The theory of coloration of texti

Second edition

Edited by Alan Johnson

1989

Society of Dyers and Colourists

on behalf of the Dyers' Company Publications Trust

Published by the Society of Dyers and Colourists, PO Box 244, Perkin House, 82 Grattan Road, Bradford, West Yorkshire BD1 2JB, England, on behalf of the Dyers' Company Publications Trust.

Typeset by Wessex Typesetters
(Division of The Eastern Press Ltd)
Frome, Somerset

ISBN 0 901956 48 1

The theory of coloration of textiles

Dyers' Company Publications Trust

The Dyers' Company Publications Trust was instituted by the Worshipful Company of Dyers of the City of London in 1971 to encourage the publication of textbooks and other aids to learning in the science and technology of colour and coloration and related fields. The Society of Dyers and Colourists acts as trustee for the fund, its Textbooks Committee acting as the Trust's technical subcommittee.

Contributors

Brian C Burdett
Business manager, Dyeing and Finishing Research and Development, British Textile Technology Group, Manchester

(the late) **Charles H Giles**
Formerly honorary professor, Department of Pure and Applied Chemistry, University of Strathclyde, Glasgow

Wilfred C Ingamells
Lecturer, School of Home Economics and Institutional Management, University of Wales College of Cardiff

Alan Johnson
Formerly senior lecturer in polymer and fibre science, UMIST, Manchester

(the late) **Francis Jones**
Formerly senior lecturer, Department of Colour Chemistry and Dyeing, University of Leeds

Paul Rys
Professor, Technisch-Chemisches Laboratorium, Eidgenössische Technische Hochschule, Zürich

Harry H Sumner
Formerly manager, Technological Services and Dyeing Research Section, ICI plc, Organics Division, Manchester

Heinrich Zollinger
Honorary professor, Technisch-Chemisches Laboratorium, Eidgenössische Technische Hochschule, Zürich

Preface

The first edition of this book was published in 1975 as the first of a series published under the auspices of the Dyers' Company Publications Trust. From the outset it accomplished all that was expected of it and has been in constant demand, especially by students.

The decision to publish a second edition was taken some seven years ago and Mr Bird agreed once again to be the editor. Unfortunately a number of unforeseen circumstances intervened which seriously delayed publication and also led to the change in editorship. I can only hope that my efforts will bear comparison with those of my predecessors.

In planning this edition it was felt that the original chapters entitled 'Influence of fibre structure on dye uptake' and 'The response of fibres to dyeing processes' should be combined into one, and this has been undertaken and the information updated by Dr Ingamells. Other changes involve Chapter 1 on physical chemistry, which has been rewritten by Mr Burdett in place of Dr Marshall who is no longer active in the subject, and Chapter 4 on thermodynamics, where Mr Sumner agreed to contribute a chapter equivalent to that written by the late Dr Peters. The remaining chapters have been updated by their original authors.

Regretfully Professor Giles died shortly after submitting the manuscript of his chapter and before it had been edited. I have therefore carried out the minimum of alterations, leaving it as far as possible as he wrote it: this is particularly obvious in the format of the references. A further cause for regret is the untimely death of Dr Jones in November of last year, but in his case he had been able to cooperate fully in the editing of his chapter. Both of our former colleagues are greatly missed.

In this edition SI units have been used throughout but c.g.s. values are often quoted as well because these are the ones found in the original references. Modern chemical nomenclature has been used where appropriate, so that a number of familiar terms appear in new guises. The most important of these is the replacement of the term molecular weight by relative molecular mass (r.m.m).

My thanks are due to my authors, to members of the Society's staff and to Dr J Macqueen who converted my final version into a form suitable for typesetting.

ALAN JOHNSON

Contents

CHAPTER 1

Physical chemistry essential to dyeing theory

BRIAN C BURDETT

1.1 INTRODUCTION

Physical chemistry is concerned with the influence of physical quantities such as temperature, pressure, concentration and electrical potential on chemical reactions and on the reactants and products of such reactions. The objective is to obtain an understanding of chemical change. Consequently a wide range of different chemical behaviour falls within the investigational methods of physical chemistry.

Initially results are obtained from experimental procedures, after which the results are placed in the form of mathematical equations known as *laws* to formulate general statements or *hypotheses*. An hypothesis attempts to interpret the experimental facts in terms of simpler or more understandable behaviour. In other words, the purpose of an hypothesis is to state the causes of the observed effects. When a particular hypothesis has withstood the test of time and other alternative hypotheses have been eliminated or proved to be unsatisfactory, we refer to it as a *theory*.

Coloration theory rests firmly on physical chemistry. Dyeing or printing of textile substrates depends on the transfer of molecular species (the majority of which are coloured) by diffusion through application media, for example dye solutions or print pastes, adsorption on molecular surfaces of substrates and final diffusion within the structures of the substrates; covalent reactions may occur during the transfer process. As a consequence the physical chemistry of coloration includes a treatment of adsorption, electrical double-layer and related interfacial behaviour, reaction kinetics and thermodynamics, all of which we shall discuss in this chapter.

Emphasis has been placed on simplicity and clarity, so that average readers can follow the presentation while those with greater ability will find that this chapter provides a useful ground work for further study.

1.2 THERMODYNAMICS

1.2.1 Introduction

Thermodynamics often seems an enigma. This need not be the case, however, if it is realised that thermodynamics is concerned with macroscopic variables such as volume, pressure, temperature and concentration and with the relationships between them. These relationships do not necessarily describe any particular kind of system – thermodynamics is of general applicability, whether for studying the formation of urea from a gaseous reaction involving carbon dioxide and ammonia, or for studying the interaction between acid dyes and wool in an aqueous environment.

It must be stressed also that thermodynamics allows us to make predictions. This is an important and valuable feature of the subject. These predictions, however, concern only the *differences* between the initial and final (or equilibrium) states of the change. Thermodynamics *cannot* predict the rate of change.

It is possible to predict recipes and the controls for a given dyeing process by calculations involving thermodynamic relationships in conjunction with colour physics. The discerning reader will appreciate that such predictions do not deal with the rate at which dye is adsorbed from solution by a substrate, but only with the end result. Relationships that describe change only are developed through thermodynamics.

In dealing with change the extent to which the change occurs must be defined. In thermodynamic terms this is equilibrium: a system is said to be *at equilibrium* when it has reached a state where no more macroscopic changes will occur.

1.2.2 The First Law of Thermodynamics

The First Law of Thermodynamics deals with three fundamental concepts, namely, energy, work and heat, and the relationships among them.

Energy

Thermodynamics is concerned with *internal energy* U, that is energy acquired by means of the mass and motion of molecules, intermolecular forces and chemical composition. But thermodynamics is aimed at the *change in energy* rather than absolute values. Thus if ΔU represents the change in the total internal energy then

$$\Delta U = U_2 - U_1 \tag{1.1}$$

where U_2 is the final value and U_1 the initial value of the internal energy of the system. The significant features of this relationship are that the energy change depends only on the initial and final states, and is independent of the path between them (Figure 1.1).

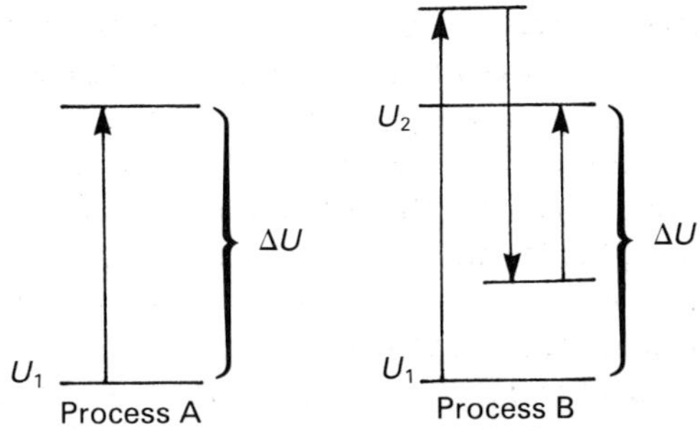

Figure 1.1 – Two paths between initial (U_1) and final (U_2) states that have the same ΔU value irrespective of intermediate stages (reproduced by permission of Prentice Hall Inc., Eaglewood Cliffs, New Jersey, from 'Introductory physical chemistry', by A R Knight, © 1970, p. 5)

It is convenient to classify all energy leaving or entering a system as either work or heat.

Work

Work w may be defined as the energy *lost* from the system other than by heat transfer. For example, when a gas is formed in a chemical reaction, work is done by the system in pushing back the surrounding atmosphere. Note that *work is done by the system* and, conventionally, it is a *positive* quantity. This type of work is known as pressure–volume or PV work, and the formation of a gaseous product results in an increased volume. If the volume change is represented by ΔV, then the work done is given by

$$w = P\,\Delta V \tag{1.2}$$

where P represents the *constant external* pressure.

For a general situation, Eqn 1.2 would be written as

$$w = \int_{V_1}^{V_2} P\,\mathrm{d}V \tag{1.3}$$

where $P\,\mathrm{d}V$ is the product of external pressure and an infinitesimal volume change at that pressure, and

$\int_{v_1}^{v_2}$ is the sum of all the $P\,\mathrm{d}V$ terms between the limits of V_1 and V_2.

If P remains constant, then Eqn 1.3 becomes

$$w = P \int_{V_1}^{V_2} \mathrm{d}V = P\,\Delta V \qquad \text{(constant pressure)} \qquad (1.4)$$

which is identical to Eqn 1.2. No PV work will ever be done if the volume does not change.

Heat

Heat q may be defined as a change in energy shown by a change in the temperature of the system or its surroundings. Thus it is possible to estimate the heat absorbed by a system from a knowledge of the change in absolute temperature ΔT, such that

$$q = C\,\Delta T \qquad (1.5)$$

The additional factor C is the heat capacity of the system. Conventionally, q is *positive* when *heat is absorbed by a system*. In Eqn 1.5 q is expressed in joules (J) and C in joules per kelvin (J K^{-1}).

The First Law

The First Law of Thermodynamics expresses the relationship between the energy change in going from one state to another state and the two mechanisms by which this energy can be changed, i.e. heat and work. The law is sometimes known as the *Law of Conservation of Energy* and states that energy cannot be created or destroyed. Thus whenever energy of one kind is converted into energy of another kind, there exists a definite quantitative relationship between them, which is

$$\Delta U = q - w \qquad (1.6)$$

This relationship expresses the statement that the net change in energy is the difference between heat gained and work lost. Note that Eqn 1.6 also states that energy, heat and work all have the same units, i.e. joules (J).

1.2.3 Applications of the First Law

In any process there are certain parameters that can be allowed to vary or that can be maintained constant. For example, if a chemical reaction is carried out in a sealed vessel, then the volume of the system will remain constant. Most chemical reactions are not carried out under constant volume, however, but at constant pressure in a reaction vessel open to atmospheric pressure. We shall consider each of these two situations.

Constant-volume processes
If a reaction is carried out at constant volume then $dV = \Delta V = 0$ and, according to Eqn 1.2, $w = 0$. Hence, from Eqn 1.6,

$$\Delta U = q_v \tag{1.7}$$

where q_v is the heat absorbed at constant volume (a subscript indicates that the property indicated by the subscript is constant). The heat absorbed is, therefore, equal to the increase in the energy of the system in a reaction, and this energy change is a characteristic of the reaction under *any* conditions; the use of constant-volume conditions is a convenient way to determine the numerical value of ΔU. Further, ΔU may be regarded as the *heat of reaction at constant volume*.

Constant-pressure processes and enthalpy
For processes carried out at constant pressure it is convenient to define a new thermodynamic quantity which involves pressure and volume. This quantity is *enthalpy* or *heat content* H and it is defined by the equation

$$H = U + PV \tag{1.8}$$

The *change in enthalpy* is of more interest than the absolute value, however. Consequently a more suitable equation is as follows:

$$\Delta H = \Delta U + \Delta(PV) \tag{1.9}$$

For a constant-pressure process, Eqn 1.9 becomes

$$\Delta H = \Delta U + P\,\Delta V \qquad (P \text{ constant}) \tag{1.10}$$

If only PV work is possible and at constant P

$$w = P\,\Delta V$$

(Eqn 1.4) which, with Eqn 1.6, gives

$$\Delta U = q_p - P\,\Delta V \tag{1.11}$$

Combining Eqn 1.11 with Eqn 1.10 gives

$$\Delta H = q_p \tag{1.12}$$

The expression indicates that ΔH is a measure of the heat absorbed in a constant-pressure process. This is a *positive* quantity of heat and for a chemical process we would have an *endothermic reaction*. When heat is liberated, we would have an *exothermic reaction* with a *negative* value for ΔH. In this context, ΔH is known as the *heat of reaction* (at constant pressure).

Note here the similarities with ΔU. As with ΔU, ΔH is the enthalpy change for *any* process, not just those at constant pressure. Any change from initial to final states will be associated with a ΔH value.

An interesting feature of ΔH refers to reactions involving liquids and solids only. If the pressure remains constant, the difference between the energy change and the enthalpy change depends on $P\ \Delta V$ (Eqn 1.10). For up to moderate pressures, if a reaction involves *only* liquids or solids, the volume change will be very small (if the temperature remains constant) so that $P\ \Delta V$ will be insignificant. Thus Eqn 1.10 reduces to

$$\Delta H = \Delta U \qquad \text{(liquids and solids only)} \tag{1.13}$$

i.e. changes in enthalpy equal changes in internal energy.

Reversible (ideal) and natural (real) processes

A careful distinction is made in thermodynamics between various kinds of process, and the most significant is that of describing a process as *reversible*. In such a process the change from initial to final state takes place by a series of infinitesimal steps. A truly reversible process would take an infinitely long time to carry out; consequently, it is an idealised process to which we can compare real processes that occur in nature. A natural or irreversible process is one that takes place by finite steps.

If we consider a gas at pressure p expanding against an external pressure P, then the process of expansion will be reversible if

$$p = P + \mathrm{d}p \tag{1.14}$$

At each value of P the system will be in equilibrium, since there is only an infinitely small difference between opposing and driving forces, i.e. internal and external pressures. If $\mathrm{d}p$ in Eqn 1.14 becomes finite, the system is no longer in equilibrium and the process is thermodynamically *irreversible*. For our expanding gas the irreversible process is represented by

$$p = P + \Delta p \tag{1.15}$$

Although it is possible to classify thermodynamic processes as reversible or irreversible, other classifications are possible on the basis of maintaining one or other parameter constant. Thus we may have the following:

isothermal processes	processes at constant temperature
isobaric processes	processes at constant pressure
isometric processes	processes at constant volume
adiabatic processes	processes in which no heat enters or leaves the system.

Isothermal expansion

When a gas expands at constant temperature, the process is said to be isothermal. For an ideal gas, where the internal energy U is dependent solely on the absolute temperature, $\Delta U = 0$. Substituting for w from Eqn 1.3, The First Law (Eqn 1.6) becomes

$$\Delta U = q - w = q - \int_{V_1}^{V_2} P\,\mathrm{d}V = 0 \qquad (1.16)$$

If the expansion is reversible (Eqn 1.14), then the ideal gas law will apply. So, for one mole of ideal gas, $p = RT/V$ and, substituting into Eqn 1.16,

$$q = w = \int_{V_1}^{V_2} P\,\mathrm{d}V = \int_{V_1}^{V_2} p\,\mathrm{d}V = \int_{V_1}^{V_2} RT\,\mathrm{d}V/V \qquad (1.17)$$

Since RT is a constant, then integrating Eqn 1.17 gives

$$q_{\mathrm{rev}} = w_{\mathrm{T}} = RT \ln (V_2/V_1) \qquad (1.18)$$

or, if initial and final pressures are more convenient, then using Boyle's Law

$$P_1V_1 = P_2V_2 \qquad (T \text{ constant}) \qquad (1.19)$$

Eqn 1.18 becomes

$$q_{\mathrm{rev}} = w_{\mathrm{T}} = RT \ln (P_1/P_2) \qquad (1.20)$$

Vaporisation and sublimation

During vaporisation or sublimation of a pure substance a phase change takes place at constant temperature. If the external pressure is constant and equal to the vapour pressure at that temperature, there will be true equilibrium between the condensed phase and the vapour, and a reversible process will occur. As heat is added to the system to maintain constant temperature, more vaporisation takes place. The temperature will increase only after all the condensed phase has vaporised or sublimed.

If P is the vapour pressure of the condensed phase at a particular temperature and ΔV is the change in volume per mole as the substance vaporises, then, using Eqn 1.10, we have

$$\Delta H_{vap} = \Delta U_{vap} + P\,\Delta V \qquad (1.21)$$

For sublimation, ΔH_{sub} and ΔU_{sub} replace ΔH_{vap} and ΔU_{vap} in Eqn 1.21.

For vaporisation, the First Law (Eqn 1.6) becomes

$$\Delta U_{vap} = q_{vap} - w = q_{vap} - P\,\Delta V \qquad (1.22)$$

which, when combined with Eqn 1.21, gives

$$\Delta H_{vap} = q_{vap} \qquad (1.23)$$

This relationship provides a definition of the heat of vaporisation ΔH_{vap}, as the heat absorbed when one mole of liquid phase is converted to the vapour phase at constant pressure. A similar definition applies for the heat of sublimation of a solid to a vapour, ΔH_{sub}.

1.2.4 Thermochemistry

Thermochemistry is that part of thermodynamics that deals with heat and energy changes in chemical processes. Different substances have different amounts of internal (chemical) energy, so that the total energy of the products of a chemical reaction is usually different from that of the reactants. Consequently, chemical change is accompanied by the liberation or absorption of energy which may appear in the form of heat. Thus we have the equations

$$\Delta U = \Sigma U(\text{products}) - \Sigma U(\text{reactants}) \qquad (1.24)$$

$$\Delta H = \Sigma H(\text{products}) - \Sigma H(\text{reactants}) \qquad (1.25)$$

for the net changes in energy and enthalpy, respectively. If heat is liberated during the reaction, the process is *exothermic* (negative value for ΔH), but if heat is absorbed it is *endothermic* (positive value for ΔH). The quantity ΔH, expressed by Eqn 1.25, is the enthalpy change or heat of reaction, and this equation forms the basis of thermochemistry.

Thermochemistry equations and Hess's Law

The changes that can take place in a chemical reaction can be represented by a thermochemical equation, which includes not only the mass balance but also the change in energy or enthalpy of the system as it is altered from the initial (reactants) to final (products) state.

The equation

$$\text{C(graphite)} + \text{O}_2\text{(g)} \rightarrow \text{CO}_2\text{(g)} \qquad \Delta H_{298} = -393.50 \text{ kJ} \qquad (1.26)$$

represents the reaction between solid carbon (graphite) and gaseous oxygen to yield carbon dioxide at 25 °C or 298 K. In other words, when 12.01 grams of solid carbon (in the form of graphite) and 32 grams of gaseous oxygen react completely at 25°C to yield 44.01 grams of gaseous carbon dioxide, at constant pressure, there is a *decrease* in enthalpy (since ΔH is negative) of 393.50 kilojoules (kJ). Notice that ΔH is expressed in *kilo*joules; it is general practice to use kilojoules because the use of joules implies an accuracy greater than is usually attainable experimentally. Further, a definite temperature should always be specified. Since the enthalpy change (or heat of reaction) depends on whether the reactants and products are solids, liquids or gases, it is necessary to specify the state of the system by indicating the phase, and concentration or pressure, and the nature of the solvent if the species is in solution. This is done by adding the state symbols s, l or g after the formula, respectively; reactions taking place in aqueous solution are indicated by the symbol aq. For example;

$$\text{HCl(aq)} + \text{NaOH(aq)} \rightarrow \text{NaCl(aq)} + \text{H}_2\text{O(l)} \quad \Delta H_{298} = -57.32 \text{ kJ} \quad (1.27)$$

Strictly the use of 'aq' implies (a) the use of dilute solutions and (b) conditions in which addition of further water causes no detectable heat change.

The information given in Eqns 1.26 and 1.27 may be used in two ways: either to indicate the heat effect of the reactions where, in these cases, heat is liberated, or to carry out thermodynamic calculations and interpret the thermodynamic change as the systems move from the initial states to the final states. Considering the large number of possible chemical reactions for which a ΔH value may be required, a complete set of data would be very formidable. However, since the path by which we proceed from reactants to products does not alter the value of ΔH, then thermochemical equations can be treated by simple algebra. Thus by combining ΔH values for a series of reactions containing common species, the enthalpy change of a reaction of interest can be calculated.

Let us consider two examples.

1. If we wish to calculate the enthalpy change at 298 K of the process

$$2\text{C(s)} + 2\text{H}_2\text{(g)} + \text{O}_2\text{(g)} = \text{CH}_3\text{COOH(l)} \qquad (1.28)$$

we can use the information given in the following three reactions:

$$CH_3COOH(1) + 2O_2(g) \rightarrow 2CO_2(g) + 2H_2O(1) \qquad \Delta H = -876.13 \text{ kJ} \qquad (1.29)$$

$$C(s) + O_2(g) \rightarrow CO_2(g) \ \Delta H = -393.71 \text{ kJ} \qquad (1.30)$$

$$H_2(g) + \tfrac{1}{2}O_2(g) \rightarrow H_2O(l) \ \Delta H = -285.77 \text{ kJ} \qquad (1.31)$$

Thermoequations 1.29, 1.30 and 1.31 require to be algebraically manipulated, *including* each ΔH value, so that summation will give Eqn 1.28. First Eqn 1.29 must be reversed so that CH_3COOH appears as product as in Eqn 1.28; the enthalpy change will then become +876.13 kJ. Then, if Eqns 1.30 and 1.31 are each multiplied by 2 and added to the reversed Eqn 1.29, Eqn 1.28 is obtained. The corresponding enthalpy changes are

$$(-393.71 \times 2) + (-285.77 \times 2) + (+876.13) = -482.83$$

so that the algebraic sum, −482.83 kJ (or kJ mol^{-1}), is the ΔH value required for Eqn 1.28, i.e.

$$2C(s) + 2H_2(g) + O_2(g) \rightarrow CH_3COOH(l) \qquad \Delta H = -482.83 \text{ kJ} \ (1.32)$$

2. The oxidation of carbon (graphite) can be represented by the three thermochemical equations

$$C(s) + O_2(g) \rightarrow CO_2(g) \qquad \Delta H = -393.50 \text{ kJ mol}^{-1} \qquad (1.33)$$

$$C(s) + \tfrac{1}{2}O_2(g) \rightarrow CO(g) \qquad \Delta H = -110.25 \text{ kJ mol}^{-1} \qquad (1.34)$$

$$CO(g) + \tfrac{1}{2}O_2(g) \rightarrow CO_2(g) \quad \Delta H = -283.25 \text{ kJ mol}^{-1} \qquad (1.35)$$

where the summation of Eqns 1.34 and 1.35 is expressed by Eqn 1.33 in terms of both chemical species and enthalpy change. Figure 1.2 illustrates Eqns 1.33 to 1.35.

These two examples illustrate the principle of additivity of chemical equations and their enthalpy values. The principle is known as *Hess's Law of Constant Heat Summation*, which states that the enthalpy change of a chemical reaction is the same whether it is carried out in one or several stages. In other words, ΔH depends only on the initial and final stages of a process, as already discussed.

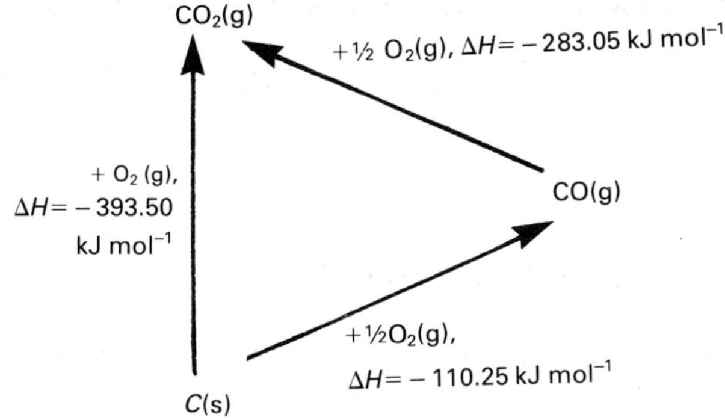

Figure 1.2 – Enthalpy change in the formation of $CO_2(g)$; the direct and two-step processes have the same value of ΔH

Standard enthalpies of formation

Although we can measure the ΔH value of a reaction such as those of Eqns 1.26, 1.27 and 1.30, we cannot measure the individual enthalpies of the components because thermodynamics deals with *change* only. To enable thermodynamic data of individual substances to be tabulated in a systematic way a *standard state* is used as a reference point for the purposes of calculation. The standard state is arbitrarily defined as the lowest enthalpy (and free energy) state of a substance at 298 K and at a pressure of one atmosphere. Although values are almost invariably quoted at 298 K, a standard state may be defined *at any temperature*. For purposes of calculation, the enthalpies of all elements in their standard states are arbitrarily chosen to be zero at all temperatures. Thus in Eqn 1.33, the enthalpy change of the reaction is equal to the standard enthalpy of CO_2 because the standard enthalpies of C and O_2 are zero. In this context, the standard enthalpy is signified by $\Delta H^\ominus$, where the superscript $^\ominus$ indicates the standard state. It should be clearly understood that since thermochemical quantities always represent *differences*, the basis of reference chosen for calculation purposes is immaterial *provided it is always the same*. The one described above is simple and convenient, but it does not imply that the enthalpies of elements are actually zero.

One of the most useful types of information, which makes many calculations possible from a small amount of reference data, is the standard enthalpy change in a reaction in which a substance is formed from its constituent elements. This is known as the *standard enthalpy of formation*, $\Delta H_f^\ominus$. The units of $\Delta H_f^\ominus$ are kJ mol^{-1} and, unless otherwise indicated, at a temperature of 298 K. Thus in Eqn 1.33, -393.50 kJ mol^{-1} is the standard enthalpy of formation of CO_2 from its elements. Likewise from Eqn 1.32, $\Delta H_f^\ominus$ for CH_3COOH is -482.83 kJ mol^{-1}.

Armed with tabulated data of standard enthalpies of formation, we can calculate enthalpy changes for reactions. For purposes of calculation let us consider the reaction

$$C_2H_4(g) + H_2O(l) \rightarrow C_2H_5OH(l)$$

From tables of standard data, we know that

$$\Delta H_f^{\ominus}[C_2H_4(g)] = +52.28 \text{ kJ mol}^{-1} \quad (1.36a)$$

$$\Delta H_f^{\ominus}[H_2O(1)] = -285.84 \text{ kJ mol}^{-1} \quad (1.36b)$$

$$\Delta H_f^{\ominus}[C_2H_5OH(l)] = -277.65 \text{ kJ mol}^{-1} \quad (1.36c)$$

and thus the reaction may be written in enthalpy terms as

$$+52.28 + (-285.84) = -277.65 - \Delta H^{\ominus} \quad (1.36d)$$

where $\Delta H^{\ominus}$ is the standard enthalpy change for the reaction. The value of $\Delta H^{\ominus}$ is therefore -44.09 kJ mol^{-1}.

For any reaction, the standard enthalpy change is the total standard enthalpy of formation of products less that of reactants, or

$$\Delta H^{\ominus} = \Sigma n\ \Delta H_f^{\ominus}\text{(products)} - \Sigma n\ \Delta H_f^{\ominus}\text{(reactants)} \quad (1.37)$$

where n is the number of moles of each compound. (Notice the similarity between this equation and Eqn 1.25.)

As a final comment, there is nothing mysterious about standard states. They are only a set of reference conditions chosen arbitrarily to simplify the tabulation and calculation of thermodynamic data.

Variation of ΔH with temperature

In general, the enthalpy change of a chemical reaction varies with temperature. If, say, the standard enthalpy of formation of liquid water is required at several different temperatures then it is imperative that a method of calculation is available. In order to determine the variation of ΔH with temperature, Eqn 1.25 is differentiated with respect to temperature at constant pressure, leading to

$$\left(\frac{d(\Delta H)}{dT}\right)_p = \left(\frac{dH(\text{products})}{dT}\right)_p - \left(\frac{dH(\text{reactants})}{dT}\right)_p \quad (1.38)$$

According to Eqns 1.5 and 1.12,

$$\left(\frac{dH}{dT}\right)_p = C_p \tag{1.39}$$

Hence Eqn 1.38 may be rewritten as

$$\left(\frac{d(\Delta H)}{dT}\right)_p = C_p \text{ (products)} - C_p \text{ (reactants)} = \Delta C_p \tag{1.40}$$

where ΔC_p is the difference in the heat capacities at constant pressure of the final and initial states (e.g. products and reactants in a chemical reaction). Eqn 1.40 is sometimes known as the *Kirchhoff equation*.

In order to make practical use of this equation it must be integrated between the temperature limits T_1 and T_2, giving

$$\Delta H_2 - \Delta H_1 = \int_{T_1}^{T_2} \Delta C_p \, dT \tag{1.41}$$

We may now consider two situations:

(a) If ΔC_p is constant and independent of temperature, over a limited range, then this term can be taken out of the integration. This results in Eqn 1.42:

$$\Delta H_2 - \Delta H_1 = \Delta C_p (T_2 - T_1) \tag{1.42}$$

(b) However, if the individual values of C_p vary with temperature, then Eqn 1.41 must be used and ΔC_p is expressed as a function of temperature such that

$$\Delta C_p = a + bT + cT^2 \ldots \tag{1.43}$$

where a, b, c . . . are constants determined by the nature of the products and reactants, and may be obtained from tabulated data.

It is evident from either Eqn 1.40 or Eqn 1.41 that variation of the enthalpy change with temperature depends on the magnitude and sign of ΔC_p. The larger the difference in heat capacities of products and reactants, the greater the variation of enthalpy with temperature. If ΔC_p is positive (i.e. the total heat capacity of the products is greater than that of the reactants) ΔH will increase with temperature. If ΔC_p is negative, then ΔH will decrease with increasing temperature.

1.2.5 The Second Law of Thermodynamics

So far, in discussing chemical processes, we have concentrated on the energetics of change. We have shown that the First Law can lead to exceedingly useful and general conclusions regarding enthalpy and energy changes that accompany chemical processes. However, the First Law gives no indication as to whether or not a reaction will actually take place, knowledge of which is of importance to a chemist. For the reaction depicted in Eqn 1.26 we would expect that under given conditions some particular equilibrium will exist between graphite, oxygen and carbon dioxide, but there is no way in which the First Law can tell us where that equilibrium lies. The First Law cannot predict whether the reaction indicated by Eqn 1.36 will ever occur. Some additional factor is thus necessary.

For a chemical reaction it would appear that ΔH would be the determining factor. It might be expected that if a reaction evolved heat through a decrease in enthalpy (an exothermic reaction) it would be *spontaneous*. Where the enthalpy change is large this rule holds reasonably well, but it does not always hold where ΔH is small. There are many exothermic reactions that are not spontaneous and many endothermic reactions that are. Thus when the enthalpy change is large, some additional factor is more than outweighed by the enthalpy value.

The concept of entropy

In examining the characteristics of a spontaneous process, it is useful to consider *disorder* or *randomness*. If we isolate a system in which the process is occurring (i.e., permit no energy or mass transfer in or out of the system), then a spontaneous process appears to be characterised by an *increase in disorder or randomness of the isolated system*. That is, the process involves a change from a more ordered to a less ordered state. The following examples illustrate such a change.

(a) As a gas expands from a volume of high pressure to one of low pressure, so that it fills all the volume available, the molecules move further apart and, consequently, the system moves from a more ordered to a less ordered condition.

(b) In diluting a concentrated solution, the solute molecules move further apart and the system is less ordered when the solute distribution is finally complete.

In physical processes of this kind the increase in randomness is obvious. In chemical reactions the change in order of the system is not always so obvious, particularly as energy transfer invariably occurs between system and surroundings, i.e. we find it more difficult to envisage an isolated system.

When dye molecules are transferred (spontaneously) from solution to a fibre in a dyeing process, they become *more ordered* because of confinement in the solid fibre phase. This apparent contradiction in our discussion of spontaneous processes becomes more sensible only when the randomness of the total system (*the whole dyeing process*) is analysed.

If we accept that an increase in randomness of an isolated system is a criterion of spontaneity, then we have a basis for the Second Law. The thermodynamic function that is a direct *measure of randomness* is *entropy*, S. As the absolute values of thermodynamic initial and final states cannot be calculated, then the *change in entropy* ΔS is more significant.

The Second Law

From the concept of entropy we may now say that the Second Law of Thermodynamics states that for a spontaneous process in an isolated system there is an increase in entropy: the entropy change occurring during the process, ΔS, is positive:

$$\Delta S(\text{isolated system}) > 0 \tag{1.44}$$

However, for a closed but *non-isolated system* (i.e. energy transfer but no mass transfer occurs with the surroundings) the restriction that ΔS must be positive for a spontaneous process does not apply. Here the entropy change may be positive, negative or zero. Thus for chemical reactions (including dyeing processes) it may be difficult to interpret the Second Law.

Notwithstanding this difficulty, it is useful to define and consequently calculate ΔS, and thus predict whether a process will occur. For a thermodynamically reversible process, the First Law as formulated by Eqn 1.6 may be written as

$$\Delta U = q_{\text{rev}} - w_{\text{max}} \tag{1.45}$$

Only a limited maximum amount of work can be obtained (w_{max}), for which there will be a corresponding limit to the value of q_{rev} (the heat absorbed); when $w = w_{\text{max}}$, q is also at its maximum value. This value of q_{rev}, therefore, places a restriction on the amount of energy that can appear as work. The remainder, which is q_{rev}, represents the unavailable energy. For an infinitesimal amount of heat absorbed, q_{rev} is related to the entropy (at constant T) by the equation

$$T\,\mathrm{d}S = dq_{\text{rev}} \tag{1.46}$$

or

$$\mathrm{d}S = \frac{\mathrm{d}q_{\text{rev}}}{T}$$

For a finite change between initial and final states,

$$\Delta S = \frac{q_{rev}}{T} \tag{1.47}$$

or

$$q_{rev} = T\,\Delta S$$

As can be seen from this equation, the units of entropy are joules per kelvin ($J\,K^{-1}$).

If the temperature varies between T_2 and T_1, then ΔS may be evaluated using the equation

$$\Delta S = S_2 - S_1 = \int_{T_1}^{T_2} \frac{dq_{rev}}{T} \tag{1.48}$$

Truly reversible processes are rare. Consider a thermodynamically irreversible process in which q_{irr} is the amount of heat absorbed. Since q_{rev} is the maximum amount of heat that can be absorbed, q_{rev} is greater than q_{irr}, so that at constant T

$$\Delta S = \frac{q_{rev}}{T} > \frac{q_{irr}}{T} \tag{1.49}$$

Since, therefore, ΔS cannot be calculated from q_{irr}/T we must devise some reversible path of going from initial to final state in order to calculate ΔS.

1.2.6 Free energy

We can predict that for an isolated system a spontaneous process will occur when $\Delta S > 0$. As indicated earlier, however, for chemical reactions which cannot be isolated the entropy change of the surroundings must also be considered such that

$$\Delta S(\text{non-isolated system}) + \Delta S(\text{surroundings}) = \Delta S(\text{total system}) > 0 \tag{1.50}$$

It is the *total entropy* that must increase for the reaction to be spontaneous. The ΔS value for the non-isolated system may be positive, negative or zero, and this explains apparent contradictions when real processes such as dyeing are discussed. An additional criterion is required if we are to predict the course of a chemical reaction.

The thermodynamic function that fulfils this criterion is the *free energy* G, sometimes known as the *Gibbs free energy* for constant-pressure processes. The free energy is defined by the equation

$$G = H - TS \tag{1.51}$$

However, as in all changes from the initial state to the final state, it is the *change* in free energy, ΔG, that is more useful, that is:

$$\Delta G = \underset{\text{final state}}{G(\text{products})} - \underset{\text{initial state}}{G(\text{reactants})} \tag{1.52}$$

and this is given by

$$\Delta G = \Delta H - \Delta(TS) \tag{1.53}$$

The units of ΔG are joules (J). For a constant-temperature process,

$$\Delta G = \Delta H - T\,\Delta S \tag{1.54}$$

Significance of ΔG

To appreciate the meaning of the free energy change, let us substitute Eqn 1.9 into Eqn 1.53 to obtain

$$\Delta G = \Delta U + \Delta(PV) - \Delta(TS) \tag{1.55}$$

which is a perfectly general equation. Restricting ourselves to constant-pressure and constant-temperature processes, then

$$\Delta G_{TP} = \Delta U + P\,\Delta V - T\,\Delta S \tag{1.56}$$

However, $$\Delta U = q_{\text{rev}} - w_{\text{max}} \tag{1.45}$$

and $$q_{\text{rev}} = T\,\Delta S \tag{1.47}$$

so that substitution in Eqn 1.56 leads to Eqn 1.57:

$$-\Delta G_{TP} = w_{\text{max}} - P\,\Delta V \tag{1.57}$$

The significance of Eqn 1.57 is twofold, namely,

(a) w_{max} represents the work done if the chemical process is reversible, and

(b) $P\ \Delta V$ is the work done against the confining pressure and is of little practical significance for chemical processes.

Consequently, the difference between w_{max} and $P\ \Delta V$ is the *useful* or *net work*, that is:

$$-\Delta G_{TP} = w_{useful} \tag{1.58}$$

Thus, the decrease in free energy in a process is equivalent to the maximum amount of useful work.

We may obtain a further insight into the significance of ΔG, by substituting

$$\Delta H = q \tag{1.12}$$

and

$$T\ \Delta S = q_{rev} \tag{1.47}$$

into Eqn 1.51, to obtain

$$\Delta G_{TP} = q - q_{rev} \tag{1.59}$$

A spontaneous process occurs irreversibly, i.e. $q = q_{irr}$, but since $q_{irr} < q_{rev}$ it follows that $\Delta G_{TP} < 0$. Thus for a *spontaneous process*, ΔG_{TP} is *negative* and the value of the free energy decrease is a measure of the tendency of a process to proceed spontaneously.

This conclusion is consistent with Eqn 1.50, as shown by the following argument. From Eqn 1.54, $\Delta S - \Delta H/T = -\Delta G/T$ where $\Delta S \equiv$ entropy change of the system and $-\Delta H/T \equiv$ entropy change of the surroundings, because the heat ΔH is lost by the surroundings at temperature T. However, because the sum of the two entropy terms must exceed zero for change to take place and tend to zero at equilibrium, $\Delta S - \Delta H/T$ must be greater than zero (at constant T and P) so that $-\Delta G/T > 0$ or $\Delta G < 0$, where $-\Delta G/T$ is a measure of the overall entropy change.

In this context, ΔG_{TP} can be regarded as a measure of the *chemical potential* of the system. As a reaction proceeds, the system moves from a high potential to a lower potential, which is the *driving force* for any process.

Returning to Eqn 1.59, if ΔG_{TP} is *positive* the process is thermodynamically impossible (or the process is proceeding in the opposite direction), because (from Eqn 1.45) $q_{rev} \equiv q_{max}$ and therefore $q_{actual} \not> q_{rev}$. Finally, if $q_{actual} = q_{rev}$ then, from Eqn 1.59, $\Delta G_{TP} = 0$ which implies that the process is *reversible* and at an *equilibrium state*.

Standard free energy changes

A knowledge of ΔG for a process is of considerable importance. The use of

Eqn 1.54 would enable values of ΔG to be calculated where values of enthalpy and entropy are known. Unfortunately this rarely occurs, with the consequence that a reference must be defined. This is taken as the free energy change in a *standard state*, denoted by $\Delta G^\ominus$, that is the difference in the free energy between products and reactants in their *standard state*. If standard enthalpy and entropy values are available, then $\Delta G^\ominus$ can be evaluated from Eqn 1.60:

$$\Delta G^\ominus = \Delta H^\ominus - T\,\Delta S^\ominus \tag{1.60}$$

As with standard enthalpy, the standard state is characterised here also by a pressure of 1 atmosphere and, by convention, the value of $\Delta G^\ominus$ of elements at 25 °C (298 K) is taken as zero. The units of $\Delta G^\ominus$ are joules per mole (at constant temperature).

The significance of the standard free energy change is that it is the change of free energy associated with a reaction taking place under such conditions that each of the reactants and products is in its standard state, so that

$$\Delta G_f^\ominus = \Sigma n\,\Delta G_f^\ominus(\text{products}) - \Sigma n\,\Delta G_f^\ominus(\text{reactants}) \tag{1.61}$$

By analogy with standard enthalpies (see Eqn 1.37) we are dealing with standard free energies of formation. If the value of $\Delta G^\ominus$ is negative the reaction will take place at the particular temperature. Reversal of a particular reaction does not alter the standard free energy change but merely reverses the sign.

$\Delta G^\ominus$ has a *definite* value at every temperature for any given reaction and it is related to the reaction equilibrium constant at the same temperature. The tabulation of standard free energies is thus equivalent to the tabulation of equilibrium constants. These features of $\Delta G^\ominus$ make it an important parameter in the study of reactions.

The dependence of free energy on pressure

We have seen previously that

$$G = H - TS \tag{1.51}$$

and

$$H = U + PV \tag{1.8}$$

Thus

$$G = U + PV - TS$$

and, for differential change,

$$dG = dU + P\,dV + V\,dP - T\,dS - S\,dT \tag{1.62}$$

Cancellation of terms occurs when dU is replaced by $T\,dS - P\,dV$, which comes from Eqn 1.6 by substitution of $T\,dS$ for heat adsorbed (Eqn 1.46) and $P\,dV$ for work done by the system (Eqn 1.3) for a reversible process. Thus

$$dG = V\,dP - S\,dT \tag{1.63}$$

When P is constant,

$$dG = -S\,dT \tag{1.64}$$

and when T is constant,

$$dG = V\,dP \tag{1.65}$$

Substituting for V by the ideal gas law, then

$$dG = RT\frac{dP}{P} \tag{1.66}$$

Integrating between two states,

$$G_2 - G_1 = \Delta G = RT\int_{P_1}^{P_2}\frac{dP}{P} = RT\ln\left(\frac{P_2}{P_1}\right) \tag{1.67}$$

For a change in pressure from the standard state at $P_1 = 1$ atmosphere to another state at $P_2 = P$, Eqn 1.67 becomes

$$\Delta G = G - G^{\ominus} = RT\ln P$$

or

$$G = G^{\ominus} + RT\ln P \tag{1.68}$$

Here G is the free energy per mole in the non-standard state where the pressure is P atmosphere. For the free energy of formation under non-standard conditions, we require the equation

$$\Delta G_f = G_f^{\ominus} + RT\ln P \tag{1.69}$$

These derivations assume ideal conditions.

The dependence of free energy on concentration

The free energy of the components of a liquid solution depends on their concentration expressed as either mole fraction x or molarity c. By analogy with Eqn 1.67 we may write

$$\Delta G = RT \ln\left(\frac{x_2}{x_1}\right) \tag{1.70}$$

and taking x_1 as unity (definition of the standard state) and $x_2 = x$, the non-standard state, then

$$\Delta G = G - G^{\ominus} = RT \ln x$$

or

$$G = G^{\ominus} + RT \ln x \tag{1.71}$$

For the free energy of formation, we have

$$\Delta G_f = \Delta G_f^{\ominus} + RT \ln x \tag{1.72}$$

If molarity is used instead of mole fraction, then by the same argument as above,

$$G = G^{\ominus} + RT \ln c \tag{1.73}$$

and

$$\Delta G_f = \Delta G_f^{\ominus} + RT \ln c \tag{1.74}$$

As with pressure, these derivations assume ideal conditions, i.e. that the molecules of solute are completely independent of each other.

The dependence of free energy on temperature

Eqn 1.54 can be rearranged to give

$$\frac{\Delta H}{T} = \frac{\Delta G}{T} + \Delta S \tag{1.75}$$

which represents a straight line for the plot of $\Delta H/T$ against $1/T$, assuming the value of ΔG remains constant over the temperature range of interest. From the slope of the line, we can obtain the value of ΔG. A more rigorous mathematical approach is necessary, however, because the free energies of individual components in a chemical reaction will vary with temperature, and consequently the overall ΔG for the process will vary with temperature. For a small change in temperature, values of ΔG may be determined as a function of temperature by the relationship

$$\frac{\Delta G_1}{T_1} - \frac{\Delta G_2}{T_2} = \Delta H\left(\frac{1}{T_1} - \frac{1}{T_2}\right) \tag{1.76}$$

where ΔH is constant between the two temperatures T_1 and T_2. Thus free energy at one temperature can be calculated from a known value of free energy at another temperature by substitution in the equation

$$\Delta G_2 = \Delta H + \frac{T_2}{T_1}(\Delta G_1 - \Delta H) \tag{1.77}$$

This assumes that the value of ΔH is known. Alternatively, ΔH may be calculated if values of ΔG are known.

1.2.7 Equilibrium and the equilibrium constant

Many chemical reactions occur in both the forward and backward directions and, at a constant temperature, such reactions eventually reach a state of *dynamic equilibrium*, at which the forward and backward rates are equal. This means that the reaction does not stop but the concentrations of the components of the system cease to change unless there is a change in the reaction conditions, e.g. a change of temperature or a product that is removed. When a given reaction reaches equilibrium at a given temperature, the ratio of the concentrations of products to those of reactants has a unique value. This is the value of the *equilibrium constant*, K. Regardless of the initial concentrations of reactants, the same numerical value of K is always obtained at the given temperature.

Experimentally, equilibrium and the value of K at constant temperature may be checked by allowing the reaction to reach equilibrium using several different initial concentrations of reactants. When no further change in concentrations is observed, the *ratio* of product to reactant concentrations should be the same in all cases.

For a chemically reversible reaction written as

$$a\text{A} + b\text{B} \leftrightarrows m\text{M} + n\text{N}$$

where the symbol $\leftrightarrows$ indicates both a forward and a backward reaction, the equilibrium constant is defined by

$$K = \frac{[\text{M}]^m\,[\text{N}]^n}{[\text{A}]^a\,[\text{B}]^b} \tag{1.78}$$

where square brackets [ ] indicate concentration in moles per litre (mol l^{-1}). For a given temperature a reaction will have a unique value of K, but the dimensions of K depend on the values of m, n, a and b. If $a + b = m + n$, then K would be dimensionless.

If the concentrations of all the components of the equilibrium reaction can be determined experimentally, then K can be calculated provided that the stoichiometric equation is known. On the other hand, it is sometimes necessary to calculate the equilibrium composition from a known value of K. For this kind of determination we must have information about the concentrations of at least some of the components, either initially or at equilibrium.

1.2.8 Equilibrium and free energy

Earlier in this discussion we concluded that if a reaction

$$\text{reactants} \rightarrow \text{products}$$

has a positive value of ΔG under a given set of conditions, then the reaction is thermodynamically non-spontaneous, i.e. thermodynamically impossible. However, the reverse action

$$\text{products} \rightarrow \text{reactants}$$

will have a negative value of ΔG and will occur spontaneously. Thus if the reaction is chemically reversible,

$$\text{reactants} \rightleftharpoons \text{products}$$

a positive ΔG indicates that the reaction will occur in reverse, *not* that nothing will happen. A non-zero value for ΔG indicates the possibility that the system can attain a state of lower free energy through the formation of more products or more reactants, depending on whether the forward or the backward reaction is the spontaneous process. If ΔG is zero, however, there is no driving force for either reaction and the system will have attained equilibrium. Thus the condition that

$$\Delta G = 0 \text{ at equilibrium}$$

enables us to derive a quantitative relationship between $\Delta G^{\ominus}$ and K, i.e. between the standard free energy change and the equilibrium constant of a given process.

Commencing with the chemically reversible reaction,

$$aA + bB \rightleftharpoons mM + nN$$

the free energy of the process, which is G(products) − G(reactants), can be expressed as

$$\Delta G = mG_M + nG_N - aG_A - bG_B \qquad (1.79)$$

But

$$G = G^{\ominus} + RT \ln c \qquad (1.73)$$

where c is the molarity of a component of a reaction occurring in solution. Thus

$$\Delta G = m(G_M^{\ominus} + RT \ln c_M) + n(G_N^{\ominus} + RT \ln c_N) - a(G_A^{\ominus} + RT \ln cA) - b(G_B^{\ominus} + RT \ln c_B)$$

$$= (mG_M^{\ominus} + nG_N^{\ominus}) - (aG_A^{\ominus} + bG_B^{\ominus}) + (RT \ln c_M^m + RT \ln c_N^n) - (RT \ln c_A^a + RT \ln c_B^b) \qquad (1.80)$$

Rearrangement of the $G^{\ominus}$ and $RT \ln c$ terms gives the equation

$$\Delta G = \Delta G^{\ominus} + RT \ln \frac{c_M^m\, c_N^n}{c_A^a\, c_B^b} \qquad (1.81)$$

If

$$\frac{c_M^m\, c_N^n}{c_A^a\, c_B^b} = Q \qquad (1.82)$$

then

$$\Delta G = \Delta G^{\ominus} + RT \ln Q \qquad (1.83)$$

The expression Q does *not* represent the equilibrium constant (see Eqn 1.78) but the ratio of concentrations, each raised to the power equal to the number of moles participating in the reaction, for which the free energy of the process is ΔG. Eqns 1.81 and 1.83 are not concerned with equilibrium, but with some state of the process before equilibrium has been achieved. At equilibrium, however, where $\Delta G = 0$, Eqn 1.83 becomes

$$-\Delta G^{\ominus} = RT \ln Q_{eq} \qquad (1.84)$$

We have already seen that $\Delta G^{\ominus}$ has a unique value for a given reaction at a constant temperature, so there can be only one value of Q to satisfy this condition as expressed by Eqn 1.84. Since we are dealing with an equilibrium situation $Q_{eq} = K$, and

$$-\Delta G^{\ominus} = RT \ln K \tag{1.85}$$

which is an important thermodynamic relationship. (Of course, K in Eqn 1.85 is the same as that in Eqn 1.78.)

The use of Eqn 1.85 enables us to calculate the equilibrium constant from thermodynamic data. More significantly, the equation connects experimentally measurable concentrations of species to the thermodynamic parameters of a process and so enables $\Delta G^{\ominus}$, for example, to be calculated. If a process is not at equilibrium, then Eqn 1.83 can be used for the calculation of ΔG. By accumulating data at different temperatures, enthalpy and entropy changes can be calculated using the appropriate equations.

It is important to appreciate the relationship between $\Delta G^{\ominus}$ and K, and the significance of $\Delta G^{\ominus}$ and ΔG. Table 1.1 summarises their differences.

TABLE 1.1

Differences between $\Delta G^{\ominus}$ and ΔG

Magnitude of $\Delta G^{\ominus}$	Significance	
$\Delta G^{\ominus} < 0$	$K > 1$	Products predominate at equilibrium
$\Delta G^{\ominus} > 0$	$K < 1$	Reactants predominate at equilibrium
$\Delta G^{\ominus} = 0$	$K = 1$	Equal distribution of products and reactants at equilibrium
$\Delta G < 0$	Process is thermodynamically spontaneous	
$\Delta G > 0$	Process is thermodynamically non-spontaneous	
$\Delta G = 0$	Process is in the equilibrium state	

Reproduced by permission of Prentice Hall Inc., Eaglewood Cliffs, New Jersey, from 'Introductory physical chemistry', by A R Knight, © 1970, p. 90.

Clearly the value of ΔG indicates whether a process is possible or whether the process has reached equilibrium, while $\Delta G^{\ominus}$ indicates the state of the process at equilibrium.

Effect of temperature on equilibrium

The qualitative effect of temperature on chemical equilibrium is described by the Le Chatelier principle, but a quantitative relationship can be obtained by using thermodynamic principles. From our previous discussion,

$$\Delta G^{\ominus} = -RT \ln K \tag{1.85}$$

which on differentiating with respect to temperature at constant pressure gives

$$\frac{d\,\Delta G^\ominus}{dT} = -R \ln K - RT\frac{d \ln K}{dT} \tag{1.86}$$

Multiplying through by T and substituting Eqn 1.85 leads to

$$T\frac{d\,\Delta G^\ominus}{dT} = \Delta G^\ominus - RT^2\frac{d \ln K}{dT} \tag{1.87}$$

However,

$$\Delta G^\ominus = \Delta H^\ominus + T\frac{d\,\Delta G^\ominus}{dT} \tag{1.88}$$

where

$$-\Delta S^\ominus = \frac{d\,\Delta G^\ominus}{dT} \tag{1.64}$$

Thus, by comparing Eqns 1.87 and 1.88, we have

$$\frac{d \ln K}{dT} = \frac{\Delta H^\ominus}{RT^2} \tag{1.89}$$

an important derivation, which is known as the *van't Hoff equation*. The relationship derived here states that if $\Delta H^\ominus$ is positive, K will increase with an increase in temperature; on the other hand, if $\Delta H^\ominus$ is negative, K will decrease with an increase in temperature. The latter applies to dyeing processes at equilibrium, because of the exothermic nature ($\Delta H^\ominus$ negative) of dyeing processes. In turn this means that, for a given mass of dye applied to a substrate, the amount of dye adsorbed at equilibrium will decrease as the temperature increases.

The equation can be used not only to calculate $\Delta H^\ominus$, but to calculate a value for K at one temperature from a known value at another temperature. To use Eqn 1.89 it is common practice to integrate over a range of temperatures and so obtain Eqn 1.90:

$$\ln K = -\frac{\Delta H^\ominus}{RT} + \text{constant} \tag{1.90}$$

A plot of the values of $\ln K$ against $1/T$ will yield a straight line of slope $-\Delta H^\ominus/R$ if $\Delta H^\ominus$ remains constant over the temperature range of interest. If this condition does not apply a curve will be obtained, the slope of which at any temperature will be equal to $-\Delta H^\ominus/R$. If only two temperatures are of interest and they are sufficiently close for $\Delta H^{\ominus}$ to remain constant, then a more conventional form of equation can be used, namely,

$$\ln\frac{K_2}{K_1} = -\frac{\Delta H^{\ominus}}{R}\left(\frac{1}{T_2} - \frac{1}{T_1}\right) \tag{1.91}$$

Eqn 1.91 can be very easily obtained if we substitute Eqn 1.85 into Eqn 1.76 and rearrange.

1.2.9 Equilibrium in ideal and non-ideal solutions

We have already derived relationships showing the dependence of free energy on pressure (Eqns 1.67, 1.68 and 1.69) and, by analogy, we have shown the dependence of free energy on concentration (Eqns 1.70 to 1.74). At this stage it must be emphasised that for ideal solutions and for the solvent of non-ideal solutions, the mole fraction of the solvent x is related to the vapour pressure of the solution P and the vapour pressure of the pure solvent $P°$ by Raoult's Law, which states that

$$P = xP° \tag{1.92}$$

If it is assumed that the vapour in equilibrium with the solution behaves as an ideal gas, then substitution of $P_1 = x_1P°$ and $P_2 = x_2P°$ into Eqn 1.67 gives Eqn 1.70 and the subsequent Eqn 1.72 for the free energy of formation.

Equations based on the use of solute molarity c instead of mole fraction are dependent on Henry's Law, which states that the solute concentration is related to the vapour pressure in equilibrium with the solution by the equation

$$P = kc \tag{1.93}$$

where k is a proportionality constant. By substitution of $P_1 = kc_1$ and $P_2 = kc_2$ into Eqn 1.67, Eqn 1.74 is obtained.

1.2.10 Activity and activity coefficient

Equations in the preceding sections that describe behaviour in solution are applicable only to extremely dilute solutions; in more concentrated solutions considerable deviations often occur. In such solutions it is convenient to express the free energies per mole of the components in terms of their *activities* a, which we define by equations of the type

$$G = G^{\ominus} + RT \ln a \tag{1.94}$$

where $G^{\ominus}$ is the free energy per mole when the activity is unity, i.e. when the particular component is in its standard state under defined conditions.

Comparison with Eqn 1.73 shows that activity a has simply replaced concentration c for an ideal solution, i.e. a solution which is so dilute that the solute particles do not interact with each other. In real solutions solute particles often do interact, however, particularly if the solute is ionised, so that a function of concentration has to be used in order for the behaviour of the solution to be accurately described by the ideal equations. The activity is this idealised function of the real concentration.

The activity of the solvent is usually taken as unity in the pure liquid. Then in Eqn 1.94, $G^{\ominus}$ is the molar free energy of the pure liquid. However, Eqn 1.67 may be written to give

$$G = G^{\ominus} + RT \ln\left(\frac{P}{P^{\ominus}}\right) \qquad (1.95)$$

from which we see that

$$a = \frac{P}{P^{\ominus}} \qquad (1.96)$$

The behaviour of an ideal solution may be described by

$$G = G^{\ominus} + RT \ln x \qquad (1.71)$$

where $x = P/P^{\ominus}$, which is a representation of Raoult's Law (Eqn 1.92). We now have two equations (1.94 or 1.95, and 1.71) that describe the behaviour of the solvent in ideal and non-ideal solutions, respectively. The ratio a/x is a measure of the deviation from ideal behaviour and it is known as the *activity coefficient* f_1, where

$$f_1 = a/x \qquad (1.97)$$

For the solvent as the pure liquid, $f = 1$.

If the activity of the solute is being described, then we may use Eqn 1.73 to represent the variation of the molar free energy of the solute with concentration in dilute solutions. Otherwise, Eqn 1.94 should be used to describe the variation if the solution is non-ideal. As before, the ratio a/c is then a measure of the deviation from ideal behaviour, and the activity coefficient f_2 is given by

$$f_2 = a/c \qquad (1.98)$$

When c is very small, f_2 may be placed equal to unity.

1.2.11 Chemical potential

Closed and open systems

So far our discussions have considered only *closed* systems, that is, systems (processes) where the mass of material is constant or there is no transfer of material during the change considered. In many processes material is transferred, however, and in these processes material is consumed, produced or exchanged. These systems are known as *open*. The most common type of open system is heterogeneous, containing two or more phases where mass and composition vary.

For our discussion, the most familiar system of this kind is the dyeing process, where solid and solution phases are in intimate contact. Here the total dyeing process is a closed system. However, both the dyebath solution (in which dye concentration is decreasing) and the fibre (in which dye concentration is increasing) are open systems. In the study of open systems it is necessary to introduce the concept of partial quantities.

Partial quantities

The properties of a system may be distinguished as being either intensive or extensive. An *intensive quantity* is one that is independent of the mass or amount of substance in a system, such as temperature, density, viscosity, surface tension and refractive index. Such quantities are characteristics of the system and are the same whether we have, for example, a gram or a kilogram of substance present. An *extensive quantity* is one that is dependent on the mass or amount of substance in a system. Such properties as volume, mass, heat capacity and free energy are extensive properties, and they change values in direct properties to the amount of substance present.

It is possible to place a quantity in either of the above two groups depending on the units that are used. For example, while volume V is an extensive quantity, volume per gram v (commonly known as density) is an intensive quantity; likewise heat capacity C_p is an extensive quantity, but the heat capacity per gram c_p (or specific heat capacity) is an intensive quantity. We should note that the intensive quantity has been obtained by dividing the extensive quantity by the *mass* of material present; in these examples mass is in grams, i.e. $v = V/m$ and $c_{\mathrm{p}} = C_{\mathrm{p}}/m$.

It is possible to obtain an intensive quantity by dividing the extensive quantity by the *amount of substance n* that is present. In these instances the word 'molar' is placed before the name of the extensive quantity. For example, molar volume is the volume divided by the number of moles of substance, i.e. $V_{\mathrm{m}} = V/n$. (Sometimes the subscript m is used to denote the molar quantity.) Likewise molar free energy is the free energy divided by the amount of substance, i.e. $G_{\mathrm{m}} = G/n$.

The extensive quantity G is determined completely by the state of the system, that is, by temperature, pressure and amounts (n_1, n_2, n_3, . . .) of various substances present. We may represent this statement by the function

$$G = \mathrm{f}\,(T, P, n_1, n_2, n_3, \ldots) \tag{1.99}$$

If T, P and all amounts n remain constant except n_i, then G will change according to the derivative

$$\left(\frac{\mathrm{d}G}{\mathrm{d}n_i}\right)_{T,P,n_1,n_2,\ldots} = \mu_i \tag{1.100}$$

The derivative is called the *partial molar quantity*, which in this case is the *partial molar free energy* for the component i and is represented by the symbol μ_i.

In any open system it is essential to use partial molar quantities, because mass and composition vary. It is necessary therefore for us to use these quantities in the thermodynamic study of dyeing systems. Dyeing takes place in a heterogeneous system containing two or more phases, and mass and composition certainly vary.

Chemical potential and its application

The quantity μ_i, defined by Eqn 1.100, is also known as the *Gibbs potential* or *chemical potential*. It is an intensive quantity and represents an increase in the quantity G of the system resulting from the addition at constant temperature and pressure of one mole of component i to such a large quantity of the system that its composition remains virtually unchanged. We must add that the concentration of any other components in the system must remain constant.

Since G is a thermodynamic function so is μ, and we may use it in place of G in equations discussed previously. For example, we may express Eqn 1.94 as

$$\mu_i = \mu_i^{\ominus} + RT \ln a_i \tag{1.101}$$

Since, by Eqn 1.98,

$$f = a/c \tag{1.98}$$

then

$$\mu_i = \mu_i^{\ominus} + RT \ln c_i + RT \ln f_i \tag{1.102}$$

If $f_i = 1$ (as in dilute solutions), Eqn 1.102 becomes

$$\mu_i = \mu_i^{\ominus} + RT \ln c_i \tag{1.103}$$

which is the alternative form of Eqn 1.73.

If we have two phases 1 and 2, and a component i present in both phases, then at the equilibrium distribution stage of i between 1 and 2

$$(\mu_i)_1 = (\mu_i)_2 \tag{1.104}$$

If the concentration of i in each phase is sufficiently small for the activity coefficients to be unity, then from Eqns 1.103 and 1.104 we may write

$$(\mu_i^\ominus)_1 + RT \ln (c_i)_1 = (\mu_i^\ominus)_2 + RT \ln (c_i)_2 \tag{1.105}$$

which on rearranging gives

$$\ln \frac{(c_i)_1}{(c_i)_2} = \frac{(\mu_i^\ominus)_2 - (\mu_i^\ominus)_1}{RT} \tag{1.106}$$

or

$$\ln \left(\frac{(c_i)_1}{(c_i)_2}\right) = p \tag{1.107}$$

where p is the partition coefficient of i between 1 and 2. This coefficient is constant because the right-hand side of Eqn 1.106 is independent of composition. A further simplification of Eqn 1.106 is

$$\frac{(c_i)_1}{(c_i)_2} = A \tag{1.108}$$

where

$$A = \exp \left\{\frac{(\mu_i^\ominus)_2 - (\mu_i^\ominus)_1}{RT}\right\} \tag{1.109}$$

Chemical potentials are applied to dyeing systems because the dyebath and the fibre are thermodynamically open systems. If in the previous discussions we replace phases 1 and 2 by ϕ (the fibre) and σ (the dye solution) respectively, and component i represents a dye species, then at equilibrium

$$\mu_\phi^\ominus + RT \ln a_\phi = \mu_\sigma^\ominus + RT \ln a_\sigma \tag{1.110}$$

or

$$-\Delta_\mu^\ominus = -(\mu_\phi^\ominus - \mu_\sigma^\ominus) = RT \ln (a_\phi / a_\sigma) \tag{1.111}$$

where $-\Delta\mu^\ominus$ is known as the *standard affinity of the dyeing process*. It is a measure of the tendency of the dye component to move from its standard state in the dye solution to its standard state in the fibre. It can be regarded as the driving force of the kinetics of the dyeing process.

Experimentally it is more useful to replace activities by concentration terms, despite errors that are produced through approximations. Eqn 1.111 may then be written as

$$-\Delta\mu^{\ominus} = RT \ln (c_{\phi}/c_{\sigma}) \quad (1.112)$$

The use of these equations for dyeing systems is discussed in Chapter 4.

1.3 ACIDS AND BASES

1.3.1 Brønsted and Lowry

Acids and bases have been defined in many ways, but one of the most useful descriptions is that associated with Brønsted and Lowry. An *acid* is defined as a substance that has a tendency to lose a proton and a *base* as a substance having a tendency to gain a proton. The 'strength' of an acid is a measure of its tendency to lose a proton, and the 'strength' of a base is a measure of its tendency to gain a proton.

There must therefore be a relationship between an acid and a base; the interaction between them is known as a *proton-transfer reaction*. We may write

$$\underset{\text{acid}}{A} \rightleftharpoons \underset{\text{proton}}{H^+} + \underset{\text{base}}{B} \quad (1.113)$$

and the acid and base that differ thus by a proton are said to form a *conjugate pair*; every acid must have its conjugate base and every base its conjugate acid. As examples we have

$$\underset{\text{acid}}{CH_3COOH(aq)} \rightleftharpoons H^+(aq) + \underset{\text{conjugate base}}{CH_3COO^-(aq)} \quad (1.114)$$

$$\underset{\text{acid}}{NH_4^+(aq)} \rightleftharpoons H^+(aq) + \underset{\text{conjugate base}}{NH_3(aq)} \quad (1.115)$$

Acids and bases may be ions or molecules, but at least one of a conjugate pair must be ionic. This is an important conclusion of the Brønsted and Lowry concept of acids and bases: ions as well as un-ionised molecules can possess acidic or basic properties.

It must be appreciated that for an acid to ionise (i.e., lose or release protons) a base of some kind must be present. For example, hydrogen chloride dissolved in benzene shows no acidic properties because the basic properties of benzene are very feeble. But if a base, such as ammonia or an amine, is added to the benzene solution then the proton from the hydrogen chloride can be transferred to the base. If hydrogen chloride is dissolved in water, however, its ionisation is practically complete because water, in common with many solvents, is itself a base and is able to take up protons. The ionisation equilibrium of hydrochloric acid (i.e. hydrogen chloride dissolved in water) may be written as

$$\underset{\text{acid}}{HCl} + \underset{\substack{\text{base}\\ \text{(solvent)}}}{H_2O} \rightleftharpoons \underset{\text{acid}}{H_3O^+} + \underset{\text{base}}{Cl^-} \qquad (1.116)$$

When ammonia is dissolved in water there is a transfer of protons from water (acting as the solvent) to the ammonia with the subsequent formation of ammonium ions; we may write the ionisation equilibrium for this transfer process as

$$\underset{\text{base}}{NH_3} + \underset{\substack{\text{acid}\\ \text{(solvent)}}}{H_2O} \rightleftharpoons \underset{\text{acid}}{NH_4^+} + \underset{\text{base}}{OH^-} \qquad (1.117)$$

which is another way of writing Eqn 1.115. Eqn 1.117 may be written in the alternative form as

$$\underset{\text{base}}{NH_3} + \underset{\substack{\text{acid}\\ \text{(solvent)}}}{H_3O^+} \rightleftharpoons \underset{\text{acid}}{NH_4^+} + \underset{\text{base}}{H_2O} \qquad (1.118)$$

These examples of proton transfer reveal three interesting points.

(a) The hydrogen ion in aqueous solution is not the proton H^+; it consists of at least one water molecule associated with a proton. Thus strictly we are dealing with the *hydronium ion* H_3O^+, although there are several other, less firmly held, water molecules in the unit (as is the case with other ions in aqueous solution).

(b) Water has the capability of acting either as an acid or as a base. Substances that can act in this way (that is, to accept or to donate a proton) are known as *amphoteric* or *amphiprotic* substances.

(c) The interaction between an acid and a base leads to the formation of another acid, conjugate to the base, and of another base, conjugate to the acid. In fact, all types of acid–base reactions can be represented as an equilibrium between two acid–base systems, viz.

$$\text{acid 1} + \text{base 2} \rightleftharpoons \text{acid 2} + \text{base 1} \qquad (1.119)$$

1.3.2 Ionisation constants

Acids

For any acid HA dissolved in water, the ionisation equilibrium is

$$HA + H_2O \rightleftharpoons H_3O^+ + A^- \tag{1.120}$$

and the equilibrium constant may be written in terms of activities as

$$K = \frac{\text{activity}\ (H_3O^+)\,.\,\text{activity}\ (A^-)}{\text{activity}\ (HA)\,.\,\text{activity}\ (H_2O)} \tag{1.121}$$

If we are dealing with dilute solutions, the activity of the water molecules is approximately that of pure water and, by convention, this is placed equal to unity. For simplicity we will represent H_3O^+ by the symbol H^+, and then Eqn 1.121 may be written as

$$K_a = \frac{\text{activity}\ (H^+)\,.\,\text{activity}\ (A^-)}{\text{activity}\ (HA)} \tag{1.122}$$

where K_a is known as the *ionisation constant* of the acid. If, as in section 1.2.10, we replace each activity term by the product of the concentration and activity coefficient, and if the concentration of each species is so small that each activity coefficient may be placed equal to unity, then Eqn 1.122 can be written as

$$K'_a = \frac{\text{concentration}\ (H^+)\,.\,\text{concentration}\ (A^-)}{\text{concentration}\ (HA)} \tag{1.123}$$

More conventionally, this equation is written as

$$K'_a = \frac{[H^+]\,[A^-]}{[HA]} \tag{1.124}$$

where square brackets [ ] indicate concentration in mol l^{-1}.

Values of K'_a for some common acids are shown in Table 1.2.

Bases

The ionisation of a base B in aqueous solution may be represented as

$$B + H_2O \rightleftharpoons BH^+ + OH^- \tag{1.125}$$

TABLE 1.2

Ionisation constants of acids and bases at 25 °C

Acid	K'_a*	Base	K'_b*
Formic acid	1.77×10^{-4}	Ammonia	1.74×10^{-5}
Acetic acid	1.75×10^{-5}	Pyridine	1.60×10^{-9}
Benzoic acid	6.46×10^{-5}	Aniline	4.10×10^{-10}
Chloroacetic acid	1.33×10^{-3}	Urea	1.5×10^{-14}
Lactic acid	1.37×10^{-4}		
Phenol	1.20×10^{-10}		

* Strictly both K'_a and K'_b should have the units mol 1^{-1}.

so that by analogy with Eqns 1.122 and 1.124 we may write

$$K_b = \frac{\text{activity }(BH^+).\text{ activity }(OH^-)}{\text{activity }(B)} \tag{1.126}$$

and

$$K'_b = \frac{[BH^+]\,[OH^-]}{[B]} \tag{1.127}$$

Values of K'_a for some common bases are shown in Table 1.2.

Ionic product of water

Since water can behave as both an acid and a base (see section 1.3.1), there must exist in all aqueous solutions, as well as in pure water, the proton-transfer equlibrium represented by

$$H_2O + H_2O \rightleftharpoons H_3O^+ + OH^- \tag{1.128}$$

where the equilibrium position is very much in favour of the 'reactants'. The equilibrium constant is given by

$$K = \frac{\text{activity }(H_3O^+).\text{activity }(OH^-)}{[\text{activity }H_2O)]^2} \tag{1.129}$$

If, as in other examples, the solution is so dilute that the activity of the water may be taken as unity, and writing H^+ (hydrogen ion) in place of H_3O^+ (hydronium ion), then

$$K_w = \text{activity }(H_3O^+)\text{ . activity }(OH^-)$$

or

$$K_w = \text{activity}(H^+) \, . \, \text{activity}(OH^-) \quad (1.130)$$

where K_w is known as the *ionic product constant* for water. Ignoring activity coefficients, then

$$K'_w = [H^+][OH^-] \quad (1.131)$$

$$= 1.008 \times 10^{-14} \text{ mol}^2 \text{ l}^{-2} \text{ at } 25 \text{ °C}$$

In a neutral solution, or in pure water,

$$[H^+] = [OH^-] = (1.008 \times 10^{-14})^{\frac{1}{2}} = 1.004 \times 10^{-7} \text{ mol l}^{-1} \quad (1.132)$$

When $[H^+] > 1.004 \times 10^{-7}$ mol l^{-1}, the aqueous solution is acidic; when $[OH^-] > 1.004 \times 10^{-7}$ mol l^{-1}, the solution is alkaline.

The equilibrium depicted by Eqn 1.130 or 1.131 has a large heat effect, and accordingly there is a large temperature coefficient for K_w or K'_w. Thus the value of K'_w is temperature-dependent, as indicated in Table 1.3 for the temperature range 0–100 °C. We see that the value of K'_w increases rapidly with increasing temperature; at the same time the concentration of hydrogen ions required for neutrality increases, a fact that is too rarely appreciated.

From Eqn 1.131, we see that once the value of either $[H^+]$ or$[OH^-$ is known the other can be calculated. For example, if $[H^+]$ is 2.10×10^{-3}mol l^{-1}, then $[OH^-] = K'_w/[H^+] = 4.80 \times 10^{-10}$ mol l^{-1}.

TABLE 1.3

The influence of temperature on the value of K'_w

Temperature/°C	$K'_w \times 10^{14}$ /mol^2 l^{-2}	$[H^+] \times 10^7$ /mol l^{-1}	pH for neutrality
0	0.114	0.338	7.47
10	0.292	0.540	7.27
25	1.008	1.004	7.00
50	5.474	2.340	6.63
70	20.416	4.518	6.35
100	54.945	7.412	6.13

1.3.3 The pH scale

The hydrogen ion, or more strictly the hydronium ion, concentration varies over a considerable range. In the titration of a weak base with a strong acid, for example, a typical variation might be between 10^{-10} and 10^{-2} mol l^{-1}, a range so wide that graphical interpretation of the results would be impracticable. However, if a logarithmic scale is used, each unit becomes a tenfold variation and, by a convenient relationship, we avoid negative exponents. The scale that has been adopted is known as the *pH scale*, and the pH value of a solution is defined by the expression

$$\mathrm{pH} = -\log\,[\text{activity}(\mathrm{H}^+)] \approx -\log\,[\mathrm{H}^+] = \log\,(1/[\mathrm{H}^+]) \qquad (1.133)$$

The relationship between pH and $[H^+]$ is illustrated in Table 1.3 for the variation of neutrality of water with temperature.

The logarithmic representation may be extended to the hydroxide ion, such that

$$\mathrm{pOH} = -\log\, a_{\mathrm{OH}^-} \approx -\log\,[\mathrm{OH}^-] \qquad (1.134)$$

In addition, this type of notation is used frequently to express the ionisation constants of acids and bases and the ionic product constant of water. Thus $pK'_a = -\log K'_a$, $pK'_b = -\log K'_b$ and $pK'_w = -\log K'_w$.

If Eqn 1.131 is written in logarithmic form, i.e. $\log K'_w = \log\,[H^+] + \log\,[OH^-]$, then we readily obtain Eqn 1.135:

$$pK'_w = \mathrm{pH} + \mathrm{pOH} = 14 \text{ (at 25 °C)} \qquad (1.135)$$

Eqn 1.135 means that pH or pOH can have any value, but their sum must be 14. It should be emphasised that the pH scale is *not* limited to the range 1 to 14, although this impression is often given. We do not meet such situations very frequently, but it is possible to have $[H^+] > 1$ mol l^{-1} and hence a negative value of pH, and pOH $> +14$.

The pH value of an aqueous solution is an important property and any changes in that value, whether deliberate or unintentional, may change drastically the reaction taking place or may initiate another reaction. Often very little meaning is attached to pH and pH changes that occur in a process, and far too often it is interpreted vaguely as 'the level of acidity'. In most cases we do not need to consider the meaning of pH in terms of solution theory, but only to accept the values as a practical scale of acidity and alkalinity. Familiarity with its use is essential, particularly the logarithmic relationship to hydrogen ion concentration.

1.3.4 Electroneutrality

In the equilibrium in pure water, the stoichiometry requires that

$[H^+] = [OH^-]$, and hence it is possible to calculate this concentration from the value of K'_w (see Eqns 1.131 and 1.132). In more complex systems, similar stoichiometric relationships are sometimes necessary for us to evaluate additional unknown concentrations. When we are dealing with ionic solutes, the solution must always remain electrically neutral regardless of the source or mechanism of production of the solute ions.

In pure water we have simply

$$[H^+] = [OH^-] \tag{1.132}$$

For an aqueous acetic acid solution, where we have the two ionic equilibria

$$CH_3COOH \rightleftharpoons CH_3COO^- + H^+ \tag{1.114}$$

$$H_2O \rightleftharpoons H^+ + OH^- \tag{1.128}$$

then electrical neutrality of the solution (or charge balance) is represented by

$$[H^+] = [CH_3COO^-] + [OH^-] \tag{1.136}$$

As a further example, in an aqueous solution containing sulphuric acid and sodium sulphate the following ionic equilibria exist:

$$\begin{aligned} H_2SO_4 &\rightleftharpoons H^+ + HSO_4^- \\ HSO_4^- &\rightleftharpoons H^+ + SO_4^{2-} \\ Na_2SO_4 &\rightleftharpoons 2NA^+ + SO_4^{2-} \\ H_2O &\rightleftharpoons H^+ + OH^- \end{aligned}$$

Electrical neutrality is then represented by

$$[H^+] + [Na^+] = [HSO_4^-] + 2[SO_4^{2-}] + [OH^-] \tag{1.137}$$

In Eqns 1.136 and 1.137, the complexity of evaluating unknown quantities may be simplified by substituting $[OH^-]$ with $K'_w/[H^+]$ (Eqn 1.131) so that, for example, Eqn 1.136 would become

$$[H^+] = [CH_3COO^-] + K'_w/[H^+] \tag{1.138}$$

In systems involving simultaneous equilibria, it is essential to represent the equality for *all* ions present. Subsequently, from a knowledge of the properties of a particular system it may be possible to neglect certain terms. In a system represented by Eqn 1.137, for example, if the pH value is 3 ($[H^+] \approx 10^{-3}$) then the value of $[OH^-]$, which would be approximately 10^{-11}, could be neglected without serious error.

1.4 ELECTROCHEMICAL PROCESSES

1.4.1 Electrode phenomena

Metals such as zinc and copper have a tendency to dissolve in water forming cations, while their valence electrons remain on the metal. The tendency of metals to dissolve is decreased, and may even become reversed, if ions of the same type as those derived from the metal are present in solution.

If a zinc rod is placed in 1.0 mol l^{-1} zinc sulphate solution, zinc will dissolve to produce Zn^{2+} ions in solution and the rod will become negatively charged with respect to the solution. A potential difference is produced between the rod and the solution, and an electrical double layer is formed. This *electrode process* can be represented by the equation

$$Zn \rightarrow Zn^{2+} + 2e$$

If a rod of copper is placed in 1.0 mol l^{-1} copper sulphate solution, ions from the solution are deposited on the copper rod, which becomes positively charged with respect to the solution. In this case the equation for the electrode process is

$$Cu^{2+} + 2e \rightarrow Cu$$

In these equations, e represents the unit of negative charge, the electron.

In general, the number of ions released or deposited is very small and these processes rapidly reach equilibrium due to the potential differences that are established. As the negative potential is developed on the zinc rod, the rate at which zinc ions leave the rod to go into solution is reduced. On the other hand, as the rod becomes negatively charged, the rate at which zinc ions become deposited on the rod is increased. Equilibrium is established when the two rates are equal. The potential difference that is obtained is unique for this process, and depends on the activity (or, approximately, on the concentration) of ions in solution and the temperature.

For any metal M we can write

$$M \rightleftharpoons M^{z+} + ze$$

to represent the electrode process, and this will be associated with a potential difference E between electrode and solution. The value of E for a single-electrode system cannot be measured because there is no method of making contact with the solution without establishing another electrode potential difference. For convenience a *reference electrode* is defined. This is the *standard hydrogen electrode*, which consists of pure hydrogen gas at unit partial pressure in contact with a solution of hydrogen ions at unit activity. The hydrogen bubbles over the surface of a platinum electrode coated with finely divided platinum (platinum black). The potential difference between this electrode and the solution is chosen as 0.000 V at 25 °C. The equation representing this electrode process is

$$H^+ + e \rightleftharpoons \tfrac{1}{2}H_2$$

By using the hydrogen reference electrode we may measure the potential difference of any other electrode system with respect to the hydrogen electrode. A *standard* electrode potential $E^\ominus$ implies unit activity of the ions and a temperature of 25 °C.

If we combine the two electrode systems described earlier so that electrical contact is made, without mixing the solutions, then electrons will flow from zinc to copper through an external circuit. The process is spontaneous and will continue until all the zinc has been oxidised to Zn^{2+} ions or until all the Cu^{2+} ions have been reduced to copper, whichever process is completed first. These particular electrode reactions form the basis of the *Daniell cell*; the standard electrode potential of the zinc oxidation process is +0.763 V, whilst that of the copper reduction process is +0.337 V, so that the standard potential of the Daniell cell is +1.100 V. The electron-transfer reaction here is one of a general category of oxidation–reduction reactions known as *redox* reactions.

1.4.2 Cells and electrodes

A cell enables a spontaneous electrochemical reaction to produce a detectable electric current. If the redox reactions contained within the cell are to be utilised, the reactions must be capable of being separated physically so that electrons lost in one part of the system are gained in another. Oxidation occurs at an electrode in one *half-cell* and reduction occurs at an electrode in another half-cell. The electrode at which oxidation occurs is the *anode*, while reduction occurs at the *cathode*. Since electrons are lost in the oxidation process they leave the cell at the anode, flow through the external circuit and re-enter at the cathode. To prevent physical mixing of the solutions containing electrolyte while still allowing for ion migration in the cell, the two half-cells are connected by a *salt-bridge*. This is usually a saturated solution of potassium chloride, which is sometimes suspended in a gel.

If A and B are pure metals and A^+ and B^+ ions are present in aqueous solution, then the half-cell processes (that is, the separated oxidation and reduction processes) are:

oxidation: $A \rightarrow A^+ + e$
reduction: $B^+ + e \rightarrow B$

and algebraic summation gives us the overall cell reaction of

$$A + B^+ \rightarrow A^+ + B$$

If we combine the first two reaction equations in section 1.4.1, then the overall cell reaction for the Daniell cell is

$$Zn + Cu^{2+} \rightarrow Zn^{2+} + Cu$$

Various types of electrode are involved in redox half-cells. The most important are discussed below.

Metal–metal ion electrodes
Here the metal is placed into a solution of its ions, with which there is a reversible reaction; sometimes these electrodes are known as *cation-reversible*. Examples of the half-cell reaction and their designations are:

(a) oxidation: $Zn \rightarrow Zn^{2+} + 2e$
[designation, $Zn \mid Zn^{2+}(x \text{ mol l}^{-1})$];
(b) reduction: $Ag^+ + e \rightarrow Ag$
[designation, $Ag^+(x \text{ mol l}^{-1}) \mid Ag$].

Metal-insoluble metal salt electrodes
The metal is in contact with an insoluble salt of the metal, which in turn is in contact with a solution containing the anion of the metal salt; the term *anion-reversible* is sometimes used to describe these electrodes because of the dependence on the anion concentration. Examples are:

(a) reduction: $AgCl(s) + e \rightarrow Ag(s) + Cl^-(aq)$
[designation, $Cl^-(x \text{ mol l}^{-1}) \mid AgCl(s) \mid Ag(s)$];
(b) reduction: $Hg_2Cl_2(s) + 2e \rightarrow 2Hg(s) + 2Cl^-(aq)$
[designation, $Cl^-(x \text{ mol l}^{-1}) \mid Hg_2Cl_2(s) \mid Hg(s)$].

Gas–ion electrodes
Mention has been made already of the hydrogen electrode, (b) below, which is typical of this class of electrodes. Although not used industrially, this

particular electrode is used as the primary zero on the electropotential scale. Another example is the chlorine electrode, (a) below.

(a) oxidation: $2Cl^- \rightarrow Cl_2 + 2e$
[designation, $Pt \mid Cl_2(y\ Pa) \mid Cl(x\ mol\ l^{-1})$];
(b) reduction: $2H^+ + 2e \rightarrow H_2$
[designation, $H^+(x\ mol\ l^{-1}) \mid H_2(y\ Pa)\ Pt$].

Redox electrodes
Bright platinum is placed into contact with a solution containing an ionic species in two valency states, such as Fe^{2+} and Fe^{3+}, Sn^{2+} and Sn^{4+}, hydroquinone and quinone, or vat dye and leuco dye. The potential developed by this type of electrode is dependent on the ratio of the concentrations of the two valency states, e.g., $[Fe^{3+}]/[Fe^{2+}]$ or [leuco vat]/[vat]. An electrode of this type may be represented thus:

reduction: $Fe^{3+}(aq) + e \rightarrow Fe^{2+}(aq)$
[designation, $Fe^{3+}(x\ mol\ l^{-1})$, $Fe^{2+}(y\ mol\ l^{-1}) \mid Pt$]; or
oxidation: $Fe^{2+}(aq) \rightarrow Fe^{3+}(aq) + e$
[designation, $Pt \mid Fe^{2+}(y\ mol\ l^{-1})$, $Fe^{3+}(x\ mol\ l^{-1})$].

The writing of a particular electrode reaction as oxidation or reduction and the resulting conventional half-cell designation is *arbitrary*.

Depending on the nature of the particular electrode comprising the second half-cell, the redox electrodes could be involved in *either* oxidation *or* reduction.

1.4.3 Thermodynamics of cells

Relationship between E and ΔG
It is possible to obtain a relationship between the electrode potential developed during an electrochemical process and the free energy change of that process. Once obtained, the relationship enables us to use a variety of thermodynamic equations. To derive the relationship let us use gas expansion as an analogy.

The work done w during gas expansion is given by Eqn 1.4:

$$w = P\Delta V \tag{1.4}$$

For electrical work w_{el} there is a similar expression based on the product of the electrical charge q and the voltage V, such that

$$w_{el} = qV \tag{1.139}$$

But
$$q = zF \tag{1.140}$$

where F is the Faraday constant and z is the stoichiometric number of e transferred in the electron-transfer (oxidation–reduction) reaction, so that

$$w_{\text{el}} = zFV \tag{1.141}$$

If we have a reversible electrochemical reaction, then Eqn 1.141 represents the maximum possible work, because V will be at a maximum value. Thus V may be replaced by E, the electrode potential, and Eqn 1.141 becomes

$$w'_{\text{max}} = zFE \tag{1.142}$$

which represents the maximum useful work. From section 1.2.6 we know that the maximum amount of useful work is equivalent to the decrease in free energy at constant temperature and pressure, i.e.

$$\Delta G_{TP} = -w'_{\text{max}} \tag{1.143}$$

Thus substitution into Eqn 1.142 gives the important equation

$$\Delta G_{TP} = -zFE \tag{1.144}$$

which is the relationship we require between E and ΔG that enables us to manage the thermodynamics of electrochemical processes.

Concentration dependence of E

We must distinguish between the reversible electrode potential of a reaction E and the *standard* reversible electrode potential $E^{\ominus}$, which is the voltage obtained when all components of the reaction are present in their standard states. In this case Eqn 1.144 is written

$$\Delta G^{\ominus} = -zFE^{\ominus} \tag{1.145}$$

To demonstrate the importance of distinguishing between E and $E^{\ominus}$ and the significance of the concentration of the reaction components, let us take the general reaction $a\text{A} + b\text{B} \rightleftharpoons m\text{M} + n\text{N}$ (cf. section 1.2.8), for which

$$\Delta G = \Delta G^{\ominus} + RT \ln Q$$

If the general reaction involves electron transfer, then substitution from Eqns 1.144 and 1.145 leads to

$$-zFE = -zFE^{\ominus} + RT \ln Q \tag{1.146}$$

which on rearranging gives

$$E = E^{\ominus} - \frac{RT}{zF} \ln Q \tag{1.147}$$

Eqn 1.147 is known as the *Nernst equation*, and it is *the primary relationship in electrochemistry*. The equation permits calculation of several important quantities from a relatively small number of experimental determinations or from tabulated data (particularly of standard oxidation and reduction potentials at 25 °C or 298 K).

Almost invariably, ionic species are involved in electrode reactions, which implies that activity coefficients (see section 1.2.10) must be evaluated. However, in calculations involving electrochemical processes, these coefficients often cancel; even so, the difference in the value of Q can be small, and the effect on E is even less because of the logarithmic relationship in the Nernst equation.

For the Daniell cell reaction $Q = [Zn^{2+}]/[Cu^{2+}]$ and, at 298 K, the Nernst equation may be written as

$$E = E^{\ominus} - 0.0128 \ln ([Zn^{2+}]/[Cu^{2+}]) \tag{1.148}$$

where $z = 2$ and $RT/F = 0.0257$ V.

If the reduction of quinone to hydroquinone is represented by the following reaction:

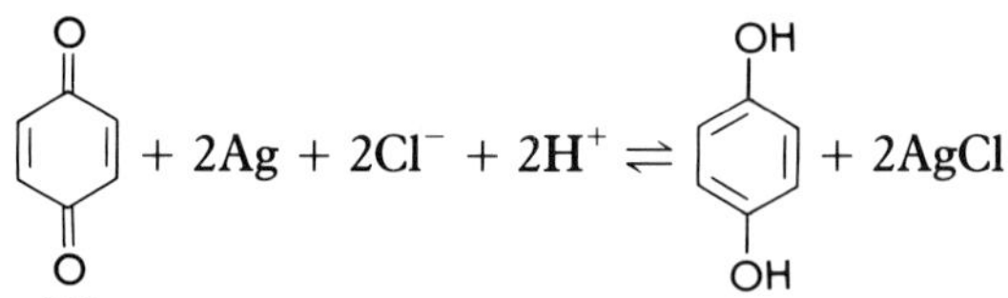

(Q) (HQ)

then

$$E = E^{\ominus} - 0.0128 \ln\left(\frac{[HQ]}{[Q]\,[Cl^-]^2\,[H^+]^2}\right) \tag{1.149}$$

If ln is converted to 2.303 log and log $(1/[H^+]^2)$ is then converted to pH (see section 1.3.3), Eqn 1.149 becomes

$$E = E^{\ominus} - 0.0296 \log\left(\frac{[HQ]}{[Q]\,[Cl^-]^2}\right) - 0.0591 \text{ pH} \tag{1.150}$$

Thus if all concentrations in the system, except $[H^+]$, are constant, the experimentally determined value of E would be a linear function of pH. The implication is that this cell (or others involving hydronium ions) could be used, in theory at least, to determine the pH value of a solution. Alternatively, if all concentrations other than [HQ] and [Q] are kept constant, then

$$E = E^{\ominus} - 0.0296 \log([\mathrm{HQ}]/[\mathrm{Q}]) + \text{constant} \quad (1.151)$$

so that $E = E^{\ominus}$ + constant when [HQ] = [Q], i.e. when the concentrations of the reduced and oxidised forms of the system are equal.

1.5 SURFACE CHEMISTRY

1.5.1 Surface tension

In the bulk interior of a liquid any given molecule is completely surrounded by other molecules and, as a result, it is attracted equally in all directions, that is, it is subjected to intermolecular forces that are spherically symmetrical. At the surface, however, these forces act only in the direction tending to pull the molecules back into the homogeneous interior of the liquid, because the number of molecules per unit volume is larger in the bulk of the liquid than in the vapour above the surface. As a result of this inward pull liquid surfaces have a tendency to contract to the smallest possible area for a given volume. An isolated drop of liquid will always take on a spherical shape which reduces its surface area to a minimum. We can envisage therefore an energy, the *surface free energy* or *surface tension*, that offers resistance to any expansion of the surface. The change in free energy dG associated with such a process at constant temperature is proportional to the change in surface area dA, and the proportionality factor is the surface tension γ, i.e.

$$dG = \gamma \, dA \quad (1.152)$$

Surface tension may be regarded as a measure of *cohesive energy* or *work of cohesion*, w_c. For a single liquid this corresponds to the work required to pull apart a column of liquid of unit cross-sectional area, and is given by the equation (see Figure 1.4(b))

$$w_c = 2\gamma \quad (1.153)$$

Table 1.4 gives values for surface tension of some liquids at 20 °C. Note the dramatic way in which γ changes with the molecular structure of the liquid.

We should note that cohesive energy is work/unit area, so that the dimensions are force × length/length^2. This simplifies to force/length, which are the dimensions of surface tension.

Substances that in solution increase the cohesive energy of a liquid (e.g. water),

TABLE 1.4

Surface tension of liquids at 20 °C

Liquid	γ/mN m^{-1}
Water	72.8
Nitrobenzene	41.8
Acetic acid	27.6
Propanone	23.7
Ethanol	22.3
n-Hexane	18.4
Mercury	485

including most salts, increase the surface tension and are excluded from the surface layer, whereas substances that decrease the cohesive energy, such as alcohols, decrease the surface tension and are concentrated in the surface.

1.5.2 Contact angles and wetting

When a liquid is placed on a flat solid surface it may spread over the surface or it may form drops having a definite angle of contact θ with the surface, as depicted in Figure 1.3.

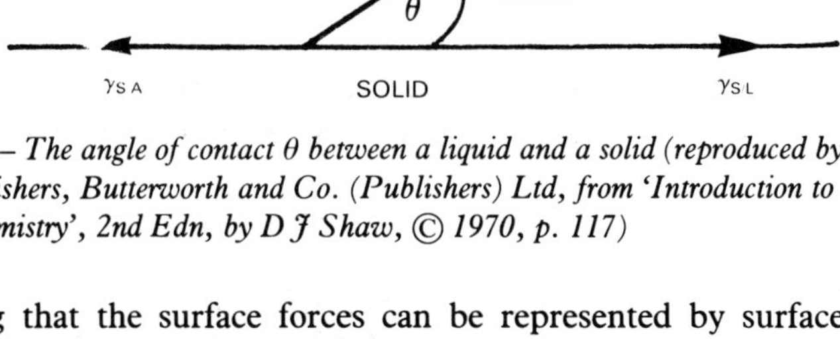

Figure 1.3 – The angle of contact θ between a liquid and a solid (reproduced by permission of the publishers, Butterworth and Co. (Publishers) Ltd, from 'Introduction to colloid and surface chemistry', 2nd Edn, by D J Shaw, © 1970, p. 117)

Assuming that the surface forces can be represented by surface tensions acting in the direction of the surfaces, then

$$\gamma_{S/A} = \gamma_{S/L} + \gamma_{L/A} \cos \theta \qquad (1.154)$$

Zero contact angle occurs when the forces of attraction between liquid and solid are equal to or greater than those between liquid and liquid; a finite contact angle occurs when the liquid adheres to the solid less than it coheres to itself.

Adhesive energy, or *work of adhesion*, w_a, between two phases A and B is the work required to separate unit area of interface and form two separate phases. Then w_a is related to the appropriate surface energies by the equation (see Figure 1.4(a))

$$w_a = \gamma_A + \gamma_B - \gamma_{A/B} \tag{1.155}$$

Figure 1.4 – Work of adhesion (a) and work of cohesion (b) (reproduced by permission of the publishers, Butterworth and Co. (Publishers) Ltd, from 'Introduction to colloid and surface chemistry', 2nd Edn, by D J Shaw, © 1970, p. 77)

For a liquid and solid in contact, the work of adhesion $w_{S/L}$ is given by

$$w_{S/L} = \gamma_{S/A} + \gamma_{L/A} - \gamma_{S/L} \tag{1.156}$$

and combining this with Eqn 1.154 gives

$$w_{S/L} = \gamma_{L/A}(1 + \cos\theta) \tag{1.157}$$

The tendency of a liquid to spread over a solid surface, or to wet a solid, may be defined by a spreading coefficient S, given by the equation

$$S = \gamma_{S/A} - (\gamma_{L/A} + \gamma_{S/L}) \tag{1.158}$$

Substituting Eqns. 1.153 and 1.156 into Eqn 1.158 gives $S = w_{S/L} - w_L$. Thus spreading, or wetting, occurs when liquid adheres to the solid more strongly than liquid coheres to itself.

Alternatively, a solid is completely wetted by a liquid if $\theta = 0$, and only partially wetted if $\theta > 0$. Complete non-wetting assumes $\theta = 180°$, which is unrealistic because it requires that either $w_{S/L} = 0$ or $\gamma_{L/A} = \infty$. There is always some solid–liquid attraction; water will adhere to a paraffin wax surface ($\theta \approx 110°$). It is not entirely satisfactory to correlate wetting with a contact angle less than 90° and non-wetting with an angle greater than 90°.

The spreading of water on a hydrophobic solid surface (loomstate cotton, wool, polyester) is considerably improved by the addition of surface-active agents. In Eqn 1.157, $w_{S/L}$ is increased and $\gamma_{L/A}$ is decreased, both of which contribute to decreasing the value of θ. Surface-active agents are used as wetting agents in many industrial situations; in the wet processing of textile materials they are of considerable importance. In addition to lowering $\gamma_{L/A}$, wetting agents must lower $\gamma_{S/L}$ so that a choice of agent to suit the particular nature of the solid surface has to be made. Unsymmetrical molecular structures, such as sodium di-n-octyl sulphosuccinate, are often good wetting agents since micelle formation is not favoured because of steric hindrance, which in turn allows the presence in solution of relatively high concentrations of unassociated surface-active molecules and hence a greater lowering of $\gamma_{L/A}$ and $\gamma_{S/L}$. Non-ionic surface-active compounds are also good wetting agents.

1.5.3 Surface adsorption

Wetting agents, and surface-active agents in general, decrease the surface tension of, for example, water by becoming concentrated at the surface. Taking the sodium salts of long-chain fatty acids as the model for the anionic type agent, then the polar carboxylate group is readily soluble in water and is immersed in the bulk of the solution, while the insoluble hydrocarbon chain is a component of the surface, along with some water molecules. In a general way, when solute molecules of a solution concentrate at the solution surface (or strictly, the solution–air interface) we can regard the process as *adsorption*. The extent of adsorption at a solution surface may be estimated from surface tension data by an equation known as the *Gibbs adsorption isotherm*. For a dilute solution of concentration c, the equation may be written as

$$\Gamma = -\frac{c}{RT}\frac{\mathrm{d}\gamma}{\mathrm{d}c} \qquad (1.159)$$

where Γ represents the excess concentration of solute per cm^2 of surface (or concentration of solute adsorbed at the surface), $\mathrm{d}\gamma/\mathrm{d}c$ the rate of change (increase or decrease) of the surface tension of the solution with the increase in concentration of the solute, R the familiar gas constant, and T the absolute temperature. According to Eqn 1.159, any solute that causes the surface tension of the solvent to decrease ($\mathrm{d}\gamma/\mathrm{d}c$ negative) will have a higher concentration in the surface than in the bulk of the solution because Γ will be positive. In other words, a solute that decreases the surface tension at an interface will concentrate at that interface. If $\mathrm{d}\gamma/\mathrm{d}c$ is positive (the solute increasing the surface tension), then Γ is negative, and the concentration of solute will be lower in the surface than in the bulk solution. Some solutes do produce this behaviour.

The excess concentration of solute at the surface is temperature-dependent, since according to Eqn 1.159 this concentration is inversely proportional to the absolute temperature. This relationship occurs for all adsorption processes.

1.5.4 Adsorption from solution

Adsorption, absorption, physical and chemical interaction

Adsorption from solution is important in many practical situations, such as those in which modification of the solid is of primary concern (e.g. dyeing of hair and textile materials, bleaching of wool and cotton) and those which involve the removal of unwanted material from solution (e.g. purification of sugar solutions and solutions containing organic compounds generally by activated charcoal or alumina). Whenever there is a solid–solution interface (and these remarks apply equally to a solid–gas interface), it is necessary to distinguish between *adsorption* and *absorption*. Adsorption refers to an excess concentration at the surface, while absorption implies a more or less uniform distribution in the solid of the solute (or gas or vapour). In these processes the solid phase is known as the 'adsorbent' or 'absorbent', and the solute (or other species) as the 'adsorbate' or 'absorbate', respectively. In some cases it may be difficult to distinguish between adsorption and absorption, or, because of lack of data, it may not be known which process is occurring; the term *sorption* is then used. It is probable that adsorption occurs at all surfaces.

If a molecular species completely saturates the adsorbent surface with a single layer of molecues (or ions), then the adsorption is known as *unimolecular* or *monomolecular*. On the other hand the formation of multilayers can occur, and we have *multimolecular* adsorption. The latter no doubt occurs in the dyeing of nylon fibres with monosulphonated acid dyes, where dye aggregation is believed to take place within the fibre.

In practice, there are two kinds of adsorption. Interaction of the adsorbate with the surface may take place simply through forces of physical attraction, i.e. intermolecular forces, and this type of adsorption is known as *physical adsorption*. It occurs on surfaces where the valency requirements of the atoms in the surface have already been satisfied by bonding with neighbouring atoms. If, on the other hand, a surface is unsaturated, that is, the atoms in the surface are not fully satisfied by bonding with neighbouring atoms, then *chemical adsorption* or *chemisorption* occurs. Here, chemical bonds form between the adsorbate and the surface.

Physical adsorption occurs rapidly, because no activation energy is required. For the same reason it is temperature-independent. Chemisorption, being a chemical reaction, may require an appreciable activation energy, in which case it will proceed at a reasonable rate only above a certain temperature. In many instances of adsorption it is difficult to distinguish between the two types by the rate of the process. If a surface is so unsaturated that the activation energy is zero or very small, chemisorption can be rapid even at low temperatures. Thus rapid adsorption need not be physical adsorption.

If the adsorbent is porous or consists of fine capillaries, penetration of adsorbate may be extremely slow. Penetration to the interior surfaces may result in chemisorption or physical adsorption. Thus a slow process need not be chemisorption. It should be noted that chemisorption ceases when the adsorbate can no longer make direct contact with the surface; it is therefore a single-layer process. No such limitation applies to physical adsorption, and multilayers are possible. It is possible for multilayer physical adsorption to occur after the formation of a chemically adsorbed monolayer.

Surfaces

A solid surface may be regarded as a plane of potential energy fluctuations, which may be either uniform or non-uniform (see Figure 1.5). The troughs, or wells, represent *adsorption sites* and the adsorption process is *localised*. If, on the other hand, the troughs are very shallow and cannot be regarded as adsorption sites, then the adsorption is *non-localised*.

In localised adsorption, two extreme conditions can be considered. If the energy fluctuations are the same between any pair of sites and all sites have the same energy, then the surface is uniform or *homogeneous* (Figure 1.5(a)).

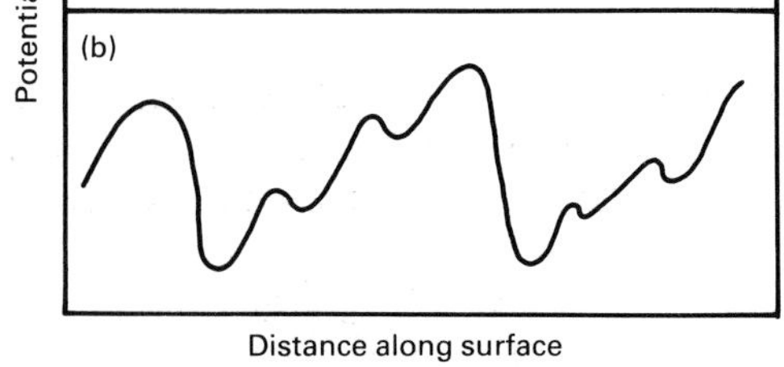

Figure 1.5 – Distinction between surfaces: (a) uniform or homogeneous surface, (b) non-uniform or heterogeneous surface (reproduced by permission of the publishers, Butterworth and Co. (Publishers) Ltd, from 'Chemisorption', 2nd Edn, by D O Hayward and B M W Trapnell, © 1964, pp. 6 and 7)

If, however, the fluctuations are irregular and the sites have different energies, then the surface is non-uniform or *heterogeneous* (Figure 1.5(b)).

Once a molecular species has been adsorbed at a site, it can migrate from site to site only if it acquires sufficient kinetic energy to overcome the 'energy barrier' between the sites. In such cases the adsorption process is said to be *mobile*. If the kinetic energy is less than the energy barrier, then *immobile* adsorption occurs. In non-localised adsorption there is little or no energy barrier opposing migration from site to site, and non-localised adsorption is therefore always regarded as mobile. Thus it is in localised adsorption only that we need to consider the kinetic energy or activation energy necessary for removal or *desorption* from a site.

Desorption is always activated. If the activation energy of adsorption is E and the heat of adsorption q, then the activation energy of desorption E' is given by

$$E' = E + q \tag{1.160}$$

Thus, for the adsorbate to migrate to a neighbouring site a minimum energy of E' is required to lift it to the energy level from which adsorption at another site can occur. According to Eqn 1.160 E' is always finite, so that, even when $E = 0$,

$$E' = q \tag{1.161}$$

The heat of adsorption is an important property of adsorption processes, and its significance in dyeing processes is discussed in Chapter 4.

At this stage, however, the dependence of the heat of adsorption on temperature should be noted. The Clausius–Clapeyron equation may be written in the form

$$\frac{\mathrm{d} \ln c}{\mathrm{d}T} = \frac{q}{RT^2} \tag{1.162}$$

where c is the equilibrium concentration of adsorbate (e.g. a dye) in solution and other parameters have these usual meanings. Assuming q is constant (between temperatures T_1 and T_2), then on integration Eqn 1.162 becomes

$$q = \frac{RT_1T_2}{(T_1 - T_2)} \ln\left(\frac{c_1}{c_2}\right) \tag{1.163}$$

where c_1 and c_2 are the aqueous concentrations at the two temperatures; experimentally, the concentration of adsorbate in the solid must be the same at the two temperatures.

Immobile adsorption of an adsorbate leads to a more ordered or less random system, with consequent decrease in the entropy of the system. At the same time adsorption involves a decrease of free energy which, by use of the thermodynamic relationship

$$\Delta G = \Delta H - T\,\Delta S \tag{1.54}$$

implies a negative value for ΔH, i.e., the heat of adsorption ($q \equiv \Delta H$) or heat of dyeing is negative. Thus by this argument adsorption is exothermic, which is the same conclusion as that from Eqn 1.163 when $T_2 > T_1$. For gas–solid interaction where no other molecular species are involved, the negative values of ΔG and ΔS (with $-T\,\Delta S$ being positive), and the subsequent negative value of ΔH, are not difficult to understand. However, the argument is more complex for adsorption from solution, since in thermodynamic terms we are concerned with *overall* changes in the solute(s), solvent and solid (including hydrated surfaces and solute–water interactions) that could lead to positive net entropy and enthalpy changes, i.e., ΔS and ΔH both positive.

In terms of sorption phenomena, the surfaces of textile fibres should be regarded as heterogeneous, although for some theoretical studies they are assumed homogeneous as, for example, in the derivation of the Langmuir isotherm equation (page 55). The adsorption sites, in many instances, are not difficult to identify. The anionic sulphonic acid and ionised carboxylic acid groups are obvious sites, as are the cationic protonated amino groups. These groups participate in ion–ion interactions between dyes and appropriate fibres, which are typical of physical adsorption. Not so easy to visualise are groups involved in non-ionic interactions such as hydrogen bonds, dipole–dipole interactions and dispersion forces. Groups acting as adsorption sites here would include hydroxyl, halide, amino and carbonyl groups. Ionic interactions have relatively strong bonding energies and must be considered to be involved in localised adsorption. At the other extreme, the weak dispersion force should be involved in non-localised adsorption; it is debatable whether this is so, however, because interactions between dyes and fibres involve π-orbitals that would increase the effective bonding energy of the dispersion forces.

Chemisorption is always localised; the reaction of triazinyl reactive dyes with ionised hydroxyl groups in cellulosic fibres fall into this category. The actual chemical reaction may not take place in a single stage, but may include transition stages. Other chapters should be studied for detailed discussions of reactive dye mechanism and intermolecular forces.

Equilibrium at surfaces; adsorption isotherms

Having obtained equilibrium at the solid–solution interface, and having experimentally determined the amount of solute adsorbed for a given solution concentration, it is necessary to represent the equilibrium data in some way. For a particular solution interacting with a given weight of a particular solid, there are only three variables: the amount of solute adsorbed x, the equilibrium solution concentration c, and the absolute temperature T. For a given weight of adsorbent,

$$x = \mathrm{f}(c,T) \qquad (1.164)$$

Experimentally, it is usual to obtain a family of curves, each referring to a fixed value of one of the variables. The most common determination is that of amount of solute adsorbed over a range of solution concentrations at a number of fixed temperatures. A family of *adsorption isotherms* is obtained, each given by

$$x = \mathrm{f}(c),\ T = \text{constant} \qquad (1.165)$$

Figure 1.6 illustrates a family of isotherms. On the other hand, if the solution concentration is determined as a function of temperature at a number of fixed adsorbed amounts of solute, the curves are *adsorption isosteres*, each given by

$$c = \mathrm{f}(T),\ x = \text{constant} \qquad (1.166)$$

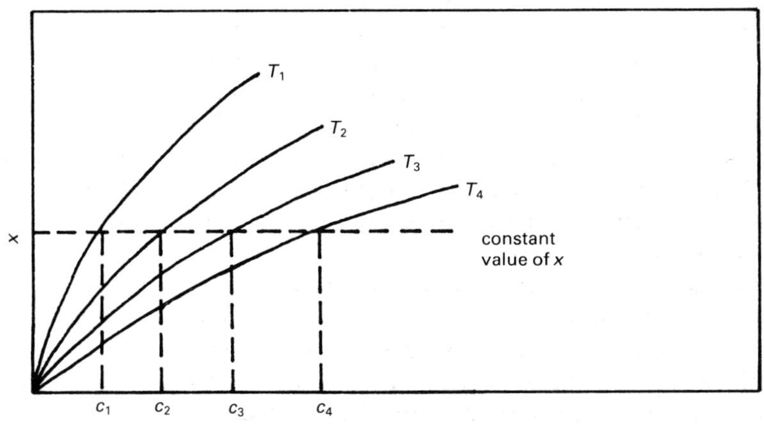

Figure 1.6 – Abstraction of isosteric data from experimental adsorption isotherms

If the amount of solute is determined as a function of temperature at a number of fixed solution concentrations, the curves are *adsorption isobars*, each given by

$$x = f(T), c = \text{constant} \tag{1.167}$$

Apart from separate experimental determinations of any curve each type of family of curves can be converted graphically to any other family. For example, the abstraction of isosteric adsorption data from a series of experimental isotherms is shown in Figure 1.6. If isosteric data is represented graphically as a plot of ln c against $1/T$, then a value of q may be obtained from the slope in accordance with Eqn 1.163.

Isotherms are the most convenient forms in which to consider adsorption equilibrium data. When related to dye–fibre systems, they have in general one of two shapes, as shown in Figure 1.7. In this diagram, concentration of dye in solution c is plotted against amount of dye x adsorbed by mass m of

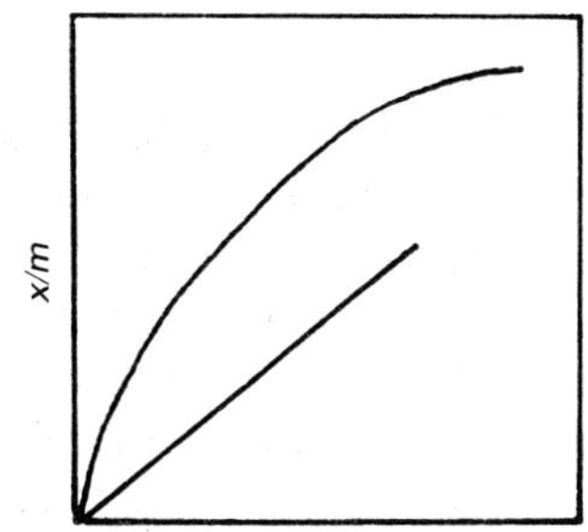

Figure 1.7 – General shapes of dye–fibre isotherms

fibre. It must be emphasised that all adsorption data, including isotherms, can be determined only by experimental procedures and can never be predicted. The shape of an isotherm reflects the mechanism of adsorption from which it arises, however, so that by determining the isotherm experimentally we have an indication of the type of adsorption occurring.

To enable us to interpret the experimental data, isotherm equations are available based on assumed mechanisms of adsorption; three such equations are cited in the literature of physical chemistry of dyeing, namely those of Langmuir, Freundlich and Nernst. Their shapes, relative to each other, are shown in Figure 1.8. In comparing experimental data with these specific isotherms it must be appreciated that where experimental data disagree with the theoretical predictions, the latter are incorrect. The theoretical isotherms are only *models* of an actual process and are therefore based on *assumptions*

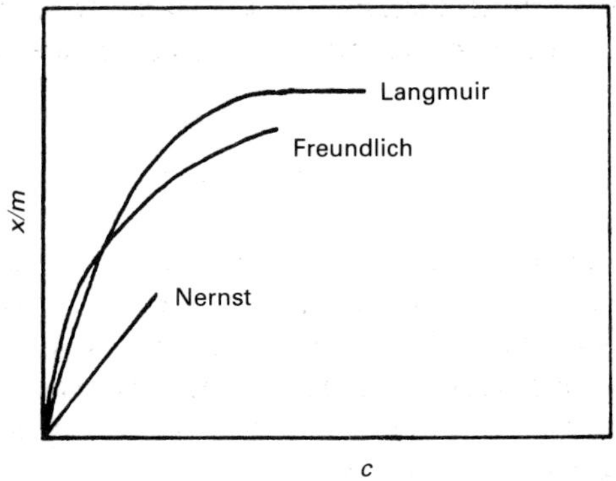

Figure 1.8 – Specific isotherms of dye–fibre systems

concerning the adsorbate, adsorbent and their interactions. They may be able to describe, but not necessarily explain, a process.

Langmuir adsorption isotherm

Langmuir's model of adsorption depends on the assumption that intermolecular forces decrease rapidly with distance, and that consequently adsorbed layers are not likely to be more than one molecule thick (mono- or unimolecular). This is no doubt true for chemisorption and for physical adsorption at low concentrations. To derive the isotherm equation it is further assumed that adsorption is localised, i.e. that it takes place at specific sites, and that the heat of adsorption is independent of surface coverage or of how many sites are occupied. This means that adsorption at one site does not influence adsorption at other sites. To obtain the Langmuir equation, a kinetic derivation can be followed in which the rates of adsorption and desorption are equated to give the adsorption equilibrium.

If θ represents the fraction of adsorption sites already occupied by adsorbate molecules, then the rate of adsorption depends on:

(a) the rate at which adsorbate molecules collide with the surface which, in turn, is proportional to the solution concentration c,
(b) the probability of striking a *vacant* site, $(1 - \theta)$, and
(c) an activation term, $\exp(-E/RT)$, where E is the activation energy of adsorption.

If it is assumed that desorption from an occupied site can occur only if the adsorbed molecule possesses the necessary activation energy, then the rate of desorption depends on:

(a) the fraction of surface that is covered θ, and

(b) an activation term, $\exp(-E'/RT)$, where E' is the activation energy for desorption.

When equilibrium is established, rate of adsorption = rate of desorption. We may then write

$$c(1-\theta)\exp(-E/RT) = k\theta\exp(-E'/RT) \tag{1.168}$$

where k is a proportionality constant.

On rearranging, we have

$$c = k\exp(-q/RT)\frac{\theta}{(1-\theta)} \tag{1.169}$$

where $q = E - E'$. If we assume that q is independent of θ, i.e., that q (and k) remain constant irrespective of the quantity of molecules adsorbed, then

$$k\exp(-q/RT) = \frac{1}{a} \tag{1.170}$$

where a is a constant at *constant* temperature. Thus

$$c = \frac{1}{a}\frac{\theta}{(1-\theta)} \tag{1.171}$$

or

$$\theta = \frac{ac}{1+ac} \tag{1.172}$$

which are forms of the Langmuir isotherm equation.

For dye–fibre systems, these equations require transposing. Let $c = [\mathrm{D}]_s$ = concentration of dye in solution, having units of mol $\mathrm{l^{-1}}$ or g $\mathrm{l^{-1}}$, and $\theta = [\mathrm{D}]_f/[\mathrm{S}]_f$, where $[\mathrm{D}]_f$ = concentration of dye on the fibre, having units mol $\mathrm{kg^{-1}}$ or g $\mathrm{kg^{-1}}$, and $[\mathrm{S}]_f$ = saturation value of dye on the fibre, also having units mol $\mathrm{kg^{-1}}$ or g $\mathrm{kg^{-1}}$. Eqns 1.171 and 1.172 may now be rewritten as

$$[\mathrm{D}]_s = \frac{1}{a}\frac{[\mathrm{D}]_f}{([\mathrm{S}]_f - [\mathrm{D}]_f)} \tag{1.173}$$

and

$$[\mathrm{D}]_f = \frac{a[\mathrm{S}]_f[\mathrm{D}]_s}{1 + a[\mathrm{D}]_s} \tag{1.174}$$

respectively. $[\mathrm{S}]_f$ represents the concentration of sites on a particular fibre, and therefore corresponds theoretically to monomolecular adsorption of dye. To check whether a set of experimental data can be represented by the Langmuir model it is necessary to plot the data in a particular way. To do this, Eqns 1.173 and 1.174 must be rearranged to give either

$$\frac{1}{[D]_f} = \frac{1}{a[S]_f[D]_s} + \frac{1}{[S]_f} \tag{1.175}$$

or

$$\frac{[D]_f}{[D]_s} = a[S]_f - a[D]_f \tag{1.176}$$

The most common procedure is to plot $1/[D]_f$ against $1/[D]_s$ which, according to Eqn 1.175, should give a straight line (Figure 1.9) with an intercept on the $1/[D]_f$ axis of $1/[S]_f$, from which $[S]_f$ may be obtained. Reciprocal plots do not give the most accurate results, however, because the plotted data tend to fall near to a straight line, and any slight variations in the slope of the line give large variations in the value of $[S]_f$.

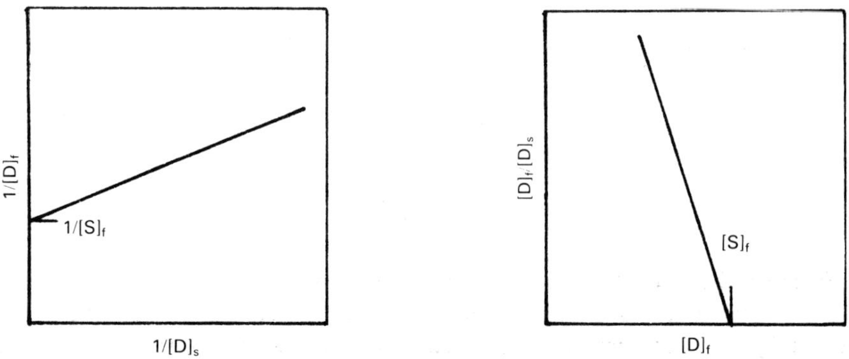

Figure 1.9 – Graphical representations of the Langmuir isotherm equation

Of more significance is the plot of $[D]_f/[D]_s$ against $[D]_f$. This is a very sensitive graphical check, and according to Eqn 1.176 a straight line should be obtained of intercept on the $[D]_f$ axis of $[S]_f$ (Figure 1.9)

Many experimental dye–fibre absorption isotherms fit the Langmuir model extremely well, but there are many that do not. For a dye–fibre system the assumptions of the model may be translated into the following implications and criticisms.

(a) The fibre surfaces are assumed to be uniform and homogeneous, with all the adsorption sites having identical potential energy. This is not true in practice because fibres have many different sites, or groups, and consequently the surfaces are heterogeneous.

(b) Adsorption is assumed to be localised, and thus the process can take place only when the adsorbate molecule collides with a vacant site. Whereas this must be true for the formation of a monomolecular layer, it cannot be true if we believe that multilayers are formed, i.e. that dye aggregation occurs by horizontal contact.

(c) A similar argument can be stated for the assumption that each site can accommodate only one adsorbate molecule.

(d) An adsorbed molecule is assumed not to interact with a neighbouring adsorbed molecule. This is another way of saying that the heat of adsorption q (which is the heat of dyeing ΔH for dye–fibre systems) is independent of θ (or of how many molecules are adsorbed). This most notable assumption is possibly true for small values of θ, but when θ is larger, and particularly when adsorption is near saturation, we might expect dye molecules to interact with each other. Any interaction between adsorbed molecules will affect E' and ΔH and, in turn, the parameter a in Eqns 1.171 to 1.176 inclusive. Thus even at constant temperature these equations may not describe (let alone explain) equilibrium dye adsorption.

(e) The interaction between adsorbed molecules can be further complicated by the geometry of the molecular orientation. Where the adsorbed molecules are extended in shape rather than spherical, horizontal or vertical adsorption can occur: a monolayer could be formed by either horizontal or vertical orientations, with the latter being more economical at large fractions of occupation. Any such geometrical consideration implies further variations in ΔH.

Figure 1.10 illustrates various forms of adsorption.

Freundlich adsorption isotherm

The variation of adsorption with the equilibrium aqueous concentration may be represented by $[D]_f = k'[D]_s^{1/n}$ where k' and n are constants at constant temperature. To check the validity of this isotherm to describe experimental data, it is necessary to take logarithms,

$$\log [D]_f = \log k' + \frac{1}{n}\log [D]_s \qquad (1.177)$$

and plot $\log [D]_f$ against $\log [D]_s$; according to Eqn 1.177 the plot should be linear. This method of representing experimental data distinguishes between the Freundlich and Langmuir models. The log–log plot is of no other value

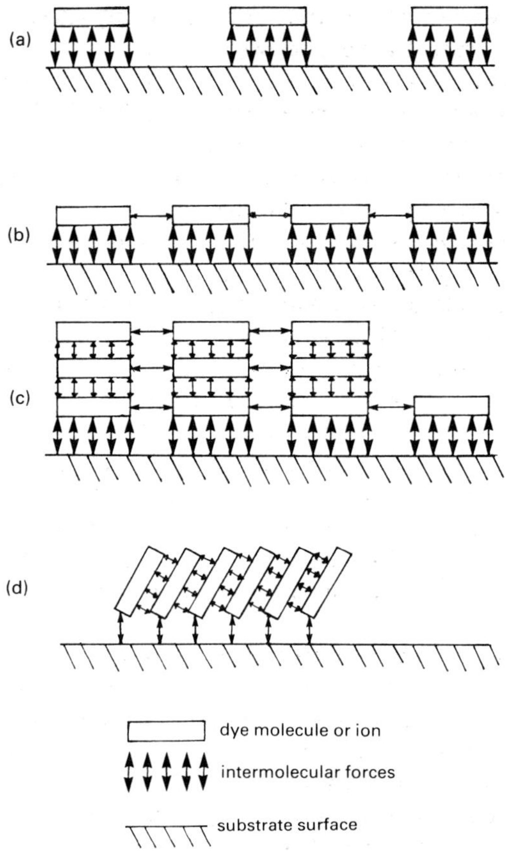

Figure 1.10 – Different forms of adsorption on a surface: (a) Langmuirian adsorption at low θ, (b) molecular interaction at high θ, (c) multilayer adsorption or aggregation, (d) non-Langmuirian monolayer causing dye interaction or aggregation; (b), (c) and (d) will produce changes in ΔH

because no other useful information can be derived, and in general this type of plot is usually linear and insensitive to small changes in data.

The Freundlich isotherm equation was originally derived empirically. However, it can be derived theoretically for an adsorption model in which the solid surface is heterogeneous or, in other words, in which the heat of adsorption decreases with increasing surface coverage. Physically this means that the higher-energy sites, giving the largest values of q or ΔH, will be occupied first, with the lower-energy sites occupied later. This in turn would cause the initial slope of the adsorption isotherm to be steeper than that predicted by the Langmuir equation (see Figure 1.8). Experimental data that can be described by such a model have been obtained in both gaseous adsorption (whether chemisorption or physical adsorption) and dye–fibre systems.

In distinguishing between the Freundlich and Langmuir models, we should remember that Freundlich does not require a limiting saturation value as in the Langmuir isotherm but, more realistically, it is based on heterogeneous surface adsorption.

Nernst adsorption isotherm

The Nernst, or linear partition, adsorption isotherm takes the form

$$[D]_f = K[D]_s \tag{1.178}$$

where K is a constant at constant temperature. A plot of $[D]_f$ against $[D]_s$ is linear according to the equation.

This isotherm equation is a mathematical representation of the *distribution law*, which states that a dissolved substance, irrespective of its total amount, distributes itself between two layers or phases in a constant concentration ratio, at constant temperature; the ratio, equal to the constant in Eqn 1.178, is referred to as the *distribution* or *partition ratio*. Theoretically the law is limited to describing the behaviour of dilute (almost ideal) solutions. The adsorption of disperse dyes from solution (not from dispersion) can be described by this law or isotherm, and ideal behaviour is often ascribed to these solutions because of the very low aqueous solubilities of disperse dyes.

If experimental data are obtained of adsorption from very dilute solutions, then in Eqn 1.174 it is possible to neglect $a[D]_s$ in comparison with unity in the denominator; under these conditions $[D]_f = K[D]_s$ where $K = a[S]_f$. Thus the Nernst adsorption isotherm is a special case of the Langmuir isotherm.

Donnan equilibrium

In previous sections, adsorption sites have been described as potential energy sources that include anionic and cationic groups within the macromolecular structure of fibres. The subsequent mathematical treatments of surface interaction neglected the influence of the macromolecular structures as polyelectrolytes affecting the equilibrium distribution of ionic adsorbate species between the two phases.

Let us consider two solutions of strong (i.e., fully ionised) electrolytes separated by a membrane that allows only some of the ions to permeate. In a system containing proteins, for example, the membrane might be permeable to small simple ions, such as Na^+ and Cl^-, but would retain the ionised macromolecular protein. When the system reaches equilibrium, the distribution of ionic species on either side of the membrane is described by the *Donnan membrane equilibrium*. This is based on two requirements:

(a) there is a free energy driving force for dilution because of the presence of a concentration gradient across the membrane, and
(b) the simultaneous maintenance of electrical neutrality in the solutions on either side of the membrane.

Firstly, let us consider a simple system where a membrane separating two solutions of sodium chloride is permeable to both Na^+ and Cl^-. These solutions, designated 1 and 2, are considered dilute so that concentration terms may be used instead of activities. The differences in free energy across the membrane for each ion are:

Membrane

Solution 1	*Solution 2*
$[Na^+]_1$	$[Na^+]_2$
$[Cl^-]_1$	$[Cl^-]_2$

$$\Delta G_{Na^+} = RT \ln ([Na^+]_2/[Na^+]_1) \tag{1.179}$$

and

$$\Delta G_{Cl^-} = RT \ln ([Cl^-]_2/[Cl^-]_1) \tag{1.180}$$

If initially $[NaCl]_1 > [NaCl]_2$, then ΔG values will be negative and sodium ions will move spontaneously from 1 to 2. Similarly, Cl^- ions will move in the same direction in response to the negative ΔG and to maintain electrical neutrality. At equilibrium, the total free energy across the membrane is zero, i.e.

$$\Delta G_{Na^+} + \Delta G_{Cl^-} = 0 \tag{1.181}$$

or

$$RT \ln ([Na^+]_2/[Na^+]_1) = -RT \ln ([Cl^-]_2/[Cl^-]_1) \tag{1.182}$$

Hence

$$[Na^+]_1 [Cl^-]_1 = [Na^+]_2 [Cl^-]_2 \tag{1.183}$$

which expresses the Donnan membrane equilibrium for the system and, while being simple to understand, is an important relationship.

Secondly, let us consider a more complex and interesting system, where one solution contains sodium chloride as before but the other contains a non-diffusable polyelectrolyte or polymer NaP (where P^- could be a protein such as wool), illustrated below.

	Membrane	
Solution 1	|	*Solution 2*
$[Na^+]_1$	|	$[Na^+]_2$
$[P^-]_1$	|	$[Cl^-]_2$

Let the initial concentrations be

$$c_1 = [Na^+]_1 = [P^-]_1 \qquad \text{and} \qquad c_2 = [Na^+]_2 = [Cl^-]_2$$

and let x mol l^{-1} of Cl^- and Na^+ diffuse across the membrane from 2 to 1 to satisfy the equilibrium of Eqn 1.183.

Although P^- cannot diffuse from 1 to 2, Cl^- will diffuse from 2 to 1 and must be accompanied by an equal number of Na^+ so as to preserve electrical neutrality. In addition to this condition the equilibrium of Eqn 1.183 must also be satisfied. If this requires the transport of x mol l^{-1} each of Cl^- and Na^+, then at equilibrium we have $[Na^+]_1 = c_1 + x$, $[P^-]_1 = c_1$, $[Cl^-]_1 = x$, $[Na^+]_2 = c_2 - x$ and $[Cl^-]_2 = c_2 - x$.

Substituting the concentration terms for diffusable ions into Eqn 1.183 we have

$$(c_1 + x)x = (c_2 - x)^2 \tag{1.184}$$

from which

$$x = \frac{c_2^2}{c_1 + 2c_2} \tag{1.185}$$

Obviously the final concentrations can be obtained for any initial distribution. The concentration of sodium chloride is *not* the same on both sides of the membrane at equilibrium, because of the presence of the non-diffusable ion. At equilibrium the counterion Na^+ is more concentrated on the polymer side of the membrane while the anion Cl^- is less. In such a system it is not necessary to have a separate membrane; the surface of a water-swollen fibre containing non-diffusable ionisable groups can act as a membrane separating internal and external solutions.

In the above argument, there must be a force opposing the concentration gradient at equilibrium restricting further diffusion of Na^+ and Cl^- from 1 to 2; this force is an electrical potential difference between the two solutions, known as the *Donnan membrane potential*. The magnitude of this potential ψ_{Donnan} is given by

$$\psi_{\text{Donnan}} = \frac{RT}{F} \ln ([Na^+]_1/[Na^+]_2)$$

$$= \frac{RT}{F} \ln ([Cl^-]_2/[Cl^-]_1) \qquad (1.186)$$

and hence

$$\frac{[Na^+]_1}{[Na^+]_2} = \frac{[Cl^-]_2}{[Cl^-]_1} = \exp(-\psi_{\text{Donnan}} F/RT) = \text{constant} \qquad (1.187)$$

where the ratio of ionic concentrations is known as the *Donnan coefficient*. Sometimes this ratio and the constant in Eqn 1.187 are given the symbol λ.

1.5.5 The electrical double layer

We have already seen that adsorption of ions from solution by solid surfaces can be complicated by the Donnan membrane phenomenon. In general, particulate matter of colloidal dimensions and indeed most solids acquire a surface charge (either net positive or net negative) when brought into contact with a polar (for example, aqueous) solution. In this section therefore we will consider in more detail the electrical behaviour of surfaces.

Origin of surface charge

The surface charge may arise from one of three causes: ionisation, ion adsorption and ion dissolution.

Ionisation Polymeric materials such as wool and silk acquire their charge mainly through the ionisation of carboxyl and amino groups giving —COO^- and —$\overset{+}{N}H_3$ groups. Carboxyl and hydroxyl groups contribute to the negative charge on cellulosic materials, where ionisation gives —COO^- and —O^-. Similar ionisation occurs with polyamide and acrylic fibres. Where weakly ionisable groups are involved, the extent of ionisation depends on the pH of the solution. For amphoteric polymers, so also does the overall charge; thus at low pH values, proteins are positively charged and at high pH values they are negatively charged.

Ion adsorption A common cause of a surface charge is the unequal adsorption from solution of oppositely charged ions. Surfaces in contact with aqueous solutions are generally *negatively* rather than positively charged, because cations are usually more hydrated than anions and thus show a tendency to remain in the bulk solution. The anions, being less hydrated and more polarising, have a larger tendency to be specifically adsorbed. The ions will bind on to the surfaces through electrostatic attraction arising from factors such as localised dipoles and induced polarisation.

Ion dissolution Ion dissolution is of little importance for polymeric materials, but of great importance for some particles of colloidal dimensions; the latter ionic substances acquire their surface charge through unequal dissolution of the oppositely charged ions of which they are composed.

The diffuse double layer

Once a surface net charge arises, it is compensated by a net equivalent amount of ions of opposite sign (*counterions*), which accumulate in solution near the surface. If the surface is net positively charged, then an equivalent amount of negatively charged ions will accumulate near the surface. This 'double layer' of ions may be considered equivalent to a parallel-plate condenser. The behaviour is more complex than this, however, and the layer may be regarded as consisting of two regions, an inner region and a diffuse region. The inner region consists of a distinct layer of adsorbed ions, while the diffuse region is a thick layer, or volume, extending some distance into the solution. In the former the fall in potential is sharp, but in the latter there is a gradual fall in potential because of thermal agitation in solution and non-uniform distribution of positive and negative charges. Overall, the layer of adsorbed ions and the diffuse volume are together known as the *diffuse (electric) double layer*.

Figure 1.11 shows a representation of the diffuse electrical double layer, assuming the surface to be negatively charged. Ions of opposite charge to that of the surface (counterions) are concentrated in the region of the surface (surface layer or volume) while ions of similar charge (*co-ions*) are repelled. As a result, close to the charged surface there is an excess of counterions over co-ions. This difference in concentration decreases with increasing distance from the surface r, until in the bulk solution the concentrations c_∞ are equivalent in order to maintain electrical neutrality. At the same time, an electrical potential ψ exists in the surface layer (due to the concentration difference), which has a maximum value at the surface ψ_0, and decreases to zero in the bulk solution.

The Gouy–Chapman Model

The simplest quantitative treatment of the diffuse layer is the Gouy and Chapman model, which is based on the following assumptions:

(a) The surface is flat and uniformly charged, and has an electrical potential of ψ_0 (usually expressed in millivolts) and a charge density of σ_0.
(b) The ions in the double layer are *point* charges; the cations have

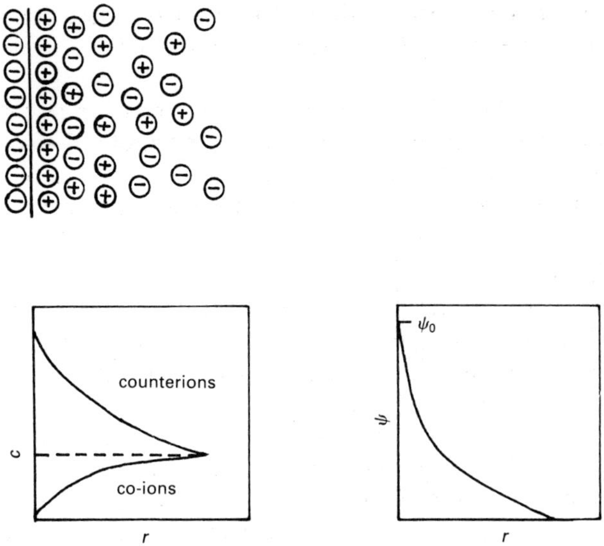

Figure 1.11 – Representation of a diffuse electrical double layer (reproduced by permission of the publishers, Butterworth and Co. (Publishers) Ltd from 'Introduction to colloid and surface chemistry', 2nd Edn, by D J Shaw, © 1970, p. 136)

concentration c_+ mol l^{-1} and charge $+e$, while anions have concentration c_- mol l^{-1} and charge $-e$. In the layer $c_+ \neq c_-$, so consequently a space charge density ϱ (rho) exists, which in turn gives rise to the electrical potential ψ.

(c) For an ion approaching the surface, work will be required to bring the ion from the bulk solution, where the concentration is c_∞, to a point near the surface. This work is dependent on ψ such that, for a single symmetrical electrolyte in which the ions carry a charge z, the electrical potential energy per ion is $+ze\psi$ or $-ze\psi$, respectively. As a consequence, the concentration of ions in the layer is dependent on the Boltzmann distribution. Thus

$$c_+ = c_\infty \exp(-ze\psi/kT) \quad \text{and} \quad c_- = c_\infty \exp(+ze\psi/kT) \qquad (1.188)$$

where k is the Boltzmann constant.

These equations imply that the concentration of ions in the diffuse layer is determined by competition between the effects of a potential field and kinetic motion. Thus, in the diffuse layer, the space charge density is given by

$$\varrho = ze\,(c_+ - c_-) \qquad (1.189)$$

Using Eqn 1.188,

$$\varrho = zec_\infty [\exp(-ze\psi/kT) - \exp(+ze\psi/kT)] \quad (1.190)$$

However,

$$\exp(x) - \exp(-x) = 2 \sinh x \quad (1.191)$$

Therefore

$$\varrho = -2zec_\infty \sinh(ze\psi/kT) \quad (1.192)$$

To obtain ψ, the Poisson equation that relates ψ to ϱ must be used. Here

$$d^2\psi/dr^2 = -\varrho/\varepsilon \quad (1.193)$$

where ε = permittivity (dielectric constant) of the solution. Elimination of ϱ between Eqns 1.193 and 1.192 gives

$$d^2\psi/dr^2 = (2zec_\infty/\varepsilon) \sinh(ze\psi/kT) \quad (1.194)$$

This equation may be integrated with the boundary conditions, $\psi = \psi_0$ when $r = 0$, and $\psi = 0$, $d\psi/dr = 0$ when $r = \infty$. The first integration then gives

$$d\psi/dr = -(8c_\infty kT/\varepsilon)^{\frac{1}{2}} \sinh(ze\psi/2kT) \quad (1.195)$$

Since $d\psi/dr$ is the slope of the potential decay curve in Figure 1.12, Eqn 1.195 informs us that at any distance from the surface the slope increases with increasing concentration c_∞, and with increasing valency on the ion z; for solutions of mixed electrolytes, c_∞ should strictly be replaced by the ionic strength of the solution I, where $I = \frac{1}{2}\Sigma c_i z_i^2$. Thus in practice, as the ionic strength of a solution increases (for example, as more NaCl is added to a dyebath for the application of direct or reactive dyes to cotton) or if electrolytes of increasing ion valency are used (for example, if NaCl is replaced by $MgSO_4 . 7H_2O$), the thickness of the diffuse layer decreases as illustrated in Figure 1.12.

So far we have concerned ourselves only with the solution component of the double layer, and have ignored the surface. It must be emphasised that the double layer as a whole must be electrically neutral. That is, the charge density on the surface σ_0 must balance the charge density in solution ϱ. Mathematically, this is expressed as

$$\sigma_0 = -\int_0^\infty \varrho_r dr \quad (1.196)$$

If this equation is combined with Eqn 1.193, then

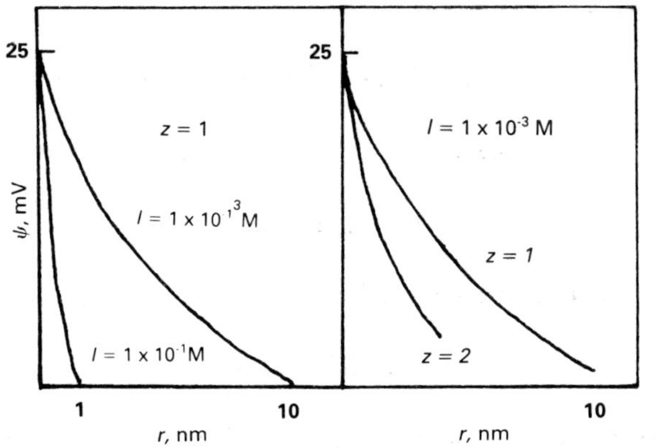

Figure 1.12 – Effect of ionic strength and valency on the diffuse layer (reproduced by permission of John Wiley and Sons Inc., from 'Physical chemistry of surfaces, 2nd Edn, by A W Adamson, © 1967, p. 213)

$$\sigma_0 = \int_0^\infty \varepsilon \mathrm{d}^2\psi_r/\mathrm{d}r^2 . \mathrm{d}r \qquad (1.197)$$

which on integration yields

$$\sigma_0 = -\varepsilon\,(\mathrm{d}\psi_r/\mathrm{d}r_{r=0} \qquad (1.198)$$

Eliminating the differential between Eqns 1.198 and 1.195 gives

$$\sigma_0 = (8c_\infty \varepsilon kT)^{\frac{1}{2}} \sinh (ze\psi_0/2kT) \qquad (1.199)$$

or

$$\psi_0 = \frac{2kT}{ze} \sinh^{-1}\left(\frac{\sigma_0{}^2}{8c_\infty \varepsilon kT}\right)^{\frac{1}{2}} \qquad (1.200)$$

Thus the value of the surface potential ψ_0 is very dependent on solution characteristics, such as z, c_∞ and ε.

The complexity of the double layer and the interdependence of the parameters can be judged by further integration of Eqn 1.194 to yield

$$\psi_r = \frac{2kT}{ze} \ln\,[1 + \gamma \exp(-\varkappa r/1 - \gamma \exp(-\varkappa r)] \qquad (1.201)$$

where $$\gamma = \frac{\exp(ze\psi_0/2kT) - 1}{\exp(ze\psi_0/2kT) + 1} \qquad (1.202)$$

and
$$\varkappa \text{ (kappa)} = \left(\frac{2e^2 c_\infty z^2}{\varepsilon kT}\right)^{\frac{1}{2}} \tag{1.203}$$

The significance of this description of the double layer can be understood best by considering the *limiting case* of low potentials, when $\psi_r \ll kT/ze$. Under these circumstances, simplifications can be made such that Eqn 1.195 becomes

$$\frac{d\psi_r}{dr} = -(8c_\infty kT/\varepsilon)^{\frac{1}{2}}(ze\psi_r/2kT) \tag{1.204}$$

or, in conjunction with Eqn 1.203,

$$\frac{d\psi_r}{dr} = -\psi_r \varkappa \tag{1.205}$$

Eqn 1.200 becomes

$$\sigma_0 = \varepsilon \varkappa \psi_0 \tag{1.206}$$

and Eqn 1.201 becomes

$$\psi_r = \psi_0 \exp(-\varkappa r) \tag{1.207}$$

We may now draw several conclusions from these equations.

(a) Eqn 1.206 indicates that $\sigma \propto \psi_0$, i.e. as σ increases, ψ_0 increases. This relationship is linear for $\psi_0 < \sim 40$ mV, otherwise the full Eqn 1.200 must be used.
(b) Eqn 1.206 indicates also that the double layer is theoretically equivalent to a parallel-plate condenser with a distance of $1/\varkappa$ between the plates. This is equivalent to assuming that all the ions in the diffuse layer are located in a single plane at a distance $1/\varkappa$ from the surface. The distance $1/\varkappa$ is known as the *thickness of the double layer*; although only a theoretical concept, it is extremely useful in describing the double layer. Figure 1.13 illustrates the difference between the distance from the surface of the exponential decay curve and the double-layer thickness as given by $1/\varkappa$.
(c) According to Eqn 1.203, as the electrolyte concentration c_∞ and the charge on the ions z increase, $\varkappa$ also increases. This is also equivalent to saying that as the ionic strength of the bulk solution I increases, then $\varkappa$ increases. Thus an increase in c_∞ will compress the double layer (i.e. $1/\varkappa$ becomes smaller), and an increase in z for equivalent concentrations dramatically compresses the double layer.

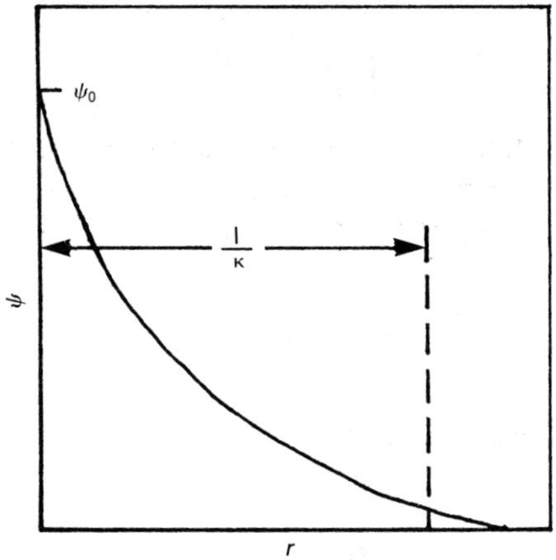

Figure 1.13 – Double layer thickness

(d) Eqns 1.204 and 1.207 confirm the behaviour of the double layer, i.e. the potential decays exponentially with distance, and this decay is very dependent upon $\varkappa$.

The compression of the double layer in the manner described above is used in those dyeing processes where electrolyte is necessary, such as the addition of sodium chloride during the application of vat, reactive and direct dyes to cellulosic fibres. In these dyeing processes, the potential barrier at the fibre–solution interface is reduced allowing dye adsorption to proceed. Details are discussed in Chapter 4.

(e) The characteristics of the diffuse double layer are determined by the surface potential ψ_0 and the ionic strength of the bulk solution I $(= 0.5\Sigma c_i z_i^2)$. When ψ_0 has a high value (e.g. 56.4 mV) the co-ions are almost expelled from the surface but the concentration of counterions is correspondingly high. However, if $\psi_0 = 4kT/e$ or $\psi = 4 \times 25.69$ mV at 25 °C, then the concentration of counterions at the surface $= c_\infty \exp(4) = 54.6c_\infty$. When $c_\infty > 0.1$M, $54.6c_\infty$ leads to unrealistically high values of surface ions. In addition, the value of $1/\varkappa$ is less than 1 nm, i.e. $1/\varkappa$ becomes of the same order as the size of small ions.

This contradiction of an otherwise useful model of the diffuse double layer lies in the assumption that the ions are regarded as point charges, with the consequent neglect of ionic diameters. Additionally, interaction between ions will no doubt occur as ionic strength increases, i.e. the solution at or near the surface will be non-ideal.

The Stern model
To take account of the finite size of ions and molecules and of their specific adsorption, we can visualise a compact layer of ions or molecules at the surface as depicted in Figure 1.14. (*Specific adsorption* implies the operation of forces between the ions and the surface other than those of electrostatic

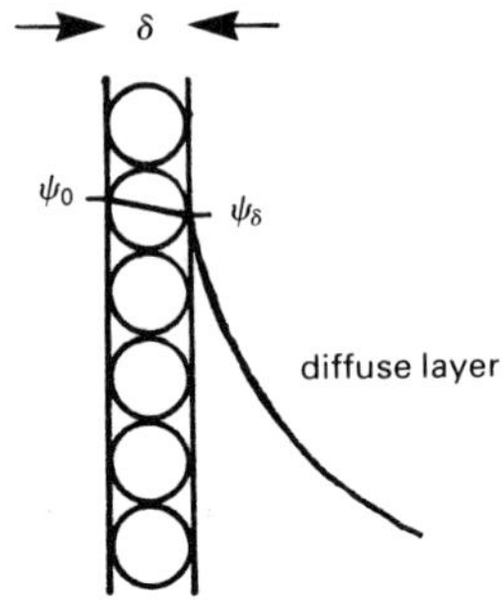

Figure 1.14 – Stern model: δ compact layer; ψ_δ potential of compact layer; $\psi_\delta < \psi_0$

origin, such as hydrogen bonds and the whole range of van der Waals forces.) This treatment of the surface is due to Stern.

We can now divide the region near the surface into two parts:

(a) a compact, immobile, surface layer of thickness δ containing a limited maximum number of ions per unit area, known as the *Stern layer*, the thickness of which will depend on the size of the ions and the aqueous environment, since ions cannot approach more closely to the surface than a distance equal to the hydrated radius, and geometry will play a significant part for large molecules such as dye ions – the ions are thus under the influence of intermolecular forces;
(b) the diffuse double layer as described by the Gouy–Chapman model, where the ions are under the influence of an electric field and thermal agitation.

In Stern's model the limitation on the number of ions in the layer immediately adjacent to the surface is defined in terms of Langmuir adsorption (see the discussion of adsorption isotherms in section 1.5.4). According to this model, if n_m is the maximum number of ions that can be adsorbed (limited by the number of sites available per unit area) and n_c is the actual number of ions adsorbed per unit area ($n_c < n_m$), then

$$n_c = \frac{n_m c_\infty \exp(-\Delta\bar{\mu}/kT)}{1 + c_\infty \exp(-\Delta\bar{\mu}/kT)} \qquad (1.208)$$

where $\Delta\bar{\mu}$ is the electrochemical potential of the interaction and involves both the electrical potential ($ze\psi_\delta$) and chemical potential ($\Delta\mu$) such that

$$\Delta\bar{\mu} = \Delta\mu + ze\psi_\delta \tag{1.209}$$

Here ψ_δ is the potential on the Stern layer. The surface charge density of this layer σ_δ is given by

$$\sigma_\delta = \Sigma zen_c \tag{1.210}$$

From this it can be seen that the value of σ_δ is limited by the individual values of n_c, unlike the value of σ_0 from the Gouy–Chapman model. Eqn 1.208 defines the parameters that determine n_c. Further, an inspection of Eqns 1.192 and 1.200 reveals that ϱ and σ_0 respectively depend only upon the appropriate electric potential ($ze\psi$), whereas Eqn 1.210 (substituted for n_c) reveals that σ_δ depends on an additional chemical adsorption potential ($\Delta\mu$).

The Donnan model

The decay of potential with distance from the surface in the Gouy–Chapman model may be represented by the line AB in Figure 1.15. At any distance r, the slope on the line is given by $d\psi/dr$ which is part of the Gouy–Chapman treatment of the diffuse layer (see Eqn 1.195). The integration of this equation will give the area under the line which is bounded by the two axes, i.e. ABO. This same area can be visualised to occupy the shaded area in Figure 1.15 (PQRO), where PQ represents the *Donnan potential* ψ_D, which remains constant and then instantaneously becomes zero at some distance R from the surface. Thus in the Donnan model, or the Donnan membrane equilibrium as described in section 1.5.4, QR in Figure 1.15 represents the outer limit of the double layer across which there is a marked change of potential. This is regarded as a membrane permeable to solute ions from the bulk solution but impermeable to the fixed charges in the surface. The volume encompassed by PQRO along the surface may be described as the surface solution, in which charges in the surface are regarded as existing in solution. Here the origin of ψ_D lies in the inability of the fixed charges to enter the bulk solution from the surface solution (see page 62).

The Donnan model is used to describe the distribution of dye ions between bulk solution and fibre (see Chapter 4). At this stage it will be sufficient to present a generalised approach.

The distribution of ions between two phases, fibre (f) and solution (s), in equilibrium is dependent on two work terms:

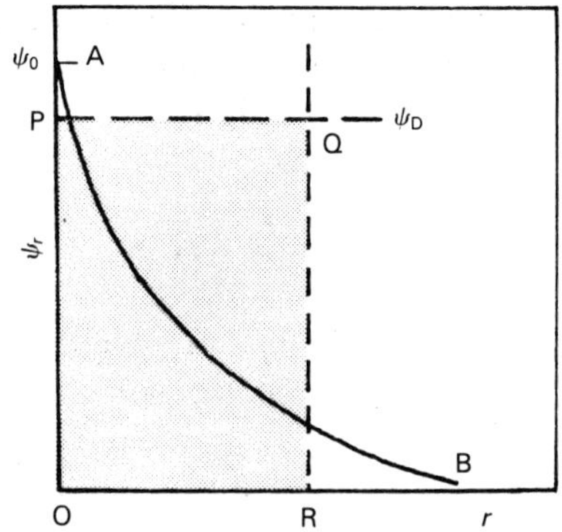

Figure 1.15 – Donnan model

(a) change in partial molar free energy (or change in chemical potential), $\Delta\mu^{\ominus}$, and
(b) change in molar electrical energy (based on the change in electrical potential), $F\Delta\psi$.

Thus the distribution of ions is dependent on two distribution coefficients, the partial molar free energy distribution coefficient given by

$$K = \exp(-\Delta\mu^{\ominus}/RT) \quad (1.211)$$

and the molar electrical energy distribution coefficient,

$$\lambda = \exp(-F\,\Delta\psi/RT) \quad (1.212)$$

At equilibrium, the usual classical relationship

$$\mu_f = \mu_s \quad (1.213)$$

must be replaced by

$$\mu_f + zF\psi_f = \mu_s + zF\psi_s \quad (1.214)$$

where μ = partial free energy or chemical potential, ψ = electrical potential, F = the Faraday constant and z = ionic charge.

In many treatments of double-layer phenomena, ψ_s is assumed to be zero. Thus if ψ_f is replaced by ψ_D, Eqn 1.214 may be rewritten as

$$\mu_f - \mu_s = -zF\psi_D \quad (1.215)$$

To develop a general distribution equation, further steps are now necessary. The chemical potential in each phase is given by

$$\mu_f = \mu_f^{\ominus} + RT \ln (c_f/V) \tag{1.216}$$

$$\mu_s = \mu_s^{\ominus} + RT \ln (c_s) \tag{1.217}$$

assuming that activity coefficients can be ignored. Here c_f represents concentration in g kg^{-1}, c_s in g l^{-1} and V (the volume of the Donnan double layer) will have units of l kg^{-1}.

Substituting Eqns 1.216 and 1.217 in Eqn 1.215, and rearranging,

$$\ln\left(\frac{c_f}{Vc_s}\right) = -\frac{zF\psi_D}{RT} - \frac{\Delta\mu^{\ominus}}{RT} \tag{1.218}$$

Therefore

$$\frac{c_f}{Vc_s} = \exp\,(-zF\psi_D/RT)\,\exp\,(-\Delta\mu^{\ominus}/RT) \tag{1.219}$$

Substituting Eqns 1.211 and 1.212 into 1.219 gives

$$\frac{c_f}{Vc_s} = \lambda^z K \tag{1.220}$$

or

$$c_f = \lambda^z K V c_s \tag{1.221}$$

which is the required distribution equation. It should be noted that in placing $\psi_s = 0$, Eqn 1.216 becomes $\lambda = \exp(-F\psi_D/RT)$. For uncharged species $z = 0$, so that $\lambda^z = 1$, and Eqn 1.221 thus becomes $c_f = KVc_s$.

The reader is reminded of the application of chemical potential in section 1.2.11.

Electrokinetic or zeta potential

If there is relative motion between a charged surface and bulk solution a plane of slip will exist in the diffuse layer, so that there is a separation of ions, and thus of charge, which in turn gives a rise to a potential difference. The plane of slip is usually known as the *shear plane* and the potential at that plane as the *electrokinetic* or *zeta* (ζ) *potential* (Figure 1.16).

So that the significance of this potential can be appreciated, let us consider the nature of each layer at the interface between a charge surface and an aqueous solution (Figure 1.17). The Stern layer may be considered immobile with regards to mobility *normal to the surface* since, if the adsorption forces

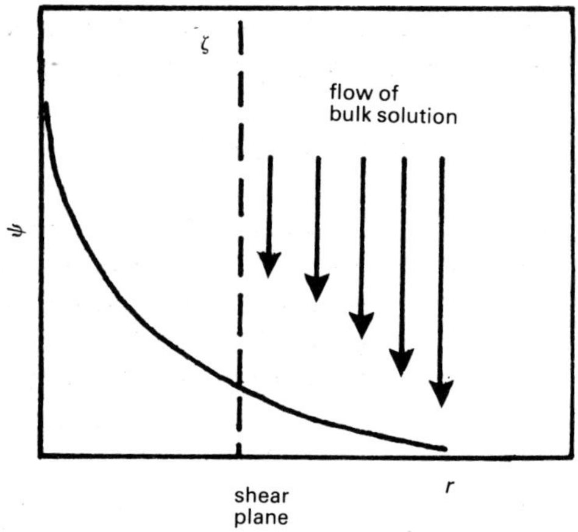

Figure 1.16 – Relative motion between bulk solution and charged surface

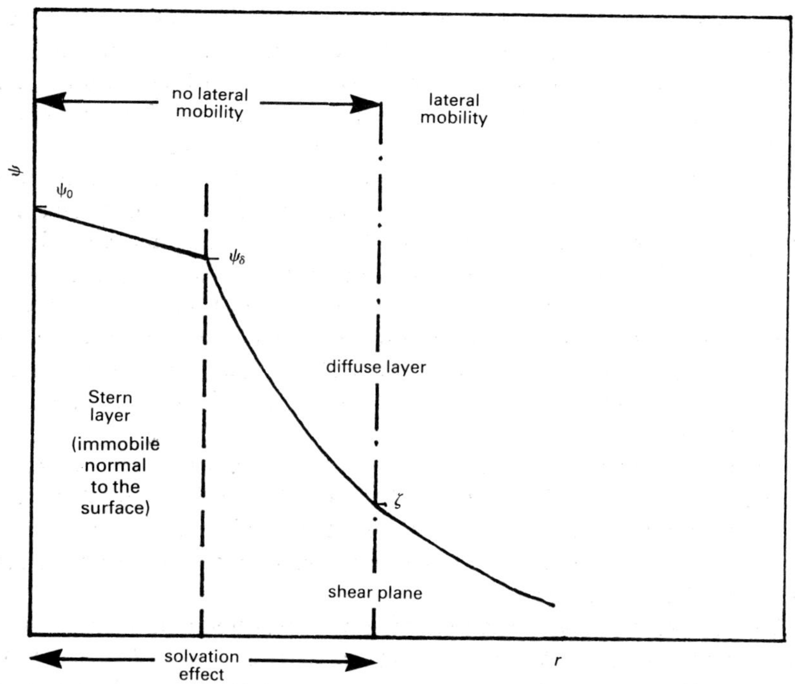

Figure 1.17 – Boundaries near a charged surface in contact with a moving solution

are strong, the time of stay of an ion in the layer will be relatively long. That is, if the average time of stay τ is given by the relationship

$$\tau = \tau_0 \exp (q/RT) \qquad (1.222)$$

where τ_0 is about 10^{-12} to 10^{-13} s and q the energy of adsorption in kJ mol^{-1}, then if q has a value of about 0.4 kJ mol^{-1}, τ will be about 10^{-13} s. This is of the order of a molecular vibration, and no adsorption would occur. On the other hand, if q has a value of about 170 kJ mol^{-1} (chemical bond energy), then τ will be about 10^{17} s, representing permanent adsorption.

In addition, *lateral mobility* or resistance to shear must be considered. The ions and surrounding medium in the Stern layer are rigidly held and the Stern layer itself is most probably immobile as regards shear. This type of immobility refers to the medium as a whole, however, and hence primarily to the solvent (e.g. water). Thus there is no reason why the shear plane should coincide exactly with the Stern layer and, as indicated in Figure 1.17, it may be located further out from the surface. The diffuse layer is a region of rapidly changing viscosity of the solution and this will add to the uncertainty as to the exact location of the shear plane and thus ζ. A situation now arises where the diffuse layer on the bulk solution side has lateral mobility (along with the bulk solution itself), and that part of the diffuse layer on the surface side of ζ, together with the Stern layer, has no lateral mobility. It should be emphasised that the ζ potential is not a phase-boundary potential, because it occurs wholly within the solution region.

In dyeing processes this potential boundary is the first at which an adsorbing dye species meets any abnormal electrical conditions. This resistance to adsorption can be reduced if the flow of the bulk solution is rapid; in turn this reduces the boundary thickness to give a lower value of ζ. Likewise an increase in c_∞ and/or z reduces the value of ζ. If, in Figure 1.18, lines 1, 2 and 3 refer to increasing concentration (c_∞) and/or ionic charge (z), then according to the Gouy–Chapman model (Eqn 1.203) the diffuse layer is compressed towards the surface, i.e. $1/\varkappa$ decreases. As a consequence, the value of ζ decreases.

Thus an increase in ionic strength has two related benefits for dye adsorption: a narrow diffuse layer and a lower ζ potential; both of these will offer a greatly reduced electrical resistance to dye adsorption. If the value of ψ_0 is reduced at the same time because of changes in surface charge density, then a similar situation arises.

The ζ potential may be evaluated from measurements carried out by electrokinetic effects. All these effects have in common the feature that relative motion is necessary between a charged surface and the bulk solution. Essentially, a charged surface is subjected to a force in an electric field and,

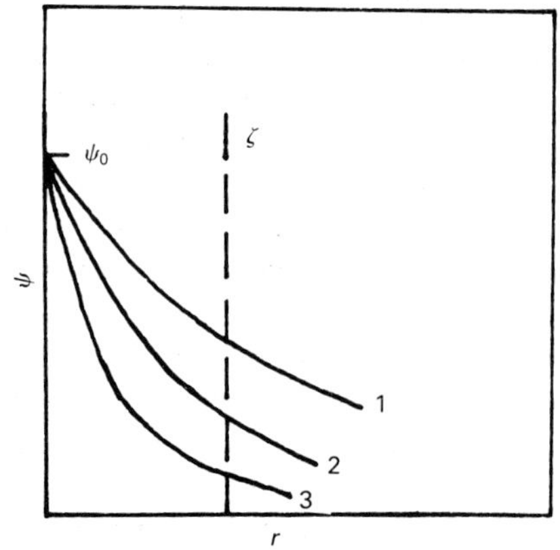

Figure 1.18 – Influence of c_∞ or z on the value of ζ

TABLE 1.5

Electrokinetic effects

Electrophoresis: solid surface moves, solution remains stationary	Sedimentation potential: solid surface moves, solution remains stationary
Electro-osmosis: solid surface remains stationary, solution moves	Streaming potential: solid surface remains stationary, solution moves
Potential *applied* externally causing movement of either surface or solution	Potential *induced* by movement of either surface or solution

at the same time, a field is induced by the relative motion of such a surface. The relationships between the various electrokinetic effects are given in Table 1.5.

Electrophoresis is the type of electrokinetic experiment most commonly carried out, because of its value as an analytical technique. It consists of applying a potential gradient (of from a few tens of volts to upwards of 10 kV) to an electrolyte solution containing also dissolved or suspended charged particles (for example, amino acids, proteins or dyes) and measuring the extent, or the rate, of movement of the particles. Since the rate of movement is directly proportional to the applied potential, the ζ potential may be calculated.

Electro-osmosis is the complement of electrophoresis. When a potential difference is applied in the latter technique, the charged particle is mobile and moves relative to the solvent which is fixed (by having it contained, for example, in a suitable strip of filter paper). If the particle is fixed, however, it is the mobile diffuse layer that moves under an applied potential difference, carrying solution with it. This is the basis of electro-osmosis; the movement of solution is measured relative to a stationary capillary or porous plug of, for example, fibres.

The *streaming potential technique* reverses that of electro-osmosis in that solution is caused to flow through a capillary or porous plug of material, and the potential induced by this flow, the streaming potential, is measured. As with all electrokinetic effects, streaming potentials are difficult to measure reproducibly; however, the technique is the one normally used to study the electrical behaviour of fibrous materials. Solution is forced under pressure P through a plug of material and the streaming potential E is measured by means of electrodes in the solution placed on either side of the plug. The ζ potential is calculated from the relationship

$$\zeta = \frac{4\pi\eta k}{P\varepsilon}.E \qquad (1.223)$$

where η is the viscosity of the solution, k its specific conductivity and ε its permittivity.

Relatively large streaming potentials can develop, particularly with solutions of low conductivity. The rapid flow of jet fuel poses a real problem in aircraft refuelling; it is essential to have an effective earth, otherwise sparks will fly!

Table 1.6 and Figures 1.19 to 1.21 illustrate results with fibres and dyes

TABLE 1.6

Surface potentials of fibres in 1×10^{-3}M KCl

Fibre	Surface potential* /mV
Acrylic (Orlon)	−44
Polyester (Terylene)	−32 to −40
Wool	−31
Secondary cellulose acetate	−28
Nylon 6.6	−22
Cotton	−17

Hexagon Digest*, **19 (1954)

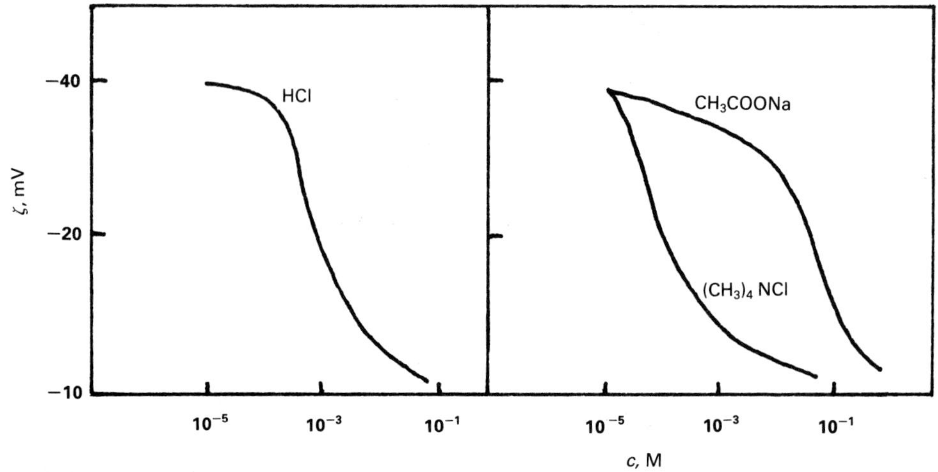

Figure 1.19 – Change in zeta potential with increasing concentration of solute (fibre: Dralon)

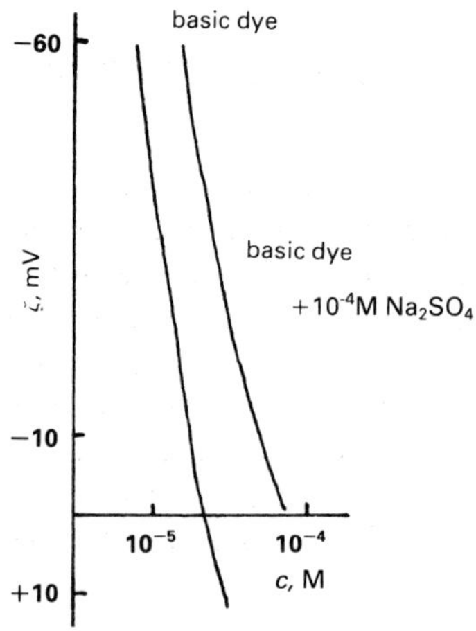

Figure 1.20 – Change in zeta potential with dye concentration (fibre: Dralon)

from the streaming potential technique. As might be expected, in general E (or ζ) varies with increasing concentration of solute, including dye, since there is a corresponding variation of ψ in the immobile layer.

The *sedimentation potential technique* is the least popular of the electrokinetic effects. Here charged particles are caused to move relative to the liquid medium by, for example, gravitational or centrifugal force. As with streaming

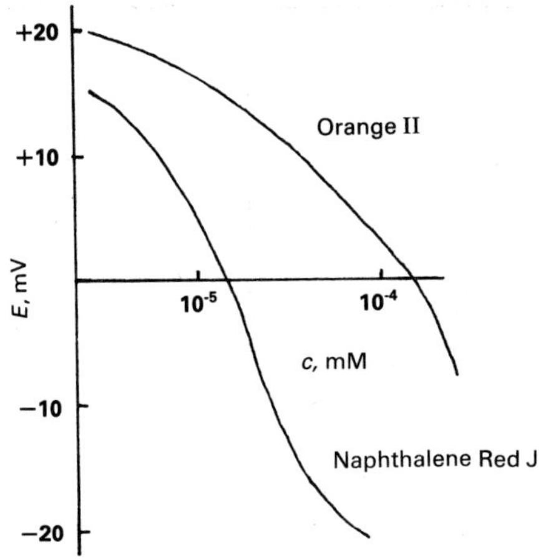

Figure 1.21 – Change in streaming potential E with dye concentration (solid: activated alumina)

potential, the induced sedimentation potential can be measured with electrodes.

1.6 KINETICS OF CHEMICAL REACTIONS

1.6.1 Introduction

Reaction kinetics is concerned with the detailed study of the rates and mechanisms of chemical reactions. The experimental part of the subject deals with precise measurement of rates of reactions at various temperatures. The interpretation of the results from the measurements leads to an understanding of the mechanisms of reactions. Combination of results from a large number of experiments gives rise to general theories of chemical reactivity.

A chemical reaction is a process of change and can be studied from one of two points of view: firstly, the ultimate extent of the change (the state of the equilibrium) and, secondly, the time taken to achieve the ultimate change.

Extent of chemical change

In principle every chemical reaction is reversible and may be written as

$$\text{reactants} \rightleftharpoons \text{products}$$

At equilibrium, the relative concentrations of reactants and products, denoted by square brackets, are determined by the equilibrium constant K, where

$$K = \frac{[\text{products}]}{[\text{reactants}]} \tag{1.78}$$

K which is related to the free energy change ΔG of the reaction at temperature T by the thermodynamic equation

$$-\Delta G = RT \ln K \qquad (1.85)$$

where R is the gas constant.

In considering Eqn 1.85 the following statements may be made:

(a) If the reaction is accompanied by a large decrease in free energy (i.e. $-\Delta G$ numerically large and therefore K large), it proceeds virtually to completion.
(b) If ΔG is numerically large but positive and therefore K very small, little or no reaction occurs.
(c) If ΔG is numerically small (whether negative or positive) and therefore K approximately unity, then the equilibrium concentrations of reactants and products are similar.

From such thermodynamic considerations, we may conclude, firstly, that the change in free energy dictates the *extent* of reaction and, secondly, that thermodynamic study *does not* indicate the rate or mechanism of reaction.

Rate of chemical change

The detailed study of chemical kinetics involves three essential stages:

(a) the precise measurement of the *rates* of reaction at different temperatures;
(b) the interpretation of results, leading to an understanding of the *mechanism* of reaction;
(c) eventually, the postulation of theories of chemical reactivity.

Before a chemical change can be studied and before a mechanism can be expressed, however, the reaction must be identified. This requires *identification of the products* and *determination of the quantities of reactants and products* during the time taken for the reaction to proceed. The quantitative relationship between reactants and products is expressed by a *stoichiometric or balanced equation*, such as

$$2NaOH + H_2SO_4 \rightarrow Na_2SO_4 + 2H_2O \qquad \text{aqueous reaction}$$

$$H_2O_2 + 2HI \rightarrow I_2 + 2H_2O \qquad \text{aqueous reaction}$$

$$H_2 + I_2 \rightarrow 2HI \qquad \text{gaseous reaction}$$

For any reaction to proceed, the molecules of reactants must *collide* with each other. The rate of reaction is proportional therefore to the frequency at which molecules collide, which in its turn is proportional to the concentrations of colliding molecules. That is,

$$\text{reaction rate} \propto \text{collision frequency} \propto \text{concentration}$$

Consider, however, the oxidation of HI: the stoichiometric equation

$$H_2O_2 + 2HI \rightarrow 2H_2O + I_2$$

suggests that

$$\text{reaction rate} \propto [H_2O_2][HI]^2$$

However, the rate is found experimentally to be proportional to [HI], not to $[HI]^2$, i.e. the reaction is

$$H_2O_2 + HI \rightarrow H_2O + HIO \qquad \text{first stage (1)}$$

and $\quad \text{reaction rate} \propto [H_2O_2][HI] \qquad \text{slow}$

The second molecule of HI is involved in the reaction

$$HIO + HI \rightarrow H_2O + I_2 \qquad \text{second stage (2)}$$

and $\quad \text{reaction rate} \propto [HIO][HI] \qquad \text{fast}$

1.6.2 Classification of chemical reactions

The oxidation of HI serves to illustrate that the stoichiometric equation does not necessarily indicate reaction mechanism. In the example, stages (1) and (2) are known as *elementary reactions* (the molecular events) and together give the *reaction mechanism*; their sum gives the overall reaction represented by the stoichiometric equation. Overall the reaction is a *complex reaction*. In any complex reaction, the *slowest* elementary reaction (or process) determines the overall rate.

A further example is the decomposition of $N_2O_5(g)$, which may be represented stoichiometrically by

$$2N_2O_5 \rightarrow 4NO_2 + O_2 \qquad \text{gaseous reaction}$$

However, the reaction is believed to take place in the following three stages, which together represent the reaction mechanism:

$N_2O_5 \rightarrow NO_2 + NO_3$	fast	stage 1
$NO_2 + NO_3 \rightarrow NO_2 + NO + O_2$	slow	stage 2
$NO + NO_3 \rightarrow 2NO_2$	fast	stage 3

Thus the rate-controlling process is stage 2.

All the reactions described so far take place in either the gaseous phase or the liquid (solution) phase. Reactions that take place in single phases only are said to be *homogeneous*. Many reactions of industrial importance involve changes at or across phase boundaries, however, and are said to be *heterogeneous*. Catalysis of gas reactions at solid catalyst surfaces, electrode reactions, many enzyme reactions and the dyeing (whether from solution or vapour) of textile materials fall into this category. There is nothing fundamentally different between these two types of process; often they take place at the same time.

In practice it may be difficult to identify elementary reactions that occur at a boundary from the rest of a heterogeneous reaction mechanism. Frequently elementary reactions are diffusion-controlled, i.e. their rates are determined by rates of diffusion either to, from or at the boundary.

The mechanism of a reaction involving a permeable solid may be subdivided into five stages:

1. Movement of reactant molecules to the surface (convection and/or diffusion).
2. Adsorption of reactant molecules onto the surface.
3. Elementary steps involving adsorbed molecules, with subsequent permeation of molecules away from the surface into the solid (diffusion).
4. Desorption of molecules from the surface.
5. Removal of desorbed molecules away from the surface (convection and/or diffusion).

Any one of these stages can be rate-controlling, and the control can pass from one stage to another during the overall reaction time. The heterogeneous nature of the dyeing process is discussed in Chapter 5.

The remainder of this section is confined to homogeneous reactions.

1.6.3 Measurement of reaction rates

In studying reaction rates, it is essential to devise an analytical technique, which may include a suitable sampling method, to enable the decrease in concentration of one of the reactants, or the increase in concentration of one of the products, to be measured from time to time during a reaction that will be occurring at constant temperature. Such concentration–time data enable kinetic behaviour to be obtained. Temperature-dependence is deduced from the results of repetitions of the procedure at various temperatures.

When using sampling techniques it must be possible to stop the reaction in the sample suddenly, so that during the time taken to carry out the analysis no further products are formed.

In any experimental technique, whether a sampling or a continuous method is used, the reaction vessel must be placed in a thermostatted bath having a uniform temperature held constant to within at most 0.1 °C.

The rate may be measured by one of the following methods:

(a) volumetric or gravimetric analysis of samples removed from a reacting solution at certain time intervals;
(b) the measurement of pressure change in a gaseous system reacting at constant volume, or of volume change at constant pressure;
(c) the measurement of change of some physical property related to the concentration of a reactant or product and which enables the reaction to be monitored without removing a sample or disturbing the reactive mixture, such as colour intensity (absorbance), refractive index, electrical conductivity, viscosity, liberation of heat, optical rotation and nuclear magnetic resonance (n.m.r.).

1.6.4 Factors determining reaction rates

The rate of chemical change depends primarily on the properties of the reacting molecules. It is also found to vary with the concentrations of reacting molecules and with temperature, so that in order to understand why molecules react at the observed rate it is necessary to express this rate in terms of quantities that are independent of both concentration and temperature.

Reaction rates may be influenced also by the presence of catalysts or inhibitors, the concentration of products, and by visible or ultra-violet radiation or ionising radiation (α-, β-, γ-rays).

1.6.5 Molecularity

Most reactions are complex and, as we have seen previously, such reactions consist of two or more elementary reactions. These are the simplest reactions by which we can represent a chemical change, and it is with these that theoretical treatments of reaction rates are concerned. They may be classified according to their *molecularity*, that is, to the *number of reactant molecules or species taking part in the chemical change*.

The molecularity is necessarily a small whole number, and is a theoretical quantity in that to evaluate it the mechanism of reaction must be known or assumed. Reactions that have a molecularity of one are termed *unimolecular*, and occur mainly in the gaseous phase. Examples are

$$N_2O_5(g) \rightarrow NO_2(g) + NO_3(g)$$

$$CH_3NC(g) \rightarrow CH_3CN(g)$$

$$\text{cyclopropane(g)} \rightarrow \text{propene(g)}$$

Bimolecular reactions (having a molecularity of two) are by far the commonest reactions in both gaseous and solution phases, and are the most important elementary reactions. Examples are

$$2HI(g) \rightarrow H_2(g) + I_2(g)$$

$$CO(g) + N_2O(g) \rightarrow CO_2(g) + N_2(g)$$

$$CH_3Br(aq) + OH^-(aq) \rightarrow CH_3OH(aq) + Br^-(aq)$$

$$\text{Dye Cl(aq)} + OH^-(aq) \rightarrow \text{Dye OH(aq)} + Cl^-(aq)$$

$$KClO(aq) + KClO_2(aq) \rightarrow KCl(aq) + KClO_3(aq)$$

Termolecular reactions are rare, since they entail the *simultaneous* collision of three molecules, and occur only in the gaseous phase.

If the mechanism of a reaction is not known, the *order* of the reaction must be used to describe the reaction. However, the order of a reaction may be stated only after *experimental work* has revealed how the rate of reaction depends on the concentrations of the molecular species, i.e. the order of a reaction is an experimental result.

Order of reaction

For the reaction

$$aA + bB \rightarrow \text{products}$$

the reaction rate may be defined in terms of concentrations of reactants, A and B, by the equation

$$\text{rate} = k_r[A]^m[B]^n \qquad (1.224)$$

This equation is known as the *rate equation* or *rate law*; the rate is found to be dependent on some power, m and n, of the reactants, and m and n are not necessarily related to the stoichiometric coefficients a and b. k_r is the *rate constant* and will be discussed later in this section.

From the rate law, the order of a reaction may be defined.

(a) The *overall order* is the sum, $m + n$, of the power terms in the rate law.

Overall order indicates how rate responds to changes in absolute concentration such as changes in pressure for gaseous reactions, or in dilution for a solution reaction.

(b) *Order with respect to a reactant* or *individual order* is the individual power of an appropriate reactant; in the example given above, order with respect to A is m, and order with respect to B is n. Individual orders indicate the sensitivity of the reaction to changes in concentration of each species and also suggest the chemical mechanism of the reaction.

In *first-order reactions*, the rate is directly proportional to the concentration of the reacting species, i.e.

$$\text{rate} = k_r[\text{A}] \tag{1.225}$$

For example, the decomposition of ethane is represented by

$$C_2H_6(g) \rightarrow C_2H_4(g) + H_2(g)$$

and

$$\text{rate} = k_r[C_2H_6] \tag{1.226}$$

Similarly for the isocyanide rearrangement,

$$CH_3NC(g) \rightarrow CH_3CN(g)$$

$$\text{rate} = k_r[CH_3NC] \tag{1.227}$$

In *second-order reactions*, the reaction rate is proportional to the square of the concentration of a single reactant or to the product of the concentrations of two reacting species, i.e.

$$\text{rate} = k_r[\text{A}]^2 \tag{1.228}$$

or

$$\text{rate} = k_r[\text{A}][\text{B}] \tag{1.229}$$

In the following two reactions,

$$2HI(g) \rightarrow H_2(g) + I_2(g)$$

$$CH_3COOC_2H_5(aq) + OH^-(aq) \rightarrow CH_3COO^-(aq) + C_2H_5OH(aq)$$

the rates of reaction are given by

$$\text{rate} = k_r[HI]^2 \tag{1.230}$$

and
$$\text{rate} = k_r[OH^-][CH_3COOC_2H_5] \quad (1.231)$$

respectively. Both are second-order reactions; in the latter (1.231), the reaction is first-order with respect to each of the reactants.

1.6.6 Concept of order

The following should be appreciated with respect to an order of a reaction:

(a) Not all reactions can be said to have an order. For example, the reaction

$$H_2(g) + Br_2(g) \rightarrow 2HBr$$

is complex, with the reaction rate given by

$$\text{rate} \propto \frac{k[H_2][Br_2]^{0.5}}{1 + \dfrac{k'[HBr]}{[Br_2]}} \quad (1.232)$$

Although this rate law is complex, an initial rate can be identified before [HBr] becomes significant. Here,

$$\text{initial rate} = k_r[H_2][Br_2]^{0.5} \quad (1.233)$$

Thus under special conditions an order can be defined.

(b) An order of reaction may be zero, integral or fractional (unlike molecularity). Order simply expresses experimental observation. Thus the overall order of the above initial reaction (Eqn 1.233), is 1.5, but it is 1 with respect to H_2, and 0.5 with respect to Br_2.

(c) Orders can be deceiving. For example, in the reaction represented stoichiometrically by

$$S_2O_8^{2-}(aq) + 3I^-(aq) \rightarrow 2SO_4^{2-}(aq) + I_3^-$$

the experimental rate of reaction is second-order, corresponding to

$$\text{rate} = k_r[S_2O_8^{2-}][I^-] \quad (1.234)$$

However, if $[S_2O_8^{2-}] \gg [I^-]$ then

$$\text{rate} = k'_r\,[I^-] \quad (1.235)$$

This is now a first-order reaction, the molecularity changing from a bimolecular to a *pseudo-unimolecular* reaction. Pseudo-molecular reactions are encountered when the concentration of one (or more) of the reactants remains essentially constant during the course of the reaction. Such reactions occur in a solvent which is itself a reactant.

The hydrolysis of reactive dyes in aqueous solution is strictly a bimolecular, second-order reaction expressed by

$$\text{rate} = k_r[\text{D}][\text{OH}^-] \tag{1.236}$$

Normally, however, the hydroxide ion concentration $[\text{OH}^-]$ is very high and is kept constant by use of a buffer, and the reaction can be considered to be pseudo-unimolecular and first-order, i.e.

$$\text{rate} = k_r'[\text{D}] \tag{1.237}$$

(d) It is not possible to deduce the order of a reaction from the stoichiometric equation. If the mechanism is simple and the stoichiometric equation is the elementary equation, then the apparent order may well be the correct one. The synthesis of hydrogen iodide

$$H_2(g) + I_2(g) \rightarrow 2HI(g)$$

is second-order as well as bimolecular. However, the same cannot be said with regard to hydrogen bromide:

$$H_2(g) + Br_2 \rightarrow 2HBr(g)$$

(e) It cannot be too strongly emphasised that the order of a reaction is a factual experimental quantity, being concerned solely with the way in which reaction rate depends upon concentration. It can be determined without prior knowledge of the mechanism of the reaction. Above all, order must not be confused with molecularity. Unfortunately some older literature and even some modern publications, refer to 'pseudo-first-order reactions' or 'pseudo-first-order rate constants'. This is to be deprecated. The use of the term 'pseudo-unimolecular' is to be preferred, as will be evident from the foregoing discussion.

1.6.7 Variation of concentration with time

For simplicity, let us consider the reaction A → products. The concentration [A] will decrease with the time of reaction t, and we can represent such changes as follows:

$$[A] = [A]_0 \text{ at zero time; } [A] = [A]_t \text{ at time } t\text{; } [A] = [A]_\infty \text{ at time } t_\infty \quad (1.238)$$

In graphical form, the simplest variations of $[A]_t$ with time are illustrated in Figure 1.22.

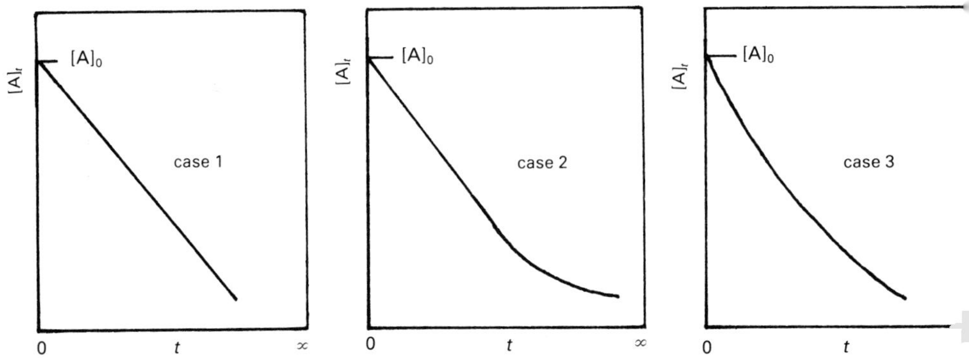

Figure 1.22 – Variation of $[A]_t$ with time t

Mathematically these relationships are defined as:

Case 1 $$[A]_t = [A]_0 - k_r t \quad \text{or} \quad k_r t = ([A]_0 - [A]_t) \quad (1.239)$$

Case 2 $$[A]_t = [A]_0 \exp(-k_r t) \quad \text{or} \quad k_r t = \ln([A]_0/[A]_t) \quad (1.240)$$

Case 3 $$\frac{1}{[A]_t} = \frac{1}{[A]_0} + k_r t \quad \text{or} \quad k_r t = \frac{([A]_0 - [A]_t)}{[A]_0\,[A]_t} \quad (1.241)$$

In each instance, the equation describes the experimentally determined data, and k_r (the rate constant) is an experimentally determined parameter.

Also in the three instances the variations of concentration with time, i.e. the reaction rates, are clearly different. Mathematically this is represented by the differential $d[A]_t/dt$, which is the slope or tangent to the plot at time t. In three cases we have the following conditions:

Case 1 $d[A]_t/dt$ is constant throughout the reaction, so that the decrease in the value of [A] is constant over equal intervals of time.

Case 2 $d[A]_t/dt$ decreases as the reaction proceeds, although in the early states it is virtually constant.

Case 3 $d[A]_t/dt$ decreases measurably throughout the whole reaction time.

Figure 1.23 gives a fuller explanation of the physical meaning of such a graph.

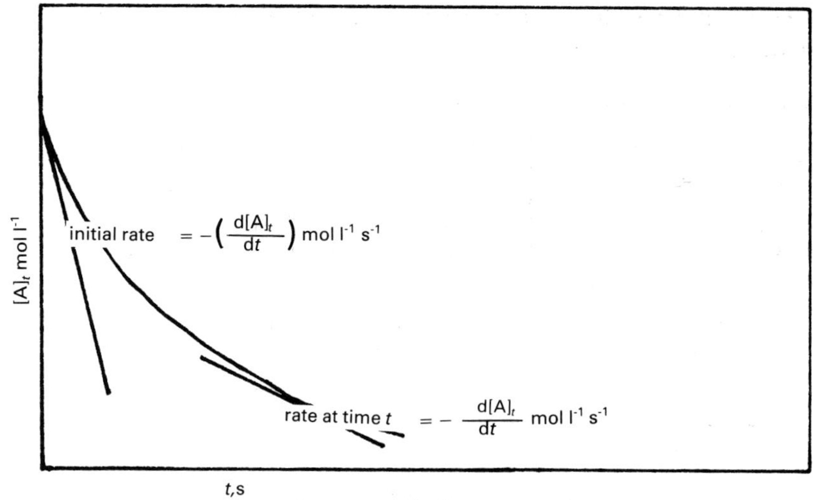

Figure 1.23 – Explanation of concentration profile; note that (1) slopes are negative, (2) the rate is defined as $-d[A]_t/dt$, *(3) the physical meaning is that the concentration of A decreases with increasing time*

More realistically, the reaction may be considered as, for example,

$$A + B \rightarrow C + D$$

and the reaction rate could be expressed in terms of any of the reactants or products as $-d[A]_t/dt$, $-d[B]_t/dt$, $+d[C]_t/dt$ or $+d[D]_t/dt$. It is conventional, however, to refer to the reactants only, since it is the reactants and not the products that actually cause the reaction to take place.

1.6.8 Variation of rate with concentration

Since the reactant concentration [A] decreases as the reaction takes place, the rate must also decrease with time according to the rate equation or rate law. Thus, for the unimolecular decomposition of A,

$$\text{rate} = -d[A]_t/dt = k_r[A]_t \text{ mol } l^{-1} \text{ s}^{-1} \tag{1.242}$$

The rate constant k_r remains unchanged throughout the reaction and provides a more convenient measure of reaction rate than does the experimentally measured rate. In general, k_r is numerically equal to the reaction rate when the reactants are at unit concentration, i.e. when $[A]_t = 1$ mol l^{-1} in the example: it is experimentally very difficult, however, to measure the rate under these conditions. The rate constant can be obtained from the initial rate if the initial reactant concentrations are known, but it is often difficult to determine the exact time at which the reaction commences.

It is more satisfactory to allow the reaction to proceed for an appropriate length of time, during which the reactants are analysed. Knowledge of the behaviour of the reaction and of k_r is obtained from an *integrated form of the rate equation*. The particular form of the integrated equation obtained depends on the number and type of concentration terms involved, i.e. it depends on the order of the reaction.

Zero-order reactions

The rate equation for a zero-order process is

$$-d[A]_t/dt = k_r[A]^0_t \tag{1.243}$$

or

$$-d[A]_t/dt = k_r \tag{1.244}$$

and this can be rearranged to

$$-d[A]_t = k_r\, dt \tag{1.245}$$

for integration with the conditions $[A]_0$ at $t = 0$ and $[A]_t$ at $t = t$, then

$$-\int_{[A]_0}^{[A]_t} d[A]_t = k_r \int_0^t dt \tag{1.246}$$

$$-\Big[[A]_t\Big]_{[A]_0}^{[A]_t} = k_r t \tag{1.247}$$

$$[A]_0 - [A]_t = k_r t \tag{1.239}$$

(see Figure 1.22). This equation indicates that a straight line will be obtained for a zero-order reaction if $[A]_t$ is plotted against t; the intercept will be $[A]_0$ and the slope $-k_r$. Thus k_r can be determined and will have units mol l^{-1} s^{-1}.

Although zero-order reactions are not common in homogeneous systems, they do occur in heterogeneous systems, particularly in catalysed reactions, including those involving enzymes.

Theoretically, a chemical reaction never ceases, and it is therefore sometimes useful to quote the characteristic time of half-change or half-reaction or *reaction half-life* $t_{\frac{1}{2}}$, i.e. the time required for the concentration of reactant to decrease to half its initial value. Thus since

$$[\mathrm{A}]_t = \tfrac{1}{2}[\mathrm{A}]_0 \qquad \text{when } t = t_{\frac{1}{2}} \tag{1.248}$$

then

$$\tfrac{1}{2}[\mathrm{A}]_0 = k_r t_{\frac{1}{2}} \tag{1.249}$$

Therefore

$$k_r = \frac{[\mathrm{A}]_0}{2t_{\frac{1}{2}}} \tag{1.250}$$

or

$$t_{\frac{1}{2}} = \frac{[\mathrm{A}]_0}{2k_r}$$

Thus, for zero-order reactions, $t_{\frac{1}{2}} \propto [\mathrm{A}]_0$.

First-order reactions

The rate equation or law is

$$-\mathrm{d}[\mathrm{A}]_t/\mathrm{d}t = k_r[\mathrm{A}]_t \tag{1.251}$$

Integration with the conditions $[\mathrm{A}]_0$ at $t = 0$ and $[\mathrm{A}]_t$ at $t = t$ gives

$$-\int_{[\mathrm{A}]_0}^{[\mathrm{A}]_t} \frac{\mathrm{d}[\mathrm{A}]_t}{[\mathrm{A}]_t} = k_r \int_0^t \mathrm{d}t \tag{1.252}$$

$$-\Big[\ln[\mathrm{A}]_t\Big]_{[\mathrm{A}]_0}^{[\mathrm{A}]_t} = k_r t \tag{1.252}$$

or

$$\ln\left(\frac{[\mathrm{A}]_0}{[\mathrm{A}]_t}\right) = k_r t \tag{1.240}$$

(see Figure 1.22). A straight line will be obtained for a first order reaction if $\ln [\mathrm{A}]_t$ is plotted against t; the intercept will be $\ln [\mathrm{A}]_0$ and the slope $-k_r$ (in seconds^{-1}). The rate constant can also be obtained from the rearranged equation without plotting data, using

$$k_r = \frac{1}{t}\ln\left(\frac{[\mathrm{A}]_0}{[\mathrm{A}]_t}\right) \mathrm{s}^{-1} \tag{1.240}$$

Once a reaction has been confirmed as first-order and k_r determined, then values of $[A]_t$ at any time t can be calculated. The time of half-reaction $t_{\frac{1}{2}}$ is given by

$$t_{\frac{1}{2}} = \frac{\ln 2}{k_r}\ \mathrm{s} \tag{1.254}$$

i.e. the half-life is independent of the initial concentration $[A]_0$. This means that, provided a reaction follows strictly first-order kinetics, the time for 50 per cent reaction cannot be changed by altering $[A]_0$; this is not so for zero-order and second-order reactions. For a first-order batch process, the batch time cannot be altered (increased or decreased) by using more or less of the initial reactants; for this, other changes must be made.

First-order reactions are common and are often observed in solution, particularly if the solvent is one of the reactants and present in large excess, in which case the reaction may be considered as pseudo-unimolecular, as stated earlier (section 1.6.6). Many gas-phase reactions and radioactive decay processes obey the first-order rate law. In decay processes the half-life is an easier concept to visualise than is the rate constant. For radium, for example, $t_{\frac{1}{2}} = 1690$ years, but k_r is given by

$$k_r = \frac{\ln 2}{1690}\ \mathrm{years}^{-1} = 1.301 \times 10^{-11}\ \mathrm{s}^{-1}$$

Second-order reactions

Second-order reactions are the most common type of reaction; in general, provided a reaction occurs in a single step and the reactants are present in roughly equal concentrations, then a second-order reaction may be expected. Reactions of this type occur often in organic chemistry. If the reaction is written as

$$\mathrm{A + B \rightarrow products}$$

then the rate law is

$$-\mathrm{d}[A]_t/\mathrm{d}t = k_r[A]_t[B]_t \tag{1.255}$$

However, consider the case where $[A]_0 = [B]_0$, i.e. $[A]_t = [B]_t$; then

$$-\mathrm{d}[A]_t/\mathrm{d}t = k_r[A]_t^2 \tag{1.256}$$

This may be rearranged to

$$\mathrm{d}[A]_t/[A]_t^2 = k_r\,\mathrm{d}t \tag{1.257}$$

which on integration with the conditions $[A]_0$ and $[A]_t$ at $t = 0$ and $t = t$ respectively gives

$$-\int_{[A]_0}^{[A]_t} d[A]_t/[A]_t^2 = k_r \int_0^t dt \qquad (1.258)$$

$$\Big[1/[A]_t\Big]_{[A]_0}^{[A]_t} = k_r t \qquad (1.259)$$

$$1/[A]_t = 1/[A]_0 = k_r t \qquad (1.241)$$

or

$$\frac{[A]_0 - [A]_t}{[A]_t\,[A]_0} = k_r t$$

(see Figure 1.22). A straight line will be obtained for a second-order reaction (where reaction depends upon the first power of the concentrations of two reactants and $[A]_0 = [B]_0$, or the second power of the concentration of one reactant only) if $1/[A]_t$ is plotted against t; the intercept will be $1/[A]_0$ and the slope will be k_r. Some workers advocate the plot of $([A]_0 - [A]_t)/[A]_t[A]_0$ against t.

Having obtained the value for k_r (in $l\,mol^{-1}\,s^{-1}$) then $[A]_t$ may be calculated at any time, using

$$[A]_t = \frac{[A]_0}{([A]_0 k_r t + 1)} \text{mol } l^{-1} \qquad (1.241)$$

The time of half-reaction is given by

$$t_{\frac{1}{2}} = \frac{1}{k_r[A]_0} \text{ s} \qquad (1.260)$$

In other words, $t_{\frac{1}{2}}$ is inversely proportional to the initial concentration of reactants, so that increasing the concentration will decrease the time taken to reach 50 per cent reaction.

In dealing with second-order reactions in which the initial concentrations are not equal, then the integrated form of the rate law is

$$\frac{1}{([A]_0 - [B]_0)} \ln\left(\frac{[A]_t[B]_0}{[A]_0[B]_t}\right) = k_r t \qquad (1.261)$$

Pseudo-unimolecular reactions

In bimolecular reactions it is often convenient to arrange that one reactant is present in large excess, so that the change in concentration of this reactant can be neglected. In other words, its concentration effectively remains constant at $[B]_0$. The rate equation (Eqn 1.224) may then be written as:

$$-d[A]_t/dt = k_r'[A]_t \quad (1.262)$$

where

$$k_r' = k_r[B]_0 \quad (1.263)$$

Eqn 1.262 is entirely analogous to Eqn 1.251, the equation for a first-order reaction, so that the value for k_r' may be readily obtained by plotting ln $[A]_t$ against t. The constant k_r' thus obtained is a first-order constant, which represents the fact that the (biomolecular) reaction is of the first order with respect to reactant A. If now the experiment is repeated several times at different values of $[B]_0$ (each large compared to $[A]_0$), and the series of values of k_r' plotted against $[B]_0$, a straight line will be obtained of slope k_r (Eqn 1.263). Alternatively, if Eqn 1.263 is put in logarithmic form, i.e.

$$\log k_r' = \log k + \log [B]_0$$

a plot of log k_r' against log $[B]_0$ should be a straight line of unit slope, having intercept log k.

This procedure is widely used in studying reactions which are thought to be bimolecular. A constant value for k_r over a range of values of $[B]_0$ is a good confirmation that the reaction is genuinely bimolecular, whilst variations in k_r with changes in $[B]_0$ are a positive indication that other processes are playing a part. Good examples of this have been found in work with reactive dyes (see Chapter 6).

1.6.9 Variation of rate constant with temperature

The rates of most chemical reactions increase as the temperature is increased. If the rate law is written as

$$\text{rate of reaction} = k_r[\text{reactant}]^{\text{order}} \quad (1.265)$$

then, of the terms on the right-hand side, only the rate constant k_r is temperature-dependent. Thus although the rate constant is independent of reactant concentration it does vary with temperature, and the variation can usually be fitted to an equation of the form

$$k_r = A \exp(-E/RT) \qquad (1.266)$$

known as the *Arrhenius equation*. In this equation, E is the *activation energy* (in J mol^{-1} or kJ mol^{-1}) and is a measure of the amount of energy required to bring about the particular reaction; A is the 'pre-exponential' or *Arrhenius frequency factor* and has the same units as the rate constant, and R is the gas constant (8.314 J mol^{-1} K^{-1}). The equation is important in reaction kinetics and can provide information about the reaction mechanism. Both A and E are virtually independent of temperature, and both are determined by the properties of the reacting molecules themselves. Once A and E are known, then for the reaction A + B → products

$$-d[A]_t/dt = A \exp(-E/RT)\,[A]_t[B]_t \text{ mol l}^{-1}\text{ s}^{-1} \qquad (1.267)$$

In logarithmic form, the Arrhenius equation becomes

$$\ln k_r = \ln A - E/RT \qquad (1.268)$$

or

$$\log k_r = \log A - \frac{E}{2.303 \times RT}$$

Thus, if a reaction is studied at different temperatures a plot of log k_r against $1/T$ should yield a straight line having slope $-E/2.303R$ and intercept of log A, so that values for E and A may be obtained. If the plot is not linear this is a good indication that other processes are intervening.

Once E and A are evaluated, and k_r of course is known, the rate of the reaction may be calculated for any conditions of concentration and temperature.

FURTHER READING

A R Knight, *Introductory physical chemistry* (London: Prentice-Hall International Inc., 1970).
A J Mee, *Physical chemistry*, 6th edn (London: Heinemann, 1962).
R Aveyard and D A Haydon, *An introduction to the principles of surface chemistry* (Cambridge: CUP, 1973).
Duncan J Shaw, *Introduction to colloid and surface chemistry*, 2nd edn (London: Butterworths, 1970).
J L Latham, *Elementary reaction kinetics*, 2nd edn (London: Butterworths, 1969).
B Stevens, *Chemical kinetics* (London: Chapman and Hall, 1961).

More advanced texts:

S Glasstone and D Lewis, *Elements of physical chemistry*, 2nd edn (London: Macmillan Press Ltd., 1960).
J Rose, *Dynamic physical chemistry – a textbook of thermodynamics, Equilibrium and Kinetics* (London: Pitman, 1961).

D O Haywood and B M W Trapnell, *Chemisorption*, 2nd edn (London: Butterworths, 1964).
Duncan J Shaw, *Electrophoresis* (London: Academic Press, 1969).
Keith J Laidler, *Reaction kinetics*, vols. 1 and 2 (Oxford: Pergamon, 1963).

Note: This list is not intended to be exhaustive; many alternative texts are available.

CHAPTER 2

Dye–fibre bonds and their investigation

CHARLES H GILES

2.1 HISTORICAL BACKGROUND

The apparently simple way in which many dyes transfer rapidly from a solution to a fibre immersed in it has a fascination to the layman and the scientist alike, and it is not surprising that it has exercised the attention of chemists throughout the whole period of the growth of modern science. As long ago as 1664 Robert Boyle, in the first scientific treatise on colour [1], surmised that 'The Liquors that Dyers imploy to tinge . . . are qualified to do so by multitudes of little corpuscles of the Pigment or Dyeing stuff, which are dissolved and extracted by the Liquor, and swim to and fro in it, those Corpuscles of Colour, insinuating themselves into, and filling all the Pores of the Body to be Dyed. . . .' We can now give a more detailed description of the dyeing process, though Boyle's picturesque sketch still holds good.

This subject attracted continued attention [2–7], but not until the mid-nineteenth century did clear theories of the mechanism of dyeing begin to be discussed. Nevertheless, the achievements of the organic chemists in producing wide ranges of new synthetic dyes far outstripped their physical colleagues' ability to interpret the dyeing process itself.

Possibly one of the earliest investigators to consider dyeing processes in some detail as a subject for chemical research was Walter Crum [7], a well-known industrialist and scientist of Glasgow, who discussed the chemistry of cotton dyeing in the 1840s. Thereafter, the subject gathered momentum and we find many authors in Britain and the European continent concerned, during the latter part of the last century, with dye–fibre reactions. Prominent amongst these were Knecht [8] and his collaborators, who laid the foundations of the modern theory of wool dyeing, and Hummel [9] and his collaborators, who made notable contributions to the understanding of the chromium mordant processes used for wool dyeing.

Little real progress could be made, however, until several landmarks in the history of chemistry had been passed, viz. the formulation of the structure of dye molecules and also of their complexes with mordanting metals, the ionic theory and, in more recent times, the concepts of the hydrogen bond and other polar bonds and of the chain-like molecules of polymers. The development of instrumentation has also played an important part, particularly the advent of photoelectric spectrophotometers.

Most of the mechanisms now believed to operate in the reactions between dyes and fibres had, however, already been suggested by the early years of this century, but the background was not rigidly quantitative [10–13]. Indeed, at least one monograph on the subject [12], published as late as 1913, is entirely descriptive, without tables or diagrams.

In the first decade of this century, Harrison [14] made some of the earliest investigations of dyeing mechanisms, employing modern physico-chemical techniques, but the greatest advances have been recorded during the past four decades. Neale [15] and his collaborators at Manchester pioneered the kinetic and thermodynamic approach to the subject, using direct dyes and cellulose. The corresponding work on wool included studies of dyeing by Elöd [16] and measurements of titration values of the fibre by acids, made by Harris *et al.* at the National Bureau of Standards [17]; this was applied, on the basis of the theory of acid dyeing of wool suggested by Knecht, to wool dyeing by Speakman and his school in Leeds [18] and by Vickerstaff and his collaborators [19] in the laboratories of ICI Ltd, Dyestuffs Division (now Organics Division), in Manchester. A significant contribution was made by Rideal and Gilbert [20] at Cambridge, who were the first to apply thermodynamic principles to an exact determination of the affinity of a fibre (wool) for acids and dyes. Their methods were later developed for cellulose fibres by R H Peters and Vickerstaff [21]. The possibility thus opened up, viz. of defining the attraction between a fibre and a dye in precise quantitative terms, led to a much more detailed study of the many factors that influence dyeing processes, and the use of such thermodynamic methods is now standard practice. Possibly the appearance of Vickerstaff's well-known monograph in 1950 [22], following an earlier work by Valkó [23], finally established dyeing studies as a distinct and important branch of physical chemistry in its own right. Its establishment was confirmed in 1953 by the holding of a Faraday Society Discussion on the subject [24]. However, in spite of the very great amount of information now available, large gaps in our knowledge and its interpretation remain, and they provide an intriguing and rewarding field for further penetration. Indeed the developing of new fibres and of new dyeing processes, which proceeds apace, continually widens the range of chemical mechanisms requiring investigation [25], so that the physical chemist in this field, like Alice, must run as hard as he can even to keep in the same place [26].

2.2 CLASSIFICATION OF FIBRES AND DYES

2.2.1 Textile fibres

For convenience in studying the mechanisms of dyeing, textile fibres may be broadly classified thus:

	Hydrophilic	
Cellulosic:	vegetable fibres, e.g. cotton and linen; also rayons (regenerated cellulose)	
Proteins:	wool, silk and hair fibres	
	Hydrophobic	
Synthetic polymer fibres:	Non-ionic	– secondary cellulose acetate, cellulose triacetate, polyester, polypropylene
	Cationic	– polyamides (nylon)
	Anionic	– polyacrylonitriles (acrylics)

2.2.2 Dyes

Dye molecules are complex aromatic structures with extended π-electron systems, polar substituents and water-solubilising groups [27, 28] (Table 2.1).

Solubility is essential to enable the dye to penetrate the intermolecular

TABLE 2.1

Solubilising groups used in dye molecules

Group	Types of dye
Permanent:	
$—SO_3Na$ (or $—COONa$)	Direct; acid; chrome mordant; reactive; 1:1 metal-complex
$—\overset{+}{N}H_2HCl^-$, $—\overset{+}{N}R_3Cl^-$	Basic dyes for cellulose, wool, silk, and acrylic fibres
$—OH$, $—NH_2$, $—SO_2NH_2$	Disperse dyes for cellulose acetate, nylon and polyester fibres; 1:2 metal-complex dyes for wool and nylon
Temporary:	
$—O^-Na^+$	Dyes for cellulose, i.e. naphthols for subsequent azo coupling on the fibre; vat dyes
onium groups, e.g. $—CH_2SC(NR_2)=\overset{+}{N}R_2$ (R=Alk or Ar)	Phthalocyanine dyes for cellulose
$—OSO_3^-Na^+$	Solubilised vat dyes

micro-pores of the fibre. The requirements of the dyer and the user of the dyed fabric are opposed, however, on a molecular level. Ideally, dyes should have very small and highly soluble molecules in solution, to ensure uniformity of penetration into the fabric, but large and insoluble molecules when inside the fibre, to ensure resistance to the manifold influences brought to bear on them in daily wear.

Except with dyes for cellulose, the nature or size of the aromatic residue is usually of less importance in determining the use of the dye for a given class of fibre than are the nature and positions of the substituent groups. The particular features of the dye molecule which determine its suitability for the respective fibres are illustrated in the following structural formulae and described below.

Direct dyes

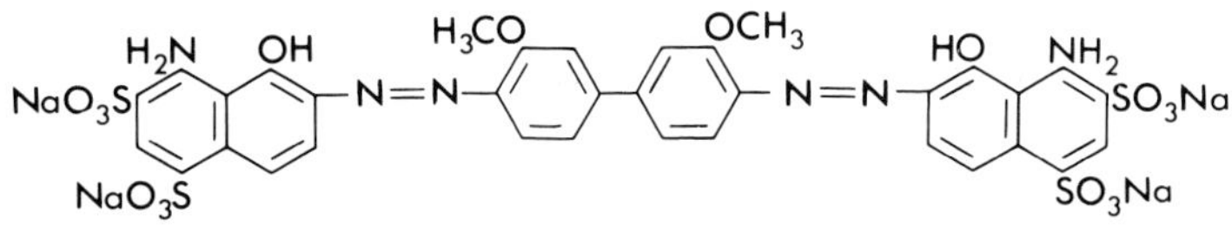

(I) C.I. Direct Blue 1 (C.I. 24410)
Sky Blue FF

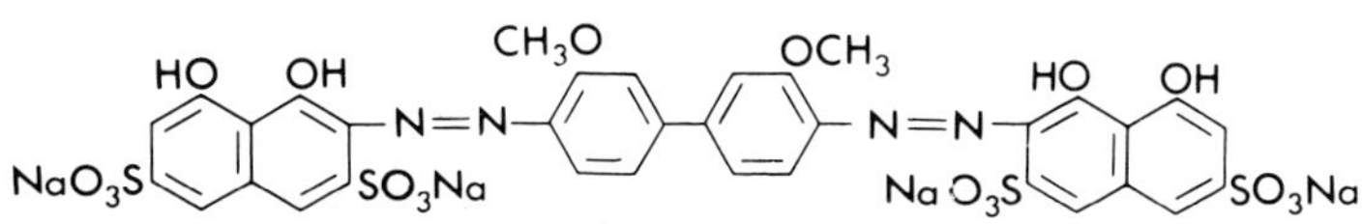

(II) C.I. Direct Blue 10 (C.I. 24340)

Vat dyes

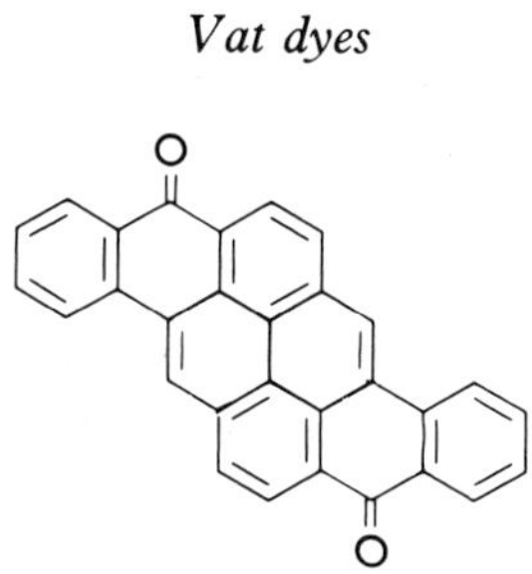

(III) An anthraquinonoid vat dye
C.I. Vat Orange 9 (C.I. 59700)

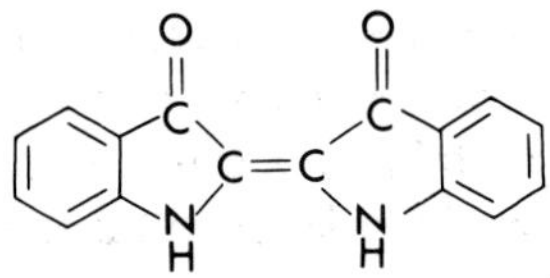

(IV) Indigo
C.I. Vat Blue 1 (C.I. 73000)

Azoic coupling components

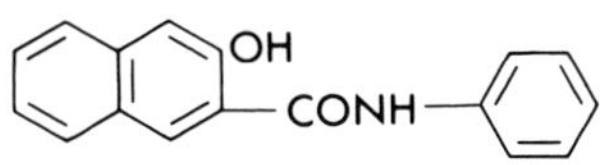

(V) C.I. Azoic Coupling Component 2 (C.I. 37505)

Acid dyes

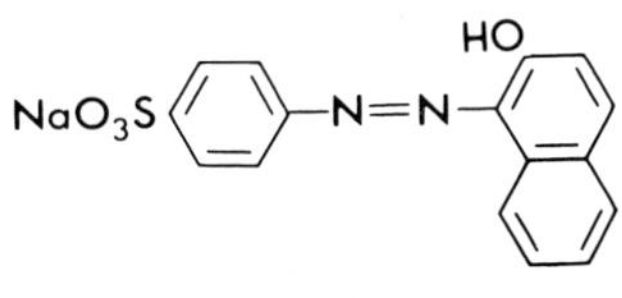

(VI) C.I. Acid Orange 7 (C.I. 15510)
Orange II

Azo metal-complex dyes

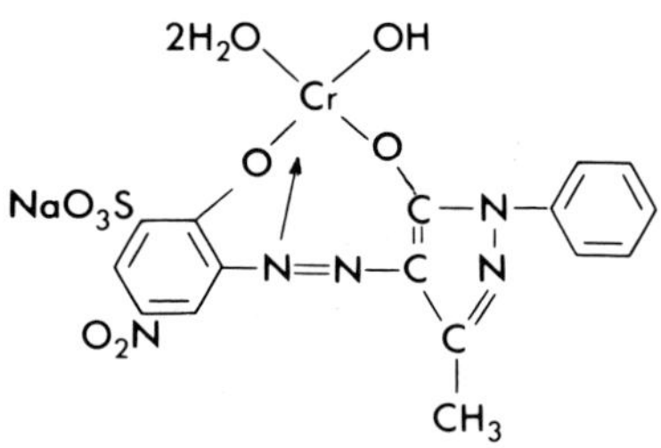

(VII) An acid-dyeing, 1:1 metal-complex dye
C.I. Acid Orange 74 (C.I. 18745)

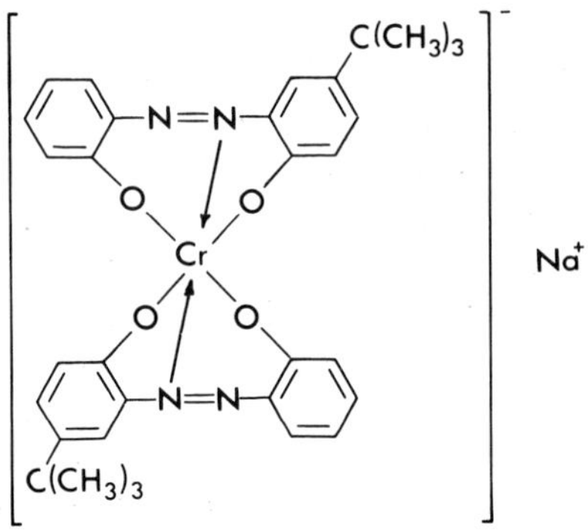

(VIII) A neutral-dyeing, 1:2 metal-complex dye
(bordeaux hue)

Basic ('cationic') dyes

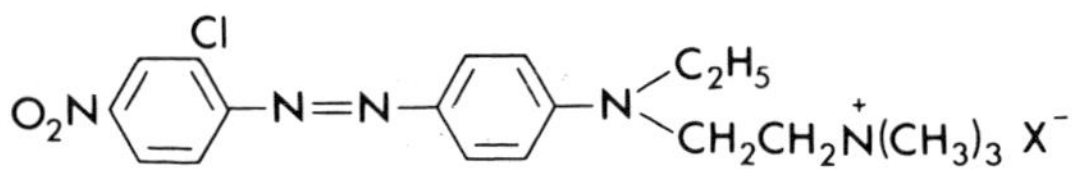

(IX) C.I. Basic Red 18 (C.I. 11085)

Disperse (non-ionic) dyes

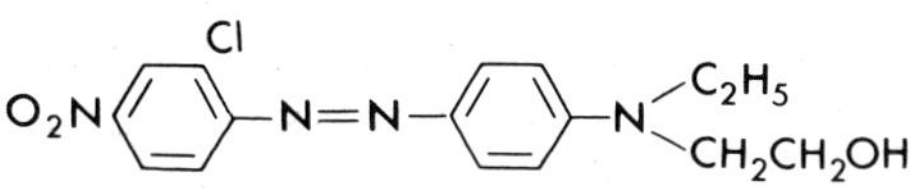

(X) C.I. Disperse Red 13
(C.I. 11115)

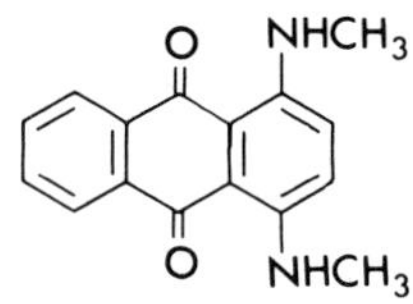

(XI) C.I. Disperse Blue 14
(C.I. 61500)

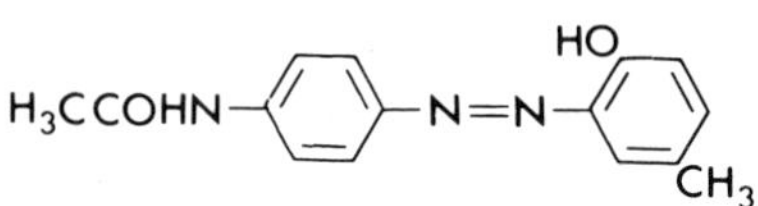

(XII) C.I. Disperse Yellow 3 (C.I. 11855)

For cellulosic fibres a long planar molecule is required, as in I and II, as well as one or more anionic solubilising groups, usually —SO_3Na, but sometimes —COONa. Vat dyes, used for cellulose, e.g. III and IV, are quinones (the quinone groups being reduced before the adsorption process and regenerated afterwards), and usually have other polar groups. The azoic dye components used for application to cellulose, and subsequently coupled on the fibre, are mainly arylamides of 3-hydroxy-2-naphthoic acid, e.g. V.

Dyes for protein fibres and nylon, e.g. VI, contain anionic groups, as for direct dyes, and if high wet-fastness is required a high relative molecular mass (r.m.m.) is necessary. Mordant and metal-complex dyes require chelating centres, which are often *o,o'*-dihydroxyazo groups; 1:1 metal-complex dyes are sulphonated (VII) but their 1:2 counterparts are not (VIII).

For acrylic fibres a cationic group is used, as in IX.

For cellulose acetate and polyester fibres, one or more polar groups must be present, e.g. —NH_2, —OH or —NHR, whilst ionic groups must be absent (X, XI and XII), and the molecule should be planar.

Dyes for man-made fibres often have a short alkyl chain, e.g. —C_2H_4—, inserted between the nucleus and a polar or cationic group, presumably to increase the substantivity for the fibre.

2.3 THE SURFACE ENERGY AND THE INTERFACIAL EFFECT

Traditionally, the chemist has tended to classify dyeing processes by the nature of the supposed dye–fibre bonding forces, and this remains a convenient procedure which is followed below. Nevertheless, evidence is growing which suggests that such forces may play a less significant role than that hitherto assumed. Fundamentally, dyeing starts at an interface – usually that between an aqueous dyebath and polymer molecules forming part of the accessible regions of a fibre – and an enquiry into its nature should first consider the behaviour of a dye molecule when it approaches this interface.

Many types of fibre have enormous surface area relative to mass (i.e. specific surface). Thus, cotton, rayon and wool have surfaces of the order of 100 m^2 per g, or more than 5 acres per lb. This surface consists almost entirely of walls of internal pores; in fact the external surface of the fibre is only about one-thousandth part of the total. When such a fibre is placed in an aqueous dyebath, water penetrates the pores very rapidly, in a time measurable in seconds. Dye takes longer to penetrate because the fibre pores exert a restraining effect on the relatively large dye molecules. Within a short time, however, a situation develops in which dye solution is in contact with the internal surface, i.e. an interface is formed between the polymer of which the fibre is composed and the solution forming the dyebath. The concept of the development of this interface is of fundamental importance to an understanding of the dyeing process.

Gibbs, in 1878 [29], first rigorously defined the consequences of forming an interface between a solution and another phase. His well-known 'adsorption equation' is described in Chapter 1 and in many text-books [30]; it defines the relation between the change in surface tension of the solution with concentration and the difference in solute concentration at the surface and in the bulk of the solution. Thus, if the surface tension of the solvent is reduced by the solute, then it follows that the solute molecules have concentrated at the surface, whereas if the surface tension is raised the solute has moved away from the surface into the bulk of the solution. In practice it is found that most water-soluble organic compounds, including dyes, reduce the surface tension of water; this demonstrates therefore that they concentrate at the surface. Highly hydrated substances, e.g. some highly ionised inorganic salts and, amongst organic compounds, glycerol and sugars, tend to increase the surface tension and therefore to move away from the surface so that they are more readily enveloped by a water 'atmosphere'. In either case the solute tends to concentrate in the region of lower energy. The interface between water and fibre-forming polymers, e.g. cellulose, cellulose acetate, nylon, polyester or protein, has lower energy than the pure water surface. This is evident from the formation by these polymers of monolayers on water [31–33], whereby the surface tension of the water is reduced, as shown by experiments with the Langmuir trough (described in section 2.4.8 below). It follows, therefore, that dyes in aqueous solution tend to concentrate at fibre surfaces. (Conversely, inorganic salts, sugars, etc., tend to move away from a fibre surface.)

The Gibbs equation can be expressed in several forms. For the present purpose it is conveniently given, for a constant surface area, thus:

$$\Gamma = -\frac{a}{RT} \cdot \frac{d\gamma}{da} \qquad (1.159)$$

where Γ is the 'surface excess', or excess concentration of solute in the surface layer, a its activity in the bulk phase, R the gas constant, T the thermodynamic ('absolute') temperature and γ the surface tension.

Using the Gibbs equation with experimental results of surface tension measurements on highly purified Sky Blue FF in water (Figure 2.1) and a modified form of the equation with surface pressure measurements on the same solution under a cellulose monolayer (the dye molecules being assumed oriented beneath the monolayer and parallel to the molecules of the cellulose chain), the following data are obtained [34]:

Surface excess (10^{14} dye molecules cm^{-2}) for 0.4×10^{-3}M dye solutions:

in water	in water under cellulose monolayer
0.48	0.056

The reduced concentration of dye produced by the presence of the cellulose

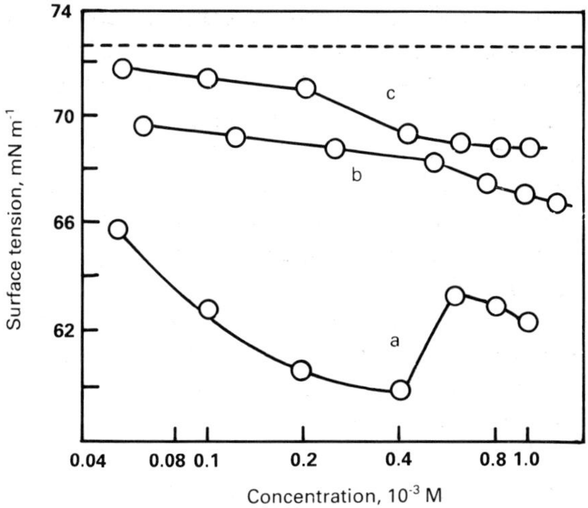

Figure 2.1 – Surface tension of solutions of Sky Blue FF: (a) purified dye; (b) as (a), after 'foaming' treatment, in 0.1M NaCl; (c) as (b), without NaCl [34a]

monolayer agrees with the observation [ref. 22, at p. 231] that highly purified Sky Blue FF is not absorbed by cellulose fibres in the absence of salt. In the case of the cellulose monolayer the dye anions are repelled by the negatively charged cellulose molecules.

In the bulk (0.4×10^{-3}M) solution of Sky Blue FF there are $0.4 \times 10^{-3} \times 6.1 \times 10^{23} \times 10^{-3} = 2.44 \times 10^{17}$ molecules of dye cm^{-3}, and therefore if there were no change in concentration at the surface, this would contain $(2.44 \times 10^{-17})^{\frac{2}{3}} = 3.897 \times 10^{11}$ molecules cm^{-2}. The above results therefore show that in the absence of salt the dye in 0.4×10^{-3}M solution is actually $0.48 \times 10^{14}/3.897 \times 10^{11} = 123.2$ times as concentrated in the air–water interface as it is in the bulk solution, and that it is $0.056 \times 10^{14}/3.897 \times 10^{11} = 14.37$ times as concentrated in the cellulose–water interface as in the bulk solution. Using a value of 3.63 nm^2 for the molecular area of this dye (measured from models), it is readily calculated that the proportion of interface covered by dye is 174.2 and 20.32 per cent; the first figure shows that the dye must be at least partly associated in the surface. (At this dilution the association is known to be low [ref. 35, Tables 1 and 2].)

There is evidence that a dye molecule 0.5 nm thick may be able to penetrate only pores that are not less than 3 nm wide [37]. Now Stone *et al.* [38, 39] find that the pores >3nm wide in wet cellulose have a total surface area of 260 m^2 g^{-1} (Figure 2.2). A layer 50 nm deep over this surface has a volume of 0.13 cm^3, or 13 cm^3 per 100 g of fibre. Suppose the dye solution under discussion, 0.4×10^{-3}M, is used as a 50-vol. dyebath, then 100 g of fibre is in 5 l of solution containing 2.0×10^{-3} moles of dye (r.m.m. 992), equivalent

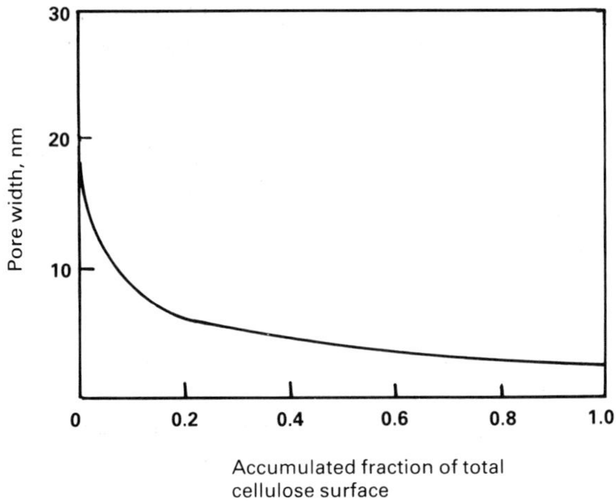

Figure 2.2 – Relation between pore width and total internal surface of cellulose [38]; (this curve is for bleached pine kraft fibres, but viscose rayon gives very similar results)

to 1.982 per cent on the mass of fibre. The interfacial layer of 13 cm^3, as just mentioned, containing approximately 123-fold excess of dye, therefore holds dye equivalent to (123 × 13)/5000 of the total 1.98 per cent, or a 0.633 per cent depth. This calculation of course refers to a bath at room temperature: nothing appears to be known of the surface tension behaviour of dye solutions at the high temperatures used for most direct dyeing. The only guide to what might be the effect of raised temperature seems to be the observation [40a] that at a water–hydrocarbon solvent interface, a rise in temperature of 20 °C reduces the rate of change of interfacial tension with concentration by only about 25 per cent; also Langmuir [40b] gives evidence that the Gibbs treatment used here is valid for hot aqueous solutions.

It is accordingly clear that, provided no extra-thermodynamic factors interfere, e.g. that there is no negative charge on the fibre, which would repel dye anions, a fair degree of exhaustion of the bath could in theory be obtained on any hydrophilic fibre without the need to postulate any force of a chemical nature between dye and fibre molecules. In practice, as will be seen later, much better exhaustion is obtained through the operation of other physical factors. Also discussed is the reason why, with a dye molecule, certain structures are most favourable for affinity for cellulose.

The effect of sodium chloride on the surface tension of solutions of direct dyes appears in one respect to be parallelled by a technical effect. The salt reduces γ with a tetrasulphonated direct dye, but has no effect on a monosulphonated acid dye. If this effect is general, then the more highly sulphonated a dye, the more should salt influence its exhaustion. In fact this hypothesis is in accordance with data given by Vickerstaff [22] for 'salt controllability', i.e. the degree of increase in adsorption caused by a given addition of salt to the dyebath:

Direct dye class	*Average salt controllability/ arbitrary units*
Disulphonated	1.1
Trisulphonated	2.3
Tetrasulphonated	>4

2.4 INTERMOLECULAR FORCES [see, e.g., 41–46]

Any discussion of the reactions between dyes and fibres must take account of the various types of force exerted by one molecule upon another, and a brief survey of these is therefore necessary. All originate in the movement of electric charges, and indeed without these movements there would be no chemistry!

2.4.1 Covalent bonds

These are the classical primary valency bonds of chemistry, e.g. the C—H bonds in methane. The two participating atoms each contribute one electron to the common linkage and the two electrons are thus shared to unite the atoms with a bond of high energy between about 84 and >840 kJ mol^{-1} (20 and >200 kcal mol^{-1}). The bonds between reactive dyes and fibre molecules are of this type; their high energy makes them difficult to disrupt and accounts for the high fastness of such dyes to wet treatments.

In terms of quantum mechanics, the covalent bond between two atoms A and B can be described by a *wave function*:

$$\psi_s = \frac{1}{[2(1+S^2)]^{\frac{1}{2}}} (\psi_{\mathrm{I}} + \psi_{\mathrm{II}}) \tag{2.1}$$

where ψ_{I}, ψ_{II} are the respective wave functions for the two ways in which the two electrons can be found, each on one of the atoms, and S is an *overlap integral*.

The wave function is the square root of the probability of finding the electron at a given point at a given time, and the overlap integral is the product of the values, at a particular point, of the wave functions extending from the two atomic nuclei, summed over the entire coordinate space occupied by the atoms. It can be shown that the expression for ψ_s predicts a minimum energy corresponding to the formation of a bond.

2.4.2 The ionic bond

When two atoms share a pair of bonding electrons but one can exert a preponderant attraction for both electrons, as with the chlorine atom in sodium chloride, the atoms may be separated as oppositely charged ions (Na^+ Cl^-) if placed in a liquid with a high permittivity, e.g. water. They then dissolve and move independently in the solution, but they exert a mutual electrostatic (Coulomb) attraction which decreases with the square of their distance of separation. Attraction of this type plays a large part in the dyeing of protein fibres and certain man-made fibres, as described later.

In the broadest sense, however, all covalent bonds are polar or partly ionic, to a degree that depends on the nature of the two linked atoms. Several different electronic structures can be written for each pair of bonded atoms; one of these has both the bonding electrons on one atom. This implies that the two atoms hold opposite charges and so can separate as ions.

One type of dyeing process which involves partly covalent and partly ionic bonds is mentioned in section 2.6.2.

2.4.3 Ion–dipole forces

As well as attracting those of opposite charge, ions in solution can exert attraction upon any polar molecule, giving rise to the so-called *ion–dipole* forces. These forces are important in dyeing, because they are largely responsible for the aqueous solubility of dyes. They originate in electrostatic attraction of an adjacent non-ionic molecule by an ion of either sign of charge. In this way an ion in water attracts around itself an 'atmosphere' of water molecules and is thereby kept in solution and separated from other ions. The 'atmosphere' is of indefinite depth, as the attraction falls off according to the inverse square law, but the central ion holds at least one layer of water molecules, and perhaps more, in an ordered manner. Various physical methods have been used to measure the number of water molecules carried by different ions in water. The results do not always agree amongst themselves, but generally cations appear to carry more water molecules than do anions, and small cations more than large cations. A theoretical treatment gives the expression for the radius r of the ionic 'atmosphere' (or the 'Debye length') as

$$r = 1/K \tag{2.2}$$

where

$$K = \left(\frac{8\pi Ne^2}{\varepsilon kT \times 10^3}\right)^{\frac{1}{2}} I \tag{2.3}$$

N being the Avogadro number, e the magnitude of the charge on the electron, ε the permittivity of the solvent, k the Boltzmann constant, T the thermodynamic temperature, and I the total ionic strength of the solution.

2.4.4 Hydrogen bonds [see, e.g., ref. 47–49]

The very small size of the hydrogen ion, or proton, and its consequent very high polarising power cause it under certain conditions to form bonds with two atoms simultaneously. The required condition is that the H atom must be covalently bound to one of high electron affinity, whereby it is positively polarised. This occurs mainly with H bound to N, O or F, but in some cases the bond is to Cl, S or C. Although the hydrogen bond is so well known and has been intensively studied, its theoretical basis remains somewhat obscure. It can act in five different ways:

(a) Intermolecularly and extending over many molecules, as in water:

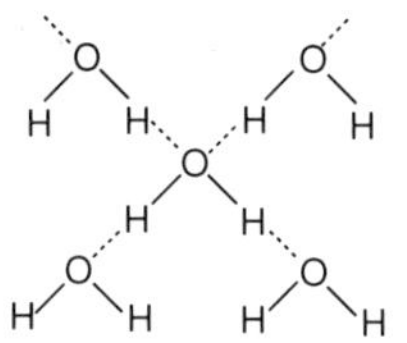

(b) Intermolecularly, whereby two molecules are united in a dimer, as in acetic acid:

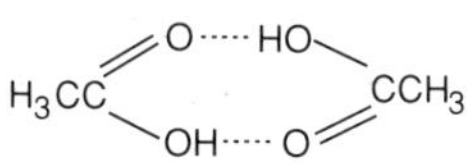

(c) Intermolecularly, where the electron donation is by a >C=C< double bond or by the π-electron system of an aromatic nucleus [50], as in benzene, which links with the hydrogen atom of a hydroxyl group in an alcohol or phenol:

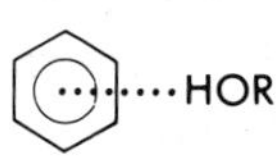

This bond may be important in dyeing, e.g. in poly(ethylene terephthalate) dyeing, where a π-bond between the aromatic nucleus and a hydroxyl group in a disperse dye may be possible.

(d) Intramolecularly, whereby a chelate ring is formed, as in Orange II (VI):

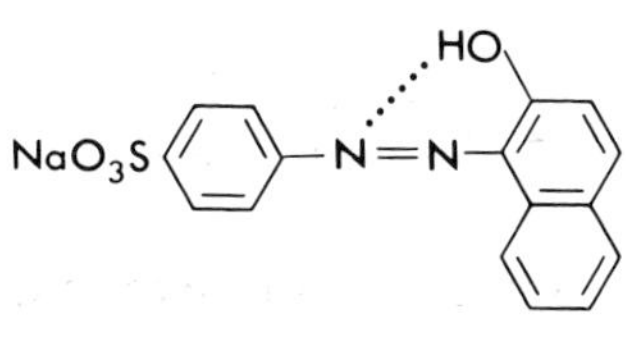

This arrangement is most stable when a six-membered ring is formed, through five- and seven-membered rings formed by hydrogen bonding have been detected.

(e) Intramolecularly in an ion, as in HF_2^-

$$(F \cdots H \cdots F)^-$$

Types (a) and (b) are responsible for abnormal physical properties, e.g. the high boiling points of water and ammonia and the high melting points of organic acids, sugars and urea.

Hydrogen bonds have some of the properties of normal valency bonds; they are directional and, as stated, they form only when the two interacting atoms can approach very close to each other. Their fundamental importance in determining the geometry of the molecules of proteins and DNA is now well known; in liquid water the molecules participate in a vast network of hydrogen bonds, since all the three atoms in each water molecule can form such bonds. The hydrogen atom forming the bond can conveniently be considered to be in a state of incipient ionisation.

The groups participating in the bond are thus in effect Lewis (or Brønsted) acids and bases respectively. Lewis acids and bases are defined as electron acceptors and donors respectively; Brønsted acids and bases are defined as donors and acceptors of a proton, respectively [51]. These two definitions are alternative views of acid–base reactions which are more widely applicable than the classical definition of acids and bases as substances giving rise in aqueous solution to hydrogen or hydroxide ions respectively. More simply, as explained in section 1.3, acids are substances that tend to lose protons and bases are substances that tend to combine with protons.

The hydrogen bond is relatively weak (*ca* 8.4–41.9 kJ (2–10 kcal) mol^{-1}) and therefore can be readily formed and readily broken – one of its important features. It can be regarded as being formed by partial *proton* donation and acceptance or, alternatively, by partial *electron* acceptance and donation by the respective groups. Thus, in the solvation of an amino group by water, say in a disperse dye, a proton of water is partly held by the nitrogen of the amino group, and the spare electrons of nitrogen are partly held by this proton:

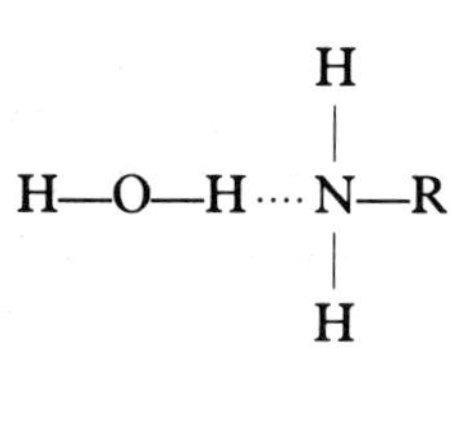

Examining of the formulae of most dyes and fibres shows that they have many groups that would be expected to form intermolecular hydrogen bonds, and consequently these bonds are often considered to take part in dyeing operations. A complicating factor, however, is the presence of the water of the dyebath, which by itself bonding strongly with these groups may inhibit their attachment to other molecules. We shall examine this matter later.

2.4.5 Charge-transfer forces

An electron-deficient (electron-acceptor) atom of one molecule can form a bond with an electron-rich (electron-donor, or nucleophilic) atom of another type of molecule, i.e. one with a lone pair of electrons, even though both are 'saturated' in the normal sense. The combination gives rise to the so-called *charge-transfer* complexes. The bond is of the dative covalent type, and it can often persist not only in the solid state but also in solution and even in the vapour state. Examples are the picric acid–aromatic hydrocarbon addition compounds and those formed by amines or ethers with a halogen molecule. They can be described also as complexes between a Lewis acid and a Lewis base, and their formation is somewhat similar to the acid–base combination of ionised compounds. The charge-transfer bond has also been described as intermediate in character between a single covalent bond and no bond at all. It is possible that a charge-transfer mechanism is involved in the bonding of disperse dyes to hydrophobic fibres.

2.4.6 Van der Waals forces

These forces are so named because they were first recognised by van der Waals in 1873 in his study of the cause of the non-conformity of gases to the ideal gas laws. They are usually divided into two main groups, known respectively as polarisation (or electrostatic) and dispersion forces. Sometimes the term is used to include hydrogen bonds, dipole–dipole, dipole-induced dipole and π-orbital forces, but this use is confusing and will not be followed here. We shall use the term in the more familiar sense.

Polarisation forces

The symmetrical distribution of electrons around atoms is disturbed when the atoms are chemically combined. In many cases this results in a permanent unsymmetrical distribution of electric charge in the resulting molecule, which therefore has a permanent electric or dipole moment. Such a type of molecule, which includes most non-symmetrical organic molecules that are not hydrocarbons, is said to be *polar*. Methane (CH_4) is symmetrical and non-polar, but if one, two or three of its hydrogen atoms are replaced by, say, halogen atoms the molecule becomes unsymmetrical and has a dipole moment. When all four are replaced by the same substituent, as in CCl_4, symmetry is restored and the dipole moment is zero. Molecules with dipole moments attract each other in two ways: by permanent dipoles and by induced dipoles, rather as one permanent magnet attracts another and as it attracts a piece of soft iron by first inducing magnetism therein.

Clearly, if a molecule with a dipole moment approaches another in such a way that the positive charge on one is nearer to the negative than to the positive charge on the other, a mutual attraction will result. But a complication arises because the molecules in a gas or liquid are in constant motion and any given pair do not necessarily approach each other orientated in a straight line and, as it were, head to tail; they approach in a completely random manner. The result is that each causes a slight disturbance of the electronic cloud of its neighbour, which thus acquires an *induced dipole moment*. The net result, however, is that the energy of the molecules becomes lower as they approach, and thus they exert mutual attraction.

Theoretical treatment of the situation for the randomised approach of all molecules gives an expression for the interaction energy per molecule, W_i:

$$W_i = -\frac{\mu^2}{r^6}\left[\alpha_0 + \frac{\mu^2}{3kT}\right] \tag{2.4}$$

where μ is the dipole moment, α_0 the *distortion polarisability*, k the Boltzmann constant and r the intermolecular distance. α_0, which has the dimensions of volume, is equal to μ_m/E, where μ_m is the dipole moment after distortion and E the applied field. In the above expression the negative sign indicates an attraction, the first term being the part due to the distortion of polarisation of one molecule by the permanent moment of the other, and the second term the part due to the orientation caused by the mutual effect of their permanent dipoles.

It is seen from Eqn 2.4 that the polarisation van der Waals forces fall off with the sixth power of the distance between molecules.

Dispersion or London forces

These forces are named from the chemist F London, who in 1930 first described their theoretical basis; they account for the fact that non-polar molecules, e.g. CCl_4 or argon, can attract one another. Consider the situation of the electron cloud around an atom of argon. The cloud of electrons is distributed spherically around a positive nucleus; there is no net dipole moment, but this distribution is an average over a period of time. At any instant, however, the electrons, which are constantly circulating, may be so distributed that there is a separation of positive and negative charge and therefore a dipole moment. This moment then has a vector which is continually changing with the movement of the electrons, and the net result is a zero moment. When two such atoms are very close to one another their momentary dipoles interact and the motions of their electrons synchronise to cause a continuing electrostatic attraction between them; the energy of the system is thus lowered and a weak bond established. The energy is proportional to α^2/r^6 where α is the polarisability of the molecules and r is their separation. The α value is greater the greater the volume of the molecule, so that with increase in molecular size the dispersion forces increase and also become relatively more effective than polarisation forces – an important consideration in the present context, since both fibre and dye molecules are large. (The above trends are very familiar in the general tendency for melting or boiling points to rise with r.m.m.) The very rapid fall-off of these forces with separation, and their increase with increasing r.m.m., should clearly cause long dye molecules to align as closely and as nearly parallel as possible with the long molecules of a fibre.

As the intermolecular distance r is decreased, the attraction changes to a mutual repulsion; this distance can be visualised as that where the electron clouds of the molecules come close enough to begin to interpenetrate.

2.4.7 Hydrophobic interaction

This is a term which is now frequently used to describe the tendency of hydrophobic groups, especially alkyl chains, to associate together and escape from an aqueous environment. Its first description is commonly attributed to Kauzmann, Némethy and Sheraga [52], in their studies of the tendency of alkyl side-chains of proteins to escape from water, although Meggy [53] had previously developed a similar concept to explain the dye–wool bond [cf. ref. 54]. The effect is really due to two causes acting simultaneously: the van der Waals forces between the hydrocarbon groups and the hydrogen bonds between the water molecules. Each set of forces causes respective assemblages of molecules or groups to associate together and to exclude the other. Hydrophobic bonding thus offers a useful new concept to unify and illuminate known diverse phenomena, but it is not strictly a new type of bond or intermolecular force.

2.4.8 Identification of dye–fibre forces

For reasons already given, the direct identification of these forces, other than those due to ionic charges, is extremely difficult and seldom attempted. Instead, studies are made of reactions between simpler compounds (usually termed *model* compounds) having some of the distinguishing features of the dyes and fibres under investigation, and the behaviour of the more complex systems is then inferred.

Since in many dyeing operations the existence of more than one type of force has been postulated, it is convenient to consider the identification of the types of force separately from the discussion of the dye–fibre systems.

In spite of more than a century of experiment and speculation, and intensive work in laboratories during the past three or four decades, there is still no unanimity of opinion regarding the nature of dye–fibre adsorption forces. One of the main difficulties in determining their nature is the relatively small quantity of dye present. Thus in a normally dyed fibre there may be less than 2 g dye per 100 g of fibre; even sophisticated modern analytical techniques are of little value in identifying bonds in such systems, since any indications of dye–fibre bonds are swamped by the vastly greater number of intermolecular bonds between the complex fibre molecules or the dye molecules themselves. We have therefore to rely on indirect evidence to give clues to dye–fibre linkages, and many physical tests on simpler systems have been so used. These include vapour-phase adsorption by fibres [55], chromatography [56], permittivity and refractive index measurements on solutions of binary mixtures [57, 58], r.m.m. measurements [59], polarography [60], tests with monolayers on aqueous solutions [61] (all carried out with either dyes or model compounds), determinations of melting points of binary mixtures of solid dyes [62] and adsorption tests of dyes on inorganic surfaces [63].

The experimental methods which appear to have been most used with model compounds are the measurement of refractive index changes in solutions, by the 'method of continuous variations' and monolayer techniques.

The method of continuous variations

In this method some physical property (spectral absorption, viscosity, permittivity, refractive index and several others have been used) is measured for a range of binary solutions, each containing as solutes the two components A and B under investigation, in a range of A:B ratios from 100:0 to 0:100; each solution has the same total molarity.

The method seems first to have been described as far back as 1911, and it has been used many times since then [64]. It is sometimes called 'Job's method', because Job and his collaborators used it to measure spectral absorption curves of dyes [64]. In the writer's laboratory, it has been used with a very wide variety of pairs of model solutes, some of which included dyes and fibrous polymers, in water and various other solvents, mainly to detect hydrogen-bonding complexes.

The plots are usually linear, and a change of slope, or an inflection, indicates a complex of A with B, in the ratio of the two components corresponding with the inflection (Figure 2.3). The theoretical basis of this correspondence has been worked out [59, 64, 65].

Apart from the rapidity and simplicity of this method which enable large

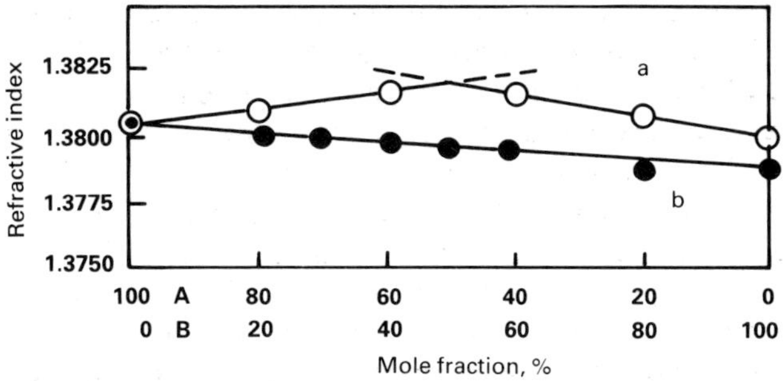

Figure 2.3 – The method of continuous variations, illustrating the use of refractive index measurements for detecting complex formation in water [36]

A, cellulose (calculated as cellobiose residues); B, (a) Sky Blue FF, (b) C.I. Direct Blue 10

Sky Blue FF forms a 1:1 complex with each cellobiose residue in the cellulose chain; C.I. Direct Blue 10 forms no complex. Strictly the square of the refractive index should be used, being linear with concentration but, over the very small ranges of value used in this type of test, the index itself can be plotted

numbers of systems to be tested, a particular advantage is that it can be used as readily in aqueous as in non-aqueous media; this can be said of few if any of the many other methods of detecting hydrogen bonds. This is obviously important in any study of the behaviour of dyes and fibres in aqueous dyebaths. A very recent development, based on earlier work by Yoshida [66], enables it to be used for quantitative measurements of thermodynamic parameters of the hydrogen bonding of model solutes in water [58].

The method has been used with many hundreds of systems, in a variety of solvents. Apart from the evidence of the simple complex ratios mentioned, there are numerous cross-checks on the inferences these give of the reactivity of various groups. Thus the number of molecules of a solute B combined with a solute A agrees with the number of free bonding centres in A, and in structures where intramolecular hydrogen bonding can occur this can be detected by the loss of complexing ability of the centres involved. Examination of the results for given systems, in several solvents respectively, shows that usually the ability for complex formation is influenced in relatively few cases by hydrogen bonding between the reacting groups and the solvent, i.e. by their protection by solvation. This conclusion is supported by the data in Table 2.2, where a non-bonding solvent, tetrachloromethane, is seen to give the same results in most cases as bonding solvents such as alcohols, aromatic hydrocarbons, dioxan, and especially water. The results with water are of considerable interest because of their relevance to dye–fibre systems, as well as to reactions in living organisms. The exceptions are also interesting. They

TABLE 2.2

Complexes detected by refractometry [65]

Type of bond	Solutes A	Solutes B	Solvent*	Total mol. concn	Temp. /°C	Mol. ratio of complex, A:B
Alk—OH + Ar—OH	D-Glucose	Phenol	W	0.25	20	0
			ED	0.1	20	1:6
>C=O + Ar—OH	Propanone	Phenol	W	0.1	20	0
	Di-isobutyl ketone		D	0.25	20	1:2
	Ethyl methyl ketone		CCl_4	0.1	14.83	1:1, 1:2
—N=N— + ArOH	Azobenzene	Benzyl alcohol	CCl_4	0.1	20	1:1
		Phenol	B	0.25	19	1:1, 1:2
—COOR + Ar—OH	Ethyl acetate	Phenol	CCl_4	0.1	13.11	1:1
—COOR + Ar—Cl	Ethyl acetate	Chlorobenzene	CCl_4	0.1	14.85	1:1
			EW	0.1	14	1:1
	Phenyl acetate		B	0.1	21	1:1
			EW	0.25	14	1:1
—COOR + NO_2	Ethyl acetate	4-Nitrophenol sulphuric ester	W	0.04	20	2:1
	Glycerol triacetate		W	0.04	20	1:1

*B = benzene; D = dioxan; ED = ethanediol; EW = ethanol–water (1:1 vol./vol.); W = distilled water.

show that, when water is the solvent, some groups lose their ability to form intermolecular hydrogen bonds; these include aldehyde and keto groups and, significantly, the hydroxyl groups in simple carbohydrates, such as glucose, sucrose and cellobiose. Phenolic groups, amino groups, azo groups and ester groups are active in water and in aqueous mixed solvents, a fact which has an important bearing on the dyeing of some man-made fibres.

An additional fact revealed is the ability of some ester groups to act as proton donors, whereby —CH ··· O— or —CH ··· N— bonds are formed. This fact is confirmed by affinity measurements in which paper chromatography is used [56], and by heat-of-reaction measurements using gas–liquid chromatography, described below.

The monolayer method [61]

This method was developed from the technique of spreading monolayers on

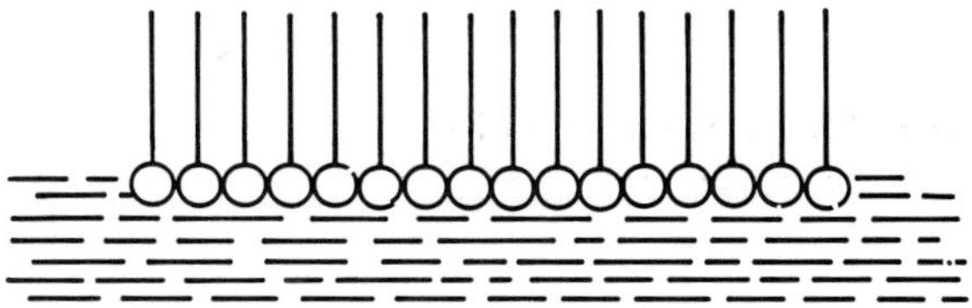

Figure 2.4 – Monolayers on water: molecules of a long-chain alkyl compound, e.g. hexadecan-1-ol (cetyl alcohol); the hydroxyl ('head') groups dissolve in the water surface, the alkyl chains do not [61i]

Figure 2.5 – Monolayers on water: a polymer molecule consisting of alternating water-soluble and water-insoluble groups [61i]

aqueous liquids (Figures 2.4 and 2.5), particularly by use of the Langmuir trough. Essentially, the principle entails measurement of the pressure exerted by the monolayer on a floating barrier (Figure 2.6). Movement of a second barrier step by step enables the monolayer to be gradually decreased in area

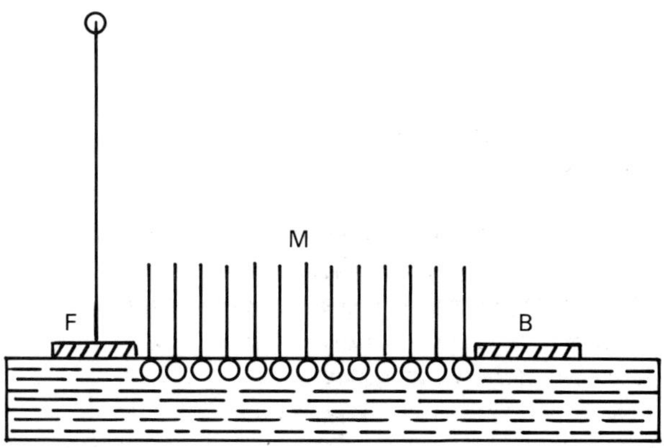

Figure 2.6 – Principle of the Langmuir–Adam balance for measuring monolayer characteristics.

This side elevation shows the trough filled just to the brim with water which is covered by the monolayer M held between the freely movable boom B and a delicately suspended float F. As B is swept across the surface towards F the monolayer molecules become more tightly packed and more regularly orientated; this reduces the surface tension of the water surface between B and F. The higher surface tension to the left of F therefore causes F to move towards the left, and its movement is measured by an optical lever [61i]

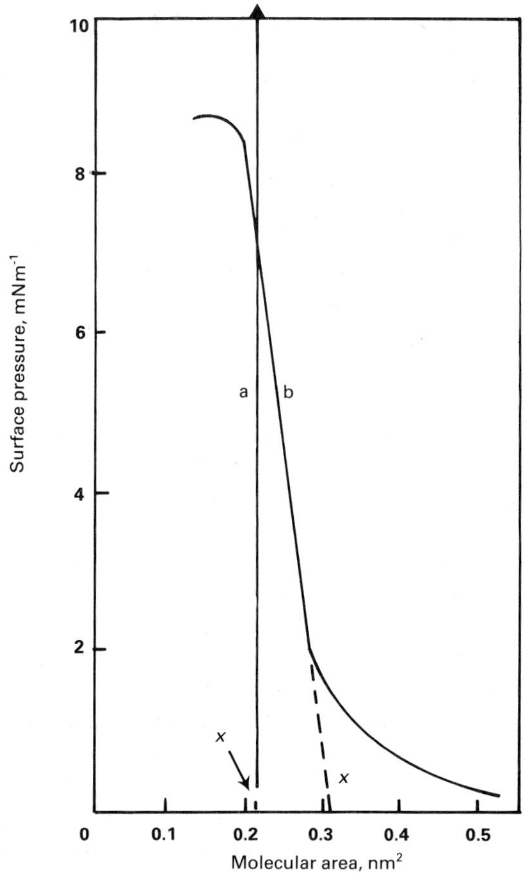

Figure 2.7 – Typical force–area curves obtained with a Langmuir-Adam film balance. (a) Curve for a long straight-chain carboxylic acid, e.g. stearic acid. (b) Curve for a disperse dye (C.I. Disperse Red 13), formula

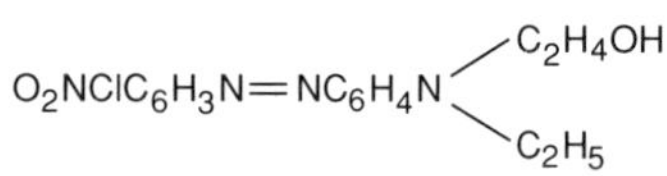

The force on the float F (Figure 2.6) is measured for a range of areas of surface, by varying the position of the boom B, and the area occupied by each molecule is calculated from the weight of substance originally placed in the film. The points x give the molecular areas for close vertical packing. That for the dye (0.31nm²) agrees well with the value (0.32nm²) measured from a molecular model [61i].

and the pressure corresponding to each value of the area to be noted and plotted on the pressure–area diagram (Figure 2.7). Other parameters of the monolayer are also often measured, especially its viscosity and changes in its electric potential.

A very important advantage of this method over all others for investigating dye–fibre reactions is that the polymers which form most of the common fibres can now be spread as monolayers. (The main exception is wool, but in its place other spreading proteins can be used.) The reactions of the polymers with dyes and model compounds can thus be studied directly, without complications introduced by the pore structure of a fibre. This important advantage is discussed later.

Surface tension of dye solutions
Mention may also be made of the determination of the surface tension of dye solutions, which has been used with a few cationic and anionic dyes [34]. The latter included a monosulphonated acid dye (C.I. Acid Red 88), a direct dye (Sky Blue FF (I), see Figure 2.1) and a tetrasulphonated acid dye with a relatively small molecule. All were found to be surface-active, i.e. they lowered the surface tension of water. The important implications of this fact are discussed later. One new discovery made in the course of this work is that dyes in their normal commercial state, or after purification by normal crystallisation, all give surface tension *vs* concentration curves with a pronounced minimum. This is indicative of the presence of a trace of some highly surface-active impurity, and it was found that elimination of this impurity by prolonged foaming of the solution with a current of air and removal of the foam (which is caused by the impurity) was necessary to produce solutions not showing a minimum. It seems probable that the impurities are highly soluble substances with small aromatic molecules (possibly naphthalenesulphonates in the case of the acid dyes), since they do not seem to influence the reactions of dye solutions with polymer monolayers spread on them.

2.5 ADSORPTION AT SURFACES

The basis of all dyeing operations is the concentration of a dye at a solid surface, and this is a particular case of the phenomenon known as *adsorption*. This phenomenon has been known for a long time [67], but modern studies of the subject began with Langmuir's concept of gas adsorption [68], viz. that gas molecules striking a solid surface are captured to form a layer one molecule deep, known as a *monolayer*. This is a dynamic process, with molecules being both captured and released into the gas phase. On this basis Langmuir formulated his famous equation:

$$\frac{p}{v} = \frac{1}{bv_{\mathrm{m}}} + \frac{p}{v_{\mathrm{m}}} \tag{2.5}$$

where v is the volume of gas adsorbed by the solid at pressure p, v_{m} is the volume in the completed monolayer, and b is a function of temperature. Thus a plot of p/v against p, or of $1/p$ against $1/v$, should be linear. Although the situation in dyeing from solution is more complex than gas adsorption, the above Langmuir equation can be applied [69] to most dyeing processes involving hydrophilic fibres, if the pressure p is replaced by the concentration $[\mathrm{D}]_{\mathrm{s}}$ of dye in solution and v is replaced by the concentration $[\mathrm{D}]_{\mathrm{f}}$ of dye in the fibre, both variables being measured at equilibrium. Thus a plot of $1/[\mathrm{D}]_{\mathrm{s}}$ against $1/[\mathrm{D}]_{\mathrm{f}}$ should be linear (cf. Figure 1.9).

In gas adsorption (or sorption), the more usual and more complex case of the building up of superimposed monolayers, termed *multilayers*, was investigated theoretically by Brunauer *et al.* [70], who developed the well-known, so-called BET equation of wider applicability, but this has not been used much in dyeing studies.

Any adsorption is usually studied by plotting a graph known as an adsorption isotherm [71]. For a dyeing process this is a plot of $[\mathrm{D}]_{\mathrm{f}}$ against $[\mathrm{D}]_{\mathrm{s}}$, and the graph may have one of the four shapes shown in Figure 2.8,

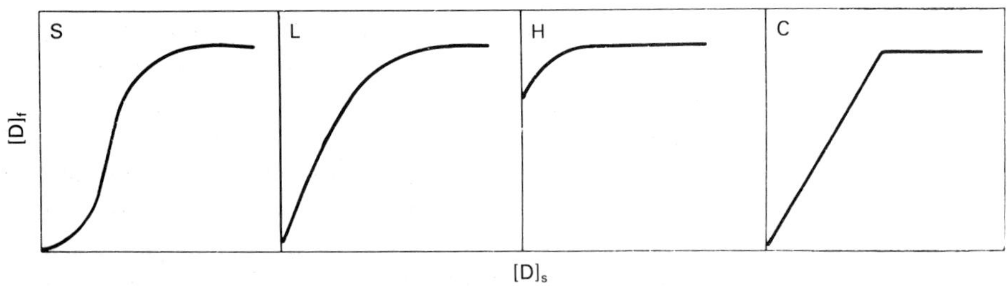

Figure 2.8 – The four classes of solute isotherm: S, with solutes having strong intermolecular attraction parallel to the surface, and/or strong competition from a second solute or both together; L, with solutes having weak intermolecular attraction; H, with solutes of high r.m.m. and high affinity for the surface, including cases of chemisorption; C, with solutes that penetrate the solid preferentially to the solvent [71]

the shape giving information about the nature of the dye–fibre forces. The theoretical treatment is described elsewhere [69].

The most common isotherm in dyeing is the L-type, which is obtained for most dyeing processes involving hydrophilic fibres. The linear C-isotherm is obtained with disperse dyeing of hydrophobic fibres. The H- and S-types are rarely found.

An alternative method of plotting is to use logarithmic instead of arithmetic scales, i.e. to plot log $[D]_f$ against log $[D]_s$ (Eqn 1.177). If the system gives the L-isotherm with arithmetic scales, it will give a linear log–log plot [67], provided that the concentration is not taken to too high a value.

(For a description of the often-quoted Freundlich adsorption isotherm and its relation to the Langmuir isotherm see, e.g., [67b] and also Chapters 1 and 4.)

2.6 DYEING MECHANISMS [72]

Dyeing mechanisms may be described by defining simply two broad classes of dye–substrate association, with the qualification that in many systems these may overlap so that more than one type of association may operate. Consequently, all dyeing mechanisms can be divided into three groups, in which the dye–substrate attraction (a) is specific, i.e. owing to covalent bonds, hydrogen bonds or other directed bonds, or (b) is non-specific owing, for example, to ion exchange or van der Waals forces, or (c) is produced by mere mechanical restraints.

2.6.1 Specific dye–fibre bonds

Covalent bonds

Apart from the application of reactive dyes, described in Chapter 6, only one normal dyeing process appears to result in covalent bonding between dye and substrate; this is the dyeing of anodised aluminium. Even in this case covalent bonding occurs only to a limited extent, much of the dye being bound by ion-exchange processes. This dyeing process differs in many ways from textile dyeing but, because of the rigidity and non-swelling nature of the substrate, some aspects of its dyeing are more readily investigated, and have special theoretical interest. Since this book is confined to the coloration of textiles a discussion of aluminium dyeing is outside its scope, but further information can be obtained by reference to the original sources and reviews [73–78].

2.6.2 Mordant dyeing of wool [79–83]

This is one of the oldest of all dyeing processes, and gives very high fastness. In ancient times, naturally occurring phenolic colouring matters were applied with the aid of, e.g., iron salts or alum, but today sulphonated anthraquinone or azo dyes with chelating centres, e.g. *o,o′*-dihydroxyazo compounds, are applied with potassium or sodium dichromate as the mordanting salt. The mordant may be applied before, during or after dyeing.

The dichromate anion is reduced in the fibre to the Cr^{III} cation, with simultaneous splitting and oxidation of some of the disulphide bonds of the keratin. The overall reduction of Cr^{VI} occurs in steps, of which the final one is an oxidation:

$$Cr^{VI} \xrightarrow[\text{reduction}]{} Cr^{IV} \xrightarrow[\text{reduction}]{} Cr^{II} \xrightarrow[\text{oxidation}]{\text{air}} Cr^{III} \quad (2.6)$$

Cr^{III} then reacts by successive addition of dye molecules to the chromium atom, to give 1:1 and 1:2 metal–dye complexes [82]. Hartley [82] and Dobozy [84] have reviewed the present knowledge of the chemistry of the many processes that occur in chrome mordant dyeing. A typical chrome mordant dye is C.I. Mordant Violet 5; its two complexes can be formulated as XIII and XIV.

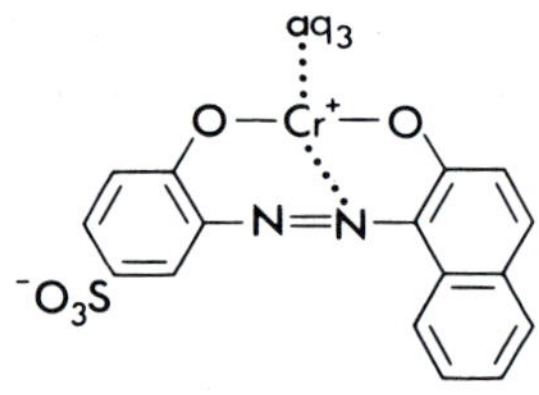

(XIII)

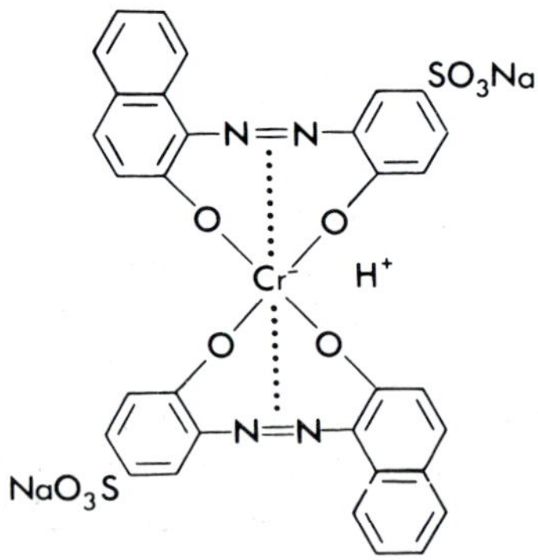

(XIV)

Some dyes appear to form the 1:1 and others the 1:2 type of complex; fall in pH tends to favour the 1:1 complex. The evidence is qualitative, mainly based on the colour of the fibre, because the complexes cannot be removed unchanged [85]. In a few cases analysis of the dyed and mordanted fibre after removal of unreacted chromium salts has appeared to show the presence only of the 1:2 complex [86]. Moreover, further reactions involving mutual interaction of hydroxyl groups attached to the chelating metal to give polymeric substances probably occur. These reactions, viz. (a) 'olation' and (b) 'oxolation', are known to occur in aqueous solutions of some trivalent chromium salts [87, 88]:

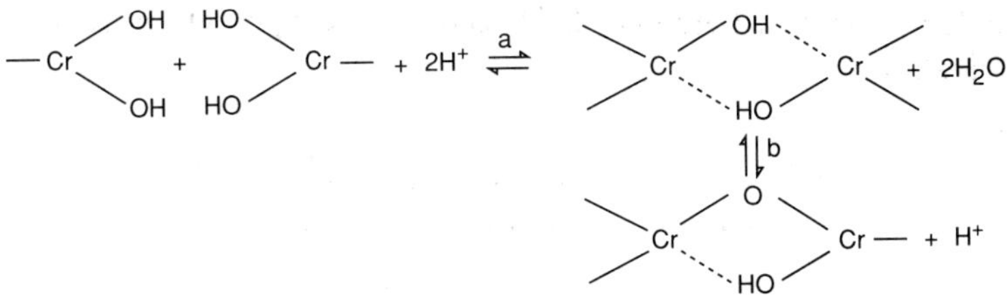

(2.7)

In fact, spectroscopic and other analytical techniques have revealed the presence of 1:1 and 1:2 complexes in the fibre during the dyeing of wool with the (unsulphonated) hydroxyquinone compound haematoxylin, the colouring principle of the logwood tree. Also, properties of the reaction products prepared *in vitro* indicated the presence of olated polymers [87, 88], e.g. analyses of the product of reaction of haematein and potassium dichromate in boiling dilute aqueous solution were consistent with the presence of olated compound:

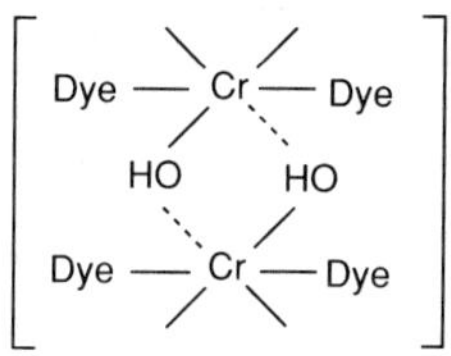

as well as unolated 1:2 metal–dye complex [80].

It is clear from the reaction sequence shown above that the olation reaction (a) will produce a rise in pH and the subsequent oxolation (b) a fall in pH. These changes are in fact observed when an aqueous solution of the 1:2 chromium–haematein lake is allowed to stand at a raised temperature [80].

The haematein colouring matter which was used for the investigations just mentioned is a natural product and therefore is not sulphonated; most of the present-day mordant dyes are synthetic sulphonated compounds. Nevertheless, the conclusions reached are considered to be applicable to the synthetic products, since sulphonated and unsulphonated azo dyes display similar complexing reactions with metals.

If only the 1:1 complex is present, the possibility of covalent linkage between the chromium atom and polar groups in the wool molecule exists, e.g.

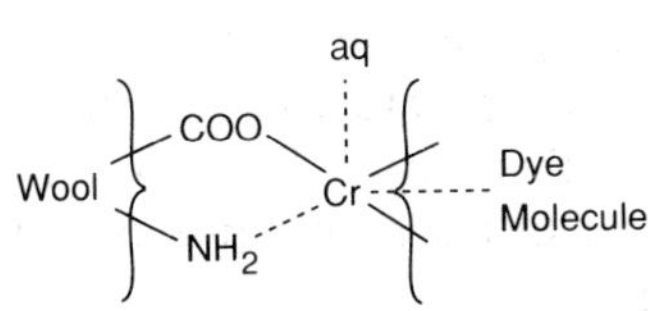

Several authors have obtained evidence that the tanning reactions which proteins undergo when treated with chromium salts can be explained by the hypothesis of chromium–protein covalent bond formation [89–91] and such bonds may thus operate in the chrome mordanting of wool. Another line of evidence that they do so is given by statistical analysis of washing fastness data of acid and mordanted dyes on wool. This shows that the mordanted dyes tend to have better fastness than would be expected from their r.m.m.s if they form only 1:2 complexes [92]. The higher resistance to washing may be due to the presence either of a proportion of 1:1 complexes, covalently bonded to the fibre, or of polymeric olated bodies in which the covalent bonds are formed between the linked dye molecules.

Against the view that dye–chromium–protein bonds are formed, however, may be set the results of studies of reactions between metal-complex dyes and monolayers of a protein and of a nylon derivative respectively, on water. With the 1:2 metal–unsulphonated dye complexes there was no evidence of protein–metal bonding, and the only type of bond detected appeared to be one between the monolayer and certain proton-donating substituent groups, especially hydroxyl groups, in the dye molecules [93a]. These conclusions were the result of a lengthy investigation, for details of which the original report must be consulted. Similar work in the author's laboratory also gave no evidence of bonds between protein and chromium in the 1:1 metal–sulphonated dye complexes [93b].

2.6.3 Hydrogen bonds: disperse dyes on hydrophobic fibres [25, 94–106]

The mechanism of the dyeing of these fibres, especially of cellulose acetates, with non-ionic (disperse) dyes has been intensively studied by several authors, particularly by Bird and the Leeds school, and a considerable body of facts

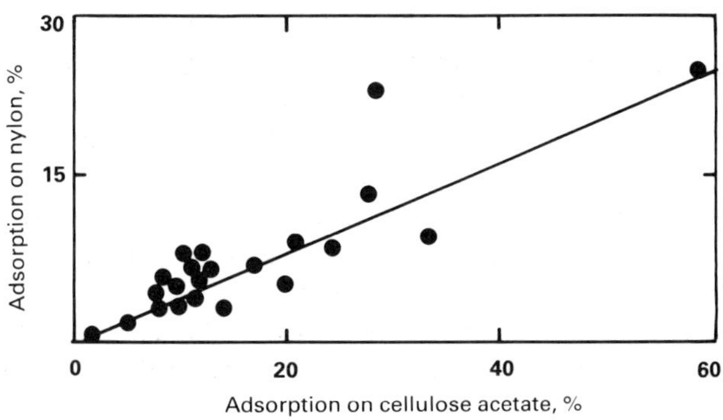

Figure 2.9 – Relation between saturation adsorption of disperse dyes on cellulose acetate and on nylon at 85°C (data from Vickerstaff) [22,100]

is available on which to base an hypothesis of the dyeing mechanism. Typical disperse dye structures (X, XI, XII) are shown in section 2.2.2. Since there are no ionic groups either in the dyes or in the fibres (with the exception of polyamide and acrylic fibres), dye–fibre attraction clearly must originate in one or more of the various types of polar and dispersion forces. An apparently linear relation (Figure 2.9) between saturation adsorption of disperse dyes on (non-ionic) secondary cellulose acetate and on (ionic) nylon suggests that the mechanism of disperse dyeing on these fibres is the same, but nylon adsorbs less dye because of its lower content of amorphous material. Simple model non-ionic proton donors, e.g. phenol, are readily adsorbed by cellulose acetate, and this must almost certainly be due to hydrogen bonding, as first suggested by Marsden and Urquhart [107]. Vickerstaff [22] later extended this concept to dyeing with disperse dyes, all of which contain simple polar groups in their structure. Most disperse dyes contain proton-donor groups, e.g. —OH or —NH_2, as well as proton acceptors, e.g. >C=O, —N=N—, —$N(CH_3)_2$ and —OCH_3.

Complexing between proton acceptors and simple organic acetates is readily demonstrated by the method of continuous variations, as already described [57, 108]. The acetates can be considered as models for cellulose acetate. Several sulphonated naphthols in water do not appear to complex with acetates, however, presumably because they are too soluble. This would help to explain the very low affinity of *sulphonated* dyes for cellulose acetate. The sulphate group, however, is apparently less solvated by water than the sulphonate group and the case of the *sulphated* Solacet water-soluble dyes, formerly used for cellulose acetate, is an interesting one [109]. These have, as a water-solubilising group, a sulphate ester group, which is probably less highly ionisable than the sulphonate group and so less effective in preventing dye–fibre bonding at the water–fibre interface. The Solacet dyes produce the L-type isotherm, indicating normal adsorption in water-filled pores, with saturation limits rather less than those of corresponding disperse dyes. The saturation increases with rise in salt (NaCl) concentration, indicating dye association, which must occur in water-filled pores.

There is little difficulty in accounting for the above complexes by proton donation from a hydroxyl group in one solute. Similarly it was shown by extensive refractometric tests that a hydrogen atom in acetate groups is sufficiently polar to take part in complexing with proton acceptors [108]. Thus isopropyl acetate complexes with anisole: 1:1 in dioxan or ethyl acetate, 1:1 with propanone in toluene and 4:1 with azobenzene in toluene. Weak chelation between 1:2-acetate groups was observed, whereby a seven-membered ring is formed. This evidence can explain the adsorption by cellulose acetate of various aromatic compounds without proton-donor groups, e.g. azobenzene, and it also shows that all the polar groups in a disperse dye molecule, and not merely those with a free hydrogen atom, can contribute to the affinity.

When monolayer tests were made with films of cellulose triacetate (CTA) mixed with two typical disperse dyes A and B respectively, A caused no change in the film area, whereas B expanded it (Figure 2.10). The interpretation is

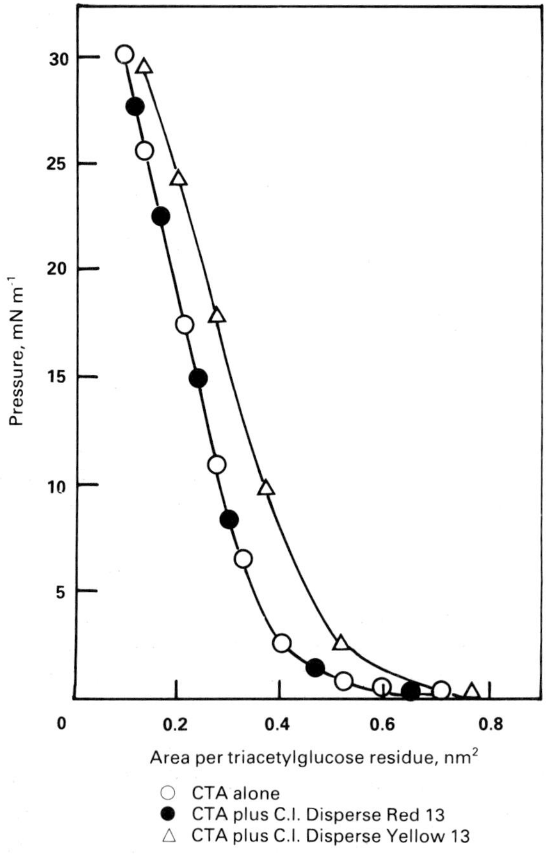

Figure 2.10 – Surface pressure–area curves for cellulose triacetate (CTA) monolayers, alone and as mixed film with disperse dye (proportion, 1 dye molecule: 1 hexa-acetyl-cellobiose unit) [61g]

that the dye molecules lie parallel, withe CTA chains and molecules of A forming face-to-face complexes in the plane of the water surface. The widths of the molecules of A and B are respectively equal to and greater than that of CTA; therefore A does not cause the CTA chains to move apart, whereas B does so.

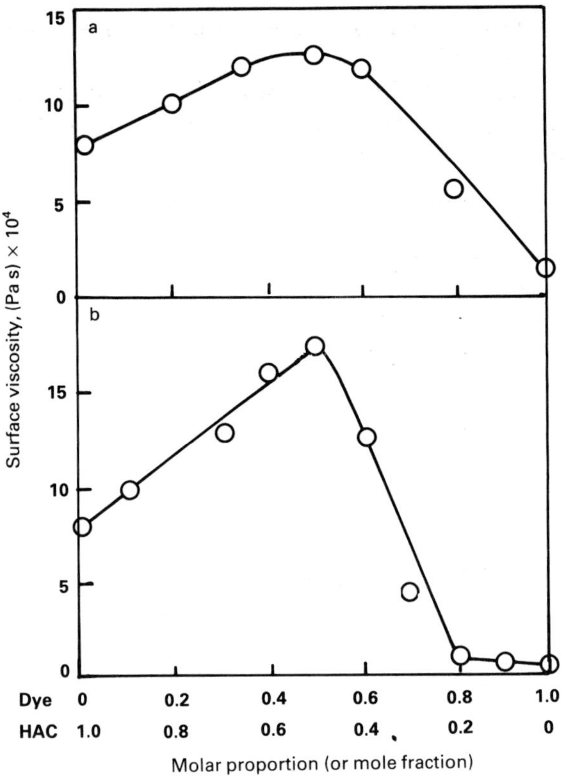

Figure 2.11 – Surface viscosity of mixtures of cellulose triacetate and disperse dye in monolayers, plotted according to the method of continuous variations: (a) C.I. Disperse Red 13; (b) C.I. Disperse Yellow 13 (HAC = hexa-acetylcellobiose residue) [61g]

Monolayer viscosity measurements by the method of continuous variations (Figure 2.11) show clear signs of a 1:1 complex between an (azo) disperse dye molecule and each hexa-acetylcellobiose unit in the cellulose triacetate molecule [61f]. This suggests, as do other monolayer experiments, that complexing is face-to-face, the molecules lying parallel [61c, 61g].

Solubility

Both aqueous solubility and saturation adsorption increase together in a series of dyes (Figure 2.12). This is the reverse of the relation shown by water-soluble dyes on a typical hydrophilic polymer, cellulose. Thus Griffiths and Neale [110] found that the saturation adsorption of direct (anionic) dyes by regenerated cellulose falls with increase in sulphonation of the dye molecule, and thus with increase in its solubility in water (Table 2.3).

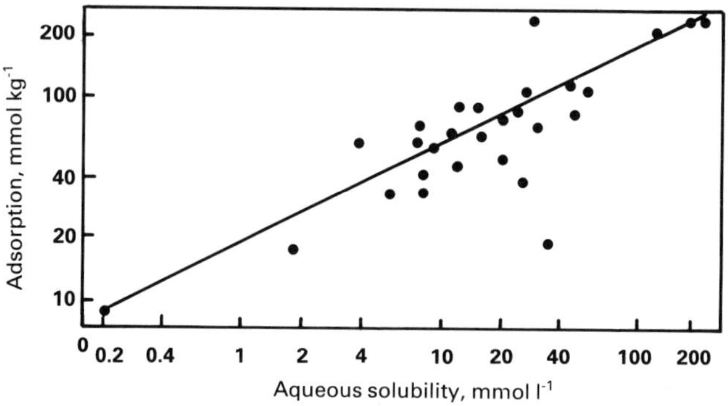

Figure 2.12 – Relation between water solubility of disperse dyes and their saturation adsorption in cellulose acetate [100] (line drawn by inspection)

TABLE 2.3

Effect of sulphonation on maximum adsorption of disazo direct dyes by regenerated cellulose [110]

Dye (C.I. no.)	Maximum adsorption /% on wt of cellulose
Disulphonated:	
24140	1.45
24205	2.10
24105	2.25
24100	2.45
23040	2.84
22120	3.10
23500	3.30
24380	3.40
Tetrasulphonated:	
24340	0.545
22610	0.610
22605	0.880
24400	0.945
23850	0.950
24270	0.990

The direct correlation between solubility and adsorption with disperse dyes shows that both are directly related to a third parameter. This must be the number and nature of polar and π-bond groups in the dye molecule.

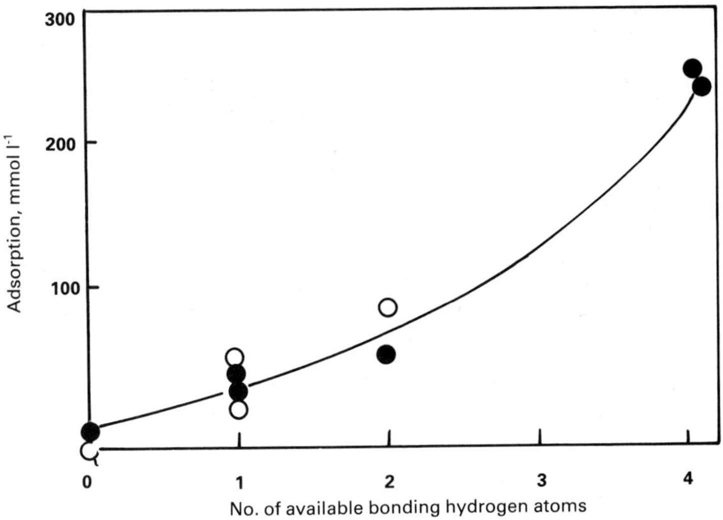

Figure 2.13 – Relation between number of hydrogen atoms available for intermolecular binding in multifunctional azobenzene disperse dyes (no 2- or 2'-subsituents) and their maximum adsorption by cellulose acetate [100]

The effect of hydrogen-bonding groups
This effect arises because the dye adsorption mechanism of polar hydrophobic fibres differs from that of all other types of dye–fibre systems in being almost entirely one of hydrogen and π-bonding. Van der Waals or 'hydrophobic' bonding has a minor effect. The subject has been discussed by several authors. Figure 2.13 illustrates the effect of the number of bonding groups [100]; apparently the more bonds that the dye can form with the fibre, the more readily can it break fibre–fibre bonds and so penetrate deeply into the polymer structure. The ability of the polymers to act as both proton donors and proton acceptors, in bonding with polar groups or the π-electron system in the dye molecule, can be demonstrated in several ways; these include monolayer experiments, and tests of various physical properties of binary mixtures of dyes and polymers in model compounds. The most recent of these methods is gas–liquid chromatography, as developed by de Vries *et al.* [111] for measuring the heat of reaction between cellulose acetate as stationary phase and a large number of model compounds. A few of these compounds and results obtained with them are listed in Table 2.4.

The conclusions reached by these authors are:

1. Phenols and aliphatic alcohols act as proton donors to the carbonyl oxygen atom of the acetate group and the hydroxyl group in secondary cellulose acetate.

TABLE 2.4

Heats of adsorption for various solutes on secondary cellulose acetate (CA) and cellulose triacetate (CTA) [101]

Solute	Temperature range /°C	$\Delta H^\ominus$ CA /kJ mol^{-1}	$\Delta H^\ominus$ CA /kcal mol^{-1}	$\Delta H^\ominus$ CTA /kJ mol^{-1}	$\Delta H^\ominus$ CTA /kcal mol^{-1}
Alcohols:					
Methanol	60–80	36.4	8.7	34.3	8.2
Ethanol	60–90	46.9	11.2	39.8	9.5
Propanol	60–80	52.3	12.5	41.9	10.0
Heptan-1-ol	120–140	53.6	12.8	44.0	10.5
Phenols:					
Phenol	150–171	56.5	13.5	55.3	13.2
4-Ethylphenol	150–171	64.9	15.5	63.6	15.2
4-Chlorophenol	150–171	63.6	15.2	73.6	17.6
Hydrocarbons:					
n-Hexane	20–40	3.4	0.8	3.8	0.9
Benzene	30–50	22.2	5.3	26.0	6.2
Toluene	30–50	30.1	7.2	32.6	7.8
Halogen compounds:					
Trichloromethane	60–80	27.2	6.5	24.7	5.9
Tribromomethane	110–130	31.0	7.4	31.0	7.4
Tetrachloromethane	30–50	23.0	5.5	19.3	4.6

2. Aromatic hydrocarbons act as electron donors and the acetate group as an electron acceptor (other hydrocarbons are adsorbed only by weak dispersion forces).

3. The halogen compounds are adsorbed by electron transfer, from the carbonyl oxygen atom of the acetate group, to the protons of the haloforms and the chlorine atoms of tetrachloromethane respectively.

[The heat values for phenols are roughly the sum of the values for those for a typical alcohol (ethanol) and a typical aromatic hydrocarbon (benzene).]

The linear isotherm (*C-curve* [69])
A feature that distinguishes the dyeing of disperse dyes on hydrophobic fibres from all other dyeing processes is the nature of the adsorption isotherm, which is linear with an abrupt change to a horizontal plateau [112] (Figure

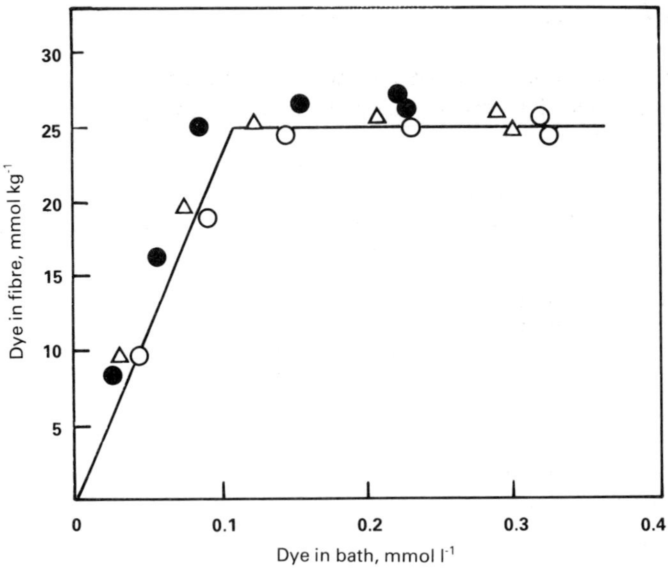

Figure 2.14 – A typical linear (C-type) adsorption isotherm, for C.I. Disperse Red 13 on nylon (90 °C) [62]

2.14). As stated previously, the disperse dye molecule may be held face-to-face with the cellulose acetate molecular chain by multiple bonding. This causes the dye molecule to leave the water of the dyebath – in which disperse dyes have low solubility – and penetrate between the polymeric chains of the fibre, beyond the regions penetrated by water. New sites for adsorption are thus developed as the solute penetrates further and further into the structure. The final result can be considered to be a form of 'solid solution', a term used to explain the dyeing of secondary cellulose acetate in the earliest days of its investigation [22, 113]. The first solute molecules entering the fibre may be likened to a wedge driven between the polymer chains of the fibre and opening them up to more solute. In such a process, the stronger the dye–fibre bonding (i.e. the larger the number of groups on the dye molecule that can bond with the fibre), the greater is its capacity to break the interchain bonds of the fibre. Thus we should expect that, with increase in number of bonding groups in the dye, more dye will be able to accommodate itself in the fibre, i.e. that the saturation limit for dye on fibre will rise with the number of bonding groups in the dye molecule. This effect, in fact, can readily be observed (Figure 2.13). As adsorption increases, more fibre–fibre bonds must be broken, and in extreme cases, e.g. phenol adsorbing from water in nylon, the whole fibre eventually disintegrates.

The C-curve condition can be considered a special case [112] of the

Langmuir monolayer hypothesis, according to which the concentration of gas in a monolayer, established by adsorption of its molecules on a solid surface, is determined by the balance between capture of molecules by the surface and their evaporation therefrom. If the dye molecules are likened to those of the adsorbate gas, then the amount y of dye adsorbed is, by the Langmuir treatment, as given in section 2.5,

$$y = \frac{by_{\mathrm{m}}c}{1 + bc} \tag{2.8}$$

where y_{m} is the amount adsorbed when the monolayer is complete, c is the concentration of dye in solution (equivalent to gas pressure in the original Langmuir treatment) and b is a constant, the adsorption coefficient, which is a function of temperature.

This treatment may be extended to the case of the present linear isotherms by assuming that y_{m} is no longer a constant, but increases proportionately with c, i.e. that the surface available for adsorption increases as each adsorbed dye molecule breaks apart a little more of the fibre structure [69]. This is like the action of a zip fastener, the slider being the entering molecule. Thus,

$$y_{\mathrm{m}} = k(1 + \alpha c) \tag{2.9}$$

where k is a constant and α is in the nature of a coefficient of expansion of the surface. Under C-curve conditions, Eqn 2.9 may be written.

$$y = \beta c \tag{2.10}$$

where β is another constant. It follows from Eqns 2.8, 2.9 and 2.10 that

$$b = \frac{\beta}{k + k\alpha c - \beta c} \tag{2.11}$$

Eqn 2.11 is valid only up to a critical value $c = c^* = k/(\beta - k\alpha)$ (i.e. $k + k\alpha c - \beta c = 0$), because if c is greater than c^* the sign of b would change from positive to negative, and this is impossible, because the components of b are positive. Thus at $c = c^*$, $b = \infty$, i.e. the rate of escape of adsorbed dye from the fibre is zero. It follows that, when the concentration c^* is reached, the value of y (Eqn 2.8) will remain fixed, *provided no fresh internal surface is formed*, and the curve will change abruptly to the horizontal. This is the condition found experimentally. If a fresh region of the fibre is opened up when $c = c^*$ and a similar adsorption mechanism applies, there will be a sharp inflection in the curve and then a second linear portion of different slope. In a few cases, e.g. benzyl alcohol on cellulose triacetate, this very rare type of curve has actually been obtained [114].

Alternatively the process can be considered as follows. The rate of solute adsorption is independent of the available area of substrate at any given time, and is dependent only on the concentration c of the solution in contact with it. The rate of desorption, however, is proportional to the area of the surface covered by adsorbed dye molecules and hence, to a first approximation, to the concentration y of adsorbed solute in the fibre. Thus rate of adsorption $= k'c$ and rate of desorption $= k''y$, where k' and k'' are constants. At equlibrium $k'c = k''y$, so y/c is a constant [69].

Conclusion

It is thus possible to account for the adsorption of all disperse dyes on all hydrophobic polymers by the combined action of weak dispersion forces together with strong hydrogen bond forces. (On polypropylene only the weak dispersion forces can operate [115].) It is clear that, for all disperse dyes and all hydrophobic fibres, every polar substituent group in the dye molecule and the aromatic system as a whole contribute to the strength of the dye–fibre bond if the molecule is planar and thus can lie flat against the polymer molecule. This accounts for the much higher substantivity which planar disperse dye molecules have, compared with non-planar ones [106], and also for the evidence given by measurement of monolayer areas and viscosities that a strong 1:1 complex is formed by face-to-face packing between planar disperse dyes and the hexa-acetylcellobiose residues in cellulose triacetate (Figure 2.11).

Most of the evidence for the hydrogen-bonding mechanism of disperse dyes has been gained from experiments with cellulose acetates, but there seems no reason to doubt that nylon and polyester adsorb these dyes by the same general mechanism. In confirmation, as stated, there is a linear relation between solubilities of such dyes in cellulose acetate and nylon respectively, and this indicates a similar mechanism. Certainly both nylon and polyester have more hydrocarbon groups than has cellulose acetate. In nylon, however, these are in the form of alkyl chains, and it follows from the very low substantivity of disperse dyes for the purely alkyl hydrocarbon polymer polypropylene that they must have a negligible effect on dye substantivity. Nevertheless, the aromatic nuclei in polyester may take part in hydrogen bonding through the π-electron system.

2.6.4 Disperse dyes on cellulose

Aspland and Bird [116] examined the adsorption of disperse dyes by cellulose. The amount taken up is very small, but the peculiar feature of the dyeing is that the adsorption isotherm is of the linear or C-type (Figure 2.15). Further,

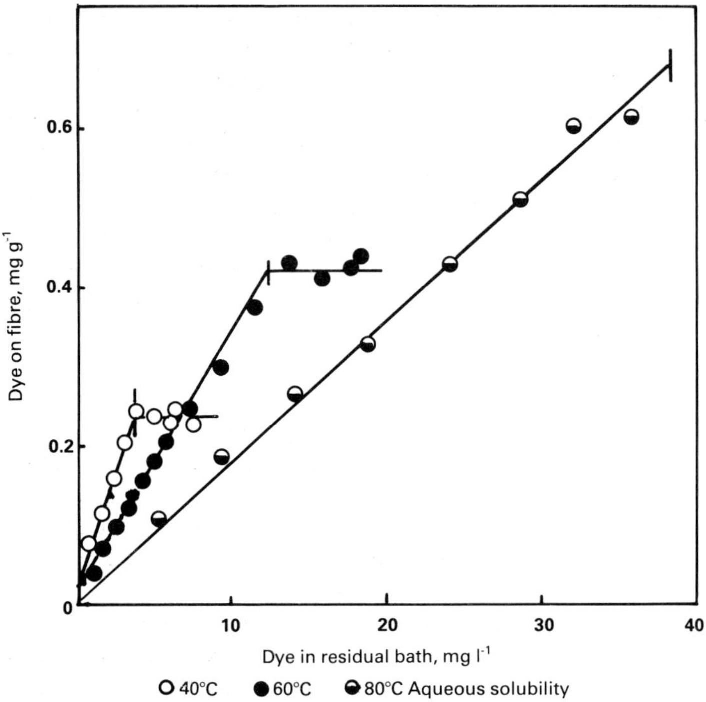

Figure 2.15 – Adsorption isotherms for a disperse dye on viscose rayon [116]

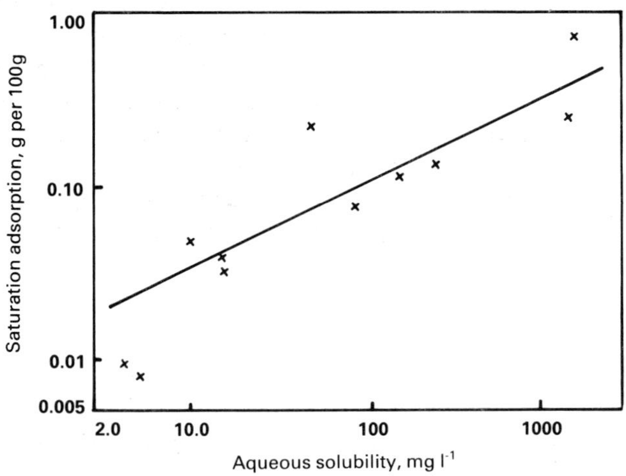

Figure 2.16 – Relation between saturation adsorption of disperse dyes on viscose rayon and their aqueous solubility, at 60 °C, plotted from data by Aspland and Bird [116] (line drawn by inspection)

an examination of the saturation adsorption for the dyes used shows that this increases regularly with their aqueous solubility (Figure 2.16); both these features resemble those for disperse dyeing already discussed and are unlike the characteristics of direct dyeing (which gives L-type [112] adsorption isotherms). This leads to the conclusion that the disperse dyes penetrate into some of the crystalline regions of the cellulose, where they unite by hydrogen bonding in absence of water. This might explain an observation that direct dyes have better fastness to light on cellulose when applied from ethanol than from water [117]; in absence of water they can perhaps penetrate the crystalline regions where they are better protected against oxygen and hence fade more slowly.

2.6.5 Protein fibres

The role of hydrogen bonding in the dyeing of protein fibres has never been very clearly defined, and indirect evidence must be used to interpret it. These fibres are usually dyed with ionic dyes, and ion-exchange and other non-specific forces mask any contribution from hydrogen bonds.

Investigations were made by the author into the adsorption of a very wide variety of organic substances, from the vapour phase and from solution in water and organic solvents, by wool, silk and nylon [118, 119]. The conclusion, supported by the monolayer tests described below, was briefly that proton donors have higher substantivity than proton acceptors, in all cases. The phenolic group (but not the phenolate ion) appeared to have much higher substantivity than other groups. In ability to bind with proteins or nylon, the groups fall roughly in the order: ArOH > $ArNHCOCH_3$ > $ArNH_2$, ArNHR, AlkOH > proton acceptors. Bird and Firth [120] found that non-ionic (disperse) dyes have very low saturation adsorption values on wool. The dyes they used have several polar groups, but no phenolic group; probably the adsorption is low because of competition by water and the inability of the dyes to penetrate beyond the water-swollen regions, as they can do in hydrophobic fibres (or the water-impenetrable region is small).

Tests made by introducing dyes beneath protein monolayers on water [93, 121] showed that of the two dyes,

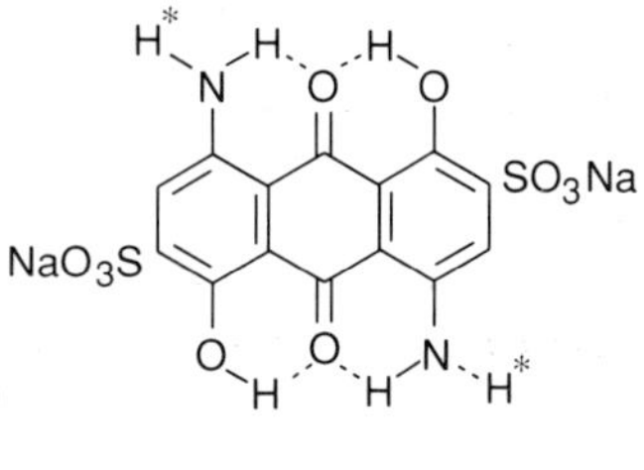
(XV)

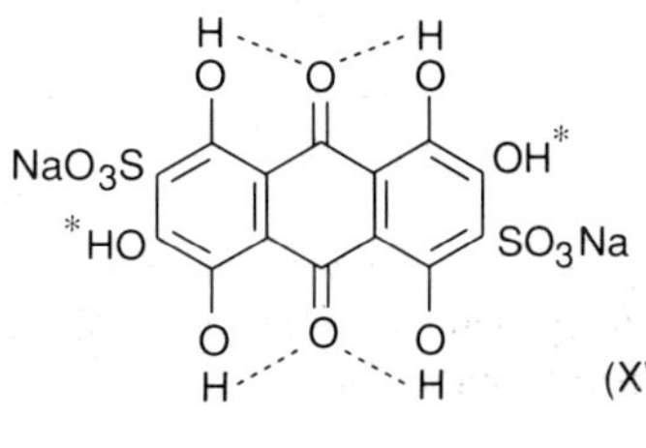
(XVI)

both of which have hydrogen atoms (marked*) free to bond, the one (XVI) with the free hydroxyl groups had a much more marked effect than the one with free >NH groups.

This effect was similar to 'tanning' by tannic acid, in that the monolayer was made more rigid and less compressible. Few dyes used for wool contain a free phenolic group, however, and the general conclusion was that hydrogen-bonding effects are usually not very important in normal wool dyeing.

Non-specific forces: ion exchange

This operates in the technically very important process of acid dyeing of anionic dyes on fibres containing amino groups (i.e. wool, silk and nylon), also with basic dyes on these fibres, on which they are not much used, and on the newer acrylic fibres, for which they are widely employed. The broad principles of the ion-exchange effects in the acid dyeing of wool, etc., have often been described [22].

In water the amino and carboxyl groups of the protein fibre (*W*) can be considered as zwitterionic pairs, and entering hydrogen ions from an acid (HX) then neutralise the carboxyl groups by back-titration:

$$\bar{O}OC—W—\overset{+}{N}H_3 \xrightarrow{H^+X^-} HOOC—W—NH_3^+ \quad X^- \qquad (2.12)$$

Thus the capacity of protein (or nylon) fibres for acids should be equivalent to their free amino-group content. That it is very closely so can be seen from the curves of Figure 2.17. It is also confirmed by studies of the temperature-dependence of the wool titration curve, which reveal that the heat of combination of hydrogen ions is almost zero in acid solution and approximately -42 kJ $equiv.^{-1}$ (-10 kcal $equiv.^{-1}$) in alkaline solutions, values which are close to those for the heats of ionisation of many organic acids and the heats of dissociation of various substituted ammonium ions, respectively [22]. At very low pH values, excess of acid is taken up. This is believed to be due to protonation or to hydrolysis of the peptide or amide groups [122]. Hydrolysis is suggested by the weakening of nylon fibres dyed at these low pH values. Rideal and Gilbert's thermodynamic treatment of the acid titration of wool was made on the assumption that all the adsorbed ions are located on specific sites. Some discussion has taken place subsequently upon the alternative application of the Donnan membrane principle, on the assumption that the adsorbed anions are not specifically located [123]. This is discussed in more detail in Chapter 4.

The titration curves show that anions differ in affinity*, which in fact rises with relative anionic mass, as shown by the shift of pH required for half-

* Affinities of some inorganic anions for silk increase in the order $Cl^- < Br^- < SCN$ [123i].

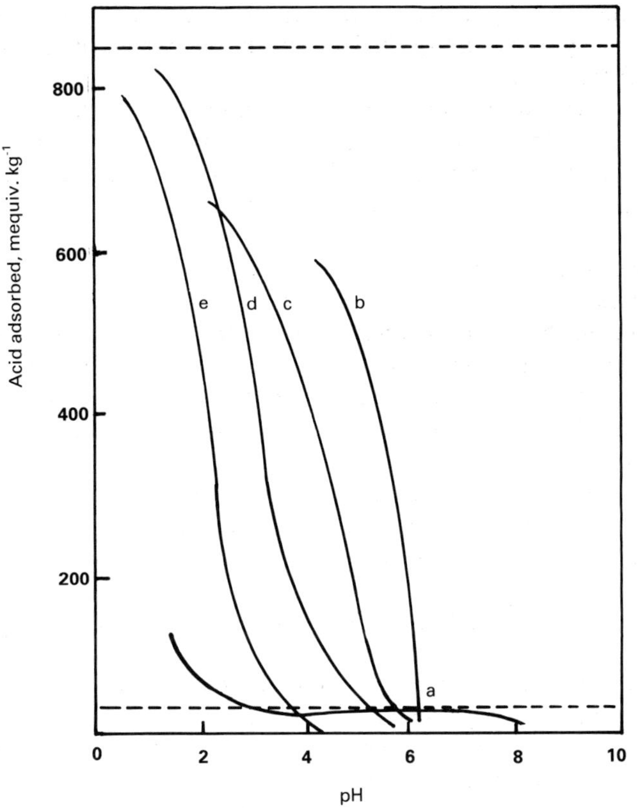

Figure 2.17 – Acid titration curves of nylon [25] and wool [22]; the broken lines represent the values corresponding to the respective free amino-group contents of the two fibres. Nylon: (a) dye C.I. 18050 (disulphonated). Wool (monobasic free acids): (b) dye C.I. 26660; (c) dye C.I. 15510; (d) naphthalene-2-sulphonic acid; (e) hydrochloric acid

saturation (Figure 2.17). The forces responsible for this change in affinity with molecular size or r.m.m. have not been very precisely defined. Hydrogen bonding probably plays some part, if the dye contains proton-donating groups, but van der Waals forces must also be involved. It had been suggested [53] that these operate between the hydrocarbon portions of the dye molecules themselves, which tend to escape from the water and seek the hydrophobic regions of the protein, with consequent destruction of the hydrocarbon–water interface ('hydrophobic bonding') as already mentioned. It might be expected that measurements of thermodynamic parameters could throw light on this matter, but they do not because water is desorbed to give place to adsorbed dye, and this dye itself may associate further after adsorption, so that all measurements give only a net result which is sometimes difficult to interpret.† Adsorption of the dye Orange II (VI), in free acid form by wool and by collagen, shows an increase in entropy [124, 125] attributed to greater order in the water originally present on the ionic sites, than in the adsorbed dye. Indeed, it has been suggested that entropy is the major contributor to affinity [54, 126] (cf. p. 273). Few data on apparent heats of dye adsorption on protein fibres have been obtained, but work on dye adsorption by ion exchange on inorganic powders shows that the apparent heat may be negative, zero or positive, depending upon changes in the state of association of the dye before and after adsorption [63], so that the actual numerical magnitude of the parameter gives little information, unless of course comparison is made directly with a value for some very similar reaction.

Two interesting researches on such lines have been described. In the first [127], comparisons were made by calorimetry of the heat of reaction in water of hydrochloric acid and the free acid of Orange II (VI), respectively, with eight amino acid constituents of wool. This was found to be less than 1 kcal mol^{-1} in every case, interpreted as evidence that the apparent heat of adsorption of this dye by wool (-9.3 kcal or -38.9 kJ mol^{-1}) is not due to direct interaction between the dye and groupings in the amino acid residues of the fibre. This agrees with the suggestion above that proton acceptors do not readily form hydrogen bonds with proteins: Orange II, because of its intramolecular bond (section 2.4.4) has no hydrogen atom free to bond, and is a proton acceptor [57].

In the second research [128], the heat of dimerisation of Orange II in water was found to be -44 kJ mol^{-1} (-10.5 kcal mol^{-1}). These results suggest that the apparent heat of adsorption of the dye is possibly due to dye–dye affinity on adsorption, rather than dye–fibre affinity.

We have already seen that the saturation value for monobasic acids and dyes with low r.m.m. corresponds to the amino-group content. This situation is difficult to reconcile with the known structure of the fibre, however. Later knowledge has done little to clarify the situation discussed by Vickerstaff [22], who remarked that, in spite of the apparent theoretical impossibility, 'all the experimental evidence leads to the conclusion that, in wool saturated with dye, the dye ions are not all in close proximity to charged amino groups' and that this conclusion raises difficulties in developing a quantitative theory.

There is in fact evidence that, at high dye contents, adsorption of dye occurs either on peptide groups with breakdown of the protein chains and collapse of the fibre into powder [22, 129] or as aggregates in cavities in the fibre [130, 131] (cf. Figure 2.18).

The data must be regarded as still open to re-interpretation; the present

† Zollinger [54] has described the relation of 'hydrophobic bonding' to wool dyeing.

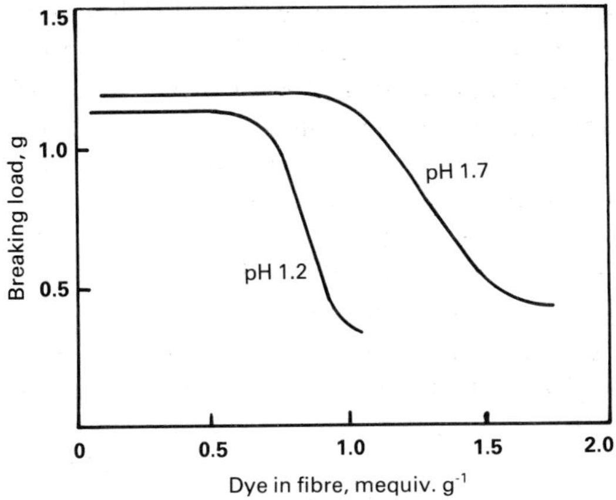

Figure 2.18 – Relation between concentration on wool of the milling acid dye C.I. Acid Blue 138 and fibre-breaking strength [19a]

author considers that the dye is probably concentrated in voids in the fibre structure, and that the correspondence between dye anion capacity and basic group content of the fibre may be partly fortuitous.

Further discussion of the adsorption of acid dyes by wool will be found in Chapter 4.

2.6.6 Dyeing with basic dyes [133–135; see also Chapters 3 and 5]

Soon after polyacrylonitrile (PAN) fibres had been introduced it was found that basic dyes could readily be applied to them [133]. This led in the early 1950s to renewed interest in this very old class of dyes, and to an outburst of research activity by the dye manufacturers which resulted in the development of several ranges of entirely new types of basic (cationic) dye of bright hue and high fastness. The 'basic-dyeing' nylon and polyester fibres also can be dyed with dyes of this type.

Basic dyes are adsorbed by PAN fibres to give Langmuir-type ('L' [112]) isotherms, and by replotting these on reciprocal scales and measuring the intercept, the number of acidic adsorption sites can be determined. The values for various PAN fibres dyed at low pH were determined by Balmforth *et al.* [134] and found to agree well with the respective values obtained by other methods (Tables 2.5 and 2.6). Moreover, no sample of PAN examined showed a dye adsorption, even at high pH, exceeding the total number of acidic groups. These facts provide excellent confirmation of the hypothesis that the dyeing mechanism is a simple one of ion exchange whereby the dye cation displaces H^+, Na^+ or K^+. Potentiometric titration shows that most PAN fibres contain not only the strong acidic groups thus revealed but

TABLE 2.5

(Strong) acidic dye sites of PAN fibres [134]

Fibre	Dye sites/mequiv. kg^{-1} from Langmuir isotherm intercept (pH 5.4 dye-bath)	by non-aqueous conductance titration (after ion exchange)*
Acrilan 16 (Chemstrand)	36	41
Orlon 42 (DUP)	54	54
Dralon (BAY)	44	44
Dralon Neu (BAY)	55	58

* Before removal of Na^+ and K^+ by ion exchange these values were approximately half those shown here, respectively.

TABLE 2.6

Acidic end-group content of PAN polymer [134]

Before ion exchange:	
Na^+	15
K^+	5
H^+	21
Total	41
After ion exchange (H^+)	41.5
Sulphur analysis	40.5
Dye adsorption (pH, 5.4)	40
Reciprocal of number-average r.m.m. ($1/\overline{M}_n$)	40

End-group content measured in mequiv. kg^{-1}. PAN polymer has a low specific viscosity (= 0.15).

additional weak groups, which can be determined by a special method of potentiometric titration. At low pH adsorption occurs mainly on the strong acidic groups (Figure 2.19). Some of the stronger sites appear to be sulphate and sulphonate groups introduced by the redox polymerisation catalyst. The weaker ones have an ionisation constant similar to that of carboxylic acids.

Basic dyes are little used on hydrophilic fibres, and consequently their dyeing mechanism thereon has received little attention. The mechanism on wool appears to be similar to that of anionic dyes [136], and the titration

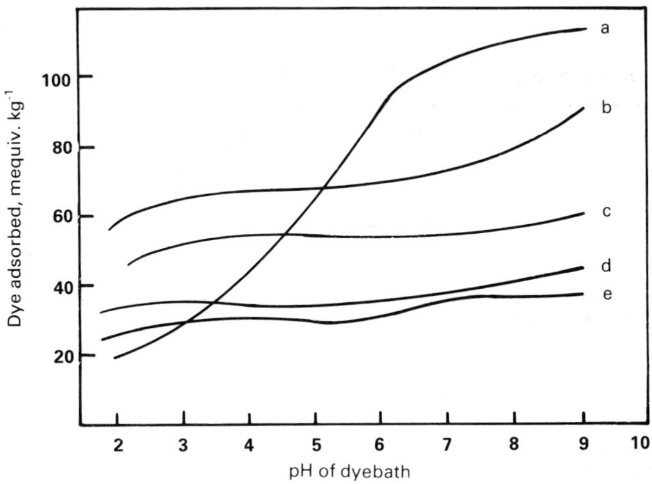

Figure 2.19 – Change of adsorption of C.I. Basic Blue 22 by various PAN fibres with pH of dyebath (100 °C, 16 h dyeing in 40-vol. bath) [134]:
(a) Courtelle E; (b) Beslon; (c) Orlon 42; (d) Acrilan 16; (e) Acribel

curves are mirror images of those for anionic dyes; the old-established basic dyes have rather low affinity and are rapid-dyeing on this fibre.

2.6.7 Other non-specific forces: cellulose dyeing

Cellulose dyeing is the most important of all dyeing processes in terms of quantity, yet it is probably the least well understood. Many theories have been proposed, and re-proposed, to account for the very strong substantivity which many dyes have for cellulose, even though there appear to be few reactive groups in the fibre [22, 34, 36, 137, 138, 139–156]. The earliest theories [11], applied to many dyeing processes, advanced by Hellot and several other authors from the mid-eighteenth century onwards, offered a purely mechanical explanation, whereby the fibre pores were swollen sufficiently by the hot dyebath to admit dye particles, which enter and are later trapped therein on cooling. We shall see later that similar ideas are again coming to the fore.

Possibly Hodgson [140] made the first real advance in clearly stating that the dye molecule must be planar for marked affinity for cellulose to be manifest. Thus of the two direct dye structures, XVII and XVIII, XVII gives a highly substantive cotton dye, but XVIII does not.

(XVII)

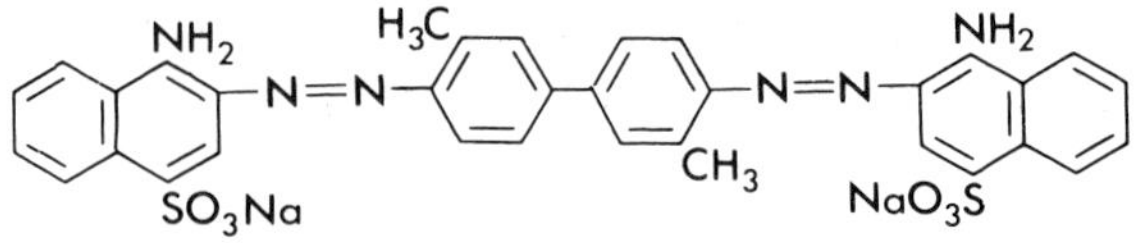

(XVIII)

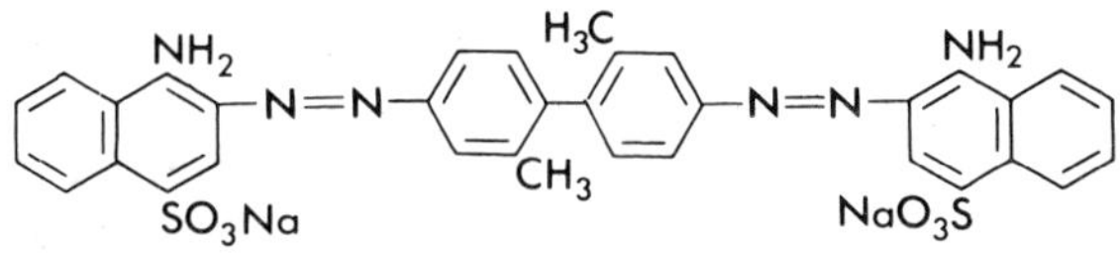

Hydrogen bonds
XVIII is non-planar, because of steric interference by the methyl groups. Consideration of such facts led to the hypothesis, advanced by several authors, that hydrogen bonds are formed between the dye and the fibre molecules [22, 141], either directly or subsequent to adlineation and close approach of dye and fibre molecules caused by strong van der Waals attraction [148].

Later evidence put forward by various authors threw doubt on the hydrogen bond hypothesis and it was suggested that these bonds are not formed, because the cellulose hydroxyl groups are too firmly attached to water to allow their bonding to polar groups in a dye. The unique ring structure of glucose, the building unit of cellulose, ensures maximum retention of an 'atmosphere' of water molecules around it, though the aldehyde group in the open-chain form, not significantly present in cellulose, can form bonds in water. Cellobiose, formed by condensation of two molecules of glucose, is the actual repeating unit of the cellulose chain.

The results of a monolayer experiment [61a] supported this suggestion. When a surface-active benzidine disazo dye is spread as a monolayer on water containing cellobiose, the area per dye molecule rises to the value required to accommodate a single layer of water molecules between each dye and cellobiose molecule, thus preventing hydrogen bonds forming between them, whereas no such layer is evident when mannitol is used instead of cellobiose. Mannitol, unlike cellobiose, does form hydrogen bonds with other solutes in water, and so can form them with dyes.

The following additional hypotheses have been proposed to account for dye–cellulose affinity.

Ion–dipole interaction Wegmann [153] postulated attraction between dipoles on the cellulose ether groups and ionic groups in the dye molecules.

This does seem a possible contribution to the attractions involved, but no extensive research has been carried out to investigate it.

Hydrogen π-bonds Supporting a suggestion of this type of bond, Yoshida *et al.* [155] found that sucrose, glucose and cellobiose form 1:1 complexes in water with simple *aromatic* sulphonate anions but not with an *aliphatic* one, the inference being that the bond involves the aromatic nucleus. Yoshida's observations have recently been confirmed, and extended to a sulphonated anthraquinone dye, which was found to form a 1:1 complex with sucrose in water [157]. The entropy of formation is high (314 J (75 cal) K^{-1} mol^{-1}, suggesting that the complex is formed by two co-operating molecules orientated parallel to each other and face to face. The evidence was in favour of the operation of this bond in cellulose dyeing, until it was found that typical sulphonated azo dyes, including Sky Blue FF (I), do not form any complex with methyl-β-D-glucopyranoside (a water-soluble model of cellulose) or with sucrose, so that its importance in cellulose dyeing is ruled out. It appears that the presence of an azo group in an aromatic structure prevents formation of the π-bond complex [157], but the cause of this prevention is not known.

The linear relation between the affinity of anionic dyes for cellulose and the (log) length of conjugate chain in their molecules (Figure 2.20) suggests that the source of affinity lies in either fibre–dye or dye–dye van der Waals forces, or both, rather than in hydrogen-bonding groups, since many of the dyes concerned do not differ in their content of the latter, yet differ markedly in affinity. Peters and Sumner [138] found a similar relation between the affinity of some of these dyes and the (log) extinction coefficient.

Acid–base bonds

Evidence of a possible bond between cellulose and free hydrogen atoms on amino groups in dyes has been obtained [36]. Refractometric tests with Sky Blue FF (I) dissolved in a solution of cellulose in alkaline Cadoxen solvent showed the presence of a 1:1 complex between the dye molecule and the cellobiose unit of the cellulose chain (Figure 2.3). This complex seems to be very weak – it has no influence on the surface viscosity of a monolayer of cellulose – and it may be the result of a weak acid–base reaction between the cellulosic hydroxyl groups and the amino groups in the dye.

This hypothesis is supported by the following observations:

1. The dye C.I. Direct Blue 10 (II), which is very similar in structure to C.I. Direct Blue 1, but has —OH and no —NH_2 groups, gives no evidence of a complex with cellulose in Cadoxen solution (Figure 2.3).

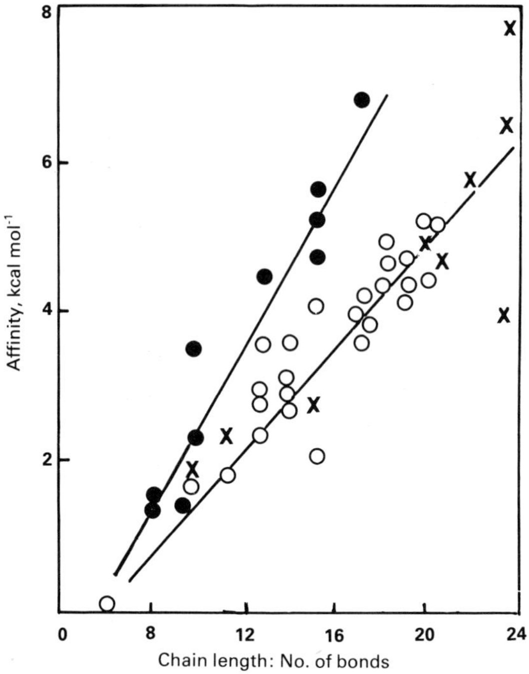

Figure 2.20 – Relationship between affinity for regenerated cellulose and molecular conjugate chain length of dye using affinity data from references 137, 138; ○, ●: affinities for cuprammonium rayon at 40 °C of planar leuco vat dyes and related compounds with respectively one and no —NHCO— group in the conjugate chain; X, affinities at 50 °C for viscose rayon of planar sulphonated azo dyes

2. Other ionic bonds do not affect surface viscosity, e.g. those between a protein monolayer on acid solutions and disulphonated acid dyes.

3. The hydrogen bonds between cellulose acetate in a monolayer and a disperse dye do considerably increase the viscosity of the monolayer, as already mentioned (Figure 2.11).

It therefore appears that a type of acid–base ionic attraction is exerted by cellulosic hydroxyl groups and free hydrogen atoms on basic groups in a dye. Some data given by Vickerstaff, for the affinity for cellulose of direct dyes containing such groups, support this suggestion (Table 2.7).

Van der Waals attraction

Further understanding of cellulose dyeing comes from monolayer experiments. Direct dyes in the water sub-phase beneath a monolayer of cellulose cause the film to expand without change in its compressibility; moreover,

TABLE 2.7

Affinities of direct dyes for cellulose (viscose rayon) at 90 °C, in presence of NaCl [22]

Dye (C.I. No.)	N-containing group(s)*	Affinity $(-\Delta\mu^\ominus)$ /kJ mol^{-1}	Affinity $(-\Delta\mu^\ominus)$ /kcal mol^{-1}
24895	none	13.0	3.1
28160	—CONH—	13.4	3.2
29185	—NHCONH—	16.7	4.0
29150	—NHCONH—	17.2	4.1
29190	—NHCONH—	19.3	4.6
29160	—NHCONH— —NHCO—	20.1	4.8
22570	>NH, >NH	18.0	4.3
30110	>NH, >NH	25.1	6.0
27720	—NH_2	15.9	3.8
22311	—NH_2	25.5	6.1
24410	—NH_2, —NH_2	26.4	6.3

*Containing unchelated H atoms.

the expansion is proportional to the width of the dye molecule [36] (Figure 2.21). Both these facts are in agreement with the hypothesis that the dye molecules are orientated flat, face to face with, and parallel to, the cellulose chains on the water surface (cf. p. 126). A similar suggestion was made by Morton [143b] to account for the dichroic effects shown by dyed cellulose. The evidence from the monolayer tests is in favour of an attraction between cellulose and dye, and not merely a chance juxtaposition of the two, each tending to float on the water surface. The question still remains, therefore, as to what is the nature of this attraction. The evidence points to the simultaneous operation of several weak forces, viz. van der Waals forces, the acid–base bonds just discussed, possibly ion–dipole interaction, and possibly a form of 'hydrophobic bonding' between the hydrophobic portion of the dye molecule and the CH groups in the cellulose molecule. This matter will be discussed again at a later stage, but first we must consider the nature of the association of dyes, both in solution and in fibres.

2.7 ASSOCIATION OF DYES

We have traced the initial concentration of dye at the fibre–solution interface, assisted as it is by forces operating between the dye molecule and the fibre polymer. Part of the technique of dyeing consists in facilitating the

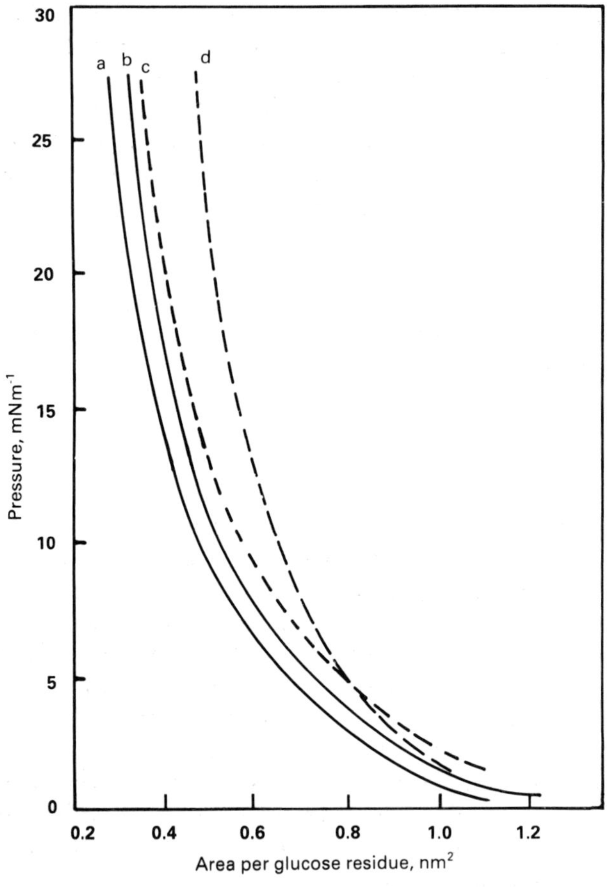

Figure 2.21 – Pressure–area curves for cellulose monolayers: (a) control (on water); (b) on solutions (10^{-5}–10^{-4}M) of azo direct or acid dyes, viz. C.I. Direct Yellow 12, Direct Red 2, Direct Blue 1, Acid Orange 7 (Orange II), Acid Red 1, Acid Red 88; (c) as (b), plus 0.1M NaCl; (d) as (b), C.I. Direct Green 26 [36]

development of such forces, e.g. by encouraging the adsorption of anionic dyes by adding dyebath assistants which either neutralise the negative charge on a fibre (salt used with cellulose fibres) or reverse it (acid used with protein fibres). Consideration of the dyeing mechanism, however, is not complete without attention being given to the subsequent behaviour of the dye molecules. There is good reason to believe that self-association into possibly quite large molecular aggregates follows their entry into the fibre. When dye molecules are held by ion-exchange forces at a *non-porous* surface, as on graphite or some forms of silica or powdered alumina, they can be almost instantly removed by cold water alonc, yet when they are held by similar forces in a fibre they are far more resistant to removal. Moreover, 4-nitrophenol, a small molecule which has little tendency to associate, enters and leaves a fibre extremely rapidly, even when the fibre contains dye particles that can adsorb it in the same way that a dye molecule is itself adsorbed by a fibre. It is therefore important to examine the evidence of dye association both in solution and in fibres.

2.7.1 In solution [158]

The absorption of light in passage through any medium is proportional to the number of absorbing molecules in its path. This law, usually known as the Beer–Lambert Law, is described by the equation

$$\log \frac{I_0}{I_T} = \varepsilon lc \tag{2.13}$$

where I_0, I_T are the intensities of light entering and leaving the medium respectively, ε is a constant (the molar extinction coefficient), I is the thickness of the absorbing medium and c is the molar concentration of the absorbing substance. Log (I_0/I_T) is a convenient practical measure of light absorption and is known as the *absorbance* (formerly optical density). Thus when l is constant, as when a cuvette of a given path length is used with a series of dye solutions, absorbance should rise linearly with dye concentration. A plot of this relation is usually known as a 'Beer's Law plot'. In fact it is often not linear, and three forms of the plot have been distinguished [159] (Figure 2.22): curve ***a*** is linear as predicted, ***b*** is concave to the concentration axis, while c is like ***b*** in the early stages, but then changes to a line of smaller curvature or to the linear form ***a***.

Curve ***b*** may be represented by the more general equation [160]

$$\log \frac{I_0}{I_T} = \frac{\varepsilon lc}{1 + gc^x} \tag{2.14}$$

where g and x are empirical positive constants. It is related to a progressive change in the distribution of size of molecular aggregates, towards a greater average size, with rise in concentration, since the larger the particle the lower is its ε value. Therefore, provided that the light is direct and monochromatic, a curvilinear Beer's Law plot for a dye indicates progressive aggregation with rise in concentration, and all such deviations from linearity are evidence of dye association.

A linear plot does not necessarily indicate that the dye is entirely monodisperse in solution, however; it may do so in some cases, but such a

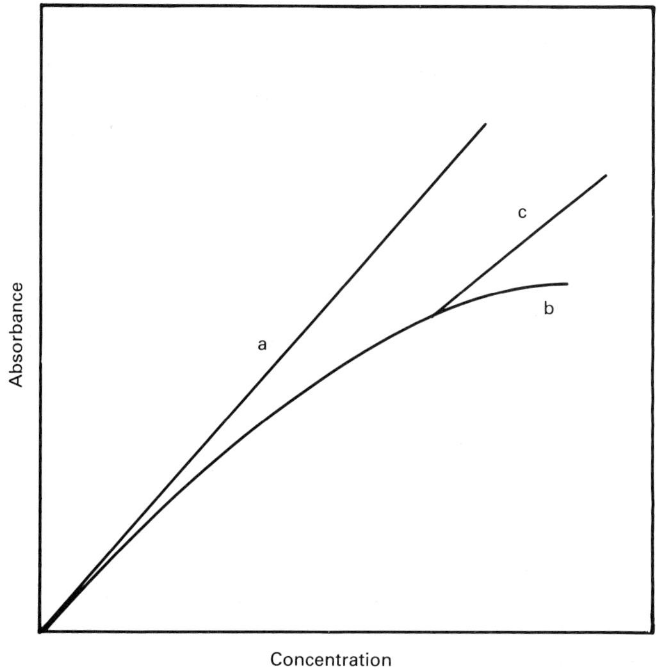

Figure 2.22 – Typical plots of absorbance versus concentration (Beer's Law plots) for dye solutions. Curve (a) is given by monodisperse dyes and in some cases by dyes associated in solution; (b) and (c) are given by associated dyes. Most dyes follow the linear pattern (a) in very dilute solution

plot can also occur with highly aggregated dyes, in which case it indicates that the aggregates are of uniform size over the range of concentrations studied. Nevertheless, most dyes appear to show non-linear Beer's Law plots over a range of very concentrated solutions.

Indeed, various types of experiment show that probably all dyes associate in aqueous solution to varying extents, according to particular features in their molecular structure; the intermolecular attraction, apparently, is mainly due to van der Waals forces. Dimers are probably formed first, and these grow by addition of further molecules (or dye ions) to give lamellar micelles in which the dye molecules are stacked up like cards in a pack [158].

Although a plot of absorbance against concentration offers, in most cases, the most ready means of detecting dye association in solution, many other techniques have been used to achieve this end, viz. measurement of diffusion rate or conductivity, ultracentrifuging, light-scattering, polarography and the plotting of spectral absorption curves. Vickerstaff [22] has surveyed this subject admirably. Most of these methods can be used quantitatively to measure N, the *aggregation number* of a dye in solution, i.e. the number of molecules in its associated particle. A more recent treatment [161] has made it possible to measure N values for dyes from the extent of deviation ΔA of the absorbance A from linearity in a Beer's Law plot. The r.m.m. M of the aggregate is given by

$$M = \frac{m}{1 - \alpha} = \frac{m\,\Delta A}{Aa} \tag{2.15}$$

where m is the relative molecular (or ionic) mass of the monomer, α is the fraction of all monomers which are associated, and a is the probability of absorption of a photon by a monomer.

Use of this expression with four different types of anionic dye gave acceptable N values, e.g. 13 for C.I. Acid Red 1 in 10^{-2}M solution and 12.5 for Sky Blue FF (I) in 1.7×10^{-2}M solution, both in absence of inorganic salts. No other published data on these dyes under similar conditions were available for comparison.

Aggregation is promoted by a high r.m.m.: ionic group ratio, by increase in length of the aromatic structure of the molecule, by rise in concentration and by the presence of inorganic salts, e.g. NaCl. It is reduced by the presence of organic solvents, e.g. ethanol, and by rise in temperature. For sulphonated dyes, which have been widely studied by various methods, N may range from two up to many powers of 10. Few measurements have been made at high temperatures, but the evidence available shows that N may fall nearly to unity in water as the temperature approaches the boiling point. Inorganic salts (NaCl has been studied most) have a marked effect in promoting association.

2.7.2 In the fibre

Hydrophilic polymers

Several methods, both direct and indirect, have been used to detect aggregation of dyes in fibres and in chemically related films (e.g. regenerated-cellulose film). Both positive and negative results have been obtained. The negative results do not necessarily prove the absence of association, however; it is probable that the method used was insufficiently sensitive. This view is confirmed in at least one case, to be mentioned, where one method gives a negative, and another method a positive result. The evidence is conveniently classified by the nature of the fibrous polymer.

Cellulosic polymers

Warwicker [162], using X-ray diffraction, detected aggregated basic dye in acidic oxycelluloses and in alginic acid. Weissbein and Coven [163] in the USA, as well as certain Russian investigators [164], obtained clear electron micrographs of aggregates of several direct dyes in cellulose. Weissbein and Coven developed a method of preparing small samples of extremely thin regenerated cellulose film suitable for electron microscopy (e.m.) and were able to correlate their results with light-fastness tests on the respective dyes in cellulose. Dyes with low fastness showed no particles, whereas those wth high fastness formed clearly visible particles (order of magnitude 1μm). The shadowing on the electron micrograph clearly revealed that the visible particles were inside the cellulose structure; moreover, the largest particles were seen to be those nearest to the outer surface, suggesting that particle growth is easiest where there is least restraint from the fibrous structure. This correlation between light fastness and particle size of direct dyes in cellulose had been predicted from statistical studies of a large volume of light-fastness data [165]. Spectroscopic tests also have provided evidence in this connection. Sky Blue FF (I) was applied to normal regenerated- and dried cellulose film and to 'never-dried' (gel) film, respectively, followed by drying. When the spectral absorption curves of a range of colour depths on the two substrates were studied, it was found that the intensity ratio of the *y*- and *x*-band peaks, respectively attributed to multimer and monomer dye (i.e the *y*/*x* absorbance ratio), was consistently higher on the film that had

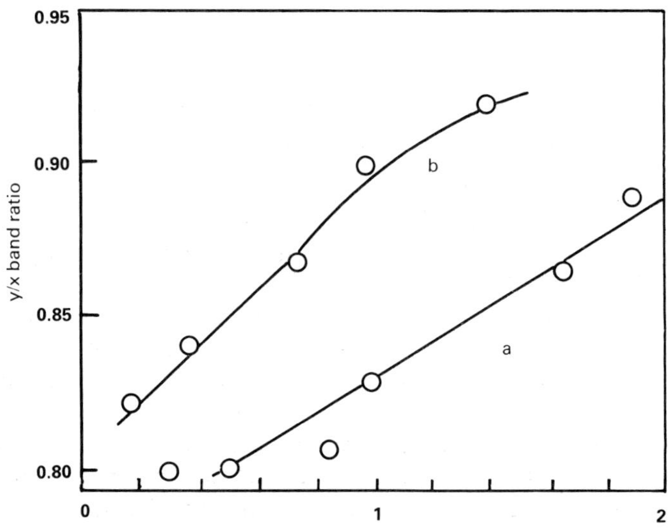

Figure 2.23 – Curves showing change in absorbance ratio (y/x bands) with rise in concentration of Sky Blue FF on regenerated-cellulose films: (a) normal film; (b) gel film. The dye is about twice as fast to light in gel film as in normal film [166]

been dyed in gel form [166] (Figure 2.23). The simplest explanation is that gel film favours dye association more than dried film does, a conclusion supported by Stone *et al.* in their measurement [39] of the porosity of these films; the gel film has the higher porosity, so the dye is under less restraint.

Recently, a new method for measuring the state of association of dyes in cellulosic substrates has been described [167]. This uses the adsorption of 4-nitrophenol from aqueous solution on to the dye. When tested on azoic and vat dyes it clearly showed the growth of particles on soaping the dyed cellulose, a phenomenon well known from other investigations. (Indeed the growth of aggregates of water-insoluble dye is an essential part of the dyeing process with both azoic and vat dyes.) It was then found possible to apply the method to direct dyes. The results with two typical dyes appear to indicate that (a) a dye of high light fastness (C.I. Direct Green 26) is present as particles comparable in size with those observed by Weissbein and Coven with this class of direct dye, whilst (b) a dye of low light fastness (Sky Blue FF, which, as stated, does not show particles either by X-ray diffraction or by e.m.) is present as multilayers, viz. superimposed monolayers up to ten or twenty layers deep. Moreover, when the same dye (Sky Blue FF) was applied to *gel* cellulose film, the method showed that more multilayers were present than in normal film, confirming the spectral tests described above. A monoazo acid dye, at a weak depth on cellulose, was found to be adsorbed as a monolayer or in only a few superimposed multilayers.

Wool

Results of several independent lines of investigation can be adduced in support of the hypothesis of dye aggregation in wool.

1. Astbury and Dawson [168] subjected wool dyed with acid dyes to X-ray diffraction tests and found that two anthraquinone dyes (C.I. Acid Blues 43 and 45) were present as crystallites in the fibre, but two azo dyes (Orange II (VI) and C.I. Acid Blue 113) apparently were not.
2. Optical microscopy shows dye-filled cavities in some samples of wool dyed with acid dyes at relatively low temperatures [131, 169].
3. As already mentioned, wool fibres in some circumstances, especially when dyed with excessive amounts of acid dye, become increasingly weaker and eventually collapse completely into powder, either by adsorption of dye on peptide groups, whereby the protein chains are separated, or through the growth of large dye aggregates in cavities in the fibre [22, 129].
4. A series of adsorption tests with vapours of water and of methanol, ethanol and propanol on samples of dyed and undyed wool gave evidence of aggregation. Adsorption of ethanol, etc., was studied in particular on fibres containing respectively benzenesulphonic acid and several acid dyes, all

applied from acid baths. On fibres containing benzenesulphonic acid the adsorption was greater than on fibres treated in a blank acid dyebath, presumably owing to distortion of the fibre molecular structure by penetration of the small aromatic anions. On all the dyed fibres, however, adsorption was less than on that containing benzenesulphonic acid, and in some cases less than on undyed fibre (Figure 2.24) [170]. The differences were attributed to blockage of fibre pores by dye aggregates.

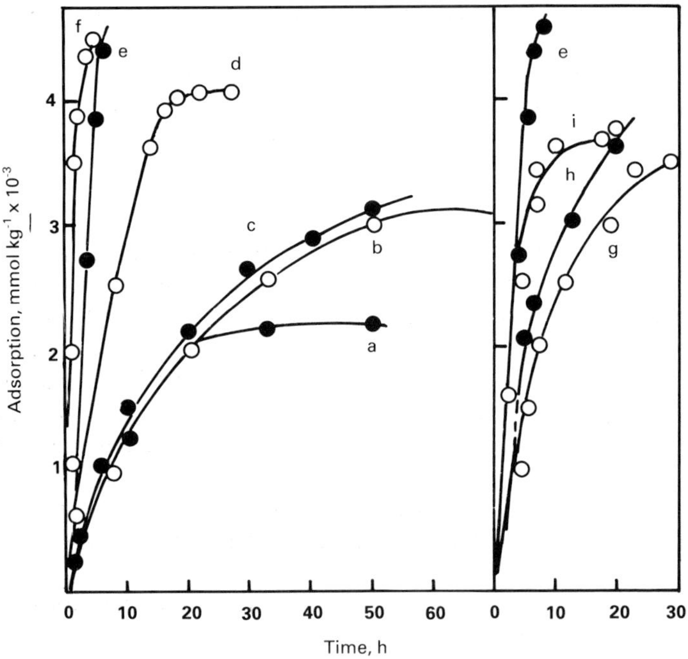

Figure 2.24 – Vapour-adsorption rate curves of untreated wool and wool containing large amounts of sulphonated dyes and other aromatic sulphonates at 35 °C for saturated ethanol vapour (except (f)). Fibre treated with: (a) C.I. Mordant Blue 1: (b) C.I. Acid Blue 45; (c) Orange II; (d) C.I. Acid Violet 1; (e) benzenesulphonic acid; (f) water vapour on wool treated in blank acid dyebath; (g) blank acid dyebath; (h) C.I. Acid Red 138; (i) C.I. Acid Red 1 [170]

5. Light-fastness tests with afterchromed dyes and some 1:2 metal–dye complexes on gelatin film (a model for wool fibres) gave fading-rate curves of zero-order type, indicative of the presence of most of the dye in the form of relatively large particles [171].

6. The few available data (Figure 5 in ref. 172) suggest that the maximum adsorption of monobasic organic acids on wool at pH 3 rises approximately with the cube of their molecular volume or r.m.m. [172]. (This is not a

molecular sieve effect, which would operate in the reverse sense.) Since mono-ionic dyes are adsorbed on non-porous surfaces as ionic micelles with aggregation numbers which also rise with the cube of their relative ionic mass [173], the inference is that organic acids are adsorbed on wool as micelles and, if so, dyes will behave similarly.

Previously, the increase of adsorption of acids or acid dyes on wool, at a given pH, with increase in r.m.m. has been attributed mainly to increased van der Waals forces between fibre and organic anions; it may now be seen as being due to increased dye–dye forces.

The evidence, therefore, is in favour of the presence of dye aggregates in wool in many cases. It is probable that all dyed wool contains them and that their size increases with the dye concentration.

Hydrophobic polymers

Under this heading are included the cellulose acetate, nylon and polyester fibres, all of which are dyed with disperse dyes. Direct evidence for the presence of dye aggregates in these substrates is less than with the hydrophilic fibres. The mechanism of their dyeing, discussed above, requires the penetration of dye directly between polymer chains and in absence of water. The dyes are possibly present as individual molecules, but more likely as small aggregates, since their molecular structure, including the absence of ionic groups, is likely to favour association. This is especially so in the dyebath which is saturated, dissolved dye being accompanied by excess solid dye in the form of a dispersion. Some azo disperse dyes are associated even in the vapour phase [174].

The conditions that favour aggregation in the dye-solution-filled pores of hydrophilic fibres are, however, largely absent. Nevertheless, there are known to be 'voids' in the hydrophobic structures, though their morphology appears to be even less understood than that of their hydrophilic counterparts [175–180]. It is believed, however, that these fibres may contain sheets of parallel or anti-parallel polymer molecules, perhaps in folded forms. Voids are detectable by low-angle X-ray scattering in cellulose acetate, in which they are said to form about 0.4 per cent of the volume of the polymer [179]. In electron micrographs of polyester fibre, after treatment with a polymerising compound and osmium tetroxide as a marker, numerous structures are evident which appear to be voids [176].

The X-ray tests that reveal voids in cellulose acetate also reveal, when used on dyed material, the presence of crystallites of disperse dye [179]. It has been suggested that acid dyes and disperse dyes occupy voids in nylon 6.6 caused by polymer-chain folding, and presumably if they do so they will be present in aggregated form [180].

Electron microscopy has failed to reveal any particles of disperse dye in cellulose acetate [103], but this is not surprising, since under the test conditions the dyes may volatilise.

Certain disperse dyes and fluorescent brighteners cause anomalous changes in specific gravity in nylon 6.6 and poly(ethylene terephthalate) in that, with increasing concentration, the specific gravity (measured by flotation in density-gradient tubes) first rises to a maximum and then falls. Other disperse dyes show no maximum or anomaly. (In all these cases, the change in light fastness with concentration follows a parallel pattern to that of specific gravity.) The inference is that these substrates contain voids that are blocked by the anomalous dyes [181]. (A similar anomaly in specific gravity is found in cellulose dyed with direct dyes, and appears to be caused by blockage of pores by aggregated dye [182].)

Spectroscopic evidence, in the form of changes in the shapes of the two peaks normally present in the adsorption spectra of anthraquinonoid disperse dyes in this type of substrate, has been adduced in support of the hypothesis just mentioned [183], but has been challenged [184]. Spectroscopy has not provided incontrovertible evidence either of the presence of aggregates of disperse dye or of the presence of monomeric dye.

It seems reasonable, therefore, to piece together these diverse elements of evidence and to postulate that disperse dyes are present at least partly as very small aggregates, i.e. as dimers, trimers, etc., in voids in the structure of all these hydrophobic fibres, though direct evidence of this hypothesis is very difficult to obtain.

2.8 SEQUENCE OF EVENTS IN THE DYEING PROCESS

We may now give greater attention to the sequence of events inside a fibre placed in a dyebath. The fibre has an enormous content of sub-microscopic pores. Into each of these the dyebath solution penetrates and forms an interface with their internal surfaces. In accordance with the laws of thermodynamics the dye – whatever its chemical class – concentrates at this interface. So-called extra-thermodynamic influences may additionally assist or retard this process. For example, the fibre may hold an electrostatic charge located at its interface with water – indeed, it usually does – and this will either assist or prevent interfacial concentration of an ionised dye. Thus cellulosic or wool fibres in water are usually negatively charged through ionisation of certain groups in their molecules, and this charge tends to repel negatively charged, i.e. anionic, dyes. With cellulosic fibres, salt is added to the dyebath to neutralise or reverse the charge and so allow direct dyes to enter the fibre; with wool, acid is added to the dyebath to give the fibre a positive charge, thus causing it to attract anionic (acid) dyes.

This account does not, however, completely describe the situation. There is not simply a large flat area of fibre substance in contact with, as it were, a layer or shallow pool of dye solution. Dyeing takes place in effect in a mass of sub-microscopic pores, and the behaviour of the dye solution in these pores is clearly important. It will be clear from the discussion in section 2.3 that the considerably increased dye concentration at the fibre–dyebath interface caused by the Gibbs effect must become more and more important as the pore in which the dye solution is held decreases in diameter; obviously the smaller the pore, the higher will be the proportion of the high-concentration interface. With rise in concentration the degree of dye association in the pores will rise, even at high temperatures, and this may account for the presence of the associated direct dyes detected in cellulose and mentioned above.

There are many unknown factors in the situation. Nevertheless, it seems fairly certain that pore structure can greatly increase the association of dyes in solution in the pores even at high temperatures, and the effect is likely to be enhanced in cellulose by the associating influence of the excess of inorganic salts used in technical practice in the dyebath.

The above discussion has been confined to cellulose because it is the only polymer for which detailed information on pore structure seems to be available, but the argument can be applied to all polymer substrates with sub-microscopic pores, including protein fibres and synthetic polymers in fibre or film form.

Dyeing may therefore entail a concentration, leading to association, of dye in the smallest accessible pores of the fibre, brought about primarily by thermodynamic and physical factors. No chemical reaction between dye and fibre need occur, though if it does, as in reactive dyeing, more dye will be fixed than otherwise. The associated dye is retained by mechanical restraint, assisted by high r.m.m. and/or a minimum number of ionic groups.

Non-ionic (disperse) dyes have low substantivity for hydrophilic fibres. The reason is probably that their very low solubility prevents a high enough concentration from developing at the internal surface, and in the crystalline regions the polymer has too firm a structure to allow penetration.

The behaviour of dye in a fibre described above implies that 'affinity' for the fibre will rise with ease of dye association, which could explain the linear relation between the length of conjugate chain in the molecules of certain series of vat and sulphonated azo dyes and their affinity for cellulose (Figure 2.20).

At this stage of the discussion two aspects of the subject require further consideration, viz. the dyeing of (a) hydrophobic fibres and (b) cellulose.

2.8.1 Hydrophobic fibres

These fibres, as their name implies, are distinguished by low moisture adsorption. The shape of their water vapour adsorption isotherms [185], which have a small 'shoulder' at very low adsorptions, shows that the water monolayer adsorption, and therefore their pore volume, is much lower than in the hydrophilic protein or cellulose fibres. The process of dye concentration and association which occurs in the hydrophilic fibre pores can therefore take place only to a very limited extent in non-ionic hydrophobic fibres, which accounts for their virtual undyeability with ionised dyes. (Ionic fibres, e.g. nylon and acrylic fibres, are clearly in a different category, because water can cluster around the ionic sites in the fibre.) 'Hydrophilic dyeing' can presumably occur in the water-filled pores, and ionised dyes will concentrate at the water–fibre interface. With few exceptions (see p. 123) ionised dyes, because of their ion–dipole attraction for water, cannot penetrate further into the polymer. If, however, an organic hydrogen-bonding agent, e.g. ethanol, is added to the dyebath for cellulose acetate, the structure is opened up and acid dyes can be adsorbed in significant quantities. Some monosulphonated acid dyes dye cellulose acetate from water alone; disulphonated ones do not. Probably this is because the former are present as large micelles and so, as in cellulose (p. 151), a deeper colour appears.

Non-ionic dyes, however, can (as already described) pass from the water phase into the fibre polymer and travel apparently beyond the water-filled regions into easily penetrated polymer. This can most readily be appreciated from the figure for normal water content of polyester fibres in air-dry conditions, which is *ca* 0.05–0.10 mol kg^{-1}, whereas simple anthraquinone disperse dyes can be absorbed (at 100 °C) to the extent of *ca* 0.18 mol kg^{-1} [186] which, having regard to the difference between the areas of a water and a dye molecule, represents a very substantially greater penetration of the polymer structure by the dye than by water vapour.

2.8.2 Cellulose dyeing

Pfeil *et al.* [154] have put forward a new hypothesis of the mechanism of cellulose dyeing which appears to explain all the facts better than any former suggestion. It recalls theories put forward soon after the introduction of direct dyes [139] and often discussed in later years, which linked substantivity with colloidal character by suggesting that colloidal particles of dye are embedded in the fibre. Pfeil *et al.* explain the difference in substantivity between a direct dye and a monoazo acid dye simply by the difference in their association abilities. Association of dye in the solution filling the fibre pores might occur when the wet fibre is removed from the hot bath and cooled, but this cannot be a complete explanation of the phenomena observed. For example, the hypothesis of Pfeil *et al.* requires that the difference between an acid dye and a direct dye on cellulose should be evident during

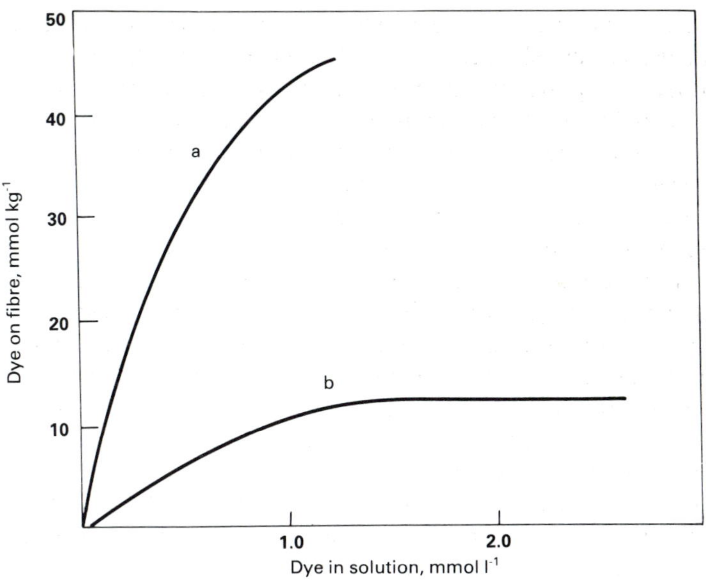

Figure 2.25 – Adsorption isotherms of (a) Congo Red (C.I. Direct Red 28) and (b) its 'half-molecule' dye, aniline→naphthionic acid, in aqueous buffer solutions at pH 9, on viscose rayon (50 °C) [137]

the dyeing process itself, for if it is not, and the dyes cover the same internal area, then there should be little difference in the amounts adsorbed.

To illustrate the principle, consider, e.g., Congo Red (C.I. Direct Red 28) and its 'half-molecule' counterpart, aniline→naphthionic acid, where the saturation adsorption values on cellulose read from the isotherm plateaux (Figure 2.25) correspond to monolayer coverages of *ca* 90 and *ca* 11 $m^2\ g^{-1}$ of surface respectively. These two dyes have identical smallest molecular cross-sections and so should enter pores equally readily. An inquiry into the large difference in their adsorptions raises the whole question of dye substantivity for cellulose, which is important in studies on all cellulose dyes, whether reactive or non-reactive.

The classical explanation is that the larger, direct dye molecule has higher affinity for sites on cellulose than has the smaller acid dye molecule. But it is difficult to imagine any type of sites that would show this difference. Moreover, if they did so, then an increase in concentration of the acid dye in solution should ultimately raise the adsorption to equal that of the direct dye. Indeed it might raise it higher than that of the direct dye, because a larger number of the smaller dye molecules than of the larger could be accommodated on the same amount of internal surface. Yet the adsorption isotherms (Figure 2.25) give no evidence that such a condition can be achieved by rise in acid dye concentration.

It seems more reasonable to postulate, with Pfeil *et al.*, that both dyes are accepted on areas of internal surface of the same magnitude, but that the larger molecule has the much greater power of self-association and so builds up multilayers.

(The question is sometimes raised, why do acid dyes of low r.m.m. dye wool but not cellulose? The short answer is that they do not! If comparison is made under neutral conditions, where both fibres hold a slight negative charge, almost no dye is adsorbed by either fibre (cf. Figure 2.17). In acid solutions, the amino groups in wool are positively charged and act as sites for anionic dyes; there are sufficient of these sites to ensure deep dyeing. Cellulose has no potentially positive sites. Furthermore, dyeing does not occur under neutral conditions in either case because small dye molecules do not exert sufficient van der Waals force, either of dye–dye nature (to associate) or of dye–fibre type.)

Pfeil *et al.* substantiated their hypothesis by determining a variety of spectral absorption curves of the dyes in cellulose film or dissolved in water or other solvents, and as solid in compressed potassium bromide, respectively. Their explanation of the need for length and planarity in a dye molecule for cellulose is that both properties allow the molecule readily to penetrate the intermicellar spaces in the fibre and there to form large aggregates. Their explanation of the aggregation is that the water of solvation carried by the dye is preferentially adsorbed on the walls of the pores, and the dye molecules are then able more readily to combine together.

Inorganic salts will greatly assist the aggregation. It may be noted that the ranking order in which the alkali-metal chlorides assist the adsorption of Sky Blue FF (I) by cellulose (Figure 2.26) [187] is almost the same as their order in promoting aggregation of the same dye in aqueous solution (Figure 2.27) [188]. As more of the unadsorbed dye becomes attached to the aggregates already present, these grow and the bath becomes more exhausted.

There are therefore two different hypotheses to account for the substantivity of direct dyes: one based on the operation of one or more of a variety of chemical attractions, as described earlier, the other based on a purely mechanical effect resulting from the pore structure of cellulose. A possibly decisive test between these hypotheses is an examination of dye–cellulose bonding when no pore structure is present. This test has been made by the author using monolayers of cellulose (spread from Cadoxen solution) on aqueous dye solutions. The results (p. 143) showed that the dye molecules are adsorbed beneath and parallel to the cellulose molecules, expanding the monolayer, but there is no difference in the effect produced by an acid dye and a direct dye, provided that their molecules are of equal width (Figure

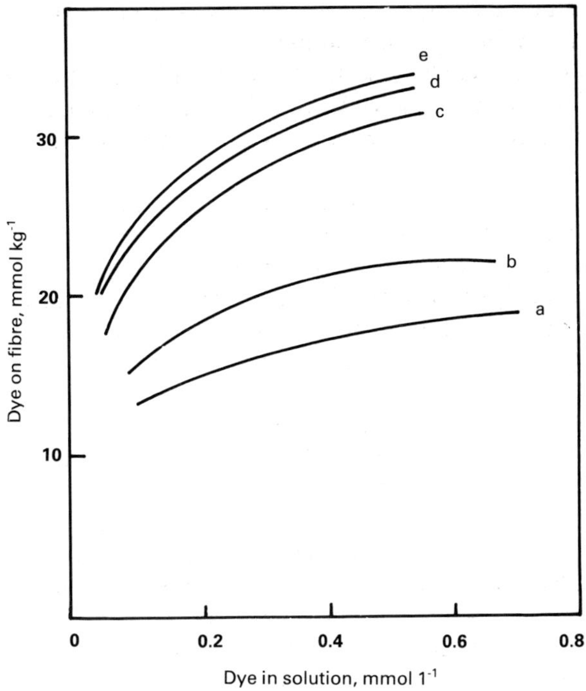

Figure 2.26 – Equilibrium adsorption isotherms (50 °C) for Sky Blue FF on cotton in the presence of 0.1M alkali-metal chlorides: (a) LiCl; (b) NaCl; (c) KCl; (d) RbCl; (e) CsCl [187]

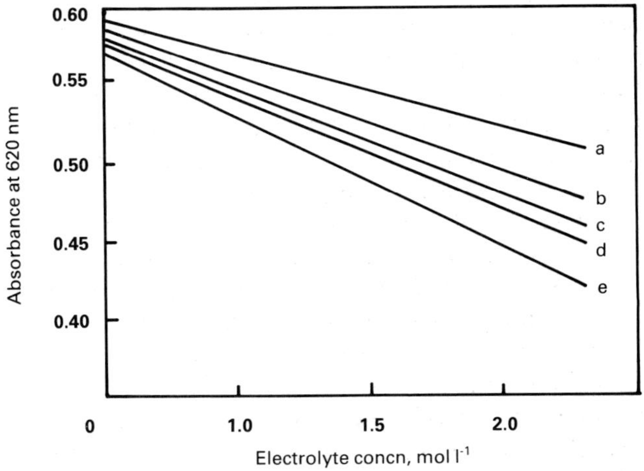

Figure 2.27 – Absorbance of aqueous solutions of Sky Blue FF (at 620 nm) in the presence of the alkali-metal chlorides: (a) LiCl; (b) NaCl; (c) CsCl; (d) KCl; (e) RbCl [188]; reduction of absorbance is an indication of increased association of dye

2.21). This result appears to confirm the importance of the morphological structure of cellulose in determining substantivity.

Further light is shed on this matter by a study of the rate curves of reactive dyes on cotton (Figure 2.28) [189]. Two effects are evident:

1. Before reaction with the fibre commences, the dyes display great differences in affinity for the fibre, comparable with those between acid dyes and direct dyes.
2. After the reaction is complete, the differences are considerably diminished, by additional uptake, mainly of the dye of low substantivity.

Before reaction starts, the dyeing mechanism is probably the same as that already described for any non-reactive dye. The additional dye adsorbed after reaction must consist of previously unadsorbed monodisperse molecules, removed unhydrolysed from solution and combined directly with cellulose. It is possible that some of this combination may take place in less ordered portions of crystalline regions of the fibre, which are swollen by the alkali used to promote reaction. For dyes with molecules 0.5 nm thick the former calculation still applies; some 260 $m^2\ g^{-1}$ of cellulose is available. If all is covered with a layer of bonded dye, which hypothetically for illustration has,

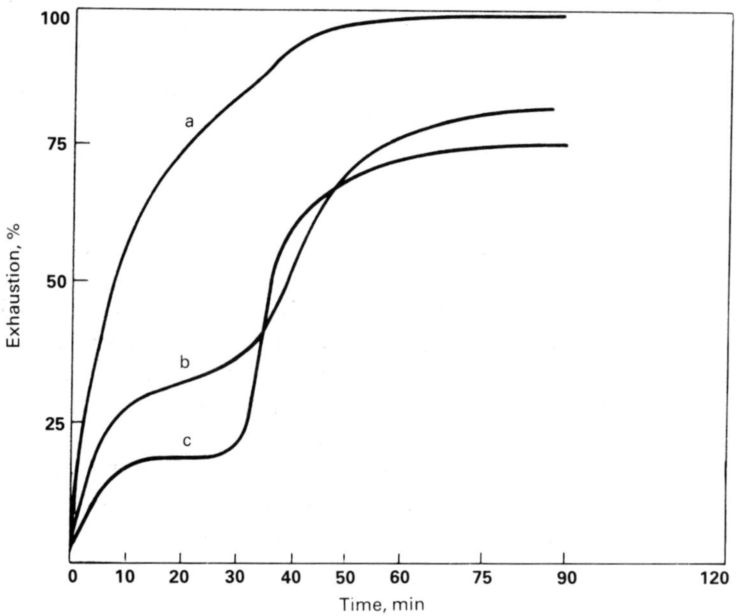

Figure 2.28 – Rate of dyeing of viscose rayon with Procion dyes at 20 °C (liquor ratio 30:1, 30 g salt per litre, sodium carbonate (1 g l^{-1}) added after 30 min; (a) Procion Yellow MXR, (b) Procion Blue MX3G, (c) Procion Red MX3B

say, the reasonable molecular area (flat) of 2.6 nm^2 and r.m.m. 610, we have:

Area of cellulose surface $= 260 \times 10^6\ cm^2$ per 100 g

Area of dye molecule $= 260 \times 10^{-16}\ cm^2$

No. of dye molecules in completed monolayer in 100 g cellulose therefore

$$= (260 \times 10^6)/(260 \times 10^{-16}) = 10^{22}$$

$$= 10^{22}/10^{23} \times 6.1 \text{ mole or } \frac{10^{22} \times 610}{10^{23} \times 6.1} \text{ g dye}$$

$$= 10 \text{ g dye per 100 g cellulose.}$$

Thus a complete monolayer would represent a dyeing of the order of 10 per cent, which is a somewhat high value, probably indicating that not all the area quoted is actually accessible to dye.

The argument from the Gibbs equation, advanced above, does not immediately seem to account for the relatively very large aggregates ($>$100 nm) of some direct dyes, visible by electron microscopy, since these are about fifty times as wide as the largest pores measured by Stone and Scallan's method. One explanation is that the aggregates formed as suggested in the small pores may, by kinetic motion, move into larger pores, where they can grow by accretion of other dye anions or micelles, under the influence of the inorganic salt present. This process could operate with all dyes.

Chitin, an abundant natural product, is essentially cellulose with an acetylamino group on most of the glucosidic residues. The monomer, acetylaminoglucose, has been shown to have much stronger hydrogen-bonding properties in aqueous solution than has glucose. Dyeing tests at pH 9, where ion exchange with the acetylamino groups and possibly consequent swelling is suppressed, showed that whereas cellulose (as stated) adsorbs much less of a typical acid dye than of a typical direct dye, chitin adsorbs both to a similar extent (*ca* 0.02 mol kg^{-1}) [137], its more highly crystalline structure restraining the build-up of large aggregates of direct dye. This restraint is also evident in tests with organic acids [172].

Conclusion

We therefore reach the tentative conclusion that all dyes associate in the hydrophilic fibres, and it can be argued that their actual usefulness as dyes, i.e. their resistance to wet treatments and to light, depends upon this fact;

any direct bond between individual non-reactive dye molecules and the fibre is then of little importance. The general tendency for affinity of dyes for wool to increase with their r.m.m., e.g. with increase in number of aromatic nuclei or in length of an attached alkyl chain, would then be attributable to greater dye–dye attraction forces in the fibre.

2.8.3 Dye aggregates and diffusion

It might be possible to gain some insight into the state of the dye in hydrophilic fibres by diffusion measurements, but little has been done on these lines. Most of the large amount of research devoted to kinetics of dyeing has been concerned with the rate of *entry* of the dye molecules into the fibre from the dyebath. They probably enter in monodisperse state, especially from a hot dyebath in which the dye is minimally aggregated. There should then be a relation between the rate of dyeing and parameters of the individual dye molecule. Merian [103, 105] demonstrated this for disperse dyes on cellulose acetate, using an equation experimentally derived by Park [190] for diffusion of organic vapours into polystyrene:

$$\log D = \log K - AV_{\mathrm{m}} - Bd \tag{2.16}$$

where D is the diffusion coefficient, V_{m} the molar volume of the vapour, d its minimum molecular diameter, and K, A and B are constants. Merian found a similar linear relation between $\log D$ for disperse dyes and a function of their molecular size and shape, based on the right-hand side of Eqn 2.16. Earlier data obtained in the present author's laboratory show that this equation can also apply to acid dyes on wool [191]. Thus rate of dyeing is probably determined by the parameters of the individual dye molecule.

Rate of desorption, e.g. into detergent solution, may not necessarily conform to the same law. In a washing test the dye molecules, probably in a mainly monodisperse form, diffuse along the fibre pores and out into the external solution but, if the dye is present as aggregated particles embedded in fine pores, these particles must first move out into larger pores where individual molecules can break away and diffuse outward. Thus the desorption sequence is:

dye aggregate in fine pores in fibre

step (a) ↓ diffusion

dye aggregate in large pores in fibre

step (b) ↓ disaggregation

monodisperse dye in solution in large pores in fibre

step (c) ↓ diffusion

monodisperse dye in external detergent solution

The rate-determining step may well be (a), which will be a function of the aggregate size. Thus in studying desorption effects of dye in fibres the term V_m in Eqn 2.16 may refer to the aggregate and not to the individual dye molecule.

On non-porous solid surfaces exposed to solutions of mono-ionic dyes, the aggregates adsorbed increase in mass with the cube of the mass of the colour ion of the dye [173]. The corresponding behaviour of dyes in fibres has not been determined, but if the above hypothesis is correct, a term proportional to log V_m, replacing V_m itself in Eqn 2.16, will possibly fit the diffusion data. Also, if wash-fastness grade (W) is a logarithmic function of D, then W may vary with log V_m and d according to a relation similar to Eqn 2.16.

2.9 CONCLUSION

The evidence given in this chapter outlines the life story of dye molecules when they are used to colour a textile material. They are of course in a state of constant motion in the dyebath. When a fibre is introduced, water from the bath immediately passes into the pores of the fibre and an interface – a very extensive internal surface – is thus formed between water and the walls of the pores. Water molecules are small and travel rapidly, in most cases filling the pores in a few seconds. Dye molecules are larger and take longer to push their way in, but their constant motion eventually carries them into all the larger pores. It carries them particularly to the walls of the pores, where at the fibre–water interface they remain closer together than in the bulk of the internal solution. This process is greatly assisted if the dye is ionic and the fibre carries an opposite charge. It is also assisted by the van der Waals forces, which are stronger with large than with small dye molecules, and which tend to align the dye molecules parallel to and in contact with the molecular chains in the fibre walls. A further action then comes into force. Because of the constriction of the smaller pores, the amount of dye in the interface layer adds significantly to the total concentration of dye in the pores. The result is that most of the dye molecules then associate into larger units. In hydrophobic fibres placed in a bath containing disperse dye, the pores are in most cases less extensive than in the hydrophilic fibres. Most of the dyeing occurs because the dye molecules possess strong affinity for the fibre polymer, arising from the formation of polar – mainly hydrogen – bonds. Consequently, they are able to push their way into the fibre substance, forcing the polymer chains apart. Some association of dye may also occur in this process.

REFERENCES

1. (a) Boyle, *Experiments and considerations touching colours; the beginning of an experimental history of colours* (London: H. Herringman, 1964); cf. (b) Giles, *J.S.D.C.*, **82** (1966) 63.

2. Petty, *An apparatus to the history of the common practices of dying*, in Sprat, *History of the Royal Society* (London: 1667), reprinted and edited by Cope and Jones (London: Routledge and Kegan Paul Ltd, 1959).
3. Hellot, *L'art de la tenture des étoffes de la laine, en grande et en petit teinte* (Paris: 1772; 1st edn, 1750) (English translation, London: R. Baldwin, 1789; reprinted (London: Scott, Greenwood and Co., 1901)).
4. Maty, *A general index to the Philosophical Transactions, vols 1–70* (London: 1787; New York: Johnson Reprint Corpn, 1963).
5. Bancroft, *Experimental researches concerning the philosophy of permanent colours*, 2nd edn (London: Cadell and Davies, 1813).
6. Laurie, *A bibliography of dyeing and textile printing* (London: Chapman and Hall, 1949) (from the sixteenth century to 1946).
7. (a) Crum, *Proc. Phil. Soc. Glasgow*, **1** (1841–44) 98; (b) *Phil. Mag.*, **24** (1844) 241; (c) *J.C.S.*, **16** (1863) **1**, 404 (this paper is notable for its excellent micrographs); (d) *On the cotton fibre, etc.* (Glasgow: Bell and Bain, 1863).
8. E.g. Knecht, *Ber.*, (a) **21** (1888) 1556; (b) **37** (1904) 3479.
9. (a) Liechti and Hummel, *J. Soc. Chem. Ind.*, **12** (1893) 240, 332; **13** (1894) 222, 356; (b) Hummel and Gardner, *ibid.*, **14** (1895) 452.
10. Pelet-Jolivet, *Die Theorie des Färbeprozesses* (Dresden: T Steinkopff, 1910).
11. Dreaper, *The chemistry and physics of dyeing* (London: J and A Churchill, 1906).
12. Wood, *The chemistry of dyeing* (London: Gurney and Jackson, 1913).
13. Sachs (a useful short review of dyeing theories, 1879–1928), *Textile Colourist, 50th Anniv. No.*, **44** (1928) 122.
14. E.g. Gee and Harrison, *Trans. Faraday Soc.*, **6** (1910) 42.
15. E.g. (a) Neale and Stringfellow, *ibid.*, **29** (1933) 1167; (b) Boulton, Delph, Fothergill and Morton, *J. Textile Inst.*, **24** (1933) 113; (c) Boulton and Morton, *J.S.D.C.*, **55** (1939) 481.
16. Elöd, *Trans. Faraday Soc.*, **29** (1933) 327.
17. E.g. Steinhardt, Fugitt and Harris, *J. Res. Nat. Bur. Stand.*, **25** (1940) 519; **26** (1941) 293; **30** (1943) 123.
18. E.g. (a) Speakman and Clegg, *J.S.D.C.*, **50** (1934) 348; (b) Speakman and Stott, *ibid.*, **50** (1934) 341; (c) Speakman and Elliott, *S.D.C. Symp. on Fibrous Proteins* (1946) 116; cf. ref. 169 below.
19. (a) Skinner and Vickerstaff, *J.S.D.C.*, **61** (1945) 193; (b) Lemin and Vickerstaff, *ibid.*, **63** (1947) 405; (c) *idem, S.D.C. Symp. on Fibrous Proteins* (1946) 129.
20. (a) Gilbert and Rideal, *Proc. Roy. Soc.*, **A182** (1944) 335; (b) Gilbert, *ibid.*, **A183** (1944) 167.
21. Peters and Vickerstaff, *ibid.*, **A192** (1948), 292.
22. Vickerstaff, *The physical chemistry of dyeing* (Edinburgh: Oliver and Boyd, 1950; 2nd edn, 1954).
23. (a) Valkó, *Kolloidchemische Grundlagen der Textilveredlung* (Berlin: J Springer, 1937); (b) see also *idem*, Coll. Chem., **6** (1946) 594 (a review).
24. 'The physical chemistry of dyeing and tanning', *Disc. Faraday Soc.*, **16** (1954).
25. See, e.g., (a) Bird, Manchester and Harris, *ibid.*, **16** (1954) 85 (cellulose acetate); (b) Peters, *J.S.D.C.*, **61** (1945) 95; (c) Remington and Gladdings, *J. Amer. Chem. Soc.*, **72** (1950) 2553 (nylon); (d) Waters, *J.S.D.C.*, **66** (1950) 609; (e) Schuler and Remington, *Disc. Faraday Soc.*, **16** (1954) 201 (polyester fibres).
26. Carroll, *Through the looking-glass* (1871), chapter 2.
27. Abrahart, *Dyes and their intermediates* (London: Pergamon Press, 1968).
28. Allen, *Colour chemistry* (London: Thos Nelson and Sons Ltd, 1971).
29. Gibbs, *Collected works* (London: Longmans, Green and Co., 1928), vol. 1 'Thermodynamics', chapter 3 (reprinted from *Trans. Connect. Acad. Arts Sci.*, **3** (1875–1878) 343).
30. (a) Adam, *The physics and chemistry of surfaces*, 3rd edn (London: Oxford University Press, 1941): (b) de Boer, *The dynamical character of adsorption*, 2nd edn (London: Oxford University Press, 1968); (c) Davies and Rideal, *Interfacial phenomena* (London: Academic Press, 1961).
31. Giles and Agnihotri, *Chem. and Ind.*, (1967) 1874.

32. Crisp, *J. Coll. Sci.*, **1** (1946) 49, 161.
33. Giles and Trivedi, *Chem. and Ind.*, (1970) 1023.
34. (a) Giles and Soutar, *J.S.D.C.*, **87** (1971) 301 (in Figure 2, symbols *a* and *b* should be transposed, and in Figure 5 symbols *b* and *c* should be transposed); (b) Giles and McIver, *J. Colloid Interface Sci.*, **62** (1977) 329.
35. Pugh, Giles and Duff, *Trans. Faraday Soc.*, **67** (1971), 563.
36. Agnihotri and Giles, *J.C.S. Perk.*, **2** (1972) 2241.
37. Giles and D'Silva, *Trans. Faraday Soc.*, **65** (1969) 1943.
38. Stone and Scallan, *Cellulose Chem. and Tech.*, **2** (1968) 343.
39. Stone, Treiber and Abrahamson, *Tappi*, **52** (1969) 108.
40. (a) Aveyard and Briscoe, *Trans. Faraday Soc.*, **66** (1970) 2911; (b) Langmuir, *J. Amer. Chem. Soc.*, **39** (1917) 1848.
41. Coulson, *Valence* (London: Oxford University Press, 1952).
42. Syrkin and Dyatkina (trans: Partridge and Jordan), *Structure of molecules and the chemical bond* (London: Butterworth, 1950), chapter 12.
43. Margenau and Kestner, *Theory of intermolecular forces* (London: Pergamon Press, 1960).
44. An entertaining, non-mathematical, but authoritative account of atomic structure and bonding is given by Hoffman, *The strange story of the quantum* (London: Penguin Books, 1965).
45. Pryde, *The liquid state* (London: Hutchinson University Library, 1966), chapter 4.
46. Castellan, Physical chemistry, 2nd edn (London: Addison-Wesley, 1971), chapter 25.
47. Pimental and McClellan, *The hydrogen bond* (San Francisco: W H Freeman and Co., 1960).
48. Murrell, *Chem. in Britain*, **5** (1969) 107.
49. Vinogradov and Linnell, *Hydrogen bonding* (London: van Nostrand Reinhold Co., 1971).
50. Dewar, *Nature*, **176** (1945) 784; *Bull. Soc. Chem. France*, (1951) C71.
51. (a) Fieser and Fieser, *Advanced organic chemistry* (London: Chapman and Hall Ltd, 1961); (b) Hassel and Römming, *Quart Rev.*, **16** (1962) 1.
52. (a) Kauzmann, *Adv. Protein Chem.*, **14** (1959) 1; (b) Némethy and Scheraga, *J. Chem. Phys.*, **36** (1962) 3382; (c) *ibid.*, **36** (1962) 3401; (d) *J. Physical Chem.*, **66** (1962) 1773.
53. Meggy, *J.S.D.C.*, **66** (1950) 510.
54. Zollinger, *ibid.*, **81** (1965) 345.
55. (a) Majury, *ibid.*, **70** (1954) 442; (b) *ibid.*, **72** (1956) 41; (c) Jones, *ibid.*, **77** (1961) 57; (d) Jones and Seddon, *Text. Research J.*, **34** (1964) 373; (e) *ibid.*, **35** (1965) 334; (f) Jones and Kraska, *J.S.D.C.*, **82** (1966) 333.
56. Suda and Ujigawa, *J. Soc. Text. Cellulose Ind. Japan*, **19** (1963) 563; through *J.S.D.C.*, **79** (1963) 557.
57. (a) Giles, Rose and Vallance, *J.C.S.*, (1952) 3799; (b) Campbell, Cathcart and Giles, *J.S.D.C.*, **73** (1957) 546.
58. Giles, Gallagher, McIntosh and Nakhwa, *J.S.D.C.*, **88** (1972) 360.
59. Arshid, Giles, McLure, Ogilvie, Rose and Eaton, *J.C.S.*, (1955) 67.
60. Yoshida, Tajiro and Oda, *Kogyo Kagaku Zasshi*, **65** (1962) 1064; through *J.S.D.C.*, **80** (1964) 224.
61. (a) Allingham, Giles and Neustädter, *Disc. Faraday Soc.*, **16** (1954) 92; (b) Harrap, *Proc. 2nd Intern. Cong. Surface Activity*, **4** (1957) 295; (c) Cameron, Giles and MacEwan, *J.C.S.* (1958) 1224; (d) Hayashi, Masako, Tachibana and Yabe, *Kogyo Kagaku Zasshi*, **63** (1960) 1545; (e) *ibid.*, **64** (1961) 1494, through *Chem. Abs.*, **56** (1962) 11840; (f) *ibid.*, **57** (1962) 3653; (g) Giles, Agnihotri and Trivedi, *J.S.D.C.*, **86** (1970) 451; for a review, see (h) Giles, *Proc. Symp. Chem. Syn. Dyes and Dyeing Mech.*, University of Bombay (1968) 14; (i) *idem, School Sci. Rev.*, **46** (1964) 3.
62. Johnson, Peters (R H) and Ramadan, *J.S.D.C.*, **80** (1964) 129.
63. Giles, Easton and McKay, *J.C.S.* (1964) 4495.
64. For references to this earlier work, see, e.g., Arshid, Giles, Jain and Hassan, *J.C.S.* (1956) 72.
65. Giles, McKay and Good, *ibid.*, (1961) 5434.
66. Yoshida and Osawa, *Bull. Chem. Soc. Japan*, **38** (1965) 140.

67. (a) Forrester and Giles, *Chem. and Ind.* (1971) 831, 1314; (1972) 318: cf. (b) Giles, *J.S.D.C.*, **89** (1973) 287.
68. Langmuir, *J. Amer. Chem. Soc.*, **38** (1916) 2221; **40** (1918) 1361.
69. Giles, Huitson and Smith, *J. Colloid Interface Sci.*, **47** (1974) 755.
70. Brunauer, Emmett and Teller, *J. Amer. Chem. Soc.*, **60** (1938) 309.
71. Giles, MacEwan, Nakhwa and Smith, *J.C.S.* (1960) 3973.
72. For a review, see, e.g., Giles, *Chem. and Ind.*, (1966) 92, 137.
73. Giles, Mehta, Subramanian and Stewart, *J.C.S.* (1954) 4360.
74. Giles, *Rev. Prog. Coloration*, 5 (1974) 49.
75. Cummings, Garven, Giles, Rahman, Sneddon and Stewart, *J.C.S.* (1959) 535.
76. Giles, Mehta, Rahman and Stewart, *J. Appl. Chem.*, **9** (1959) 457.
77. Giles, *Proc. Conf. Anodising Aluminium* (London: Aluminium Development Association, 1962) 174.
78. Giles and Datye, *Trans. Inst. Metal Finishing*, **40** (1963) 113.
79. Giles, *J.S.D.C.*, **60** (1944) 303 (review).
80. Arshid, Connolly, Desai, Fulton, Giles and Kefalas, *J.S.D.C.*, **70** (1954) 402.
81. Arshid, Desai, Duff, Giles, Jain and Macneal, *ibid.*, 392.
82. Hartley, *ibid.*, **85** (1969) 66.
83. Bird, *The theory and practice of wool dyeing*, 4th edn (Bradford: Soc. Dyers and Colourists, 1972).
84. Dobozy, *Amer. Dyestuff Rep.*, **62** (1973) 36.
85. Schetty, private communication.
86. Race, Rowe and Speakman, *J.S.D.C.*, **62** (1946) 372.
87. Merry, *The chrome tanning process* (London: A Harvey, 1936).
88. Rollinson, in *The chemistry of the co-ordination compounds* (ed. Bailar) (New York: Reinhold Publishing Corporation, 1956).
89. Schulman and Dogan, *Disc. Faraday Soc.*, **16** (1954) 158.
90. Pouradier, *ibid.*, **16** (1954) 180.
91. Gustavson, *ibid.*, **16** (1954) 185.
92. See Figure 12 in ref. 72 above.
93. (a) Giles and MacEwan, *J.C.S.* (1959) 1791; (b) Giles, MacEwan and McIver, *Text. Research J.*, **44** (1974) 580.
94. Bird, *J.S.D.C.*, **70** (1954) 68.
95. Bird, Harris and Manchester, *ibid.*, **71** (1955) 139.
96. Bird and Manchester, *ibid.*, **71** (1955) 604.
97. Bird, Manchester and Scott, *ibid.*, **72** (1956) 49.
98. Bird and Harris, *ibid.*, **73** (1957) 199.
99. Bird, *Palette* (1964) (no. 15) 25 (Basle: Sandoz Ltd) (review of cellulose acetate dyeing).
100. Giles, *Text. Research J.*, **31** (1961) 141.
101. Giles, *Brit. Polymer J.*, **3** (1971) 279; in Table IX the temperatures for phenols were 150–171°C and not as given; (ref. 100 and 101 give reviews of hydrophobic fibre dyeing, theory and practice).
102. Glenz, Beckmann and Wunder, *J.S.D.C.*, **75** (1959) 141.
103. Merian, *Text. Research J.*, **36** (1966) 612.
104. Peters, Petropoulos and McGregor, *J.S.D.C.*, **77** (1961) 704.
105. Husy, Merian and Schetty, *Text. Research J.*, **39** (1969) 94.
106. Daruwalla, Rao and Tilak, *J.S.D.C.*, **76** (1960) 418, 680.
107. Marsden and Urquhart, *J. Textile Inst.*, **33** (1942) T105.
108. Arshid, Giles and Jain, *J.C.S.* (1956) 1272.
109. Balmforth and Bird, *J.S.D.C.*, **80** (1964) 534.
110. Griffiths and Neale, *Trans. Faraday Soc.*, **30** (1934) 395.
111. (a) de Vries and Smit, *J. S. Afr. Chem. Inst.*, **20** (1967) 11; (b) Smit, Smith and de Vries, *ibid.*, **20** (1967) 144; (c) de Vries, Smit and Raubenheimer, *ibid.*, **21** (1968) 47.
112. (a) Giles, in *Sorption and transport processes in soils*, S.C.I. Monograph no. 37 (1970) 14; (b) Smith, private communication (1965); cf. (c), ref. 71 above; (d) Cruickshank, private communication (1973).

113. Burr and Burr, *J.S.D.C.*, **50** (1934) 42.
114. Gokhale, Peters (L) and Stevens, *J.S.D.C.*, **74** (1958) 236.
115. Bird and Patel, *ibid.*, **84** (1968) 560.
116. Aspland and Bird, *ibid.*, **77** (1961) 9.
117. Yabe, private communication (1969).
118. Chipalkatti, Giles and Vallance, *J.C.S.*, (1954) 4375.
119. Chipalkatti, Chipalkatti and Giles, *J.S.D.C.*, **71** (1955) 652.
120. Bird and Firth, *J. Textile Inst.*, **51** (1960) T1342.
121. Cameron, Giles and MacEwan, *J.C.S.*, (1959) 1791.
122. Greenhalgh, Johnson and Peters (R H), *J.S.D.C..*, **78** (1962) 315.
123. See, e.g., (a) Peters (L) and Speakman, *ibid.*, **65** (1949) 63; (b) Olofsson, *ibid.*, **68** (1952) 506; (c) *ibid.*, **72** (1956) 19; (d) Peters (L), *ibid.*, **70** (1954) 436; (e) *idem*, *J. Textile Inst.*, **51** (1960) T1290; (f) White, *J. Physical Chem.*, **59** (1955) 528; (g) Delmenico and Peters (R H), *Nature*, **193** (1962) 1252; (h) Alexander and Hudson, *Wool: its chemistry and physics* (London: Chapman and Hall, 1954); (i) Mitsuishi and Aida, *J. Soc. Fibre Sci. Technol., Japan*, **20** (1964) 482; through *J.S.D.C..*, **80** (1964) 618.
124. Peters (L) and Lister, *Disc. Faraday Soc.*, **16** (1954) 24.
125. Coward and Nursten, *J.S.D.C.*, **80** (1964) 405.
126. Peters (L), *ibid.*, **71** (1955) 725.
127. Derbyshire and Marshall, *Disc. Faraday Soc.*, **16** (1954) 140.
128. Derbyshire, *Trans. Faraday Soc.*, **51** (1955) 909.
129. Barr, Rowe and Speakman, *J.S.D.C.*, **58** (1942) 52.
130. Speakman and Smith, *ibid.*, **52** (1936) 121.
131. Royer and Maresh, *Amer. Dyestuff Rep.*, **32** (1943) 181.
132. Lucas, *Nature*, **210** (1966) 952.
133. Laucius, Clarke and Brooks, *Amer. Dyestuff Rep.*, **44** (1955) 362.
134. Balmforth, Bowers and Guion, *J.S.D.C.*, **80** (1964) 577.
135. Cegarra, *ibid.*, **87** (1971) 149.
136. Bird and Stancey, *ibid.*, **77** (1961) 244, and references therein.
137. Giles and Hassan, *ibid.*, **74** (1958) 846.
138. Peters (R H) and Sumner, *ibid.*, **71** (1955) 130.
139. Krafft, *Ber.*, **32** (1899) 1608.
140. Hodgson, *J.S.D.C.*, **49** (1933) 213.
141. Paine (1933), Rose (1935), quoted in ref. 22.
142. (a) Ruggli, *J.S.D.C.*, Jubilee Issue (1934) 77 (a useful review of prior knowledge); (b) Schirm, *J. für Prakt. Chem.*, **144** (1935) 69.
143. (a) Willis, Warwicker and Standing, *Trans. Faraday Soc.*, **41** (1945) 506; (b) Morton, *J.S.D.C.*, **62** (1946) 272.
144. Bamford, *Disc. Faraday Soc.*, **16** (1954) 229.
145. Giles, (a) *Compt. rend.*, XXVIIe Cong. Chim. Ind., Brussels (1954) 610; (b) *Compt. rend.*, XXXIe Cong. Chim. Ind., Brussels (1958) 484.
146. Robinson, *Disc. Faraday Soc.*, **16** (1954) 125.
147. Wahl, *Teintex*, **19** (1954) 667.
148. (a) Zollinger, *Disc. Faraday Soc.*, **16** (1954) 123; (b) *Textil-Rund.*, **14** (1959) 113; (c) *Amer. Dyestuff Rep.*, **49** (1960) 29.
149. Derbyshire and Peters, *J.S.D.C.*, **71** (1955) 530.
150. Arshid, Giles and Jain, *J.C.S.* (1956) 559.
151. Lead, *J.S.D.C.*, **73** (1957) 464.
152. Valkó, *Text. Research J.*, **27** (1957) 883.
153. Wegmann, *Amer. Dyestuff Rep.*, **51** (1962) P276.
154. Bach, Pfeil, Phillipar and Reich, *Angew. Chem.*, **75** (1963) 407.
155. Yoshida, Osawa and Oda, *J. Physical Chem.*, **68** (1964) 2895.
156. (a) Tewksbury, *Dyestuffs*, **45** (*ca* 1965) 72 (New York: Allied Chem. Corpn) (an interesting review); (b) Daruwalla, *Palette*, (1967) (no. 25) 30.
157. Giles and McIntosh, *Text. Research J.*, **43** (1973) 489.
158. Coates, *J.S.D.C.*, **85** (1969) 355 (a useful review).

159. (a) Giles and Duff, *ibid.*, **86** (1970) 405; (b) Duff and Giles, *ibid.*, **88** (1972) 181.
160. Giles, Rahman and Smith, *J.C.S.* (1961) 1209.
161. Pugh, Giles and Duff, *Trans. Faraday Soc.*, **67** (1972) 563.
162. Warwicker, *J. Textile Inst.*, **49** (1958) T148.
163. Weissbein and Coven, *Text. Research J.*, **30** (1960) 58, 62.
164. Mel'nikov, Blinicheva and Zabrodin, *Izv. Vyssh. Vcheb. Zaved., Tekhnol. Tekstil. Prom.*, (1968) (no. 2) 99; through *Chem. Abs.*, **69** (1968) 3470.
165. Giles, *J.S.D.C.*, **73** (1957) 127.
166. Giles and Rahman, *Text. Research J.*, **31** (1961) 1012.
167. Giles, Haslam, Hill and Trivedi, *J. Appl. Chem. Biotech.*, **21** (1971) 5.
168. Astbury and Dawson, *J.S.D.C.*, **54** (1938) 6.
169. Speakman and Smith, *ibid.*, **52** (1936) 121.
170. Giles, Elder and Tolia, *Text. Research J.*, **34** (1964) 839.
171. Giles, Baxter and Rahman, *ibid.*, **31** (1961) 831.
172. Giles, Hassan, Laidlaw and Subramanian, *J.S.D.C.*, **74** (1958) 647.
173. Giles, Easton, McKay, Patel, Shah and Smith, *Trans. Faraday Soc.*, **62** (1966) 1963.
174. Green and Jones, *ibid.*, **63** (1967) 1612.
175. See, e.g. *Supermolecular structure in fibres*, ed. Lindenmeyer (*J. Polymer Sci., Part C*, Polymer Symposia) (Easton, PA: Interscience Publishers, 1967).
176. (a) Catoire, Hagege and Meimoun, *Bull. Inst. Text. France*, **23** (1969) 521; (b) Sotton, Hagege and Jacquemart, *ibid.*, **24** (1970) 763.
177. Warwicker, *J.S.D.C.*, **86** (1970) 303.
178. Sotton and Vialand, *Compt. Rend. Hebd. Séance Acad. Sci.*, **272** (1971) 1381.
179. Kratky, Mittelbach and Sekora, *Kolloid-Z.*, **200** (1964) 1.
180. Brody, *Text Research J.*, **35** (1965) 895.
181. Giles, Yabe and Shah, *ibid.*, **38** (1968) 467.
182. Daruwalla and D'Silva, *ibid.*, **33** (1963) 40.
183. Giles and Shah, *J.S.D.C.*, **85** (1969) 417.
184. (a) Hoffman, McDowell and Weingarten, *ibid.*, **84** (1968) 306; (b) *idem, ibid.*, **85** (1969) 418.
185. Chabert, *Bull. Inst. Text. France*, **20** (1966) 553.
186. Schuler and Remington, *Disc. Faraday Soc.*, **16** (1954) 201.
187. Iyer, Shrinivasan and Baddi, *Text. Research J.*, **38** (1968) 693.
188. Bubser and Eichmanns, *Palette* (1965) (no. 20) 20.
189. Vickerstaff, *J.S.D.C.*, **73** (1957) 237.
190. Park, *Trans. Faraday Soc.*, **47** (1951) 1007.
191. Giles, Montgomery and Tolia, *Text Research J.*, **32** (1962) 99.

CHAPTER 3

The influence of fibre structure on dyeing

WILFRED C INGAMELLS

3.1 INTRODUCTION

The successfully applied skill of the colourist is the essence of consumer satisfaction with coloured textile goods. Goods that meet the expected standards of colour fastness and uniformity of colour distribution provide the most visible effects of all textile-finishing processes. Unfortunately, the apparent ease with which satisfactory results are obtained masks the skill and effort required.

Dyeing is usually among the last of the long line of manufacturing operations which lead to the end product and it is not uncommon to find faults, such as dark or lightly coloured stripes in the dyed cloth, which have their origin further back in the yarn or cloth manufacturing processes. Even slight variations of intensity or hue can be detected by the human eye, and their occurrence on textile material can harm the aesthetic appeal of a garment or other article for which the goods are subsequently used.

The number of factors that may conspire to influence dyeing properties is large, and explanations of their effect can be only as complete as allowed by current understanding of the chemical structure and molecular arrangement of the fibrous polymer molecules. Assuming freedom from detrimental colloidal properties of the dye solutions, such variations in dyeability are linked to differences in fibre structure either as inherent properties of the particular fibre type, or as changes in structure occurring during processing, or as changes occurring during dyeing. Simple unambiguous quantitative relationships between dyeing properties and fibre structure are therefore difficult to obtain [1]. Nevertheless investigation has provided a framework within which useful observation may be made.

The molecular mechanisms proposed for the dyeing process are often based upon the assumption that the fibre plays a passive role during dyeing. Consequently, a good deal of the fundamental work described in the literature deals with the physical forces of attraction and the chemical reactions responsible for the retention of dye in the fibre.

The dyeing process is divided into three steps [2, 3]: (1) movement of dye through the dyebath to the fibre surface (which falls outside the scope of this chapter), (2) adsorption of dye at the fibre surface (the thermodynamics of which are discussed in Chapter 4) and (3) penetration of dye through the fibre. Due to geometrical factors, the fibre structure imposes immediate resistance to the movement of the dye molecules over and above the forces of attraction between dye and polymer.

It is therefore convenient to review the situation with regard to fibre structure under the following headings,

(a) the chemical differences between fibres,
(b) fibre properties related to their inherent physical structure,
(c) fibre properties influenced by processes before dyeing,
(d) fibre properties influenced during dyeing.

The ease with which different classes of dye penetrate fibres is closely related to the interaction between fibre and dyebath medium. As most dyeing processes are carried out in an aqueous environment the capacity of a fibre to absorb water and swell is particularly important, because water absorption is an essential prerequisite for the use of water-soluble dyes. Where the water absorption is low, as with some synthetic fibres, non-ionic dyes of low aqueous solubility are required.

The water content and the absorption of dye are both governed by the polarity of the chemical groups in the fibre. Thus chemical differences between fibres determine the need to use different dye classes with different fibres, and a useful summary of the situation is presented in Table 3.1.

Those fibres that have highly polar groups absorb most water, but their accessibility is also an important factor and is discussed later. Hydrophobic fibres cannot be dyed with dyes of high aqueous solubility. These dyes are generally anionic and are repelled by the negative surface potential which the hydrophobic fibres develop when immersed in water. Values for the surface potential of some fibres, obtained in neutral aqueous potassium chloride solutions by a streaming method, together with values of zeta potentials obtained by an electro-osmosis method are given in Table 3.2. Although the two procedures produce results of different magnitude, the fibres fall into the same order whichever method is used. The hydrophobic fibres have a high negative potential, which creates the potential barrier at the fibre surface and is thus responsible for preventing the adsorption of dye anions. The high negative value of the potential for wool is associated with the hydrophobic character of the epicuticle, but since the potential is reversed or reduced in acidic dye solutions the value for wool is of lesser importance. The same applies to nylon and acrylic fibres at the pH values used for applying anionic and cationic dyes respectively [3, 4].

TABLE 3.1

Relation between chemical structure of some fibres, their water absorptivity and dyes used

Fibre	Functional groups	Water absorptivity[a] Regain (21 °C, 65% r.h.)/%	Swelling[c] in water/%	Dyes used[b]
Wool	NH_2,COOH, CONH	17.0	32–38	Acid, metal-complex, chrome, reactive
Silk	NH_2, COOH, CONH	11.0	30–41	Acid, metal-complex, basic, vat
Cuprommonium rayon	OH, COOH	11.0	99–134	Direct, vat, sulphur
Viscose rayon	OH, COOH	11.5–16.6	45–82	Direct, vat, sulphur, reactive
Cotton	OH	8.5	44–49	Direct, vat, sulphur, reactive
Secondary cellulose acetate	OH, $OCOCH_3$	6.3–6.5	6–30	Disperse
Cellulose triacetate	$OCOCH_3$	4.5[d]	Very slight	Disperse
Nylon 6.6 and 6	NH_2, COOH, CONH	4.0–4.5	2	Acid, metal-complex, direct, disperse, reactive
Acrylic	COOH, SO_3H, OSO_3H	1–2	Very slight	Basic, disperse
Polyester	OH, COOH COO	0.4	Virtually nil	Disperse
Polypropylene		0.01[d]	Nil	(Disperse)

[a] Most of these values have been obtained from *Handbook of textile fibres*, ed. Milton Harris (New York: Interscience, 1954).

[b] Dyes are listed according to their dyeing classification and the more important groups used for a particular fibre.

[c] Swelling is expressed as % increase in cross-sectional area of the fibre, which is essentially equivalent to volume swelling. Many other values quoted in the literature refer to % increase in fibre diameter.

[d] These values have been obtained from Hall, *The standard handbook of textiles*, 7th edn (London: Heywood, 1969).

TABLE 3.2

Surface and zeta potentials of fibres in 1 × 10^{-3}M KCl

Fibre	Surface potential[a] /mV	Zeta potential[b] /mV
Acrylic (Orlon)	−44	
Polyester (Terylene)	−32 to −40	
Wool	−31	−60
Poly(vinyl chloride)		−50
Secondary cellulose acetate	−28	−45
Silk		−42
Nylon 6.6	−22	
Nylon 6		−39
Cotton	−17	−31
Viscose rayon		−30

[a] *Hexagon Digest*, **19** (1954).
[b] Stackelberg, Kling, Benzel and Wilke, *Kolloid Z.*, **135** (1954) 67.

Fibres with low molecular orientation or low crystallinity generally swell easily during dyeing, because the polar groups are readily accessible. They absorb water-soluble ionisable dyes and the maximum amount adsorbed is related to the concentration of functional groups in the fibre. Although many of these groups may ionise during dyeing and interact with other ions, their specific interaction with hydrogen ions is of greatest importance [5]. For example, the adsorption of negatively charged dye ions is enhanced by the net positive charge which appears on wool or nylon fibres due to protonation of basic functional groups in an acidic dyebath. The saturation levels of dyes on nylon can greatly exceed the concentration of ionisable functional groups, however, due to non-ionic interactions such as dipole–dipole interactions, hydrogen bonding, dispersion forces and hydrophobic interactions. The surface potential is particularly significant for the adsorption of anionic dyes by negatively charged cellulose fibres where like-charge repulsion occurs at the fibre surface. This is overcome by using electrolyte to increase the ionic strength of the dyebath, thus reducing the surface potential, and by choosing dyes with flat molecules of large molecular area to facilitate non-ionic dye–fibre interactions.

In Table 3.3 the various dyeing systems are classified for regular fibres according to the type of interaction involved between dye and fibre, and it will be observed that the ionic systems are divided into those involving respectively like charges and unlike charges between fibre and dye.

TABLE 3.3

Classification of dyeing systems[a]

Fibre	Dye
1. Non-ionic systems:	
Secondary cellulose acetate	Disperse
Cellulose triacetate	Disperse
Nylon	Disperse
Acrylic	Disperse
Polyester	Disperse
Polypropylene	Disperse
2. Ionic systems:	
(a) Like charges	
Native and regenerated cellulose	Anionic
(b) Unlike charges	
Protein	Anionic
Nylon	Anionic
Acrylic	Cationic

[a] After McGregor and Peters [1].

3.2 CHEMICAL INFLUENCES

The dyeability of fibres can be changed deliberately by chemical modification, which either changes the nature of the functional groups (e.g. as in the conversion of cellulose to cellulose acetate) or introduces new functional groups during manufacture (e.g. as with basic-dyeable polyester fibres).

The chemical changes are usually accompanied by alteration of important physical properties. Consequently, the total effect of the chemical modifications is a composite of both the chemical and physical structure of the fibre.

3.2.1 Acetalisation of poly(vinyl alcohol) fibres

Treatment of poly(vinyl alcohol) fibres with formaldehyde (methanal) is used to make them less susceptible to hot water. The hydroxyl groups in the amorphous regions of the fibre are changed to acetal groups and some cross-linking also occurs, with an inevitable change in dyeing properties as exemplified in Figure 3.1 for dyeing with a direct dye [6]. Although the hydroxyl groups are being removed and the fibre is becoming more hydrophobic, the initial disruptive effect of the treatment allows more water-soluble dye to be adsorbed; the situation is reversed, however, as the extent of acetalisation increases. The latter effect is in keeping with the decreasing number of free hydroxyl groups.

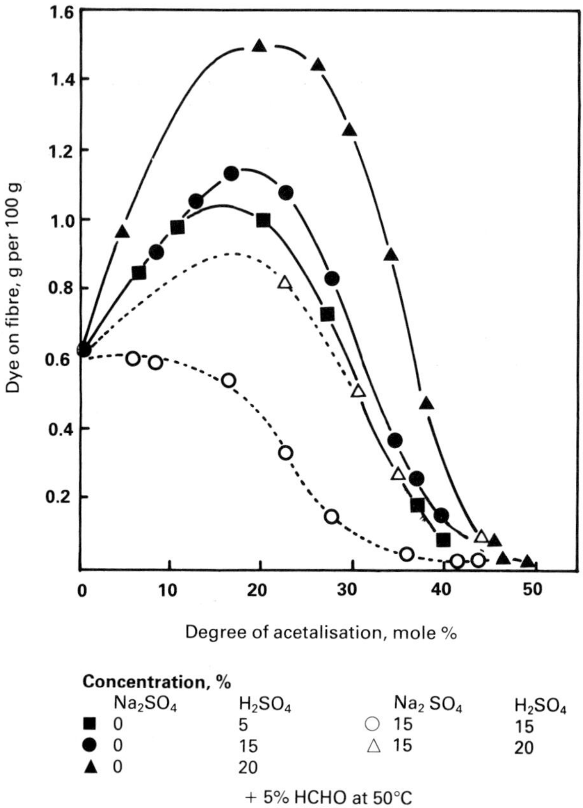

Figure 3.1 – Effect of concentration of sulphuric acid on relation between degree of acetalisation and dye adsorption

3.2.2 Acetylation of cellulose

The conversion of hydrophilic cellulose fibres to hydrophobic cellulose acetate fibres is a further example of a chemical modification which produces marked changes in dyeing properties. The increasing hydrophobic nature of the fibres with increasing degree of acetylation is shown in Table 3.4, which gives the changes in percentage water imbibition accompanying acetylation together with the changes in the most appropriate dye class. Direct dyes may still be used whilst the degree of substitution is less than 1, but as acetylation proceeds their reduced affinity becomes negligible and is superseded by an increasing affinity for disperse dyes. The change is reflected by the correlation between regain and the saturation value for the disperse dye shown in Figure 3.2.

The change in chemical composition is not the only influence on dyeing

TABLE 3.4

Influence of acetylation of cellulose on regain and water imbibition [66]

Sample no.	Acetic acid content/%	Regain at 65% r.h. and 22 °C/%	Water imbibition/%
1	54.0	7.07	25.8
2	58.53	5.96	22.5
3	60.14	4.51	17.6
4	61.6	4.05	15.4
5	62.1	3.90	15.0

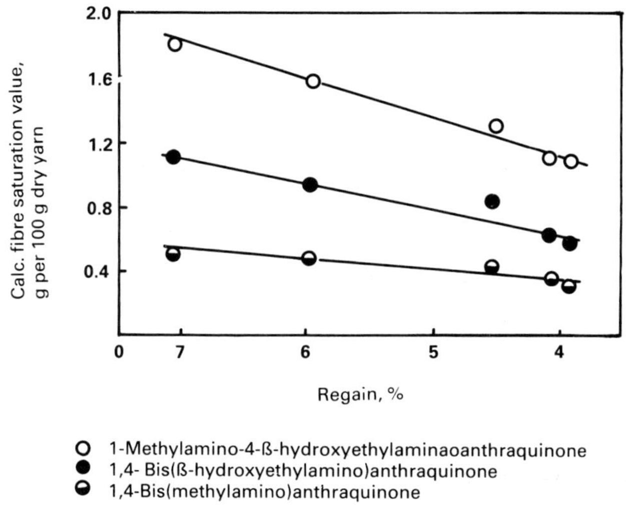

Figure 3.2 – Relation between regain and fibre saturation value

properties. There is also a recognisable disruptive influence of acetylation on fibre structure, which appears when dyeing properties of the newly acetylated 'never-dried' fibres are compared with the fibres after drying. Thus, 'never-dried' monoacetate will accept direct dyes and secondary acetate will accept selected direct and acid dyes, but not after being dried [1, 6, 7].

The changes in the molecular order of the fibrous polymers accompanying acetylation of cellulosic fibres go through stages from a high degree of order before acetylation to the largely amorphous structure of cellulose secondary acetate (2.3 acetyl groups per glucose residue; acetyl content 54–55 per cent) imparted by the bulk of the acetate groups, and then to the highly ordered crystalline arrangement of cellulose triacetate [8]. The combined effect of

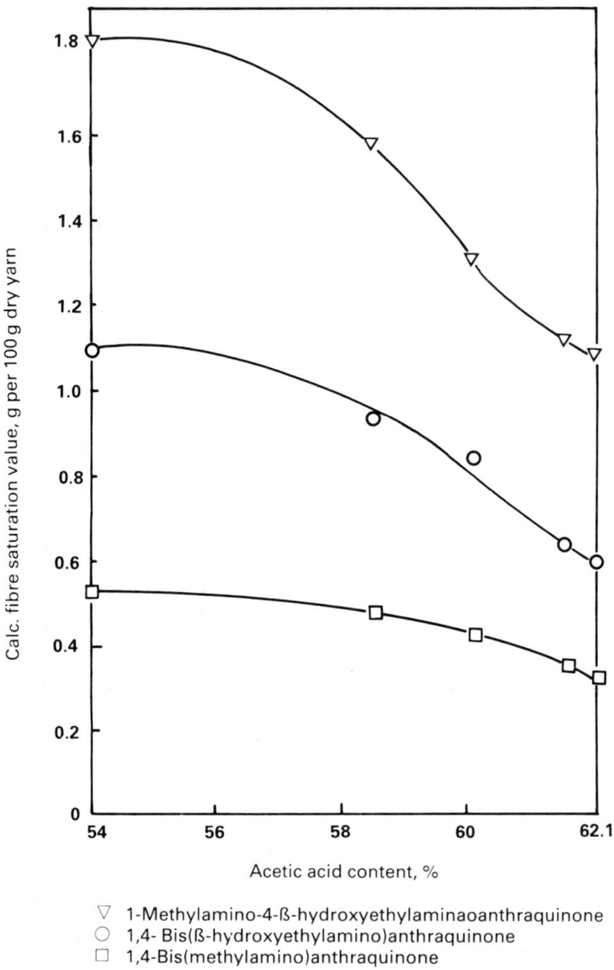

Figure 3.3 – Effect of acetic acid content on the fibre saturation value

the degree of acetylation and of changes in molecular order influence both the fibre saturation values for disperse dyes (Figure 3.3) and their dyeing kinetics (Figure 3.4).

3.2.3 Differential-dyeing or dye-variant fibre

Another approach to fibre modification is the use of deliberate chemical modification of the man-made fibrous polymer or special extrusion conditions [9–12, 13], in either case producing fibres which are very similar to the standard fibre in properties such as handle, tenacity and bulkiness, but very

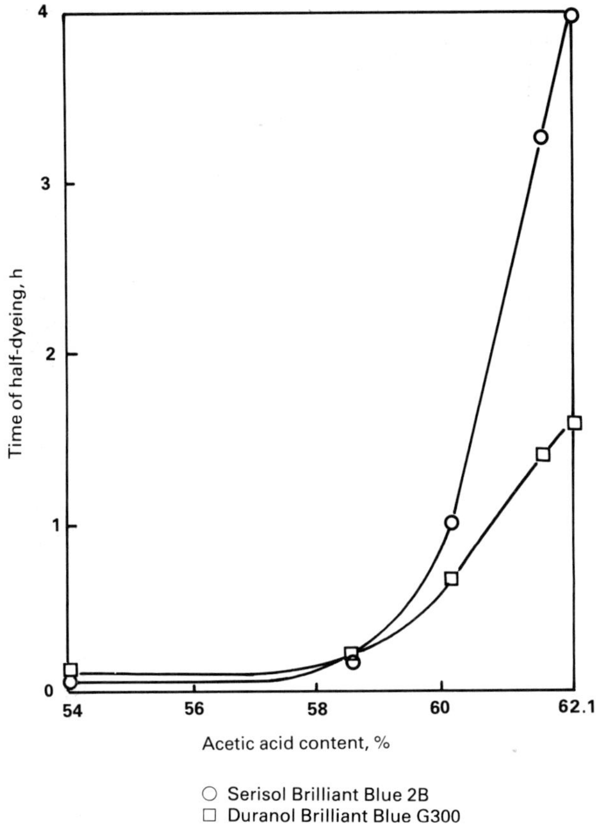

Figure 3.4 – Effect of acetic acid content on the time of half-dyeing

different in their dyeing properties. Normal and modified fibres may be blended together and dyed using a single dyebath to produce multicoloured effects.

The dye-variant fibres now technically possible encompass physical differences which influence the intensity of the colour and chemical modifications which influence the dye class which the modified fibre will accept. The possibilities are listed below, although not all may be commercially available.

Nylon:
- (a) acid-dyeable (standard type),
- (b) basic-dyeable;

Polyester:
- (a) disperse-dyeable at different levels of adsorption,
- (b) basic-dyeable;

Viscose rayon:
(a) direct-dyeable (deep-dyeing and standard type),
(b) acid-dyeable.

3.2.4 Nylon fibres

Developments in the chemistry of nylon fibres and improved yarn-spinning techniques have led to the production of (a) ultradeep-dyeing, deep-dyeing, regular- (normal-) dyeing and light- or low-dyeing yarns for the application of anionic dyes and (b) basic-dyeable yarns for the application of basic dyes [13–17].

Deliberate variation of the substantivity of acid dyes can be achieved by the control of the concentration of basic groups within the fibre. Thus

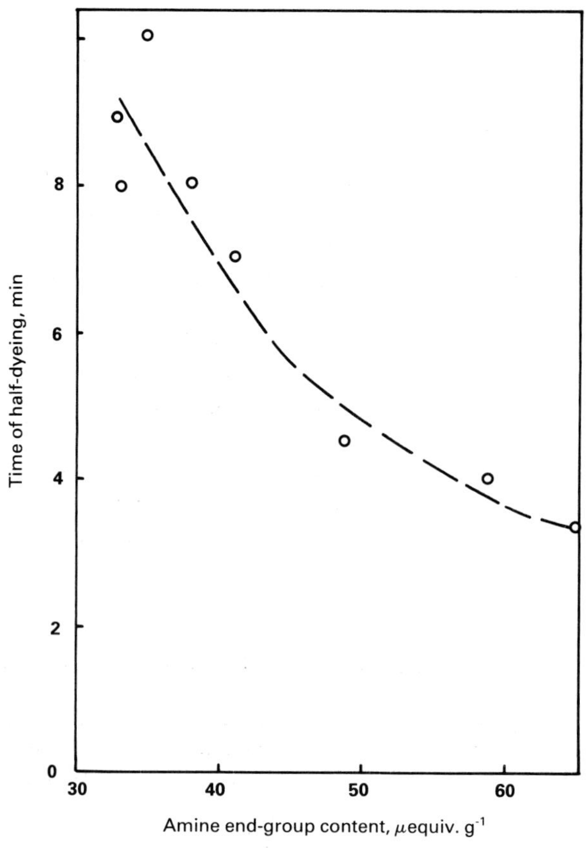

Figure 3.5 – Relationship between amine end-group content and time of half-dyeing in a Solway Blue B dyebath

increased affinity is obtained by providing more dye sites through increasing the amine end-group content of the fibre [18]. The influence on the rate of dyeing is shown in Figure 3.5, and although the equilibrium uptake also increases with increasing amine end-group content it is possible that some acid dye can be adsorbed in excess of the end-group content, through non-ionic interactions [19]. A specific example of this behaviour is shown in Figure 3.6 for a range of Monsanto fibres. The behaviour of C.I. Acid Yellow

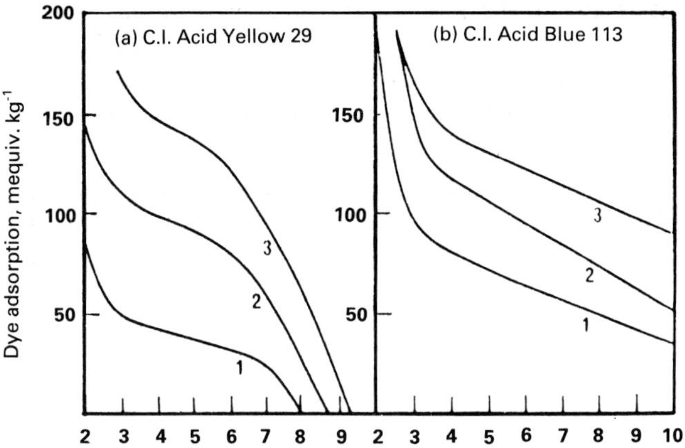

Figure 3.6 – Relationship between maximum dye adsorption and pH for differential-dyeing nylon fibres: (1) regular, (2) deep dye, (3) ultradeep dye

29 on regular- and deep-dyeing fibres is consistent with the amine end-group content shown in Table 3.5, but excess is taken up by the ultradeep-dyeing fibre.

The maximum adsorption for C.I. Acid Blue 113 is in excess of end-group content for all three fibres, and this behaviour is typical of those dyes whose adsorption by nylon fibres is more strongly dependent upon non-ionic

TABLE 3.5

Amino end-group content (in terms of acid adsorption) of some differential-dyeing nylons [16]

Monsanto fibres	Mequiv. HCl per kg fibre	DuPont fibres	mequiv. HCl per kg fibre
Regular	41	Light	32
Deep	89	Regular	46
Ultradeep	95	Deep	80

interactions. Increases in the amine end-group contents may be obtained through the reactions represented by the equations (a) and (b) [20, 21]:

$$\underset{\text{diamine}}{H_2N\text{—}X\text{—}NH_2} + \underset{\text{nylon molecule}}{HOOC\text{—}Ny\text{—}NH_2} \rightarrow \underset{\text{nylon with additional amino end-group}}{H_2N\text{—}X\text{—}NH\text{—}CO\text{—}Ny\text{—}NH_2} \qquad \text{(a)}$$

$$\underset{\text{triamine}}{H_2N\text{—}X\text{—}NH\text{—}Y\text{—}NH_2} + \underset{\text{nylon molecule}}{HOOC\text{—}Ny\text{—}NH_2} \rightarrow H_2N\text{—}X\text{—}Y\text{—}NH\text{—}NH\text{—}CO\text{—}Ny\text{—}NH_2 \qquad \text{(b)}$$

A reduction in acid dye adsorption through amide group formation is achieved through the use of reaction (c):

$$\underset{\text{nylon molecule}}{HOOC\text{—}Ny\text{—}NH_2} + \underset{\text{secondary amino-carboxylic acid}}{HOOC\text{—}X\text{—}NHR} \rightarrow HOOC\text{—}Ny\text{—}NH\text{—}OC\text{—}X\text{—}NHR \qquad \text{(c)}$$

Alternative methods for varying the adsorption of acid dyes include the introduction of derivatives of phenyl-phosphinic or -phosphoric acid and of *N*-(2-aminoethyl)piperazine [22], toluene-4-sulphonic acid [23], tributyl phosphate and hexamethylenediamine [24], and other related compounds [25]. Non-ionic interactions still operate, however, so that the adsorption of acid dyes cannot be reserved completely either by using this approach or by making the fibre more hydrophobic through acetylation [26]. The introduction of acidic groups will also suppress the adsorption of anionic dyes through an increase in the negative potential of the fibre, and consequently the magnitude of the effect will be related to the degree of sulphonation of the dye and the nature of other groups in the dye molecule [14].

3.2.5 Acrylic fibres

Normal acrylic fibres generally contain 5–10 per cent of other comonomers, such as methyl methacrylate or vinyl acetate, to give a more easily dyed fibre. The dyeing properties can therefore be subject to some control during manufacture. For example, with fibres of the same draw ratio, increasing the content of vinyl acetate monomer increases the rate of dye diffusion [27] without significant effect on the equilibrium adsorption. Increasing the molecular order in the fibre by drawing reduces both the rate of dye diffusion and the equilibrium uptake, however. Monomers with acidic groups are often incorporated into acrylic fibre polymers to enhance their affinity for basic dyes, but different comonomers which are basic in nature (e.g. vinylpyridine and substituted vinylpyridines) may be used for conferring substantivity for anionic dyes.

Other polar non-ionisable monomers (e.g. alcohols, ethers and ketones) increase the general level of adsorption through non-ionic interactions with the dye, and some monomers serve simply to produce modifications to the fibre structure which are advantageous to dye adsorption [10, 28].

The most familiar differential-dyeing fibres available are either basic-dyeable (Acrilan 16 (Monsanto), Orlon 42 (DuPont)) or acid-dyeable (Acrilan 32 and Orlon 44) [13, 29].

3.2.6 Polyester fibres

Most modified polyester fibres are based on poly(ethylene terephthalate), physical modifications being achieved during manufacture by using shorter molecular chains, changing the conditions of spinning or including supplementary condensation reactions, whilst chemical modification is achieved by copolymerisation with poly(ethylene terephthalate) as the main component.

Physically modified fibres

Dacron 35 (DuPont) shows a decreased adsorption of dyes with a better resistance to pilling than regular Dacron 54 fibre, whilst the structure of Trevira WA(FH) is more open and gives a rapid rate of dyeing which almost eliminates the need for carriers.

Chemically modified fibres

The large group of chemically modified polyesters has been produced by incorporating different comonomers into the polymer chains. A wide range has been used including isophthalic acid, 5-sulphoisophthalic acid, 4-hydroxybenzoic acid, vanillic acid and 4,4′-sulphonyldibezoic acid. Comonomers with strongly acidic groups (e.g. 5-sulphoisophthalic acid (Dacron T64)) give basic-dyeable polyester fibres, whilst alkylene oxides introduce non-ionic residues into the molecule leading to an increased adsorption of disperse dyes (Fibre K (Monsanto)). The use of small amounts of long-chain aliphatic carboxylic acids produces an open structure with improved abrasion resistance and affinity for disperse dyes [10], particularly for dyes of high r.m.m. [30].

More than one additional comonomer may also be used in one fibre to produce fibres very different from the original polyester fibres; Vycron (Beaunit), for example, contains 10–15 mol per cent of acid (including isophthalic and terephthalic acids). Grilene (Inventa) has a total acid content of 10–15 mol per cent (including 4-hydroxybenzoic acid). Both show improved dyeing properties with disperse dyes and increased resistance to pilling. Although by using cationic compounds, particularly those containing nitrogen, acid-dyeable polyester fibres are formed, the most interesting fibres in this group are those containing sulphonate groups which impart substantivity for basic dyes [10, 31].

3.2.7 Other polyester fibres

Various polyesters other than poly(ethylene terephthalate) have been used as fibres; for example, the Kodel (Eastman) range of fibres is based on 1,4-dimethylcyclohexyl terephthalate in which the effect of the cyclohexane ring is to increase chain stiffness, chain packing and interchain forces, giving rise to a fibre of better stability to hydrolysis, higher melting-point and higher second-order transition temperature than poly(ethylene terephthalate).

Modifications to the dyeing properties are affected by methods similar to those used for poly(ethylene terephthalate). Thus regular-dyeable, light-dyeable and basic-dyeable fibres are available as differential-dyeing carpet yarns [32]. The fibres appear to be more accessible to disperse dyes than are regular polyester fibres, since the use of carriers is needed, if at all, only for full depths of shade [10].

3.2.8 Polypropylene fibres

For all practical purposes unmodified polypropylene is undyeable, due to its highly compact structure. The linear isotactic non-polar polymer chains have a helical configuration which produces a highly crystalline fibrous substrate, with about only one-third of the total volume being accessible to dyes [33]. Where accessible regions exist in polypropylene, however, they allow much more rapid diffusion of disperse dyes than is found in secondary cellulose acetate (Table 3.6) although the saturation value is much lower (Table 3.7).

TABLE 3.6

Comparison of diffusion coefficients of disperse dyes in polypropylene and secondary cellulose acetate film

Dye	Temperature /°C	Diffusion coefficient /$m^2\ s^{-1} \times 10^{14}$	
		Polypropylene	Acetate
Dimethyl Yellow	60	4.81	2.3
	80	33.7	2.9
1-Anilino-anthraquinone	80	6.74	1.12

TABLE 3.7

Comparison of saturation values at 80 °C of disperse dyes on polypropylene and secondary cellulose acetate film

Dye	Saturation value /g per 100 g	
	Polypropylene	Acetate
Dimethyl Yellow	0.67	2.0
1-Anilinoanthraquinone	0.32	0.6

It has also proved difficult to improve the dyeing properties of polypropylene by the introduction of polar functional groups, but this has been achieved by incorporating polymers containing functional groups as additives to the polypropylene polymer.

The most successful commercial methods entail the incorporation of functional groups by blending suitable additives with the polypropylene polymer before melt extrusion. Organometallic compounds that form chelates with selected dyes or polymers with highly polar electron-donating groups are widely used, the most successful so far being cationic polymers (polyamides, polyvinylpyridines, polyvinyl-*N*-heterocyclics, polyamines) which increase the substantivity for disperse dyes and confer substantivity for anionic dyes [34]. Incorporation of metals such as aluminium and nickel into the polypropylene enable some disperse dyes to become chelated in the fibre to give advantages in printing. An increase in the substantivity of disperse dyes in general is achieved by the incorporation of polymers containing suitable functional groups in their structure. However, the most popular development has been the introduction of acid-dyeable polypropylene. Substantivity for acid dyes on polypropylene is realised only when the incorporation of functional basic groups is associated with an increase in the permeability of the polymer matrix. This is usually achieved by treatment with an acid before dyeing. This causes the incorporated microfilaments containing the basic groups to swell and thereby allows an increase in permeability [34–36].

3.2.9 Regenerated cellulose fibres

The high-wet-modulus viscose fibres were born out of the need to improve the low wet strength of viscose rayon. Due to the method by which they are coagulated the high-wet-modulus fibres are formed with a near-circular cross-section and a compact randomly oriented structure, which on drawing gives way to a three-dimensional array of highly oriented crystalline units with lateral order, higher than in normal viscose rayon. The interconnected crystalline units are in general aligned parallel to the fibre axis.

Swelling is lower than with normal viscose rayon. Consequently it is easier to obtain uniform penetration and distribution of dye in closely woven cloth or cheeses. Differences in dyeing behaviour are markedly dependent on the individual dye and dyeing conditions.

Differential-dyeing fibres are also found amongst the viscose rayons; they are produced by incorporating suitable components in the dope before extrusion. For example, polyvinylpyrrolidone [9] or an amino-ether of cellulose [37] will confer substantivity to acid dyes. Polyvinylpyrrolidone will also modify the substantivity to direct dyes and this has been used by Courtaulds in the production of deep-dyeing varieties of Sarille, Evlan and Vincel. The dyeing of these fibres resembles that of both viscose rayon and wool. The application of direct dyes is favoured by the net positive charge on the fibre (due to the presence of the basic groups) and consequently dyeing will take place in the absence of electrolyte. When acid dyes are applied to deep-dyeing fibres in the presence of acetic acid the increased positive charge allows very deep dyeings to be obtained. Although dye selection is important, as in all differential-dyeing systems, a wide range of multicoloured effects can be obtained on 100 per cent viscose rayon fabrics. Maximum colour contrast between normal and deep-dyeing fibres is achieved by the use of acid and the omission of electrolyte, and a subdued contrast by the use of electrolyte and the omission of acid [38].

3.3 DYEING PROPERTIES RELATED TO THE INHERENT PHYSICAL STRUCTURE OF THE FIBRE

For natural fibres the nature of the functional groups in the fibrous polymer and the permeability to dyes are determined during growth, but for man-made fibres they are determined during manufacture, since finishing operations such as the rate of cooling from the melt, drawing and heat-setting affect dyeing properties through the changes in the molecular organisation that they initiate. Many different influences of molecular organisation within the fibre therefore require attention. Detailed accounts of both the chemical and the physical structure of fibres have been given [8, 10, 39, 40, 41, 42] which reveal the incomplete understanding in molecular terms. As a useful generalisation, however, fibres may be regarded as structures in which there is a spread of molecular order, ranging from highly ordered crystalline domains to disordered amorphous regions [43]. The strength originates in the crystalline material whilst the amorphous material provides the flexibility, porosity and the regions generally accessible to liquids, dyes and other reagents. The fibre properties, including dyeing properties, vary depending upon the relative degrees of order and disorder in the structure (often described loosely as the *crystalline/amorphous ratio*).

Both orientation and crystallinity are significant in that both strongly affect the kinetics and equilibria of sorption of solvent and dye molecules by fibres. Orientation, or molecular alignment, refers to the extent to which the molecular chains are directionally correlated over the entire fibre. In some cases this is very high, with most of the molecular chains lying with their axes within a few degrees of the fibre axis, as indicated in Figure 3.7(c). In

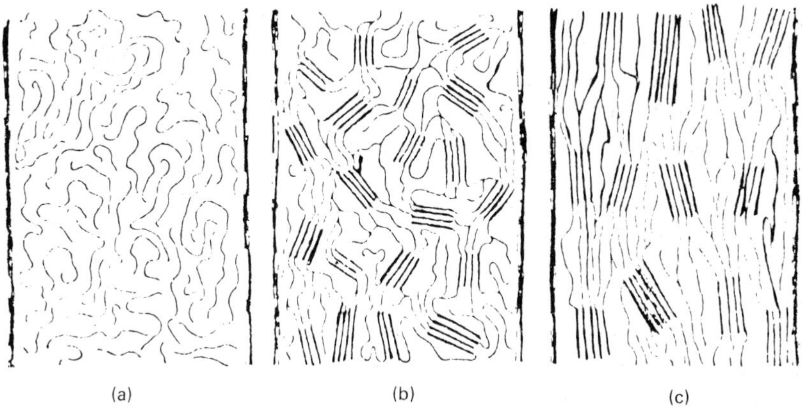

Figure 3.7 – Schematic representation of a polymer in three states

some natural fibres, such as cotton and to a lesser extent wool, the chains are helically orientated. The molecular alignment in a natural fibre is an inherent characteristic of that fibre and is for most practical purposes unalterable. Within the group of cellulosic fibres, ramie and jute both have extremely high degrees of orientation, whereas that of cotton is much lower. In man-made fibres, the degree of orientation can often be controlled at the initial extrusion stage, but more significantly at the subsequent drawing stage. In extruded undrawn filaments, the molecular chains may be packed together into crystallites but, apart from orientation at the filament surface producing a skin effect, the crystallites are in a state of random orientation (Figure 3.7(b)). Subsequent stretching or drawing orientates the crystallites along the fibre axis, to an extent that depends upon the amount of draw imparted. At the same time the molecules in the non-crystalline regions assume a certain degree of crystallinity. The resulting degree of orientation is likely to be unequally distributed between various structural units, since orientation proceeds with crystalline as well as non-crystalline regions and the orientation of the former will be different from that in the latter.

Segments of the fibre molecules tend to be not only parallel to the fibre axis, but also arranged side by side with the full three-dimensional order associated with crystalline materials. Both man-made and natural fibres show some crystallinity, although a high degree of crystallinity is not essential.

For instance, whereas nylon and polyester fibres are highly crystalline, poly(vinyl chloride) and polyacrylonitrile fibres have a low degree of crystallinity. The last two substrates, however, exhibit a high degree of molecular orientation in the drawn state, so fibre formation is assured. The degree of crystallinity, or volume fraction of the ordered regions or units of the fibre, is one of the more important characteristics of crystallinity, since it affects the maximum draw ratio of man-made fibres, dyeability and many other properties.

It is thus possible for fibres of identical chemical composition to display very different dyeing properties, because of disparities in the distribution of molecular order within the fibres. For example, the dyeing properties of different cellulosic fibres have been interpreted in terms of variations in the proportions of amorphous to crystalline material in the fibre, in the accessibility of internal adsorption sites and in the pore size of the fibres. Such models imply that the dye moves along discrete water-filled capillary channels, a portion of the dye being immobilised on the walls of the channels [44, 45]. It must be assumed that the pores are tortuous and wind their way through the amorphous regions, but are located between the crystalline regions, since it is difficult to visualise dye penetrating into highly organised crystalline regions of the fibre.

Models based on this system tend to be over-simplified, but they do permit some appreciation of the state of affairs inside a fibre within the bounds of existing knowledge. It is questionable whether any one model is likely to be suitable for explaining dyeing properties of all fibres, but they make a useful starting point. For example, natural cellulose fibres possess a highly crystalline fibrillar structure and it is easy to appreciate the existence of a system of interconnecting water-filled channels through which the dye can travel. The situation is less clear-cut with regenerated cellulosic fibres. In a viscose rayon fibre a highly ordered 'skin' surrounds the less ordered 'core' region, and cannot be penetrated by dyes unless the fibre is swollen. By suitable selection of dyes it is possible to obtain results which show that the average pore size of water-swollen viscose rayon is less than that of cotton [46]. Moreover, it is possible that dye may be transported through the two types of fibre by different mechanisms. This may be inferred from the work of Ticknor [47], who demonstrated that the rate of flow of water through viscose sheet is greater than that to be expected from a process of random diffusion, so the water must be flowing through capillary channels. His view is reinforced by measurements of the penetration of organic molecules, which suggest that diffusion through the structure becomes random when the size of the diffusing molecules reaches that of the size of the pores.

Thus it may be expected that diffusing water-soluble dye molecules will follow the same path as the water which causes the swelling up to the point at which the size of the dye molecules approaches that of the pore diameter, when they will be forced into weaving their way between polymer chains. By implication only unordered amorphous regions of the fibre can be involved, and consequently the speed of movement of the dye through the fibre will become restricted when the orientation of the polymer molecules is increased by processes such as drawing. It is well known that drawing affects the dyeing behaviour of man-made fibres and regenerated cellulose fibres, but changes in orientation and rearrangements of internal molecular structure are brought about by other factors as well. For example, the action of dry or steam heat can cause changes which are reflected in dyeing properties and in the occurrence of Barré dyeings. Furthermore, such molecular rearrangements may also take place during dyeing operations. For example, it would be surprising if the high temperatures to which polyester fibres are subjected during some fixation processes did not cause structural changes in the fibre.

Some of the earliest data reported on the effects of changes in physical structure were for work carried out on viscose rayon [48, 49, 50]. Increasing

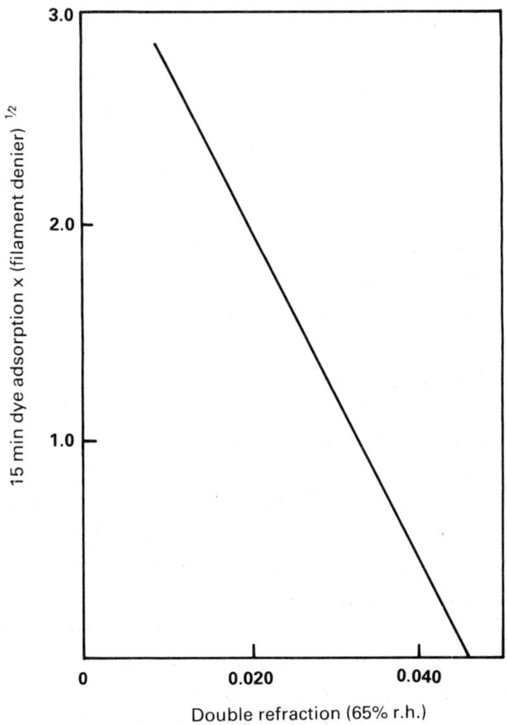

Figure 3.8 – Effect of increasing draw ratio on rate of dyeing

the degree of orientation by drawing decreases the rate of dyeing of Chlorazol Sky Blue FF (C.I. Direct Blue 1) over a range of deniers as shown in Figure 3.8. The equilibrium adsorption values are also affected (Figure 3.9). Since the moisture adsorption and density of the fibres are little affected by drawing, the decreased equilibrium uptake must be associated with the dye molecule being prevented by its large size penetrating the more orderly arrangement of the non-crystalline regions of the fibres.

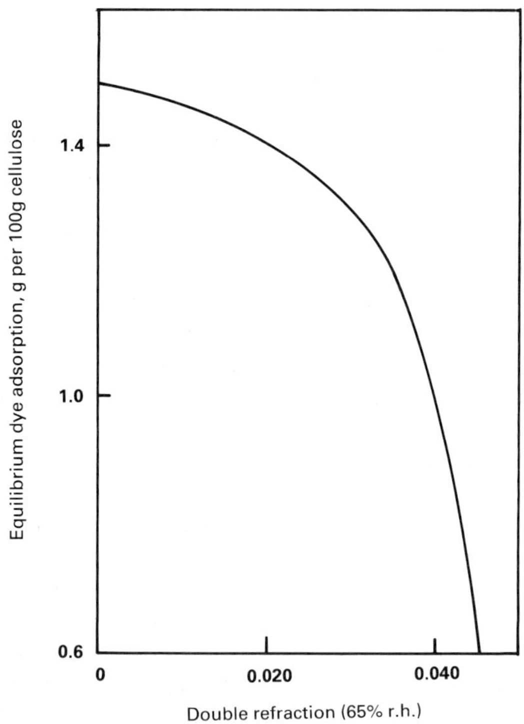

Figure 3.9 – Effect of increasing draw ratio on equilibrium dye adsorption

In Figures 3.8 and 3.9 the orientation of the fibrous molecules is conveniently represented by the birefringence (double refraction). This optical property is a very valuable measurement of the average degree of orientation of molecular chains in both the crystalline and amorphous domains. The larger the value of birefringence, the greater is the degree of orientation of the molecular chains along the fibre axis.

The simple concept of pores within a structure of variable crystalline/-amorphous molecular order has been developed to incorporate the fact that in some cases the polymer chains may crystallise in a folded-chain conformation. Evidence for this is based on X-ray crystallographic data on polymer crystals [51–54] and on other results concerning the deformation of polymers and the response of fibres to heat-setting [55]. For example, chain-folding in nylon 6.6 fibres has been investigated by Dismore and Statton [56]. In a drawn fibre the molecular chains are highly extended and essentially parallel to each other, but chain-folding exists to produce a low overall statistical order (Figure 3.7(a)). In this state there is broad distribution of intermolecular bond distances and bond directions, with a consequent broad distribution of bond energies. During heating, the least stable intermolecular bonds (i.e. those possessing the lowest bond energies) break or melt and the released segments of the molecular chains tend to move to positions that are more favourable for the formation of new intermolecular bonds of higher stability. It is suggested that the melting of a sufficient number of bonds in this way leads initially to a more random state and subsequently to a more energetically favoured state, by recrystallisation in the folded form (as represented by Figure 3.7(b)). Recrystallisation, of course, takes place in three dimensions.

This model has been used to explain the variations in the shrinkage and tensile properties of nylon 6.6 fibres with temperature, and has also been used in similar studies relating to polyester [54] and acrylic [57] fibres. It is believed that chain-folding provides a structure which is the main contributor to the fluid-like properties of fibres at elevated temperatures. The folded structure is regarded as being a flexible one whose tensile strength may be accounted for either by a small percentage of the chains in the cross-section traversing the length of more than one folded region or by bonding between the folded ends of the chains. The regularity of the folded structure in poly(ethylene terephthalate) fibre is disrupted by mechanical stress, but the folds are restored by subsequent annealing. Any attempt to counteract the shrinkage that occurs during annealing by the application of tension affects the chain-folding process, chain-folding taking place more extensively when the annealing is carried out in the absence of tension. The subsequent ability of the fibres to undergo further thermally initiated shrinkage is therefore directly related to the physical conditions that prevail at the annealing stage; thus the higher the regularity of the chain-folding that occurs, the higher is the temperature required to initiate further shrinkage.

Thus the history of the fibre can have a marked influence on the molecular organisation. It seems likely therefore that the concept will be recognised as one of the many influences on dyeing properties and the results of one investigation show a connection between chain-folding and the effects of carrier [58].

3.4 FIBRE PROPERTIES INFLUENCED BY PROCESSING BEFORE DYEING

Alterations in rates of dyeing brought about by changes in physical structure are clearly of practical importance. Variations in tension, either within or between yarns, can produce permanent deformations during fibre processing or finishing and thus affect the dyeing properties. The yarn on the inside of a package may dye at a different rate from that on the outside because of differential shrinkage, or variations may occur during heat-setting due to variations in tension or temperature. The diffusion behaviour of gases and liquids in polymer films indicate that the rate of diffusion is much more sensitive to structural changes in the polymer than is equilibrium adsorption [59]. Both aspects have been studied for the dyeing of polyester [60, 61, 62], acrylic [27, 63], nylon [64, 65], cellulose acetate [59], cotton [66, 67] and viscose rayon fibres [48, 49].

Careful comparison of the dyeing properties of cellulose acetate, cellulose triacetate and polyester fibres [67] with disperse dyes confirms that very high molecular order in the form of crystallinity is responsible for a reduced rate of diffusion [68].

Changes in the molecular order of the fibres may be modified before dyeing in two ways: by drawing the fibres and by subjecting them to some form of heating treatment. The influence of each will now be considered separately.

3.4.1 Drawing

The strength, the elastic modulus and the extension at break for man-made fibres are all brought to an acceptable level by stretching (drawing) the filaments. The drawing process enhances the orientation of the molecular chains and causes an increase in close-packing or crystallinity. The relative importance of increased crystallinity and increased alignment in the non-crystalline regions depends upon the particular fibre under examination. For example, the drawing of poly(ethylene terephthalate) produces a considerable degree of crystallinity, since after the initial manufacturing stage the oriented polymer is largely amorphous. The drawing of nylon 6.6 fibre induces a relatively small increase in crystallinity, however, since this fibre is highly crystalline in the undrawn state [8]. Nevertheless, marked changes in orientation occur, some crystallites being pulled into alignment, whilst others are sheared and smeared out (Figure 3.10) [41]. Despite the differences in physical structure between polyester and nylon fibres in the undrawn state, there appears to be no significant structural difference between them after drawing [40].

Nylon 6 is of particular interest, however, because increasing the draw ratio does not give the regular decrease in the rate of diffusion of dyes expected from an increasing degree of molecular order (Figure 3.11). Instead

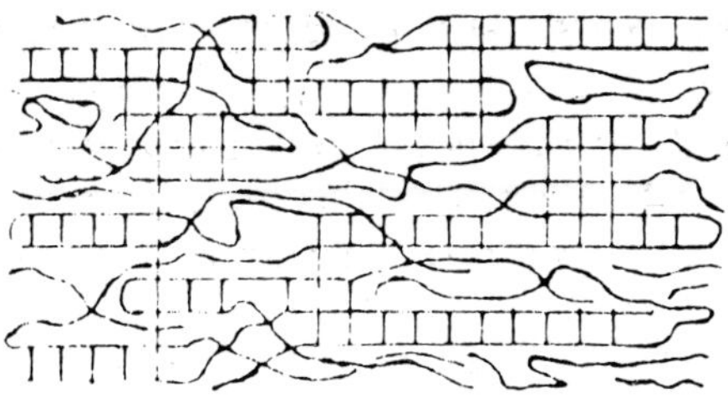

Figure 3.10 – Schematic representation of drawn structure of nylon 6.6

the radial diffusion coefficient D_r of C.I. Acid Blue 82 first increases with increasing draw ratio, then reaches a maximum and thenceforth decreases rapidly [64]. The results support the hypothesis that in the initial stages movement of crystallites causes defects which permit easier access through the non-crystalline material, but once the structure approaches a more stable

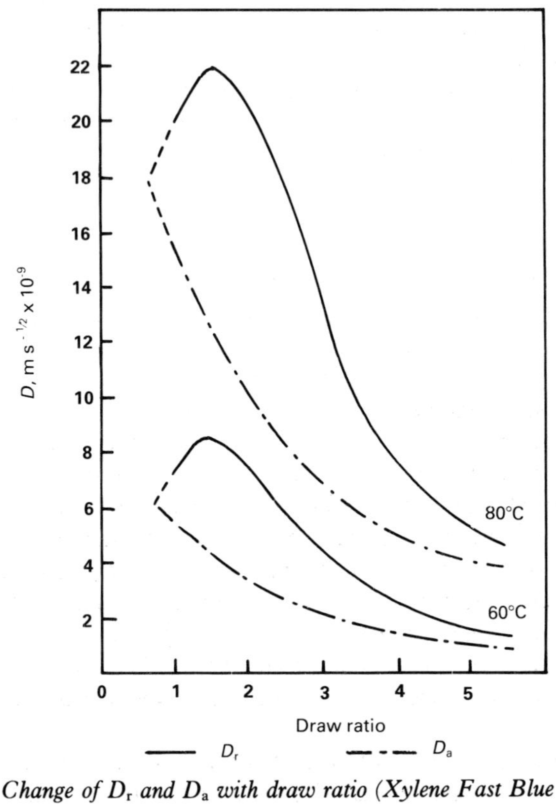

Figure 3.11 – Change of D_r and D_a with draw ratio (Xylene Fast Blue 2P on nylon 6) (D_r = radial diffusion; D_a = axial diffusion)

configuration further drawing causes the expected ordering of structure and reduction in the diffusion of the dye. With nylon 6.6 the rate of dyeing is markedly affected by drawing [65, 69] without changing the equilibrium adsorption, and evidence suggests that this is due to the increasing tortuosity of the diffusion paths in the radial direction.

3.4.2 Diffusion along the major axis of the fibre

Diffusion customarily takes place along the radial axis of the fibre during dyeing. Inspection of Figure 3.11 [64], however, reveals that diffusion coefficients along the axial direction D_a and the radial direction D_r are affected very differently by the drawing process. Diffusion along the fibre axis begins to drop as soon as structural order starts to increase, indicating differences in the stability of the structure in the two different directions. Similar experiments with acrylic fibres show that diffusion along the fibre axis stops completely when the draw ratio reaches 15 (Figure 3.12) [63]. Benzidine-based direct dyes do not diffuse along the axis of cellulosic fibres though much smaller molecules are able to do so, thus providing evidence for the existence of much smaller diffusion paths along the fibres than across it [70].

With polyester fibres both the equilibrium uptake and diffusion coefficient of disperse dyes are reduced by the increased molecular order imparted by the drawing process [61], and the influence of drawing also persists when polyester is heated in the presence of organic solvents [71].

3.4.3 The effect of heat treatments on fibre properties

The degree of drawing also affects the extent to which other changes in molecular order occur during heat treatments such as heat-setting, steam and

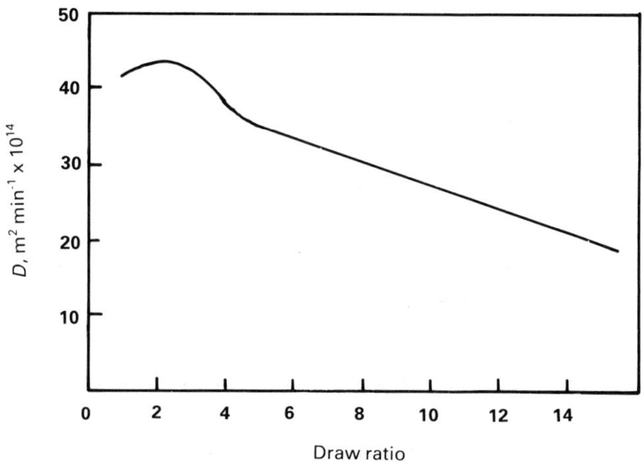

Figure 3.12 – Variation of D_a with draw ratio (Malachite Green on acrylic fibre)

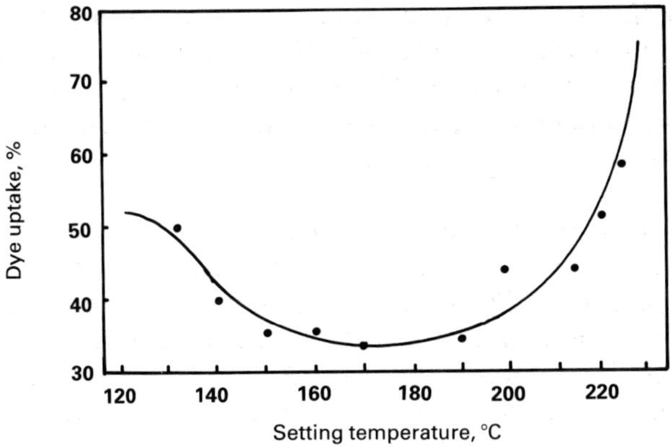

Figure 3.13 – Effect of setting temperature on dye uptake (2% Dispersol Fast Scarlet B150 at 100 °C for 90 min)

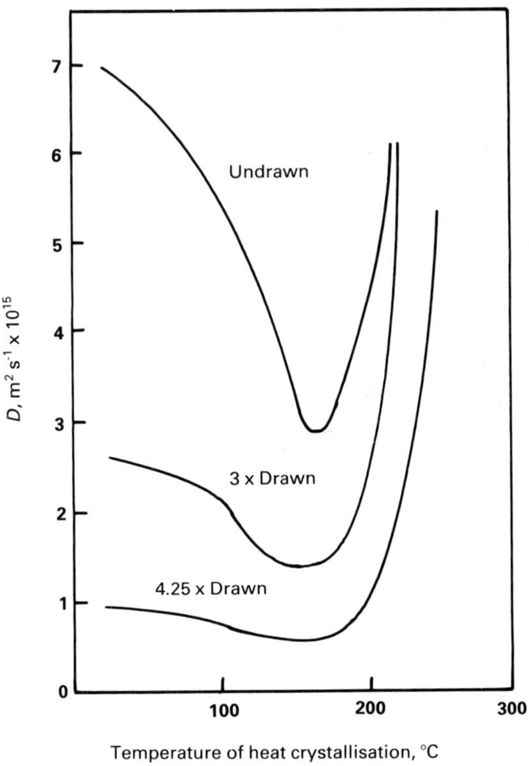

Figure 3.14 – Variation of D with annealing temperature (dyeing carried out at 100 °C)

boiling water. The receptivity of polyester fibre to disperse dyes is dependent on both the draw ratio and the heat-setting temperature in a very complex way. An increase in the temperature at which dry heat-setting is carried out causes a preliminary decrease in dye uptake. After passing through a minimum at a heat-setting temperature range of 170–200 °C the uptake of dye increases. Figure 3.13 illustrates the behaviour of C.I. Disperse Red 1 when applied to Terylene fabric at 100 °C for 90 minutes [72]. Except at temperatures of 220 °C and above, the uptake of dye is lower than that of the unset fabric. Dumbleton, Bell and Murayama [60] carried out a similar investigation which included a study of the influence of drawing on both the diffusion coefficient (Figure 3.14) and the equilibrium uptake (Figure 3.15)

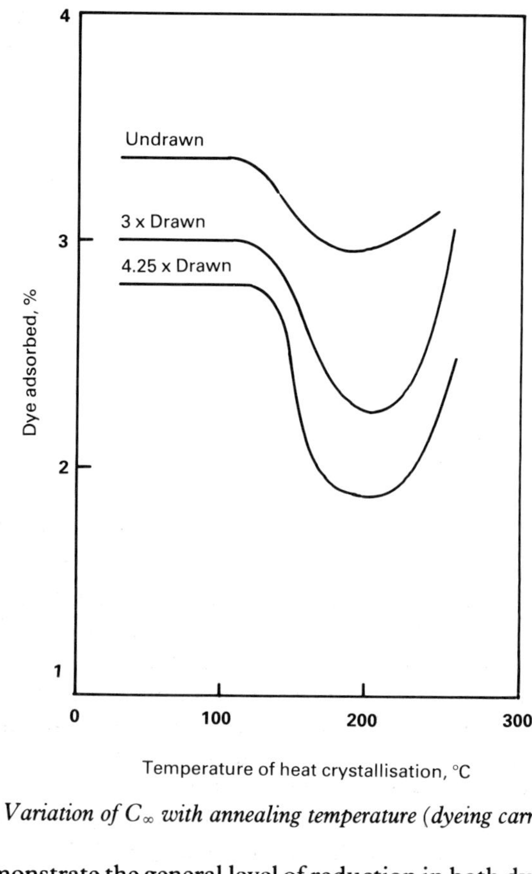

Figure 3.15 – Variation of C_∞ with annealing temperature (dyeing carried out at 100 °C)

of dye, to demonstrate the general level of reduction in both dyeing parameters and the surprising dramatic increase in dyeability which appeared once the heat-setting temperature rose above about 170 °C.

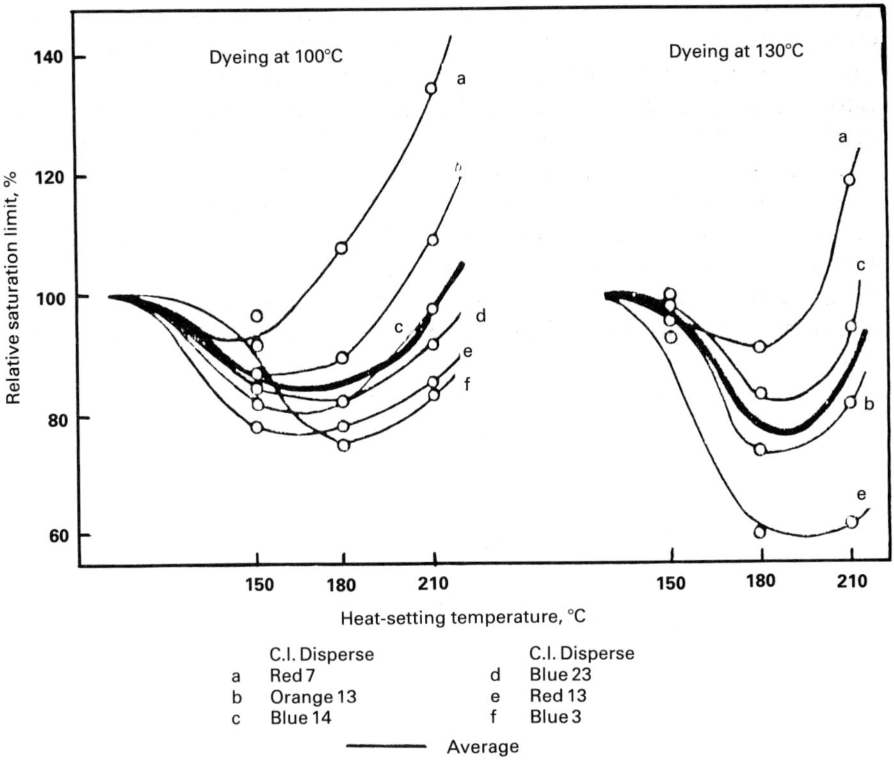

Figure 3.16 – Dependence of saturation limit of Dacron 54 (as percentage of that of unset fibre) on heat-setting temperature

This behaviour has been observed with a range of disperse dyes using Dacron 54 dyed at 100 °C and at 130 °C (Figure 3.16) [75]. The values of the diffusion coefficient also show a minimum at a given heat-setting temperature [73]. The decrease in the saturation levels and in values of the diffusion coefficients suggests that the physical structure of the fibre has compacted and made the matrix less accessible to dye molecules. At the higher heat-setting temperatures, however, further changes must take place since both the saturation levels and the diffusion coefficients then show an increase; the fibre structure has become more accessible to dye molecules. Steam-heating polyester material brings about similar changes in accessibility. In addition, yarn held under tension in either dry-heat- or steam-setting tends to show increased accessibility to dye molecules [74].

This unusual behaviour has been explained by postulating that the physical structure consists of rod-like crystalline units separated by voids, the number, size and shape of which govern the accessibility to dye molecules [74, 75].

The units are not necessarily circular in cross-section, and their outer layers consist of polymer chains that have less order than those of the bulk of the units. The segmental mobility of these chains, together with the segmental mobility of chains in the voids (non-crystalline regions), directly governs accessibility [61]. As the temperature of setting increases the accessible space between the crystalline units is reduced, owing to an increase in the tortuosity of the voids in the direction of the fibre axis and to shrinkage or annealing of the yarn. At the same time, heating increases the perfection and orientation of the units, which in turn leads to an increase in accessible space between the units. It also allows the mobility of the chain segments to increase, thereby facilitating dye accessibility. Depending upon the temperature used for heat-setting, one or other of the above mechanisms will generally predominate, resulting in the occurrence of minima for saturation levels and for values for the diffusion coefficient. When tension is applied during setting the tortuosity factor is reduced, causing the changes in orientation to predominate to a marked extent. Accessibility to dye molecules is increased accordingly.

A rigorous appraisal of the changes in structure of nylon fibres during setting processes has also been undertaken. In general, with both nylon 6 and nylon 6.6 the equilibrium adsorption and rate of dye uptake initially decrease as the temperature of pretreatment is increased, and then increase to values higher than those for the untreated fibre. Yet although the effect of heat treatment, whether by dry heat or saturated steam, is to increase the density of the fibre and thus the degree of crystallinity, the orientation of the crystalline units remains virtually unchanged [74, 76, 77]. Such behaviour would suggest a gradual decrease rather than an increase in dye adsorption.

This apparent contradiction can only be explained by once again postulating the presence of two opposing mechanisms: one decreasing dye uptake by reducing the accessible spaces or voids between the crystalline units available for diffusion and thereby increasing the tortuosity of the voids, and the other increasing dye uptake by increasing the relative size or total void content of the polymer matrix owing to an increase in crystalline perfection of the morphological units. It has been suggested that the matrix contains highly crystalline sheet-like units [78], possibly held together by tie molecules, and that changes caused by heat treatments increase the lateral order of these units. Thus at a particular temperature the polymer matrix will become more accessible to diffusing dye molecules because of the increase in void volume between the units, especially at the higher temperature of pretreatment. In other words, crystalline perfection results in an increase of internal space despite the increase in density of the morphological units. Furthermore, there is a decrease in water regain with the increase of temperature of heat treatment despite the increase in void volume. An increase in regain would be expected in this situation, but the availability of main-chain amide groups, which are responsible for adsorption of water, is decreased because of their involvement in the increased crystalline perfection. Thus the surfaces of the voids become more hydrophobic in nature, leading to a decrease in regain, even though there is an increase in dye uptake. The increase in hydrophobic character of the void surfaces increases the possibility of interaction between dye and fibre by dispersion forces and hydrophobic interaction.

Conditions under which changes in the physical structure of a fibre occur are not confined to the high temperatures of heat-setting; they are also present at the lower temperatures used in treatments carried out in aqueous solutions. It has already been noted that the rate of uptake of disperse dyes by secondary cellulose acetate is dependent upon the nature of the treatment given to the fibre before dyeing. Figure 3.17 illustrates the differences in diffusion behaviour that occur when fibres of different draw ratios are pretreated for 2 h at the temperature of dyeing, as compared with unpretreated fibres (Figure 3.18).

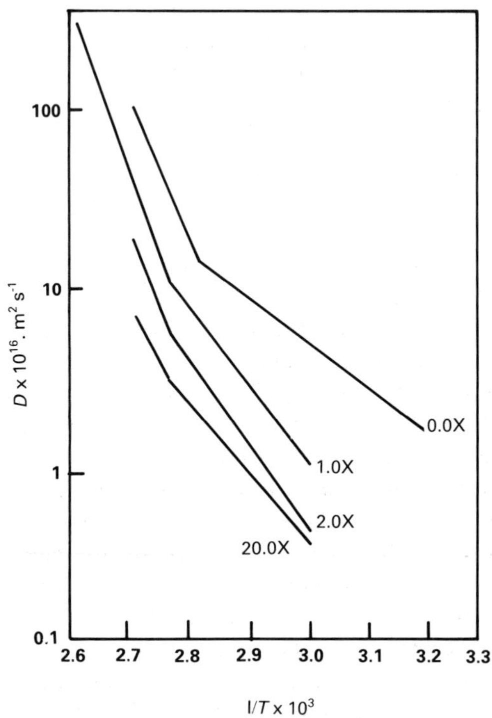

Figure 3.17 – Effect of temperature on diffusion rate of C.I. Disperse Blue 3 in acetate (pretreatment 2 h at dyeing temperature)

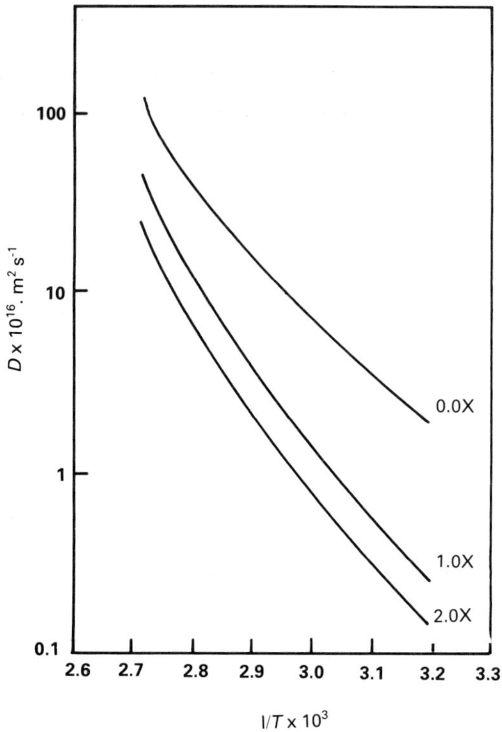

Figure 3.18 – Effect of temperature on diffusion rate of C.I. Disperse Blue 3 in acetate (no pretreatment)

When the fibres are pretreated for 2 h at the boil, the value of the diffusion coefficient shows a greater dependence on temperature, as indicated by the generally large values for the activation energy of diffusion (Table 3.8). This change in behaviour may be due to an increase in the crystallinity of the polymer matrix. Certainly, a permanent change occurs in the structure of the acetate, since the transition effect at about 87 °C disappears and the change is not particularly sensitive to the draw ratio imparted to the fibre. As would be expected the rate of dye uptake, as reflected in the diffusion behaviour, decreases with an increase in the degree of orientation and crystallinity but, more significantly, both the presence of water and the temperature of pretreatment are important in modifying the fibre structure.

Changes also occur in the structure of nylon 6.6 when the fibre is treated in boiling water. There is an increase in the density of the matrix above that for untreated fibre, indicating an increase in crystallinity (Table 3.9) [65]. The increase in crystallinity might be expected to decrease fibre accessibility,

TABLE 3.8

Effect of draw ratio on the activation energy of diffusion of C.I. Disperse Blue 3 in secondary cellulose acetate

Draw	Activation energy /kJ mol^{-1}	/kcal mol^{-1}
Pretreatment, 2 h at dyeing temperature:		
Undrawn	46.0 and 144.3	11.0 and 34.5
1.0 ×	83.7 and 171.5	20.0 and 41.0
2.0 ×	91.2	21.8
20.0 ×	70.7	16.9
Pretreatment, 2 h at boil:		
Undrawn	114.2	27.3
1.0 ×	113.0	27.0
10.0 ×	122.2	29.2

but the relation between draw ratio and the values of the diffusion coefficient at zero concentration indicates a more complex relationship (Figure 3.19). At 102 °C the boiled material has a lower rate of dyeing, whereas at 60 °C the position is reversed. It would seem that treatment with boiling water produces non-crystalline regions having a more open structure than that of the untreated fibre; hence dyeing is more rapid with the unboiled fibre [1, 65].

3.5 INFLUENCES ON FIBRE PROPERTIES DURING DYEING

The most simplistic view of a dyeing process is that of dyes travelling through water-filled capillary channels. Explanations included in earlier pages alluded

TABLE 3.9

Densities of nylon 6.6 fibres and activation energies for diffusion of Orange II (C.I. Acid Orange 7) in the fibres

Draw ratio	Diameter /cm × 10^4	Density /g cm^{-3} Untreated	Boiled	Activation energy Untreated /kJ mol^{-1}	Untreated /kcal mol^{-1}	Boiled /kJ mol^{-1}	Boiled /kcal mol^{-1}
1.00	60.7	1.141	1.146	90.4 ± 11.3	21.6 ± 2.7	74.9 ± 3.4	17.9 ± 0.8
2.02	40.3	1.142	1.151	95.0 ± 8.4	22.7 ± 2.0	68.2 ± 20.5	16.3 ± 4.9
2.99	26.0	1.142	1.152	109.2 ± 11.3	26.1 ± 2.7	75.3 ± 14.6	18.0 ± 3.5
3.99	31.7	1.147	1.154	109.2 ± 9.2	26.1 ± 2.2	77.0 ± 10.8	18.4 ± 2.6
5.02	28.9	1.148	1.153	119.7 ± 11.7	28.6 ± 2.8	85.8 ± 17.6	20.5 ± 4.2
5.99	26.4	1.149	1.157	119.7 ± 32.0	28.6 ± 7.6	81.2 ± 14.2	19.4 ± 3.4

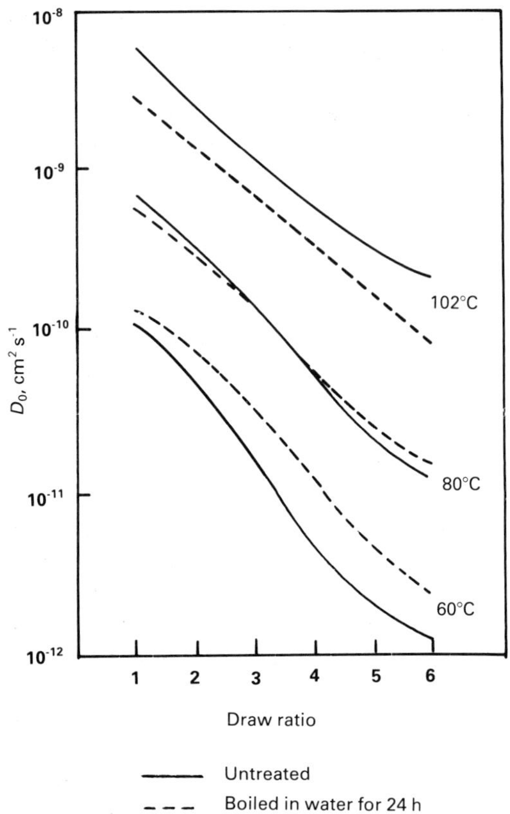

Figure 3.19 – Variation of diffusion coefficient of Orange II at zero concentration with draw ratio of nylon 6.6

to the different pore dimensions in different cellulosic fibres and changes in the relative size and volume of the void during the heat-setting of man-made fibres.

One particular limitation of using the pore model as the only mechanism contributing to dyeing is that dyes are required to follow the path of the water contained in the pores. For most hydrophobic fibres this model is not the best to use, because these fibres take up only small quantities of water and they are usually dyed with dyes of very limited aqueous solubility which form solid solutions in the fibre. For these dyeing systems theoretical considerations based on the concept of dye weaving its way between polymer chains provide an alternative. Any permanent or temporary structural changes that take place during dyeing may then be regarded as changes producing a situation in which the fibre dyeing properties are affected.

The importance of fibre structure in governing dyeing behaviour has been pointed out by Majury [79], who studied the diffusion of three compounds resembling disperse dyes in secondary cellulose acetate and cellulose triacetate respectively. Using Eqn 3.1 to represent the variation of measured diffusion coefficient with temperature, it is possible to obtain a value for the activation energy of the diffusion process:

$$D_m = D_m^0 \exp(-\Delta E_m/RT) \qquad (3.1)$$

where D_m = measured diffusion coefficient, D_m^0 = constant, ΔE_m = activation energy and R and T have their usual significance.

Majury regarded the experimentally determined activation energy ΔE_m as being divisible into two terms, i.e.

$$\Delta E_m = \Delta H_i - \Delta H_a \qquad (3.2)$$

where ΔH_i = activation enthalpy for diffusion in pores and ΔH_a = enthalpy of adsorption. ΔH_a was determined from equilibrium measurements, ΔE_m from a plot of log D_m versus $1/T$, and ΔH_i was obtained using Eqn 3.2. The results given in Table 3.10 confirm that the enthalpy of adsorption is identical for both fibres, but values of ΔH_i, representing the activation enthalpy for movement of dye in the fibre pores, are noticeably higher for cellulose acetate. This shows that the fibre with the more compact and highly organised structure offers a greater resistance to the movement of dye.

The physical properties of all fibres are related to the crystalline order and the various degrees of molecular order in the non-crystalline regions. Diffusion of dye cannot take place in the rigid structure of the crystalline regions, however, and for a clearer understanding of the mechanism of the

TABLE 3.10

Comparison of activation energies of diffusion in secondary cellulose acetate and cellulose triacetate /kJ mol^{-1} (values in parentheses are in kcal mol^{-1})

	Secondary			Triacetate		
Compound	ΔE_m	$-\Delta H_a$	ΔH_i	ΔE_m	$-\Delta H_a$	ΔH_i
4-Nitroaniline	63 (15)	29 (7)	33 (8)	109 (26)	29 (7)	80 (19)
NN-Dimethyl-4-nitroaniline	54 (13)	33 (8)	21 (5)	84 (20)	33 (8)	50 (12)
Aminoazobenzene	75 (18)	38 (9)	38 (9)	130 (31)	38 (9)	92 (22)

dyeing process it is necessary to consider what molecular rearrangements of the polymer chains govern the dyeing process. Since the molecular rearrangements will also affect the physical properties of the fibres, the question arises as to whether or not there is a connection between the variation with temperature of the physical properties of the fibre and its dyeing properties.

3.5.1 The connection between the physical properties of the fibre and its dyeing behaviour

The effect of temperature on the physical properties of man-made fibres has received a great deal of attention, but has generally been studied under dry conditions. The dyer is obviously interested in the behaviour of the fibre under the hot, wet conditions of dyeing, but experimentation under these conditions is more difficult. Nevertheless, sufficient progress has been made to show how variations with temperature of physical properties affect the dyeing properties of a fibre.

One of the earlier studies carried out by Meldrum and Ward [80] originated from the difficulties encountered when dyeing acrylic fibres. Under hot, wet conditions these fibres become dimensionally unstable and consequently, unless due care is taken, they become deformed during processing. The experiments demonstrated that the ease of compression changes markedly

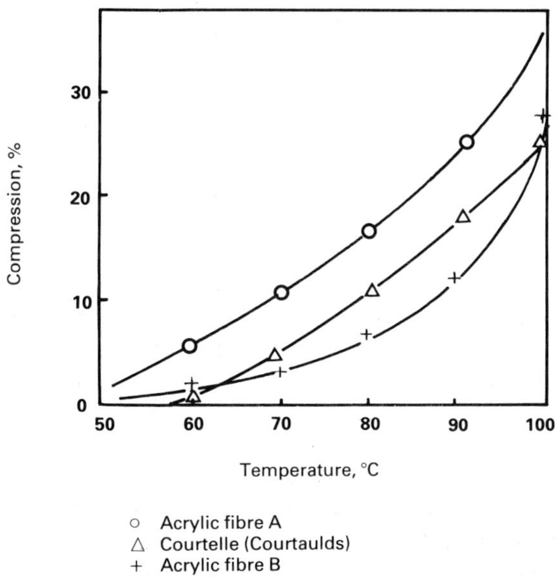

Figure 3.20 – Effect of temperature on the compression of a pack (model system) (applied load = 1.30 × 10⁴ Pa (1.88 lb in⁻²); all fibres 3 denier) [80]

over the temperature range 50–100 °C (Figure 3.20). Compressive forces in the dyebath lead to consolidation of the fibre mass and penetration difficulties in subsequent dyeing. Over the same temperature range, the response to bending and tensile forces becomes more pronounced and since the mechanical properties return to normal on cooling, the effects of such stress appear as permanent deformations.

It is not immediately obvious why this thermoplastic behaviour is related to the marked temperature sensitivity of the dyeing properties of acrylic fibres. To understand this relationship, it is necessary to consider the cause of the thermoplastic behaviour and this entails the use of the 'free volume' concept.

3.6 THE RELATIONSHIP BETWEEN TEMPERATURE AND THE PHYSICAL PROPERTIES OF MAN-MADE FIBRES

An alternative to the pore theory for representing the unoccupied space within a man-made fibre is based upon the concept of *free volume*. The free volume is defined as that volume associated with the substance which is not occupied by the constituent atoms, thus

$$V_f = V - V_0 \tag{3.3}$$

where V = specific volume of a liquid (volume g^{-1}) at temperature T, V_0 = specific volume of the liquid extrapolated to 0 K, and V_f = free volume at temperature T.

By treating amorphous polymers as viscous liquids, it has been possible to use the concept of free volume to describe the variation of physical properties with temperature [81] and, as will be seen below, to derive a correlation between the physical properties of fibres and their dyeing properties.

The glass transition temperature

If the temperature of any amorphous polymer is sufficiently low, each of the constituent polymer chains is prevented from changing its position by the presence of its neighbours; consequently, the only motions that are possible are vibrational motions of constituent atoms. In this 'frozen' state the polymers resist deformation and tend to be brittle and, generally, their properties are described as being those of a glassy polymer.

As the temperature rises, the amplitudes of the vibrations increase and sufficient energy is introduced into the system to cause rotation about some of the covalent bonds in the backbone of the polymer chain. At this stage it becomes possible for the dispersed and limited regions of free volume to coalesce and form larger spaces, thus allowing a change of position of polymer segments which happen to lie between two simultaneously rotating bonds in the same polymer chain. Movement of the polymer segment continues until it is sterically hindered by other polymer molecules. The change of position is referred to as a *segmental jump*, and once this is accomplished another polymer segment is able to move into the vacated space and so on throughout the whole polymer mass. The energy barrier to this kind of cooperative segmental mobility is overcome over a narrow temperature range, and the result is a marked increase in free volume and an equally pronounced change in the physical properties from those of a glassy state to those of a rubbery state. The polymer is then more susceptible to deformation, because its molecules can respond to an applied stress by changing to a new equilibrium position. The sudden change in properties may therefore be detected experimentally, and the temperature at which the changes take place is referred to as the *glass transition temperature* (T_g). Further increases in temperature above T_g lead to an exponential increase in free volume and a corresponding increase in the frequency of the segmental jumps. The increase in free volume above T_g may therefore be regarded as the increase in volume over and above that expected from the thermal expansion. Thus

$$V_f = V_f' + \alpha V_l(T - T_g) \tag{3.4}$$

where V_f = free volume at temperature T, V_f' = free volume at temperature T_g, V_l = specific volume of the polymer. $\alpha = \alpha_L - \alpha_g$ where α_L = coefficient of thermal expansion of the liquid (the polymer is regarded as a viscous liquid above T_g) and α_g = coefficient of thermal expansion of the polymer in its glassy state.

3.6.1 The glass transition temperature and dyeing properties

There is now convincing evidence available to support the theory that the segmental mobility in the polymer chains has a controlling influence on the rate of diffusion of dyes into fibres, since the onset of dyeing and the glass transition temperature of the fibre appear to coincide. With polyester and acrylic fibres, diffusion into the substrate does not take place at any appreciable rate until the dyebath reaches 70–80 °C. Then the increase in the rate of diffusion becomes so marked that workers have referred to a dyeing transition temperature, which generally is not far removed from the T_g as detected by physical means [82, 83]. Further indications of this behaviour are shown in Figure 3.21 [68], where the exhaustion of basic dye on acrylic fibres, basic-dyeable nylon and polyester are compared. The acrylic and basic-dyeable nylon fibres have a lower T_g than has the basic-dyeable polyester, and the rapid exhaustion begins at a correspondingly lower temperature. Sprague [84] observed a transition in the dyeing behaviour of cellulose triacetate, which he was unable to link with any detectable change

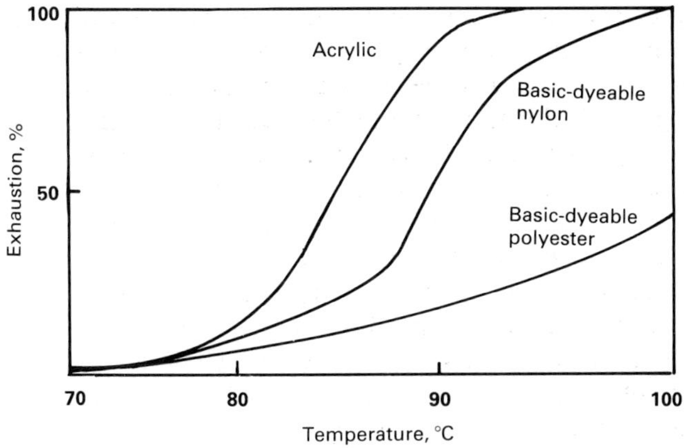

Figure 3.21 – Effect of temperature on the exhaustion of a basic dye on different fibres

in the physical properties of the substrate; in this case the transition occurred at a higher temperature for dyes of larger molecular size. Similar observations for dyes of different molecular sizes have been observed for non-ionic dyes on nylon fibres [85], but in this case the transition in dyeing behaviour was less dramatic and occurred above the glass transition temperatures of the fibres in water, which for nylon 6 and 6.6 fall in the region 0 °C to −10 °C, approximately 60 °C lower than in the dry state. Figure 3.22 serves to illustrate how increases in crystallinity increase the T_g of poly(ethylene terephthalate) [86] and by implication slower diffusion of dye [60]. If the

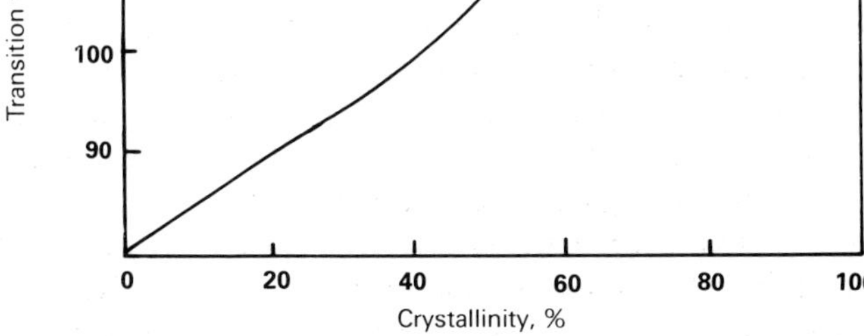

Figure 3.22 – Plot of transition temperature against crystallinity for poly(ethylene terephthalate)

molecular regularity of the system is disturbed (for example, by introducing ionic groups into the structure by copolymerisation) then the glass transition temperature decreases and the dyeing properties are improved [10], although the permeability of basic dyes is still lower than on unmodified polyacrylonitrile (Figure 3.21). It is apparent therefore that any structural changes that could occur above T_g during dyeing are also likely to influence dyeing behaviour.

A more definite connection between the glass transition temperature and the diffusion of basic dyes into acrylic fibres has been reported by Rosenbaum [82]. The theoretical principles derived are based on the concept of free volume and are the same principles that are used to explain the variations of the viscosity of polymer solutions with temperature. These theoretical considerations were originally intended for use with completely amorphous polymers in which the only forces of attraction between the polymer chains are van der Waals forces. Since man-made fibres deviate markedly from this ideal state of affairs, it is perhaps surprising that the relationship can be used to represent dyeing behaviour. Nevertheless, the correlation is good and parallels the relationship obtained by Fujita *et al.* for the diffusion of small molecules through polymers at different temperatures [89].

The Williams, Landel and Ferry (WLF) equation

The semi-empirical equation proposed by Williams, Landel and Ferry [90] for describing the variation with temperature of those polymer properties dependent upon the frequency of the segmental jumps in the polymer is

$$\log \frac{\eta_T}{\eta_{T(\mathrm{ref})}} = \log a_T = -\frac{A(T - T_{\mathrm{ref}})}{B + (T - T_{\mathrm{ref}})} \qquad (3.5)$$

where η_T and $\eta_{T(\mathrm{ref})}$ represent respectively viscosities (or other suitable properties) at a temperature T and at T_{ref}, a reference temperature often taken for convenience as T_g. A and B are empirical constants: $\log a_T$ is referred to as the *shift factor*. Clearly, once the value of the property at T_{ref} is determined it becomes possible to calculate a value of $\log a_T$ for any other particular temperature T and hence to determine the value of the property at this temperature. The use of Eqn 3.5 is therefore equivalent to monitoring the change in the frequency of the segmental jumps with temperature, through the variation in appropriate physical properties.

Rosenbaum [82] has connected such changes in the physical properties of acrylic fibres with their dyeing properties, using the WLF relationship, and consequently it has been concluded that the diffusion of basic dyes through the fibre structure is governed by the segmental mobility of the polymer chains: the more easily they can move and leave a space for a dye molecule to enter, the faster the dye molecules can move through the structure. Indications that the dyeing process is regulated in this way are obtained when attempts are made to determine the activation enthalpy of diffusion, using the Arrhenius relationship

$$\log D = \log D_0 - \frac{\Delta H}{RT} \tag{3.6}$$

Eqn 3.6 implies that a plot of log D versus $1/T$ will show a straight line relationship; in fact, however, when the experimental determinations cover a sufficiently wide temperature range, it is apparent that this is not so. Rosenbaum's results are shown in Figure 3.23, the graph in which exhibits an obvious curvature. If tangents are drawn at different parts of the curve in an attempt to measure the activation enthalpy at each point it is found that the activation enthalpy increases exponentially as the temperature is reduced. Below the break in the curve which coincides with T_g, as determined by

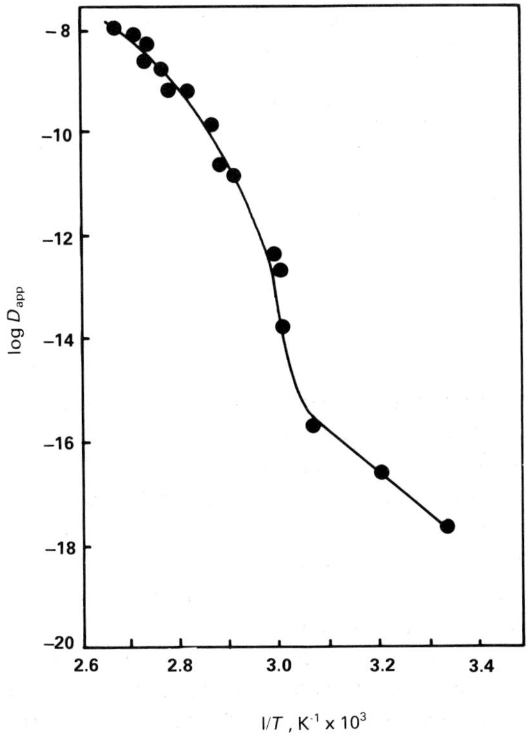

Figure 3.23 – Arrhenius plot of apparent diffusion coefficient D_{app} of Malachite Green in experimental polyacrylonitrile fibre [82]

physical measurement, the rate of diffusion becomes much less and the activation enthalpy is constant. The dependence of this diffusion behaviour on segmental mobility is confirmed by adherence of the results to a WLF relationship expressed in the form

$$\log \frac{D_T}{D_g} = -\log a_T = \frac{A(T - T_g)}{B + (T - T_g)} \tag{3.7}$$

The coincidence of experimental points with the relationship calculated from Eqn 3.7 is shown in Figure 3.24. It may therefore be concluded that diffusion is governed by the activation energy of the molecular processes taking place in the polymer structure and, in the absence of other complicating factors, this should be the same for all dyes studied. Further correlations, based

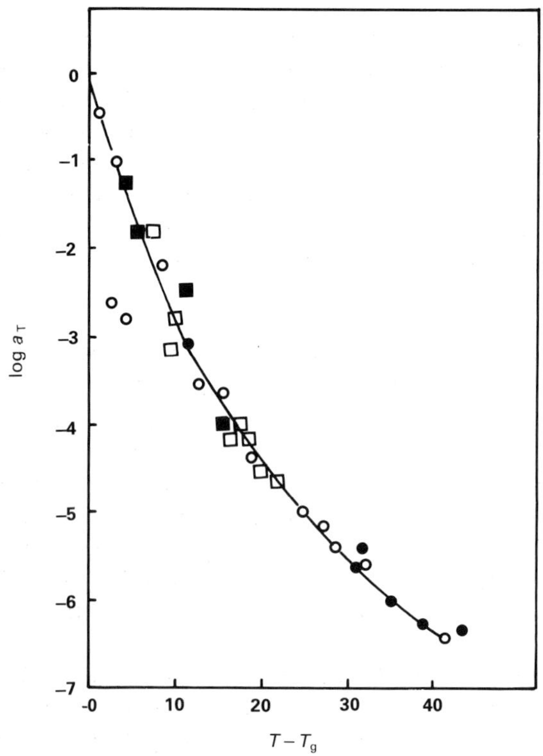

Figure 3.24 – Effect of temperature on the compression of a pack of acrylic fibres (model system) (applied load = 1.30×10^4 Pa (1.88 lb in^{-2}); all fibres 3 denier) [80] for the diffusion of Malachite Green into various polyacrylonitrile fibres, to the curve for log a_T versus ($T - T_g$) calculated using Eqn 3.7

upon the concept of free volume (and hence segmental mobility), have been derived for disperse dyes on acrylic fibres [27, 87] and on polyester substrates [60, 83] and for acid dyes on nylon 6.6 [88]. The correlations should not be taken as an indication that segmental mobility alone governs the dyeing processes, however, since other changes in molecular order which could influence dyeing, such as changes in crystallinity, may also be related to polymer chain mobility. The physical properties of semicrystalline polymers such as nylon 6.6 may also be discussed in these terms, even though they do not conform to the WLF relationship [91]. Investigation concerning the importance of molecular organisations in relation to the dyeing properties of nylon fibres have demonstrated the effects of various changes caused by dry and wet heat-settings [92, 93].

Under the conditions of dyeing used by Schreiner and Kemter [94], the dyeing of polyamide fibres with anionic dyes behaves according to the free volume model, whilst Sada *et al.* consider that both surface and pore diffusion take place [95]. Hori *et al.* [96] have discussed the combined influence of T_g, fibre swelling and other structural features on the diffusion of dyes in poly(ethylene terephthalate), whilst Rhoner and Zollinger [97] and Meyer [98] studied the kinetics of dye sorption and desorption with carefully selected acrylic fibres and conclude that both the pore and the free volume mechanisms operate simultaneously, the relative contribution from each depending upon the structure of the fibres used. Clearly, a fully comprehensive theoretical treatment explaining the complexities of practical dyeing is still awaiting development.

3.7 THE SOLUBILITY PARAMETER CONCEPT AND DYEING

It has been demonstrated in the preceding section how the free volume concept provides a theory of dyeing for man-made fibres in which ease of dyeing is related to the frequency with which segments of the constituent polymer chains in the fibre change their position and leave a space into which the dye molecule may pass. This movement of polymer segments will also reflect the ease with which the polymer chains may be separated. Consequently, it is not unreasonable to examine the possibility that ease of dyeing may also be related to the tendency of a dye-liquor to dissolve the fibre, because dissolution also requires separation of polymer molecules. Obviously, fibres do not disintegrate or dissolve under the conditions of commercial practice. Furthermore, they will not readily go into solution under other conditions, since the individual polymer molecules are held together by, for example, polar attractions and hydrogen bonds giving regions of highly organised molecular structure (crystalline regions), as well as by intermolecular attractions arising from dispersion forces. It is more usual to think in terms of the ability of the dyebath medium to penetrate and swell a fibre. Therefore, in order to assess the sensitivity of a polymer to swelling in different solvents it is necessary to have some idea of the strength of the interactions between the polymer molecules being separated.

Most dyeing is carried out in water with the aid of dyebath additives, and interaction between water and fibrous polymer is an intrinsic part of the dyeing reaction. Such interactions are involved in all textile wet-processing but discussion here is related to specific effects relevant to dyeing and the reader is directed to more extensive treatises for a fuller insight [99].

A property often used as a measure of the strength of such interaction is the cohesive energy density. All molecules in a liquid possess a potential energy E which is equal in magnitude (but opposite in sign) to the energy required to separate the molecules by an infinite distance. It originates in the attractions between molecules and, for the liquid state, it is overcome by evaporation, because during evaporation the potential energy of individual molecules is reduced to zero. The cohesive energy is expressed as E/V, where V is the molar volume, and is numerically equal to the potential energy per cm^3 of material.

The energy required to separate molecules in this way may be determined experimentally from the enthalpy (heat) of vaporisation of the liquid ΔH_v, which compensates for both the potential energy change and the work done in expansion. For one mole of a vapour obeying the ideal gas laws, the latter is equal to RT; therefore

$$\text{cohesive energy density} = \frac{\Delta H - RT}{V} \quad (3.8)$$

It is convenient for experimental purposes to represent the cohesive energy density (c.e.d.) in terms of the solubility parameter δ, where $\delta^2 = \text{c.e.d.}$ Thus the cohesive energy density can be measured for all compounds which can be vaporised. Obviously, δ cannot be measured in this way for polymers, but indirect methods of assessment are possible.

The importance of this quantity lies in its usefulness in predicting whether or not two substances will mix. When two substances are mixed to form a solution, the greater the potential energy difference between the two species, the greater will be the potential energy change on mixing. This change is equal to the enthalpy of mixing. If the potential energies of the two types of molecule in their new surroundings are similar or identical to their potential energies in the pure state, then the enthalpy of mixing will be small and mixing will take place easily. However, if the potential energies are very different in the mixture (because the attractions between like molecules are much stronger than those between unlike molecules) then mixing, if it occurs at all, will take place only with difficulty.

In thermodynamic terms, two components will mix only if the free energy decreases as a result, and the following equation must be satisfied:

$$\Delta G_m = \Delta H_m - T\Delta S_m \quad (3.9)$$

where ΔG_m = free energy change on mixing, ΔH_m = enthalpy change on mixing (heat of mixing), and ΔS_m = entropy change on mixing.

The entropy term is a measure of the disorderliness and the freedom of motion of the molecules. Mixing tends to produce conditions in which there is greater randomness of arrangement and molecular motion. For example, when fibrous molecules are dissolved in a solvent they have much greater freedom of movement than when in the fibre, where their positions are fixed by the orientated surrounding molecules. The entropy change that occurs when a polymer is mixed with a solvent is therefore always positive, and the free energy term in Eqn 3.9 remains negative. Increases of temperature increase the entropy term, so ΔG_m is reduced and mixing is facilitated. Further, ΔH_m must be negative or zero or, if positive in value, it must be numerically smaller than $T\Delta S_m$, otherwise ΔG_m will not decrease and mixing will not take place. In real systems ΔH_m is usually positive, however, so the best mixing conditions will occur when ΔH_m reaches a minimum value.

Hildebrand [100] and Scatchard [101] have attempted to express this enthalpy of mixing in terms of the solubility parameters of the two components, by means of the following equation:

$$\Delta H_m = V_A(\delta_A - \delta_B)^2 V_B{}^2 \quad (3.10)$$

where V_A and V_B are the molar fractions of the two components. Thus ΔH_m will be at a minimum when the solubility parameters of the two components are equal.

Although Eqn 3.10 was first intended for predicting the ease of mixing of liquids, its use has since been extended to mixtures of amorphous or semicrystalline polymers with liquids or dyebath additives [102]. When using the criterion of equality of solubility parameters, however, it is important to bear in mind the simplifying assumptions that were made in the original theoretical treatment and the effects of the deviations from these assumptions that occur in real systems. Failure to do this can lead to misleading deductions. These deviations have received more detailed treatment elsewhere [103, 104] and only an outline is given here to aid a better understanding of the possible effect of dyebath additives on the swelling and plasticisation of fibres.

The first assumption is that the axes of interaction of the molecules involved connect the centres of the molecules. However, it is not easy to visualise how this could be applied easily to a polymer–solvent system. It is also assumed that interaction between pairs of molecules is not influenced by the surrounding molecules and, furthermore, that there is no volume change on mixing. Finally, it is assumed that the mixing process is completely random with no specific interactions between the different molecular species other than van der Waals forces. For a dyeing system this would imply a completely amorphous fibre being dyed in a solvent which formed no hydrogen or polar bonds with the fibre molecules – a situation that is far from being realistic. It is possible to extend the predictions to real systems with some success, however, provided that the types of deviation to be expected are known.

The discrepancies are hidden in the values of the solubility parameter obtained experimentally because, although the measured enthalpies of vaporisation can indicate the intensities of the interactions between the molecules, they give no information as to the nature of the interactions. Furthermore, values for the solubility parameter of the polymers are not obtainable directly, but are deduced from swelling measurements made with the polymer immersed in different solvents of known solubility parameter. A broad maximum is obtained when the swelling of cross-linked polymers is plotted against the solubility parameter of the solvent. It is assumed that this maximum coincides with the solubility parameter of the polymer. Difficulties arise in interpretation, however, since solvents that are more or less polar than the polymer give different curves, and differences in the likelihood of hydrogen bonding in the solvent or between the polymer and the solvent also lead to different results. The existence of crystallinity in the polymer is a further complicating factor. The size and distribution of the crystalline regions will affect the swelling power of the solvent, and penetrations will also be affected by the molecular size of the penetrant. It is suggested that the effect of crystallinity can be represented by introducing a further enthalpy of fusion term (ΔH_f) into Eqn 3.9 giving, after substituting for ΔH_m from Eqn 3.10,

$$\Delta G = V_A(\delta_A - \delta_B)^2 V_B{}^2 - T\Delta S + \Delta H_f \qquad (3.11)$$

Thus, since the thermal history of the polymer affects the crystallinity, it will also influence the ease with which the fibre will dissolve (through the ΔH_f term and the change in ΔS that will occur on dissolution of the crystallites), and this will affect the solubility parameter value obtained. It is not therefore an easy matter to obtain the solubility parameter of a fibre by experimental methods. An alternative method due to Small [105] entails no experimentation, but uses molar attraction constants for specific groups, which are deduced from known solubility parameters of liquids. The values for the different groups in a polymer repeat unit are taken to be additive, and the solubility parameter is calculated to give values in close agreement with experimentally determined values. This method has been applied only with hydrocarbon, ester-containing and halogen-containing polymers, however, and cannot be used with confidence when polar groups or hydrogen-bonding groups are present.

It is clear, therefore, that measured values of the solubility parameter for either polymer or liquid must take into account all the different possible interactions, and it is necessary to consider what different interactions are involved in the cohesive energy density determinations. Unless this is done the coincidence of solubility parameters for solvent and polymer may not be sufficient grounds for expecting mixing to take place.

Contributions to the cohesive energy density, other than from dispersion forces, can be expected from the interaction of permanent dipoles in the molecules. Furthermore, the permanent dipoles may induce dipoles in other polarisable groups present in the system. If hydrogen bonding occurs, there will be a further contribution to the cohesive energy from this source. The total solubility parameter δ_t may therefore be given by

$$\delta_t^2 = \delta_d^2 + (\delta_h^2 + \delta_p^2) \tag{3.12}$$

where δ_d^2 = contribution from dispersion forces, δ_h^2 = contribution from hydrogen bonding and δ_p^2 = contribution from the polar nature of the molecules present. This is written more simply as

$$\delta_t^2 = \delta_a^2 + \delta_d^2 \tag{3.13}$$

where δ_a^2 represents all the polar forces which cause molecules to associate, and is called the *association solubility parameter* [102].

The conditions for mixing therefore depend not merely on the match of the solubility parameters of the two components, but on a match of their polar and hydrogen-bonding properties. Consequently, non-polar polymers will be swollen less by polar solvents than by non-polar solvents having equal values for solubility parameter and volume fraction. Table 3.11, which is compiled from results published elsewhere [104] for polyacrylonitrile, lists (a) compounds having solubility parameters similar to that of the polymer but different polarities, which do not act as solvents, and (b) compounds that are similar in both polarity and solubility parameter, which do act as solvents. Obviously there will be intermediate stages between the two extremes, and consequently precise prediction of the effect of additives will be difficult.

Several attempts have been made to predict the effects of hydrogen bonding in mixtures, but the magnitude of the contribution from hydrogen bonding

TABLE 3.11

The effect on polyacrylonitrile of compounds having a similar solubility parameter to that of the polymer [104]

Compound	Solubility parameter $J^{0.5}\ cm^{-1.5}$	$cal^{0.5}\ cm^{-1.5}$	Comments
Phenol	60.7	14.5	The polarity of these compounds
Ethanediol	65.7	15.7	is too low
Glycerol	69.0	16.5	to permit
Methanol	60.7	14.5	them to act as
Ethanol	53.1	12.7	solvents
Dimethylformamide	50.6	12.1	The polarity of these
Dimethylacetamide	46.8	11.1	compounds is high
Dimethylsulphoxide	56.1	13.4	and they are solvents
Dimethylsulphone	61.1	14.6	for polyacrylonitrile

depends upon the surrounding medium. Thus, when two proton donors or two proton acceptors are mixed, or when proton donors are mixed with a second component which is incapable of forming hydrogen bonds, it is safe to assume that hydrogen bonding will not interfere when applying the simple theory outlined above.

3.8 THE INTERACTION BETWEEN DYES AND FIBROUS POLYMERS

The reasoning above may be applied to both solvent–polymer and dye–polymer interactions, since in both instances the forces governing the interaction control the end result, and the concept has been considered in relation to dyes and auxiliaries for the disperse-dyeing of poly(ethylene terephthalate) [106]. The fastness to sublimation of disperse dyes on polyester fibres has been considered in terms of molar attraction constants [107], but this cannot be extrapolated directly to aqueous dyeing since sublimation relates to the vapour phase. Ibe [108], however, has produced a correlation between the solubility of disperse dyes in cellulose acetate and δ_a of the dye, which is an improvement on the poor correlation obtained with the corresponding values for the total solubility parameter.

A further correlation was obtained by Ingamells and Thomas [109] between the maximum equilibrium dye uptake (C_∞) of a range of selected azo disperse dyes (Table 3.12) and the calculated δ_a values where all the dyes were applied at the same molar concentration from an infinite bath (Figure 3.25). Although the δ_t values of the dyes fall within a narrow range (Table 3.13), the δ_a values cover a wider spread and show that the maximum equilibrium dye uptake

TABLE 3.12

Disperse dyes used in Figure 3.25 [109]

Dye	Structure	(a)	(b)	(c)	(d)
D	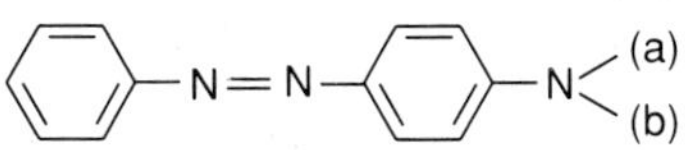	CH_3	CH_3	—	—
E		C_2H_5	C_2H_5	—	—
A	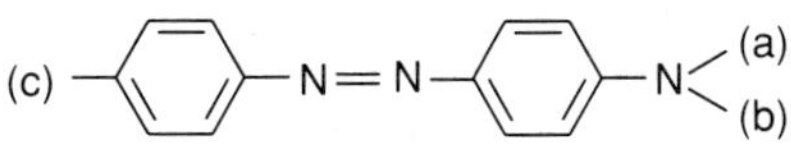	CH_3	CH_3	OCH_3	—
B		C_2H_5	C_2H_5	CH_3	—
C		H	C_2H_5	CH_3	—
F		CH_3	CH_3	Cl	—
G		C_2H_5	C_2H_5	Cl	—
H		C_2H_5	C_2H_5	NO_2	—
L		H	CH_3	NO_2	—
I	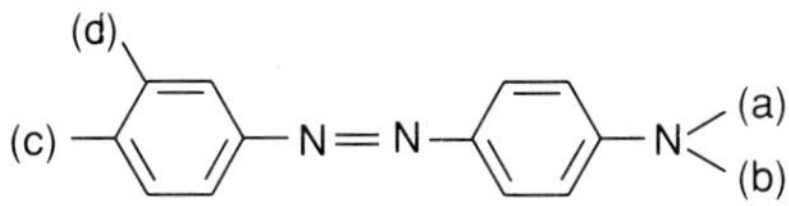	C_2H_5	C_2H_5	NO_2	CH_3
J		C_2H_5	C_2H_5	NO_2	Cl
M		C_2H_5	C_2H_5	NO_2	NO_2

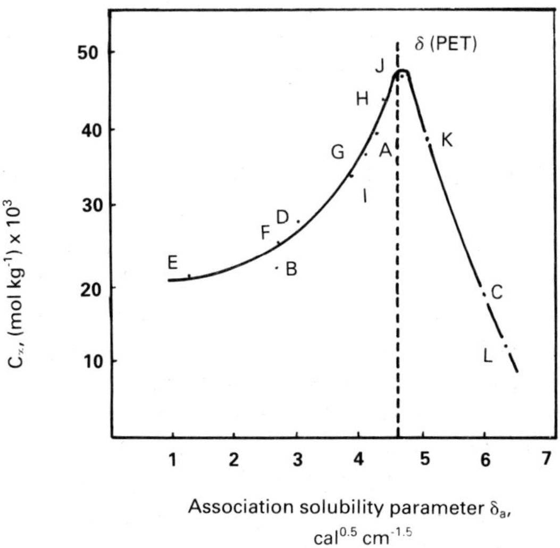

Figure 3.25 – Variation of C_∞ with δ_a [109]

occurs with the dye having a δ_a value of 4.8, close to values published for poly(ethylene terephthalate). No correlation is possible when using δ_d. The interpretation given is that the maximum affinity of the dye occurs when the relative contribution of the dispersion and association forces to the cohesive energy density of the dye matches those of the polymer. Supporting evidence

TABLE 3.13

Data for dyes in Table 3.12 [109]

Dye	Melting pt/°C	Relative molecular mass	Density /g cm^{-3}	Solubility parameter δ_t	δ_d	δ_a
D	115	225.24	1.2033	10.34	9.86	3.08
E	98	253.26	1.1361	9.85	9.77	1.30
A	132	255.25	1.2417	10.79	9.91	4.27
B	113	267.27	1.1370	10.07	9.69	2.71
C	117	239.25	1.2883	11.88	10.19	6.09
F	102	259.69	1.2126	9.88	9.43	2.92
G	112	287.71	1.3079	10.81	10.01	4.06
H	151	298.25	1.2892	10.75	9.77	4.47
L	133	256.22	1.3692	11.72	9.87	6.33
I	147	312.26	1.2168	10.23	9.45	3.92
J	143	332.70	1.3618	10.80	9.68	4.77
K	162	343.24	1.3524	10.81	9.50	5.16

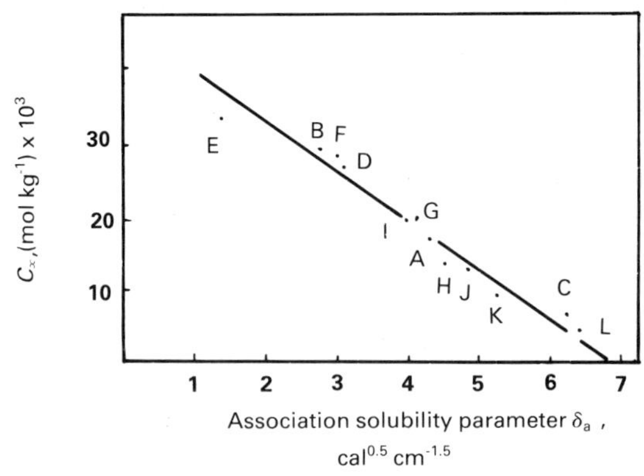

Figure 3.26 – Variation of C_∞ with δ_a for disperse dyes on polypropylene [109] (for key see Table 3.12)

is presented for a similar experiment on polypropylene (Figure 3.26), where there is no possibility of polar interaction between dye and fibre and accordingly the affinity of the dyes decreases as its polar character increases.

3.8.1 Swelling of fibre and plasticisation

One belief that many dyers hold seems more appropriate for hydrophilic than for hydrophobic fibres, viz. that dyeing is preceded by swelling of the fibre in the dyebath medium, and that if swelling is not achieved dyeing takes place only with difficulty. There is a tendency to assert also that plasticisation of fibres (e.g. by carriers) goes hand in hand with swelling. In this case a swelling action could be anticipated from a reduction in the T_g of the fibre caused by an additive. The danger of making such assumptions, however, is illustrated in Table 3.14, which gives the effects of aqueous

TABLE 3.14

The effect of various compounds on the physical properties of Acrilan (Monsanto) [110]

Surrounding medium	Solubility parameter of the compounds /$J^{0.5}$ $cm^{-1.5}$	/$cal^{0.5}$ $cm^{-1.5}$	Reduction of T_g relative to T_g in water as determined by physical properties /°C	% Increase in diameter of the filament relative to the diameter in water
Air			T_g in air = 90 °C	
Water	96.2	23.0	T_g in water = 57 °C	
Water containing 0.1 mol l^{-1} of the following:				
o-Toluidine	46.4	11.1	37	10.3
2-Ethylaniline	45.2	10.8	22	8.2
2-Ethylphenol	57.8	13.8	18.5	9.9
m-Cresol	59.4	14.2	13.5	7.0
Aniline	45.2	10.8	12.5	2.6
Phenol	60.7	14.5	10.0	
Benzyl alcohol	51.9	12.4	6.0	0.5
Amylamine	35.6	8.5	5.5	81.4
Amyl alcohol	45.6	10.9	3.0	10.0
Benzylamine	43.1	10.3	2.7	5.1
1,5-Diaminopentane	39.8	9.5	−1.5	75.4
Glycerol	69.0	16.5	−2.0	12.1
1,5-Pentanediol	61.5	14.7	−3.0	9.9
Diethylene glycol	61.1	14.6	−4.0	9.7
Ethanediol	65.7	15.7	−5.0	6.8
Ethylenediamine	47.7	11.4	−4.0	13.2

solutions of various additives on the T_g of Acrilan fibres (Monsanto), together with their effect on the radius of the fibre [110].

The plasticising effect of water on Acrilan filaments is shown by the fall in T_g to 57 °C, a reduction of 33 °C when compared with the value obtained in air. The solubility parameters of most of the listed additives fall within the same region as the values quoted in the literature for polyacrylonitrile

(53 and 65 $J^{0.5}$ (12.7 and 15.4 $cal^{0.5}$) $cm^{-1.5}$), but there is no clear correlation to be found between the change in diameter of the fibre and the reduction in T_g. Those that cause some reduction of T_g tend to cause some swelling, but the magnitude of the effect is not proportional to the degree of swelling as shown by the change in diameter. In dilute solutions of *o*-toluidine and amyl alcohol, for example, filaments show a diameter swelling of about 10.0 per cent, whilst the corresponding reductions in T_g relative to the T_g in water are both 37 °C. On the other hand, the reduction of T_g caused by dilute solutions of benzyl alcohol (2.7 °C) and amylamine (5.5 °C) bear no relation to the change in diameter of 0.5 per cent and 81.0 per cent respectively. Moreover, some of the additives (from 1,5-diaminopentane downwards) increased the swelling, but caused a slight increase in the T_g. It is therefore clear that the swelling of a fibre does not necessarily parallel its plasticisation.

The extent to which any reagent will react with and swell a fibre will depend upon the strength and nature of the interactions between the fibrous polymer and the reagent and on the amount of reagent occluded. Thus by examining the effects of equimolar concentrations of a variety of agents in acrylic fibres, it has been demonstrated that the polar component of

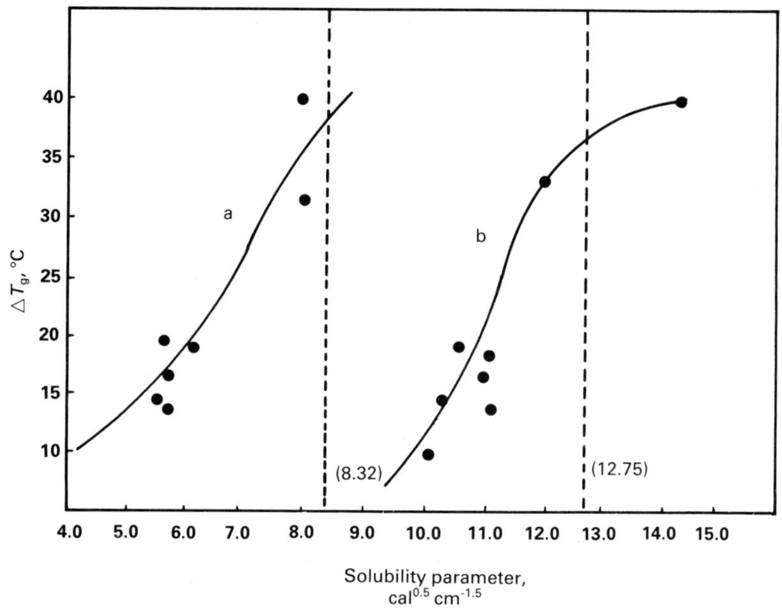

Figure 3.27 – Reduction in T_g (ΔT_g) caused by equimolecular concentration of plasticiser in Acrilan filaments related to (a) total solubility parameter δ_t, (b) association solubility parameter δ_a; the broken lines represent the total solubility parameter (53 $J^{0.5}$ (12.7 $cal^{0.5}$) $cm^{-1.5}$) and the association solubility parameter (35 $J^{0.5}$ (8.32 $cal^{0.5}$) $cm^{-1.5}$) for polyacrylonitrile

the polymer–solvent interactions governs the reduction in glass transition temperature [111] (Figure 3.27). As the total solubility parameter of the solvent approaches that of the fibre the effect is maximised. The more significant fact is hidden in this relationship, however, because the values for the dispersion solubility parameters of the solvents used are all very close and the true correlation is found by using Eqn 3.13 to obtain the association solubility parameter. When this is done the correlation is retained between ΔT_g and δ_a but not between ΔT_g and δ_d. Variations in fibre diameter are also encountered and in general tend to be smaller as the difference between the solubility parameters of solvent and fibre diminishes (Figure 3.28). In the case of poly(ethylene terephthalate), dispersion forces have been considered to be more important [112]. Unfortunately the number of compounds that can be used to provide experimental support in an aqueous system is limited by the narrow spread of the values of δ_d for those compounds examined which have adequate solubility (Table 3.15) and deductions are better made from the effects of pure solvents.

In this way Moore and Sheldon [113] obtained evidence from swelling measurements for the existence of two total solubility parameters for

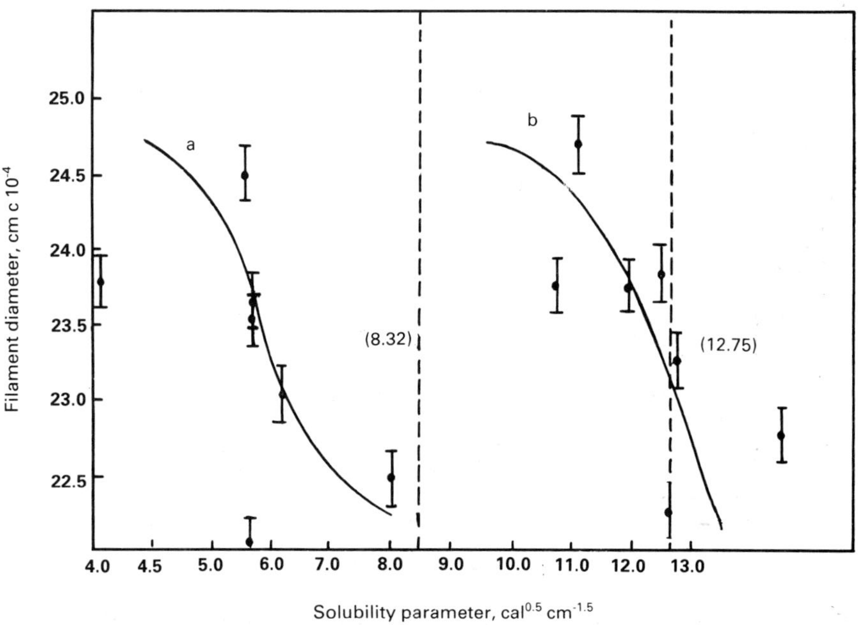

Figure 3.28 – Diameter of Acrilan filaments containing 1 mol kg^{-1} of plasticisers, related to (a) the total solubility parameter of the plasticiser δ_t, (b) the association solubility parameter δ_a; the bars represent the limits of the experimental errors [111]

TABLE 3.15

Solubility parameters at 25 °C/J$^{0.5}$ cm$^{-1.5}$ (cal$^{0.5}$ cm$^{-1.5}$)

Carrier	δ_t	δ_d	δ_a
Aniline	22.60 (11.04)	19.50 (9.53)	11.44 (5.59)
Benzoic acid	21.18 (10.40)	18.44 (9.01)	15.66 (7.65)
Benzyl alcohol	24.22 (11.97)	18.50 (9.04)	16.06 (7.85)
Benzylamine	21.08 (10.30)	17.78 (8.69)	10.95 (5.35)
m-Cresol	22.72 (11.11)	19.36 (9.46)	11.98 (5.85)
Nitrobenzene	21.74 (10.62)	19.91 (9.73)	8.72 (4.26)
Phenol	29.68 (14.50)	24.80 (12.12)	16.28 (7.95)
2-Phenoxyethanol	20.19 (9.87)	17.86 (8.73)	9.25 (4.52)
Poly(ethylene terephthalate)	21.90 (10.70)	19.52 (9.54)	9.82 (4.80)

poly(ethylene terephthalate) which are believed to represent values for different parts of the repeat unit. Similar observations [114], also based on the use of the total solubility parameter, have been used to support the view that some solvents react with the aliphatic part of the polyester molecule and others the aromatic portion. Weigmann and his co-workers [115] used the reduction in T_g caused by the solvent as a measure of the fibre-solvent interaction and obtained a diffuse correlation with the total solubility parameter of the solvent. Ingamells and Yanumet, however, used the reduction in T_g of poly(ethylene terephthalate), obtained on immersing the polymer in each of a range of solvents, as a measure of the interaction between polymer and solvent. They found a clearer correlation by considering the dispersion solubility parameter (Figure 3.29) instead of the total solubility parameter. Two clear peaks are present, one for the cyclic and aromatic solvents (δ_d = 9.15) and another for aliphatic solvents (δ_d = 8.45). Each numbered point in the figure represents a different solvent, details for which are given in the original reference [116].

Solvents still exert an influence on the polyester when applied in an aqueous medium, their effect on T_g being proportional to their concentration in the fibre. But because the values for the dispersion solubility parameter of the compounds used fall within a narrow range, they all appear to be equally effective. Similar reasoning may be applied to the increase in diameter swelling, increased equilibrium shrinkage and reduced birefringence of drawn filaments [117] which accompany an increase in the concentration of solvent introduced into the fibre.

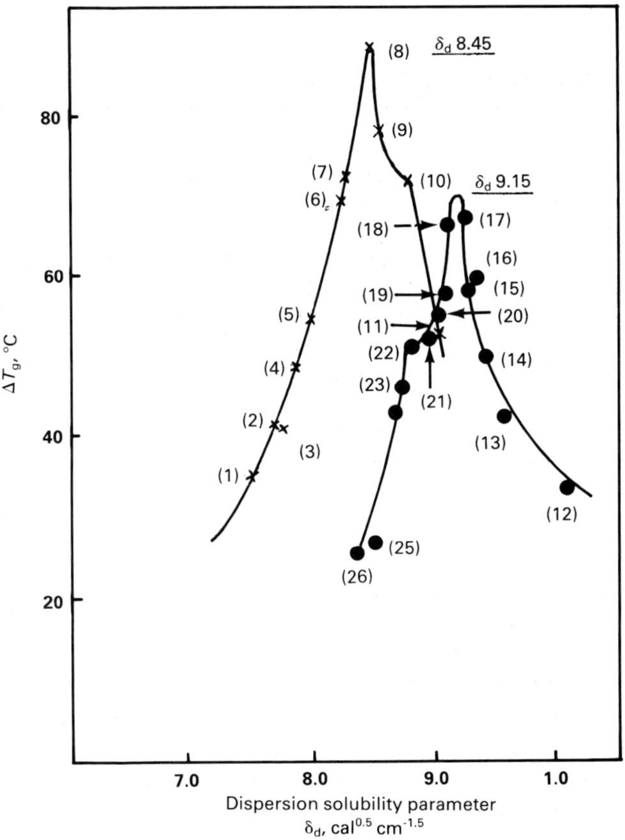

Figure 3.29 – Effect of solvents on T_g of poly(ethylene terephthalate) [116]

3.8.2 The influence of solvents on fibre structure

The influence of solvents on polymers is not limited to a reduction in the T_g of the fibre, and interpretation of their effect in terms of molecular relaxations has been considered [118–123]. The onset of segmental mobility can also be a prelude to crystallisation [124–132], surface and internal cavitation [133] and shrinkage of the fibres, which has been examined in detail for polyester [42, 134–136] and nylon filaments [137–140]. The dyeing of Nomex aramid fibres may be quoted as a recent example [141]. Pretreatment in highly polar fibre-interactive solvents causes molecular changes [142, 143] which eliminate the need for carriers in dyeing. Pretreatment of nylon 6 and 6.6 with organic compounds (e.g. solutions of phenol) also induces changes that modify the dyeing kinetics to the benefit of the dyer, and other similar treatments have been examined for polyester [42, 144, 145]. Those dyeing processes that involve the use of solvents may therefore cause changes in fibre properties which have their origin in the reaction between fibre and solvent, and a recent review of appropriate reactions is given elsewhere [99].

3.9 PRACTICAL DYEING SYSTEMS

3.9.1 The batchwise dyeing of acrylic fibres

The high sensitivity to temperature of the rate of dyeing of basic dyes on acrylic fibres can be readily understood when it is considered in relation to the free volume and the related segmental mobility of the polymer chains. Water is a very effective plasticising agent and is responsible for lowering the T_g of the fibres in the dyebath by 30–35 °C [84]. At or near the T_g the rate of diffusion of dye into the fibre increases suddenly, owing to the onset of segmental mobility. Further increases in temperature causes an exponential increase of the diffusion coefficient and therefore a great increase in the rate of dyeing. The high affinity of the dyes, which become bound to the fibre by active ionisable groups, together with the sudden increase in dye uptake once the dyeing transition temperature is exceeded, demand careful control of application conditions if level dyeings are to be achieved.

Various methods of application have been developed and have been reviewed elsewhere [146]. They all aim at regulating the uptake of dye. Levelling by migration is not feasible because of the high affinity of the dyes. Some temporary modification of the affinity is obtained by acidifying the dyebath to suppress the ionisation of the acidic groups in the fibre. For this reason the dyebath is buffered to a mildly acidic pH value, using sodium acetate and acetic acid. Further control may be gained by using a deliberately low rate of increase of temperature of the dyebath – around 0.2 °C min^{-1} – over the temperature range at which the rapid increase in diffusion is encountered. Even after taking this precaution, however, the effect of the high temperature sensitivity is not entirely counteracted. Moreover, such a gradual rise of temperature is very difficult to control in large-scale dyeing. An alternative method consists in maintaining a constant temperature of 90–95 °C and allowing the dye to enter the bath gradually; after an initial rapid uptake of dye, the rate of dye uptake during the remainder of the operation is more gradual.

Alternative approaches for regulating dye uptake make use of the dyebath additives. Neutral salts, e.g. sodium sulphate, reduce the rate of dyeing by reducing the initially negative surface potential as well as by providing metal ions which compete with dye for sites in the fibre. Colourless cationic organic compounds (*retarders*) may also be used to provide competition for the dye sites at a concentration appropriate to the concentration of the dye applied. Since the affinity of the dyes is related to the number of anionic dye sites in the fibre [147], this temporary blocking action of the retarder reduces the affinity of the dye and hence the rate of dyeing. Anionic retarders are also available; they associate with the dye, thus blocking its positive charge temporarily and limiting the concentration of dye available for dyeing at any particular time. Such compounds permit an extension of the range of application temperatures, but their effectiveness varies from dye to dye (as does that of cationic retarders) and in some cases is associated with a deterioration in fastness to rubbing.

Compatibility

The high affinity of the dyes for the fibres and the high negative zeta potential (cf. p. 77) which acrylic fibres acquire when they are immersed in water together lead to rapid saturation of the dye sites on the fibre surface. Once the surface is saturated, the rate of dyeing becomes controlled by the rate of diffusion of dye from the surface to the interior of the fibre. Provided that the amount of dye present does not exceed the total number of dye sites available in the fibre, exhaustion always approaches completion. Consequently, the rate of accumulation of dye in the fibre is controlled by the rate of diffusion into the fibre and is independent of the concentration of dye in the dyebath, so full depths take longer to dye than pale depths. This state of affairs is illustrated by Figure 3.30 [148].

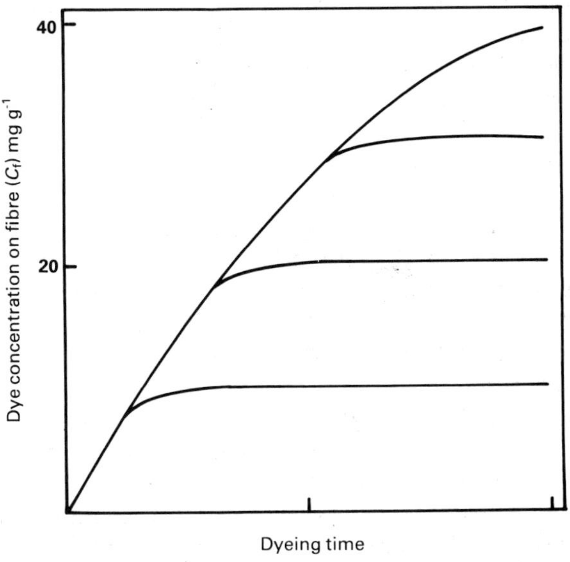

Figure 3.30 – Rate of uptake of dye (hypothetical curves for different initial concentrations of dye [148]

Difficulties can arise when using mixtures of dyes, firstly because of the limited number of dye sites available at the fibre surface and, secondly, because the rate of uptake of one dye in the presence of another is lower than when it is applied alone at the same concentration. Competition from the second dye for the limited number of dye sites available at the fibre surface retards the uptake of the first dye, and consequently can lead to problems of compatibility if the dyes chosen do not have similar dyeing characteristics. For the reasons outlined above, it is difficult to characterise the rate of dyeing of basic dyes in a quantitative way. Times of half-dyeing and exhaustion curves are no guide to the behaviour of dyes in admixture, and calculation of absolute values of the diffusion coefficient from Fick's equations (section 5.1) is too complicated for application in dyeing practice, since it requires knowledge of how the coefficient varies with concentration, which can only be obtained by measurement of dye distribution profiles in cross-sections of the dyed substrate. Moreover, the use of Hill's equation (Eqn 5.30) for diffusion into a semi-infinite cylinder can only produce values for the apparent diffusion coefficient; in any case, the calculations assume that the fibre is circular in cross-section, which it is not.

Beckmann and Glenz [148, 149] have discussed the nature of the dye–fibre bond in an attempt to gain a clearer insight into the nature of the dyeing behaviour of basic dyes. Meyer *et al.* [150, 151] have described exhaustion curves for basic dyes on acrylic fibres [146b] in terms of experimental parameters which help in selecting compatible dyes.

The slope of the plot of dye uptake *vs* the square root of the dyeing time is taken as one parameter and designated as tan α. As the concentration of dye in the liquor increases, the slope reaches a limiting value and is represented as

$$\tan\alpha = kMD^{\frac{1}{2}} \qquad (3.14)$$

where k is a constant depending on the fibre, M is the r.m.m. of the dye and D is the diffusion coefficient.

At a concentration of dye in the dyebath c_1, which is lower than that required to produce the limiting rate of dye uptake, the situation is represented by

$$\frac{\tan\alpha_{c_1}}{\tan\alpha_{\infty}} = \theta c_1 \qquad (3.15)$$

where θ is the fractional saturation of anionic groups at the fibre surface. Thus when all the surface sites are occupied $\theta c_1 = 1$, and when they are all unoccupied $\theta c_1 = 0$.

The slope of the curve for Eqn 3.15 is represented by

$$\tan \beta = \frac{K_{ads}}{M} \tag{3.16}$$

where K_{ads} is the equilibrium adsorption constant.

The expressions resulting from combining tan α and tan β can then be used to characterise the compatibility of two dyes, i.e.

$$\tan \alpha . \tan \beta = kK_{ads} D^{\frac{1}{2}} = Z \tag{3.17}$$

K_{ads} gives some measure of the affinity of the dye for the fibre, and D (the diffusion coefficient) indicates the rate of diffusion of the dye into the fibre. Thus a study of the Z values can give information about the compatibility of the dyes [146b] which it is impossible to obtain by inspection of the exhaustion curves for the individual dyes. If the dyes have the same value of the diffusion coefficient and different affinities, the dye with the higher affinity will dye preferentially. Similarly, if dyes have the same affinity but different diffusion coefficients, the dye with the higher diffusion coefficient will dye faster. In both cases incompatibility will result. For dyes with similar Z values, however, the balance of affinity and diffusion properties will be such that compatible dyeings are obtained.

More recently [152], the Society of Dyers and Colourists has proposed a compatibility test based on the K values described by Beckmann [153]. This is a simple test requiring only the production of series of dyeings. The dye under examination is dyed in combination with each of five reference dyes of known behaviour which form a scale of compatibility values ranging from 1 to 5. The compatibility value of the dye under test is determined by examining visually the five sets of dyeings and selecting the set in which build-up of colour is on tone. The compatibility value of the dye under examination is the number of the standard dye used in this set of dyeings. Two series of standard dyes, are specified, one of yellow and the other of blue dyes, the corresponding dyes in each scale being compatible. Selecting the appropriate set, e.g. the blue series with a yellow dye, maximises the effect of difference in dyeing behaviour on change in hue with build-up, and allows accurate determination of the compatibility value of the dye under test.

A test for determining the saturation characteristics of basic dyes on acrylic fibres has also been proposed by the Society of Dyers and Colourists [154].

3.9.2 Carrier dyeing

Many theories of carrier action have been proposed [155, 156], mostly on the basis of experimental observations. Because of the multitude of interrelated and complicating factors, which determine the effects observed in practical dyeing, it is very difficult to identify a mechanism unambiguously. Examination of the carrier-dyeing system involves two different aspects of the same phenomenon: the effects which the carrier may exert on the dye in the dyebath, and the effects of absorbed carrier in the fibre itself. At present, it appears that the latter effects exert the greater controlling influence.

Effect of carriers in the dye in the dyebath

One of the earlier hypotheses [157] suggested that carrier and dye form a complex that is more readily transported into the fibre than are dye molecules alone, but consideration of possible groups in the molecules of both carrier and dye which are likely to show such activity, together with spectrophotometic analysis [158], has so far failed to provide positive evidence in support of this theory. On the other hand, Zimmermann *et al.* [159] took the view that, in the presence of carrier, more disperse dye exists in true solution and, consequently, at any instant more dye is available for diffusion into the fibre. The effect with the fibre is a further aspect of the problem, which led one author to regard the adsorbed carrier as a co-fibre that enhances the solubility of the dye in the fibre [155]. Arguments of this nature tend to be confusing, however, since comparison of the partition coefficients of carriers between water and undrawn polyester fibre by Balmforth *et al.* [160] shows that, whilst uptake by the fibre of some active carriers leads to an increase in the partition coefficient of the dye between fibre and dyebath, one active carrier (benzoic acid) reduces the partition coefficient. Balmforth *et al.* suggest that this result reflects two different aspects of the carrier-dyeing mechanisms, viz. that some carriers function by causing an increase in both the equilibrium uptake of dye and the rate of dyeing, whilst others function solely by increasing the rate of dyeing. The relative importance of the two factors is difficult to determine, but it seems likely [159] that it is partly connected with partition coefficient of the carrier relative to the dye. The partition coefficient of the dye should be increased by carriers with a partition coefficient higher than that of the dye, but decreased by those with a partition coefficient lower than that of the dye, because they will make the dyebath a more favourable medium for dissolving the dye. This explanation also fits the broad generalisation that the less soluble carriers are more effective when added to the dyebath, since they tend to have a higher partition coefficient. This observation should not be confused with the effectiveness of carriers in molecular terms, because it has been demonstrated, although only with a limited number of carriers [161–163], that equimolar proportions of carriers in the fibre produce equal effects on dyeing properties.

It seems to the present author that it is dangerous to draw conclusions about the mechanism of carrier action from these generalisations, because they take no cognisance of the role of other specific features such as polarity, polarisability and molecular flexibility [164, 165], all of which influence the ease with which polymers and additives will mix. The significance of such features has not yet been assessed. Rosenbaum [165] compared the plasticising effect on acrylic fibres of various molecules with their hydrogen-bonding properties, solubility parameters and dipole moments; the evidence obtained so far indicates that dipole moment is the only parameter that predicts plasticising action, but the predictions were only tenuous.

One further aspect of carrier activity is the possibility of the formation of a surface layer of carrier around the fibre. Balmforth's investigations [160] have demonstrated that when the carrier concentration is in excess of the amount required to saturate both fibre and dyebath, a third phase of carrier forms; dye dissolves in this preferentially, thus reducing the colour yield. This has some bearing on the earlier investigations of Millson [166] who, with the aid of a microscope, was able to observe the presence of droplets of carrier on the surface of polyester fibres in a dyebath. A carrier film of thickness 10 nm on the surface of the fibre has also been reported [167]. This evidence supports an earlier suggestion of Peters and Sumner, discussed by Vickerstaff, that carrier dyeing takes place from a dye-rich surface layer of carrier [168].

Effect of carriers in the fibre

An alternative theory of carrier action attempted to relate carrier uptake with an increase in water imbibition of the hydrophobic fibre, thus creating a medium for the acceptance of the sparingly soluble disperse dye. The effect was attributed to the van der Waals attraction between water molecules and the hydrophilic substituent groups in the carrier molecule, e.g. —OH and —NH_2. However, although the water imbibition of polyester fibres increased [155], some effective carriers do not possess hydrophilic groups. Moreover, effective carriers for cellulose triacetate do not increase the water imbibition of this fibre.

Rochas [162, 169, 170] determined the extent of carrier adsorption gravimetrically and found that, even after treatment of fibres with ice-cold propanone, some carrier remained and could not be removed. Rochas attributed the origin of carrier activity to this portion and suggested that some permanent physical modifications had taken place in the fibre structure. Some doubt is indicated in the literature as to the responsibility of permanent modifications for the change in the rate of dyeing, however. Brown and Peters [171] demonstrated that although some change in the X-ray photograph of polyester fibres could be observed when trichlorobenzene was used as the carrier, the concentration of carrier required to produce carrier activity was lower than that needed to produce permanent modification of the fibre structure.

3.9.3 Swelling effects

Some modification of the fibre structure is indicated, and it is possible that carriers cause a temporary modification. For example, by analogy with the dyeing of hydrophobic fibres, the assumption may be made that the increased ease of dyeing is accompanied by an increase in fibre swelling. This assumption leads to confusion, however; although swelling has been observed with phenolic compounds [172], a clear correlation between swelling and carrier action has not yet been reported [158b, 173, 174]. This is well illustrated by the use of monochlorobenzene, which causes negligible swelling but exhibits strong carrier activity on polyester [173], and by Figure 7 in ref. 174, where the swelling effect of 2-phenoxyethanol on drawn and undrawn poly(ethylene terephthalate) is the reverse of the variation in diffusion coefficient caused by carrier action. The results confirm earlier conclusions of Waters [175] that the effectiveness of a compound in causing swelling is not the prime criterion in carrier efficiency. Attempts have been made to describe the effects of swelling in a more fundamental manner using the concept of solubility parameters. This concept, as outlined earlier, gives an indication of the likelihood of interaction between a polymer and a particular solvent. For carrier dyeing, this leads to an anticipated correlation between the solubility parameters of the carrier and the fibre. If they coincide there should be disruption of the fibre structure, which may be responsible for the increased dyeability. The concept is therefore potentially useful for describing the mechanism of carrier dyeing in terms of loosening the fibre structure or plasticisation of the fibre.

In general, it can be stated that most active carriers for polyester fibre do have solubility parameters close to that of polyester [171], but there are other compounds which have similar solubility parameters but which are not effective carriers. Moreover, the most active carriers do not appear to be those with solubility parameters closest to that of polyester, suggesting that factors other than modification of fibre structure are operative [171], such as effects on the solubility of the dye. Furthermore, a plasticising effect on acrylic fibres has been reported for compounds with solubility parameters that do not coincide with the values reported for the polymer [165]. The conflicting evidence from this approach has now been given a more rational explanation, however. For example, in section 3.8 it was shown both that the swelling of acrylic fibres is governed by association forces through the use of δ_a, and also that the influences of dispersion forces cannot be represented in a coherent manner. The closer the match between δ_a(solvent) and δ_a(fibre), the less is the swelling and the greater the reduction in T_g

(i.e. the greater the plasticisation of the fibre). Thus two solvents m identical values of δ_t, but different values for its component parts and δ_d). In this way discrepancies in the effect of particular solvent identical δ_t may easily arise. If δ_a of the solvent is close to that of the fibre little swelling and marked plasticisation is apparent. On the other hand, if δ_a is very different from that of the polymer there is less plasticisation and greater swelling.

Similar arguments can be provided for the interaction of polyester and solvents as judged by their plasticising action. Dispersion forces have been linked with the plasticisation of polyester fibres [112b] as opposed to the association forces in polyacrylonitrile [111]. If it is assumed at this stage that plasticisation governs the carrier action (the evidence for which is presented below), then the fact that dispersion forces control the plasticisation of polyester affords an explanation of some of the apparent anomalies reported by workers attempting to relate carrier activity with the δ_t of the carrier.

For example, dimethylformamide (DMF) (compound 9 in Figure 3.29) reduces the T_g of the polyester by 77 °C, and it is well known that pretreatment of polyester fibres with DMF increases the uptake of disperse dyes [127]. DMF is ineffective as carrier when used in an aqueous bath, however. Alternatively, methylnaphthalene (compound 12 in Figure 3.29) has little influence as a pure solvent but has significant carrier activity in an aqueous bath.

The evidence obtained with these and other compounds [109] suggests that the correct balance of dispersion and association forces is required for effective carrier action. Thus δ_d should be in the correct region to cause plasticisation, but if δ_a is too high (i.e. if the molecule is too polar) the concomitant higher solubility in water will prevent the accumulation of a significant concentration within the fibre. Under these conditions it cannot act on the fibre, even though as a pure solvent it has a strong plasticising action.

The plasticisation theory of carrier dyeing

The evidence so far accumulated tends to favour mechanisms in which the carrier acts through modification within the fibre rather than the external dyebath. This view is emphasised by the experiments of Rawicz *et al.* [176] who claimed that swelling of polyester filaments could be detected after immersion in an aqueous solution of compounds showing carrier activity. Their criterion for swelling was the shrinkage of the polyester filaments. To take such measurements to represent an implied increase in volume due to swelling is not clearly established as valid, but the effects produced highlight the important observation that effective carriers initiate some physical change in the fibre which is not encountered with other compounds. Thus a connection between carrier activity and a change in physical properties of the fibre can be obtained.

Other workers [177, 178] have extended the argument to show that the presence of effective carriers in the dyebath reduces the temperature at which shrinkage is first detected. They suggest also that this lowering of the shrinkage initiation temperature is equivalent to a reduction in the glass transition temperature of the fibres. Similar investigations have been made into the behaviour of acrylic fibres [179]. Consequently, the influence may be drawn that carriers lower the temperature at which the fibres become plastic, and this supports the theory that the carrier loosens the molecular structure of the fibre, thus enabling dye molecules to penetrate more readily [180].

Proof of this theory requires a correlation between plasticisation of the fibre and increase in the diffusion coefficient of the dye in the fibre. Such a correlation has been obtained by Ingamells, Peters and Thornton [83] for the dyeing of acrylic fibres with disperse dyes in the presence of phenol as carrier. Reference has already been made to the relationships between segmental mobility, physical properties, temperature and the diffusion of basic dyes into acrylic fibres, to show how diffusion of the dye through the fibre is controlled by the segmental mobility of the polymer chains. In order to extend the theory to the carrier-dyeing process two requirements need to be fulfilled. Firstly, compounds that act as carriers must lower the T_g of the fibre in the dyebath; secondly, the variation in both the diffusion coefficient of the dye and the viscoelastic properties of the fibre with temperature should be in accordance with the WLF relationship (Eqn 3.7).

The change in the glass transition temperature of acrylic filaments was detected by measuring the ease of deformation of the filaments when immersed in a blank dyebath containing carrier. The filaments were first conditioned in the appropriate solution of carrier at 95 °C, in order to avoid possible complicating effects arising from the diffusion of carrier into the filaments while the measurements were being made. The glass transition temperatures were then determined by measuring the rate of change in length of the filaments (i.e. rate of creep) when they were subjected to a light stress at different temperatures. The fractional change in length ultimately attained at each temperature was then plotted against the temperature of measurement, with the result illustrated in Figure 3.31. The onset of segmental mobility in the immersed filament takes place at 57 °C in water, at which point dramatic changes in extensibility begin. The intercept of the linear portion of the curve on the temperature axis is taken as the glass transition temperature. Addition of phenol to the system lowers the T_g so obtained, in proportion to the amount of phenol in the system. A reduction in T_g was also found using other soluble compounds with carrier activity, including nitrobenzene,

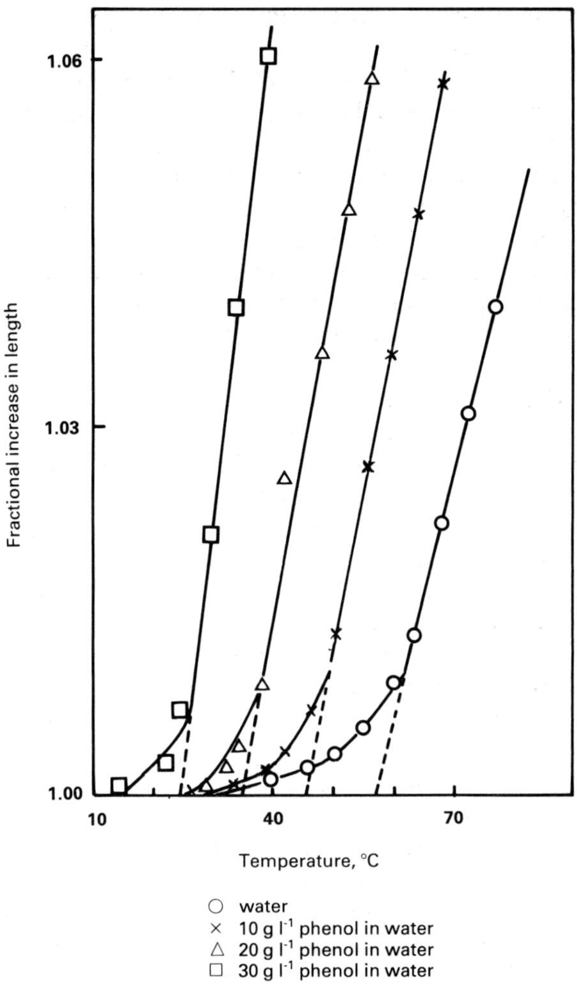

Figure 3.31 - Effect of temperature on fractional increase in length of Acrilan filaments immersed in aqueous solutions of phenol

dimethylformamide and *p*-toluidine; Kulkarni and Mashelkar [181] have made similar observations.

Determination of the appropriate shift functions from the physical measurements demonstrated the close fit of the experimental points to the relationship represented by Eqn 3.7 (Figure 3.24). The theoretical curve is calculated by substituting numerical values for the constants A and B and different values of $(T - T_g)$ in Eqn 3.7. Values of log a_T thus obtained are then plotted against $(T - T_g)$. The experimental points can be fitted to this curve, provided that their value is directly related to the segmental mobility of the polymer chains. In order to fit the experimental points to Eqn 3.7 when other polymers are used, it is sometimes necessary to change the numerical values of the constants A and B. It seems probable that this change is connected with structural features that cause the polymer properties to deviate from those expected of a truly amorphous polymer. No physical significance has yet been placed on their value, however.

The partition coefficient of the dye did not vary significantly over the range of phenol concentrations used and, by using Hill's equation (p. 383), the apparent diffusion coefficient of the disperse dye used in the various experiments was calculated from the rate of dyeing curve. From Eqn 3.7 it can be deduced that, if the diffusion of dye is governed by the segmental mobility of the polymer chains, the logarithm of the ratio of the diffusion coefficient at T_g to the diffusion coefficient at the temperature of dyeing will vary with $(T - T_g)$ in the same way as the corresponding ratio of the appropriate physical properties of the polymer. In practice, the value of the diffusion coefficient of the disperse dye at or near T_g is too small to measure and consequently, in the absence of a suitable value, it is assumed that $1/D_a$ is proportional to log a_T and the results for log D *versus* the corresponding experimental values of $(T - T_g)$ are all shifted by the same amount, in a vertical direction only. When this is done (Figure 3.32) the experimental

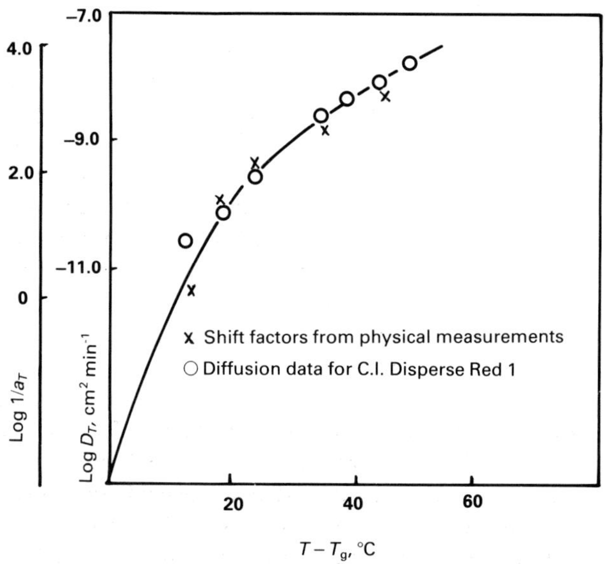

Figure 3.32 – Matching diffusion data and physical properties to the WLF equation (Eqn 3.7) using $A = 10.8$ and $B = 28.5$

points fit the calculated curve, and this can only happen if the diffusion of the dye is controlled by the segmental mobility of the polymer chains. Consequently, it can be concluded that the effect of carrier in a dyebath is equivalent to an increase in temperature and becomes apparent as an increase in the rate of dyeing. The increased rate of diffusion is equivalent to the effect which would be produced by raising the temperature by an amount ($T_g - T_{g(c)}$) where T_g and $T_{g(c)}$ are the glass transition temperatures of the fibre in water and in carrier solution respectively.

Another approach to improving the dyeability of fibres via temporary reduction in T_g is to pretreat fibres in a plasticising solvent and allow some solvent to be retained, thus lowering T_g before dyeing. Reiker and Terlinden [182] used retained chlorinated solvents to reduce the temperature of the onset of aqueous dyeing in poly(ethylene terephthalate) to 50 °C, and good dyeing is obtained at 98 °C. Chlorinated hydrocarbons are retained at even higher temperatures [183–185] and will speed up the rate of dyeing, but there is also the danger of molecular relaxations and crystallisation interfering with the improvement.

A similar approach has been used to introduce carrier into the fibre using organic solvents [186–188] rather than aqueous solutions [189]. All the processes depend on temporarily lowering T_g.

3.9.4 Solvent-assisted processes

Increasing the rate of dyeing by the use of organic dyebath additives has also been proposed for the dyeing of wool and silk [190]. Sparingly water-soluble additives, such as benzyl alcohol, butan-l-ol or various amines [191], when incorporated into dyebaths of neutral-dyeing metal-complex dyes or selected milling acid dyes, enable wool to be dyed in a shorter time and at a lower temperature than usual. The effectiveness of various accelerators has been examined [192, 193] but benzyl alcohol is the solvent usually quoted. These milder dyeing conditions reduce the risk of losing the desirable properties of the fibre [194]. Commercial methods based upon this principle are referred to as ‘solvent-assisted’ processes. Practical recommendations are available for extending the principle to the coloration of man-made fibres in general, and acrylic fibres in particular [195].

Although solvent-assisted processes have been in use for some time, there is still speculation concerning the dyeing mechanism. The first investigations of Peters *et al.* [196, 197], concerning the action of benzyl alcohol in wool dyeing, led to the suggestion that dyeing takes place from a layer of solvent covering the fibre surfaces. The dye is more soluble in this layer than in the water phase. Various experimental facts are available which support the theory that dyeing takes place from the dye-rich solvent layer, the high concentration of dye in the solvent leading to more rapid dyeing. For example, it was shown that benzyl alcohol increases the rate of dyeing of milling acid and neutral-dyeing metal-complex dyes which are soluble in the solvent, whilst dyes whose dyeing rate shows no increase can be made to respond by enhancing their solubility in the alcohol by blocking their sulphonic acid groups by conversion to quaternary ammonium or pyridinium salts. Furthermore, the magnitude of the effect of benzyl alcohol on the behaviour of an individual dye appears to correlate with the partition coefficient of the dye between benzyl alcohol and water. The more the dye favours the solvent phase, the greater is the effect of the alcohol on the rate of dyeing. Suggestions that benzyl alcohol acts as a swelling agent have been shown to be without foundation, since no significant increase in the fibre diameter can be detected when the fibres are immersed in solutions of the alcohol. Although benzyl alcohol has a low affinity for wool, its rate of uptake from the dyebath is very rapid, and this fact, together with the results of surface tension measurements, leads to the conclusion that the sorption of benzyl alcohol is associated mainly with its hydrophobic nature and that the process is one of simple physical adsorption, with little intrinsic affinity arising from specific interactions. All the above evidence supports the 'solvent layer' concept.

It may be significant, however, that solvent-assisted dyeings show many features in common with those already described in connection with man-made fibres under the heading of carrier dyeing. For example, Gokhale, Peters and Stevens [178] concluded that benzyl alcohol does not cause swelling in cellulose acetate, but increases the rate of dyeing without affecting the equilibrium absorption. This behaviour is in accordance with the plasticising effect of carriers described earlier (p. 229). Large increases in the concentration of alcohol lead to a reduction in the amount of disperse dye taken up by the fibre. Similarly, Balmforth [160] demonstrated that excessive amounts of carrier reduce the uptake of dye (but not the rate of dyeing), because the resulting increased solubility of the disperse dye alters the partition of dye between fibre and dyebath in favour of the dyebath phase.

With hydrophilic fibres it is unconventional to think in terms of carrier dyeing, but this may be because the role of the behaviour of the fibre molecules in dyeing is not yet fully understood. Speculation on the basis of available evidence can lead to the view that some form of temporary disruption in the polymer structure may be responsible for the enhanced rate of dyeing of wool in the presence of benzyl alcohol. Elliot *et al.* [241] demonstrated that wool fibres impregnated with aqueous solutions of benzyl alcohol responded more readily to an applied stress than fibres impregnated with water alone, and the slight permanent changes in physical properties caused by aqueous treatments occurred at lower temperatures. Further evidence for such a mechanism is provided by the results of Speakman *et al.* [242], who found that the resistance to extension of wool fibres stretched in aqueous solutions of propan-1-ol at room temperature is lower than that of fibres stretched in water alone. Thus with both alcohols the changed physical properties indicate more efficient bond-breaking. If this occurs during solvent-assisted dyeing, it is to be expected that the resulting temporary loosening of the fibre structure will facilitate dyeing. Lokhande has also demonstrated the practical role of water-miscible organic solvents in reducing the rate of hydrolysis in reactive dyeing and printing [195].

Medley and Ramsden have suggested that the action of the solvent takes place within the fibre [198]. Positive evidence for the importance of the plasticising action of benzyl alcohol in the dyeing of acrylic fibres has been obtained by Gur-Arieh and Ingamells [199], showing that benzyl alcohol affects the diffusion coefficient of disperse dyes in a manner predicted by the free volume theory. The alcohol reduces the T_g of the fibre and the diffusion coefficient of the dye varies in a manner predicted by Eqn 3.7. Similar evidence has been presented by Alberti *et al.*, who used a range of solvents [200].

Basic dyes are also applied with advantage to acrylic fibre [201] by the solvent-assisted dyeing process, and the high permittivity of benzyl alcohol (13.1) [202] allows the dissolution of basic dyes without the use of auxiliary (booster) solvents or emulsions which are necessary when dyeing from perchloroethylene (tetrachloroethylene). Consequently experimentation is possible using either benzyl alcohol in water (as in the normal carrier-dyeing system) or small quantities of water in benzyl alcohol (representing the situation when a surface layer of alcohol surrounds the fibre in an aqueous system saturated with the alcohol). The plasticisation mechanism applies in both cases [203], but the differences in behaviour of the dye in the aqueous and benzyl alcohol dyebaths are related to differences in the mechanical response of the fibres and in the degree of dissociation of the dye and dye sites in the two media.

3.9.5 Changes in the accessibility of dye sites during plasticisation

The equilibrium uptake of the basic dyes is also dependent on $(T - T_g)$ (Figure 3.33), a fact that supports the view [204] that the availability of the dye sites increases with increasing segmental mobility. Similar observations have also been reported for the variation with temperature of the saturation values for basic dye on reprecipitated basic-dyeable polyester fibres [205]. The removal of the effects of heat-setting on the diffusion of disperse dyes into polyester fibres also appears to be part of the same phenomenon [127, 206]. The effect does not appear to be a function of the size of the penetrant molecules, since the exchange of sodium ions in the basic-dyeable polyester

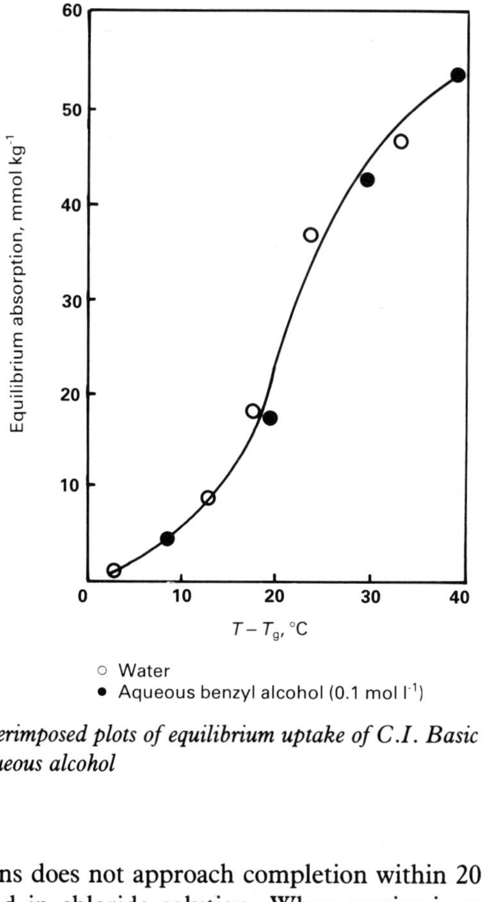

Figure 3.33 – Superimposed plots of equilibrium uptake of C.I. Basic Green 4 by Acrilan from water and aqueous alcohol

for potassium ions does not approach completion within 20 hours when the fibre is immersed in chloride solution. When carrier is present, however, exchange is rapid [207].

3.10 SOLVENT DYEING

Pressures to reduce pollution of the environment, together with the high cost of clean water and the need to conserve it, are factors which prompted an examination of the possibility of using recoverable organic solvents as alternatives to water in the wet-processing of textile goods. Unfortunately the nature of the most appropriate solvents mitigates against the desired improvements, due to ecological and toxicological problems. Consequently development activity has dwindled to a negligible level. Reviews of the subject are available, however [208–215], and various developments serve to illustrate useful principles which are relevant to any application methods involving the use of organic additives in processing.

3.10.1 The choice of solvent

Several requirements must be met by the solvent before it can be considered to be of potential use for solvent dyeing. It must be inexpensive, readily available, readily recoverable, non-toxic and non-flammable. It should also be chemically stable under the conditions of use, non-corrosive and chemically inert towards textile fibres. At present chlorinated hydrocarbons are regarded as the most suitable solvents for the purpose. Those usually considered are perchloroethylene and trichloroethylene.

Most commercial development work is centred upon perchloroethylene, because it has sufficient chemical stability under the conditions of processing and solvent recovery. Commercially available perchloroethylene contains small amounts of trichloroethylene (0.3–0.4 per cent) and hydrochloric acid (1–2 mg dm^{-3}) [215] that originate from slight decomposition, which also occurs with other chlorinated hydrocarbons during their preparation and use. For this reason the commercial solvents also contain small quantities of stabilisers, e.g. amines, phenols or crotonaldehyde dimethylhydrazone [215]. Textile-finishing processes may be modified by their presence, so the possibility of interaction between stabiliser and fibre (or other dyebath components) influences the choice of stabiliser.

Other properties of perchloroethylene

Certain physical properties of organic solvents, in particular their thermal properties, are more favourable than those of water for dyeing processes. The specific heat of perchloroethylene is 0.86 kJ (0.205 kcal) kg^{-1} K^{-1}, and the latent heat of vaporisation is 210 kJ (50.1 kcal) kg^{-1}, so heating is less expensive than with aqueous processing. It is also claimed that fabrics are wetted out more efficiently by organic solvents than by water.

The marked differences between the properties of water and perchloroethylene (or other organic solvents) can be expected to produce a marked difference in dyeing mechanism between the two systems. Water is a polar amphiprotic substance with a defined scale of pH values, a high permittivity, a high dipole moment and a unique structure which leads to what are termed hydrophobic interactions (cf. p. 113). On the other hand, perchloroethylene is a non-polar solvent to which the pH scale does not apply, and it does not possess any of the properties peculiar to water. There is a need, therefore, for lines of thought different from those that are accepted as being appropriate for aqueous dyeing systems, and various approaches have been used in attempts to develop satisfactory solvent-dyeing processes.

The solubility of dyes in perchloroethylene

The non-polar character of perchloroethylene and trichloroethylene obviously limits the solubility of ionic dyes, and the easiest dyes to dissolve are non-ionic dyes. Linear isotherms are obtained when disperse dyes are applied from perchloroethylene [210], but the partition coefficients between fibre and solvent are much lower than for those for aqueous systems [216]. Consequently the colour yields on both nylon and polyester are correspondingly lower, but the rate of attainment of equilibrium is greater [216]. Non-ionic 'basic' dyes also have been applied to acrylic fibres from perchloroethylene, and non-ionic reactive dyes can be used for nylon.

It is to be expected that the exhaustion of dye will be related to the partition coefficient of the dye, dyes with low partition coefficients giving the lowest colour yields. The partition coefficient will be increased if the solubility of the dye in the dyebath is reduced; there is a practical limit to the reduction of the solubility, however, because a very low solubility leads to an extremely low rate of dyeing. The use of a low liquor ratio with dyes of high solubility improves colour yield, but attempts to dye polyester with disperse dyes at temperatures around 120 °C have been unsuccessful, because the solubility of the dyes in the solvent is too high at these temperatures [217]. It is clear, therefore, that a suitable level of solubility must be achieved for the particular dyeing system being considered.

The scope of solvent processing has been extended by the use of dyebath additives. Some highly polar compounds which are surface-active in aqueous systems form micelles in organic solvents [215] without exhibiting very marked surface activity. These micelles have the useful property of solubilising ionic dyes in perchloroethylene, and the use of this property has been demonstrated with an acid dye applied to nylon from perchloroethylene. The size of the micelles can be varied by adding small quantities of water [215], and water may also be incorporated in the system to increase the solubility of other additives or dyes. In fact, the use of emulsions of water in perchloroethylene has been recommended for the application of both acid and basic dyes. The effect of added water is too marked, however, for an explanation of the dramatic increase in rate of dyeing to be given solely in terms of increased solubility of the dye. Apparently the presence of surface-active agent is not essential [218], since without it perchloroethylene will dissolve approximately 1 per cent of its mass of water, an amount that is sufficient to produce the desired increase in rate of dyeing; furthermore, amounts in excess of 1 per cent do not, apparently, give further technical advantages [218].

The use of secondary or 'booster' organic solvents in both batchwise and continuous methods has been described [216, 218]. These solvents increase the solubility of the dye in the dyebath. The most satisfactory solvents, used at the appropriate concentration, exert little effect on the colour yield but increase the rate of dyeing. The difficulty associated with their use lies in their recovery from the perchloroethylene after dyeing rather than in the production of technically desirable effects. 'Booster' solvents, e.g.

dimethylacetamide, methanol, benzyl alcohol and dimethyl sulphoxide, are more polar than is perchloroethylene and care must be taken to avoid increasing the solubility of the dye too much and thereby reducing the colour yield.

Again, it is not unreasonable to expect that booster solvents may act by affecting the fibre, and it is useful to consider the mechanism of solvent dyeing in terms of the theoretical principles already outlined.

3.10.2 The solubility parameter in solvent dyeing

If dyeing is to take place, space has to be available into which the dye molecules can enter, so either discrete spaces must exist in the fibre structure (which may be regarded as capillary channels down which the dye can pass) or space has to be created by the separation of polymer chains. This is true for both aqueous and organic solvent systems. It is therefore important to know how easily such spaces may be created, and accordingly knowledge of the cohesive energy density is desirable. For a solvent to be able to penetrate a polymer structure and separate the polymer chains, its solubility parameter, as well as its hydrogen bonding and polar characteristics, must match that of the polymer. To choose one particular solvent for general use must therefore lead to difficulties with some fibres, since not all fibres have the same solubility parameter. For example, perchloroethylene and trichloroethylene have solubility parameters of 39.1 and 38.5 $J^{0.5}$ (9.35 and 9.2 $cal^{0.5}$) $cm^{-1.5}$ respectively, but the values quoted for polyacrylonitrile lie in the range of 53–64 $J^{0.5}$ (12.75–15.4 $cal^{0.5}$) $cm^{-1.5}$ [105, 218] whilst for nylon 6.6 a value of 57 $J^{0.5}$ (13.6 $cal^{0.5}$) $cm^{-1.5}$ has been given, and for poly(ethylene terephthalate) 45 $J^{0.5}$ (10.7 $cal^{0.5}$) $cm^{-1.5}$ [105]. Since the value for the last fibre is closest to that for perchloroethylene, it is to be expected that when this solvent is the dyebath medium (no additive or booster solvent being present), poly(ethylene terephthalate) will be the most readily dyeable of the three fibres.

The data given above indicate why it may be necessary to modify properties of the solvent to make the solubility parameters of the dyebath and the fibre correspond. In the simplified theory of solubility parameters the effect of second solvents on the solubility parameter is expressed by Eqn 3.18:

$$\delta_{\text{mixture}} = V_1\delta_1 + V_2\delta_2 \tag{3.18}$$

where V_1 and V_2 are the volume fractions of the two components and δ_1 and δ_2 are the corresponding values of their solubility parameters. It is theoretically possible that two separate solvents which will not mix with the polymer may be made to do so if used together, provided that the solubility parameters of the mixture and polymer coincide. This argument suggests a possible second role for the booster solvents used in dyeing: in addition to improving the solubility of the dye in the dyebath, they may also facilitate better mixing of dyebath and fibre.

It is difficult to balance the requirements for a suitable solvent for dyeing synthetic fibres with disperse dyes. As indicated earlier (p. 214), the disperse dye should have an association solubility parameter close to that of the fibre, but this same requirement is expected from a solvent which will facilitate dyeing. Consequently the most suitable dye will have good solubility in the most suitable solvent, the partition coefficient of the dye will be low and the dye transfer inadequate, a situation depicted for a range of solvents in Figure 3.34 by Harris and Guion [219]: the solvent with the solubility parameter

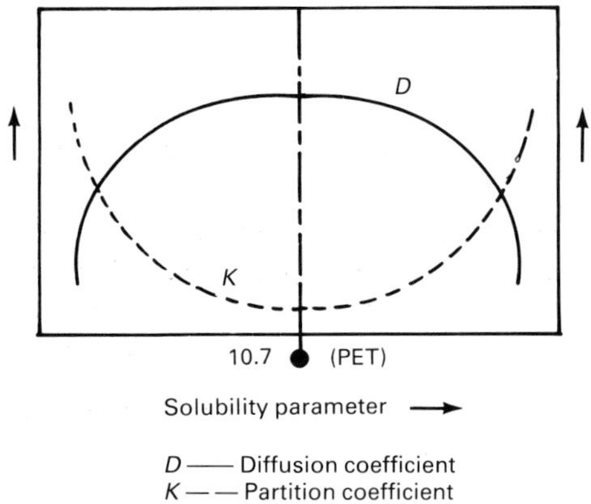

Figure 3.34 – Schematic representation of diffusion and partition coefficients as a function of solubility parameter of solvent [219]

producing the maximum diffusion rate of the dye produces its lowest affinity.

An attempt has been made by Sidiqui [220] to strike a balance of all the components of the cohesive energy of the dye, fibre and solvent when choosing an appropriate combination, and Skelly [213] suggested that the association solubility parameter of the solvent controlled the uptake of disperse dyes from an aqueous bath. Ingamells and Thomas argue that in a dye–fibre combination the relative magnitude of the polar (δ_a^2) and non-polar (δ_d^2) cohesive forces involved will influence both the affinity of the dye and the effect of solvents on the dyeing process [221]. With this in mind the dyeing transition temperature of a disperse dye in an aqueous bath was measured in the presence of a range of solvents, and the relative intensity of the association and dispersion cohesive forces in each solvent was represented by (δ_d^2/δ_a^2). The results (Figure 3.35) demonstrated that as (δ_d^2/δ_a^2) increases (i.e. as dispersion forces of the solvent become more dominant) the dyeing

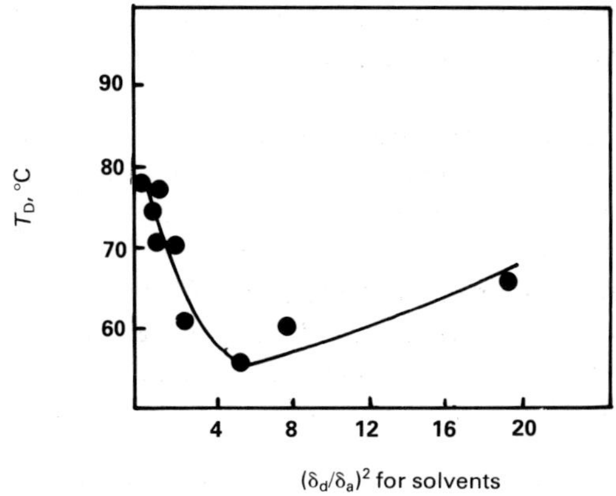

Figure 3.35 – Effect of solvents on T_g [221]

transition temperature reaches a minimum located at the value of (δ_d^2/δ_a^2) for the polyester substrate.

The relation between equilibrium dye uptake C_∞ and (δ_d^2/δ_a^2) for each dye used (Figure 3.36) remained the same (see Table 3.12), the only influence of changes of solvent being to shift the curve up or down the vertical axis (C_∞). Clearly the pattern of results is the same for each solvent and the only

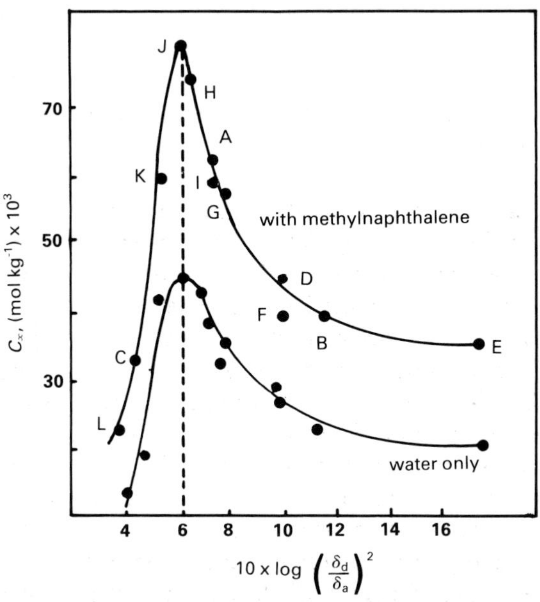

Figure 3.36 – Disperse dyes on poly(ethylene terephthalate) [221]

difference between the effects of individual solvents is one of magnitude. Figure 3.36 therefore represents the combined influence of variations in the relative intensities of polar and non-polar cohesive forces in both dye and fibre on the dyeing behaviour of polyester fibres.

3.10.3 The case of water

One outstanding discrepancy arises when attempts are made to apply the solubility parameter concept to dyeing: the case of water. Water is universally used as a solvent for dyeing both natural and man-made fibres, but the value for the solubility parameter is 96 $J^{0.5}$ (20.3 $cal^{0.5}$) $cm^{-1.5}$, a value which is very different from those given for the various fibres. The theoretical approach should not be discarded because of this discrepancy, however, since very little is understood about the precise role of water in dyeing mechanisms. Water has a strong plasticising action on man-made fibres (i.e. it lowers their T_g) and is known to swell natural fibres. In this respect it is producing the effects already discussed for various organic compounds that facilitate dyeing. Its effect is particularly noticeable in solvent-dyeing processes, where the introduction of small amounts of water have a marked effect upon the rate of dyeing.

Water has a considerable plasticising effect on nylon fibres. Data from the literature show that the T_g of dry nylon fibres is in the region of 60–80 °C, but when measured at 100 per cent relative humidity the T_g falls into the region of -10 to 0 °C. Consequently consideration of the outlined theoretical aspects of dyeing leads to the expectation that in the absence of any plasticising effect arising from water or booster solvents, the T_g of nylon fibres in perchloroethylene will be lowered only slightly and consequently dyeing will be more difficult than from water. This inference is supported by experimental observation, since small amounts of water in a perchloroethylene dyebath containing acid dyes cause a considerable increase in the penetration of dyes into nylon fibres [215]. It also seems possible that the reduction in the T_g of other man-made fibres by water may be connected to the dramatic change in dyeing rate which occurs when water is added to an organic solvent dyebath. It is interesting to note that in this context successful methods for dyeing nylon have been developed using emulsions of water in the solvent, or using 'boosters' or 'co-solvents' [222], so in each case the conditions are potentially those favouring segmental mobility.

3.10.4 Solvent mixtures

One approach to the use of ionic dyes in the non-polar perchloroethylene dyebath is to incorporate a small quantity of polar solvent to facilitate dissolution of the dye, and then gradually increase the distribution of the dye in favour of the fibre by distilling off the polar solvent very slowly. This is the principle of the STX process [223], in which nylon can be dyed with acid dyes from a perchloroethylene dyebath using methanol as co-solvent.

Dodd and Ingamells [224, 225] have also demonstrated the ability of acrylic fibres to take up basic dyes from an anhydrous perchloroethylene dyebath through the careful choice of a secondary solvent. Attention was given to plasticising the polar fibre and dissolving the polar (basic) dyes in the non-polar perchloroethylene dyebath.

It was established from measurements of the change in filament length that perchloroethylene alone caused no plasticisation. Benzyl alcohol alone has a significant plasticising action, but when used as the minor component in a perchloroethylene–benzyl alcohol mixture it causes no change in the T_g of the filaments. A dramatic reduction is readily achieved, however, by the further addition of small quantities of ethylene carbonate (as a plasticising agent) to the solvent mixture; this step also facilitates the dissolution of basic dyes in the dyebath. It is clear from Figure 3.37 that the lower the concentration of benzyl alcohol, the more efficient is the plasticising action

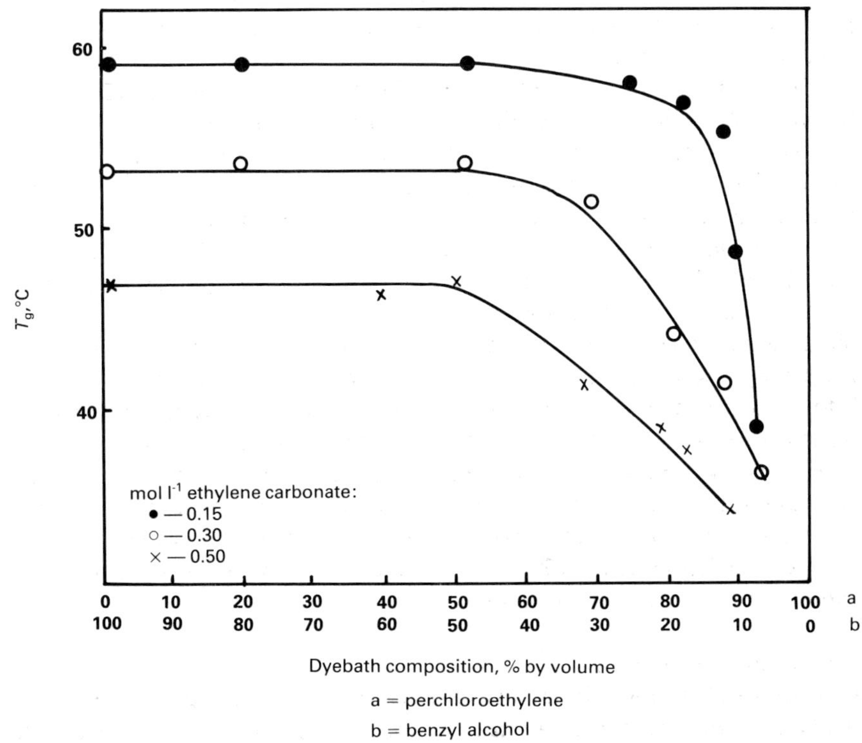

Figure 3.37 – Effect of composition of a dyebath containing ethylene carbonate, benzyl alcohol and perchloroethylene on the T_g of Acrilan 16 filaments

of the ethylene carbonate. This means that if the alcohol is removed during dyeing the dyebath phase becomes increasingly less polar, the plasticising agent moves into the fibre causing a marked drop in T_g and the dye is encouraged to move from the dyebath into the acrylic fibre. Unfortunately the boiling point of benzyl alcohol (206 °C) is higher than that of perchloroethylene; however, it can be replaced by a lower-boiling alcohol, such as ethanol. The drop in T_g as this distils off can be detected from the response of the fibres to a slight stress (Figure 3.38), and the exhaustion of the basic dyes proceeds with an efficiency matching that of an aqueous dyeing system.

An investigation of Chavan into cotton dyeing using solvents takes a

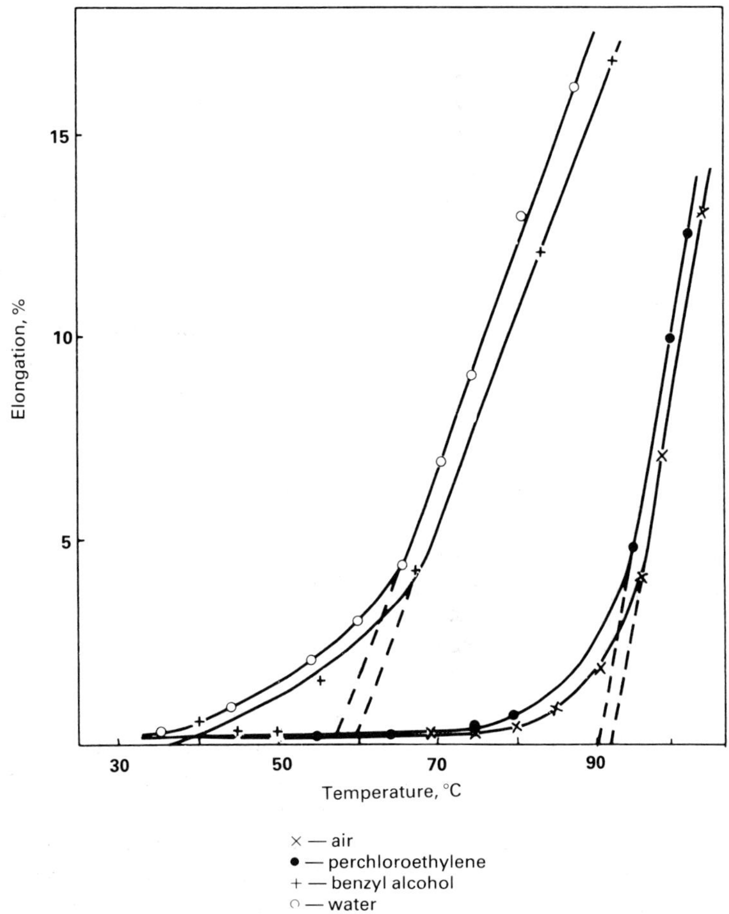

Figure 3.38 – Effect of temperature on the change in length of Acrilan 16 filaments surrounded by different media

different approach [226, 227]. In this case, dyeing with anionic dyes was aided by displacing the water from water-swollen cellulose using a mixture of DMF and trichloroethylene. Increases in dye sorption were greatest for those with lowest solubility.

Lokhande *et al.* [228–231] have given detailed attention to the different structural modifications of cellulose which occur due to the action of inorganic and organic swelling agents. Of the organic swelling agents examined morpholine had the greatest effect on structure, and the authors relate this finding to the hydrogen-bonding and polar contributions to the total solubility parameter of the swelling agent. Morpholine was also found to be the most effective reagent in enhancing the uptake of direct and reactive dyes [195, 232].

3.11 DRY HEAT-FIXATION

Interest in the dyeing of polyester fabrics by the thermofix process, in which the fabric is subjected to dry heat at about 200 °C after applying the disperse dye by padding, has been extended to the simultaneous dyeing of fabrics constructed from blends of polyester and cotton by simultaneous application of dyes suitable for both components. The fabric is padded in a liquor containing either vat dyes or reactive dyes for the cellulosic portion and disperse dyes for the polyester [233]. After impregnation, the fabric is dried and then heated to a much higher temperature to effect fixation of dye on the polyester portion. It is surprising that the process works as well as it does, because during the impregnation stage the hydrophilic cellulosic fibres pick up much more pad-liquor than do the hydrophobic polyester fibres. Consequently, at this stage, more disperse dye is present on the cotton than on the polyester where it is needed. This disperse dye must therefore transfer from the cellulose to the polyester during the baking stage. Several investigations have been concerned with the mechanism of this transfer, and three main possibilities have been considered. Firstly, dye may transfer from one type of fibre to the other where the two are in intimate physical contact. Secondly, an elaboration of the first idea suggests that transfer may take place through a medium of molten dyebath additives which are assumed to envelop the fibres during the baking stage. The third alternative, investigated by Datye *et al.* [234] and by Bent, Flynn and Sumner [235], postulates transfer through the vapour phase, when the dye sublimes off the cellulose during baking. It is the last theory that has proved to be the correct one.

The mechanism of transfer was established by a series of experiments in which a piece of cotton impregnated with pad-liquor containing disperse dye was mounted over a piece of uncoloured polyester fabric in such a way that an air gap separated the two. The composite arrangement was held firm in a suitable frame and the amount of dye transferred was determined as a

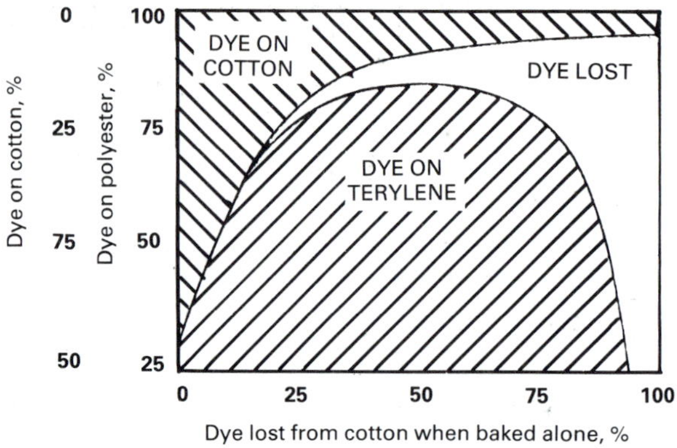

Figure 3.39 – Dye distribution versus volatility (dye, 40 g l^{-1}; baking, 1 min at 210 °C) [236]

function of time and temperature of baking, for dyes of different fastness to sublimation, as determined by their rate of sublimation from cotton. The fastness figures obtained represent the total effect of many different factors, e.g. the surface area of the dye particle per unit mass, the morphology of the dye particles, the rate of release and diffusion of dye away from the dye particles, the morphology and surface area of the cellulose, and the substantivity of the dye for the cellulose. Nevertheless, they give a measure of the ease of sublimation during baking, and this is the important consideration.

The relationship between the ease of sublimation and transfer of dye to the polyester is not simple, however. It is illustrated in Figure 3.39 [236], which indicates that dyes that do not readily sublime do not transfer to polyester easily. As sublimation increases there is first of all an increase in the amount of dye transferred, but as sublimation becomes still easier the proportion of dye that finally reaches the polyester begins to diminish and finally falls to a very low level. This behaviour may be related to the general classification of disperse dyes into groups A, B, C and D for the batchwise dyeing of polyester according to dyeing and heat fastness properties [237]. When this is done (Figure 3.40) it is apparent that group D dyes, which have the lowest dyeing rates and highest sublimation fastness, are all grouped together in the region where the transfer of dye to polyester is low. Group C dyes have transferred easily. Group B dyes, however, lie in the region in which transfer begins to fall off, whilst the rapid-dyeing group A dyes with poor sublimation fastness fall in the region where volatilisation from cotton is very easy, but transfer is again low. The complex pattern of results can be readily explained in terms of transfer of dye through the vapour phase. Thus,

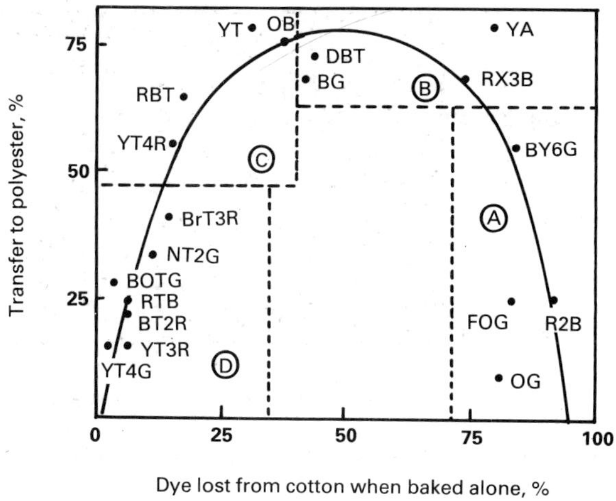

Figure 3.40 – Dye properties in relation of ABCD classification (dye, 40 g l^{-1}; time, 1 min; temperature, 210 °C) (the letter code used for each dye is given in ref. 236)

dyes with high sublimation fastness (group D) do not volatilise enough to transfer efficiently in the short time of the experiment. Where volatilisation becomes very rapid, as with the group A dyes, the dye does not diffuse into the polyester either because it is removed from the vicinity by some means (e.g. by air flow), or because it is lost by sublimation out of the polyester. Whatever the cause, the net result is loss of dye to the surroundings, and it is this which leads to contamination of equipment and staining of the fabric (Figure 3.39). Obviously, the dyes most suited to the thermofix process are those for which transfer is nearest to 100 per cent, and in general these are found within groups C and D. On the basis of other experimental evidence, the authors [235] concluded that even when the polyester and cotton were in contact the contribution to dye transfer from any additional mechanism was negligible.

Consideration of poly(ethylene terephthalate) fibres in terms of the theories outlined earlier in this chapter suggests that the conditions encountered in the dry heat-fixation of dyes should be favourable for inducing chain-folding (cf. p. 194), but the matter has yet to be investigated. Dumbleton and Bell [60] have demonstrated that the diffusion of non-ionic dyes into polyester fibres is influenced by segmental mobility but, as with nylon 6.6 [90], a marked difference is found in the relationship between diffusion coefficient and temperature compared with that expected from an amorphous polymer. The difference is believed to be due to the restricting influence on segmental mobility arising from crystallinity and orientation, and consequently the deviation will vary with the conditions of drawing and annealing. It is not easy to decide how the appropriate equations (e.g. Eqn 3.7) should be modified in order to take these changes into account. Furthermore, in aqueous dyeing systems the highest possible temperature of operation at atmospheric pressure is 100 °C. This restricts the temperature range of experiments to 30 °C (since the dyeing of polyester fibres below 70 °C is too slow to measure), whilst the effects of any physical transitions in the fibre above 100 °C cannot be easily investigated with the fibres immersed in the dyebath. For these reasons, Ito *et al.* [238] developed a technique for measuring the rate of sublimation in vacuum of simple dyes out of polyester film over a wide range of temperatures. Their results reveal the presence of more than one transition in the diffusion behaviour, one corresponding to the T_g of the polyester and the other attributable to changes in crystallinity.

3.12 COACERVATION

Other dyeing methods are available, which are based on special physical properties of the dyebath assistants. One property of commercial significance is the ability of certain compounds to form coacervates. Various physical changes are encountered when colloidal solutions are subjected to changes in pH, temperature, electrolyte concentration or the addition to the solution of other substances. Generally, the result is a reduction in solubility and the colloid separates from the solution. The form in which it separates depends upon the nature of the colloid. Sometimes it separates as a solid in the form of a dispersion of crystalline particles or as a suspension of very fine particles of colloid which cannot easily be separated from the liquid. Sometimes a liquid precipitate is formed, the droplets of which coalesce to form a liquid layer containing most of the colloid. This liquid is the *coacervate phase* and the remaining solution is referred to as the *equilibrium liquid*. It is the formation of coacervates that interests the dyer [239].

Where solubility is conferred on the molecules in the colloid by ionisable groups, coacervation can be brought about by a rise in temperature or by the addition of a less polar substance. On the other hand, the ionisation of those colloidal molecules that depend on ionic groups for solubility is reduced by electrolytes, and coacervation may therefore be induced by addition of an electrolyte. Macromolecules with ionic groups of ionic charge opposite to those in the colloid may act in a similar way.

Coacervates differ from other two-phase systems in that both the coacervate and the equilibrium liquid contain large amounts of water. Once the concentration of the colloid reaches a critical level for the formation of a coacervate under the prevailing conditions, the coacervate immediately forms a significant proportion of the total volume. Coacervates for dyeing have been produced from the products of condensation of aliphatic acids with alkylolamines. Commercial products are available, e.g. the Cibaphasols and Irgapadols (CGY) and Ethlan D (Lankro Chemicals).

The principal advantage of the coacervate system for wool is the resulting uniform application of dye. Normal pad-stream procedures for wool lead to uneven distribution of liquor in the fabric. Migration of liquor to the interior of the fabric during processing results in surfaces depleted of dye solution, leading to a skittery appearance and a higher dye content in the interior of the fabric. The overall effect is referred to as a 'sandwich' effect, because of the layered appearance of a cross-section of the dyed cloth. Earlier products such as Cibaphasol C were used to assist the application of dyes with wool by padding processes [240]. The coacervate system is prepared by mixing a surface-active agent (Cibaphasol C), an electrolyte, a thickening agent such as sodium alginate, and the dye. When the coacervate separates out it tends to float to the surface of the liquor at low concentrations of dye, because it is less dense than water. If the concentration of dye is sufficiently high, the density of the coacervate becomes greater than that of water and it sinks to the bottom of the liquid. In both cases the separation tends to be too rapid to allow the uniform application of padding liquors. For this reason the addition of thickening agents to the liquor is essential, since they reduce the rate of coalescence of the droplets of liquid precipitate.

The coacervate phase has the remarkable ability to take up either water-soluble dyes or solid particles of pigment; once on the fibre it spreads over the surface to form a uniform layer, thus evenly distributing the colouring matter. At the padding stage, no dye transfers to the fibre and the coacervate can be washed off, leaving the fibres uncoloured. Dye transfer takes place during the subsequent steaming operation.

Cibaphasol AS represents a technical improvement over Cibaphasol C, because it may be used to assist the coloration of acrylic and polyester fibres as well as of wool. The physical stability of the coacervate is also much improved and is not greatly affected by the presence of acids, electrolytes or dispersing agents. Since it forms a stable emulsion of coacervate as soon as it is introduced into water, it does not separate out in the padding liquor. Its use also prevents damage of the wool during steaming, which may occur with Cibaphasol C at the required pH of 8.9 because Cibaphasol AS can be used under mildly acidic conditions.

REFERENCES

1. R McGregor and R H Peters, *J.S.D.C.*, **84** (1968) 267.
2. *Idem*, *ibid.*, **81** (1965) 393.
3. S M Neale and R H Peters, *Trans. Faraday Soc.*, **55** (1959).
4. O Glenz and W Beckmann, *Melliand Textilber.*, **38** (1957) 296.
5. R McGregor, *Text. Research J.*, **42** (1972) 536.
6. B S Sprague, *J. Polymer Sci.*, **C20** (1967) 159.

7. A Rheiner, *Z. Angew. Chem.*, **46** (1933) 675.
8. R H Peters, *Textile chemistry*, vol. 1 (London: Elsevier, 1963).
9. I Holme, *Rev. Prog. Coloration*, **1** (1969) 31.
10. *Man-made fibres; science and technology*, vols. 2 and 3 (ed. Mark, Atlas and Cernia) (New York: Interscience, 1968).
11. H Berg, *Chemifasern*, **22** (1972) 215.
12. H G Elias, *Textilveredlung*, **4** (1969) 663.
13. C Shork, *ibid.* **2** (1967) 659.
14. M D Chantler, G A Partlett and J A B Whiteside, *J.S.D.C.*, **85** (1969) 621.
15. C Shork, *Textil Praxis*, **22** (1967) 725.
16. H Egli, *Textilveredlung*, **2** (1967) 856.
17. *Idem*, *Amer. Dyestuff Rep.*, **57** (1968) 1099.
18. W DeWinter and A Decorte, *Text. Research J.*, **41** (1971) 726.
19. H J Palmer, *J.S.D.C.*, **67** (1951) 609.
20. H Berg, *Melliand Textilber.*, **52** (1971) 448.
21. D Dierkes and E Lux, *Textil Praxis*, **22** (1967) 650.
22. Monsanto, BP 1,072,098 and 1,079,964, USP 3,304,289.
23. *Idem*, USP 3,235,533.
24. Glanzstoff, BP 1,125,633.
25. Monsanto, BP 1,072,098; 1,079,964; USP 3,304, 289; Glanzstoff, 1,100,239.
26. R H Peters, *J.S.D.C.*, **61** (1945) 95.
27. J P Bell and T Murayama, *J. Appl. Polymer Sci.*, **12** (1968) 1795.
28. P H Hobson, *Textil Chem. and Colorist*, **4** (1972) 232.
29. H Beutler, *Chemifasern*, **19** (1969) 108.
30. C Pernetti, *Amer. Dyestuff Rep.*, **61** (1972) 44(6).
31. J Pfleger and P Silvan, *Chemfasern*, **72** (1972) 1032.
32. R C Forrester and C LeBlanc, *Amer. Dyestuff Rep.*, **61** (1972) 29(6).
33. C L Bird and A M Patel, *J.S.D.C.*, **84** (1968) 560.
34. M Farber, *Soc. Plastic Eng. J.*, **24** (1968) 82.
35. *Idem*, *Amer. Dyestuff Rep.*, **55** (1966) 536.
36. M Levine and R P Weimer, *Text. Chemist and Colorist*, **2** (1970) 269.
37. M Fox, *Rev. Prog. Coloration*, **4** (1973) 18.
38. J S Ward and R Hill, *Text. Inst. Industry*, **7** (1969) 274.
39. *Fibre structure* (ed. J W S Hearle and R H Peters) (London: Butterworths, 1963).
40. J W S Hearle and R Greer, *J. Text. Inst.*, **61** (1970) 240.
41. L Rebenfield *et al.*, *J. Macromol. Sci., Rev. Macromol. Chem.*, **C15**(2) (1976) 279.
42. *Handbook of fibre science and technology*, vol. 3 (ed., M Lewin & J M Preston) (Marcel Dekker Inc., 1975).
43. Alfrey, Bohrer and Mark, *Copolymerisation*, vol. 8 in *High polymers* (New York and London: Interscience, 1952).
44. H A Standing, J O Warwicker and H F Willis, *J. Text. Inst.*, **38** (1947) T335.
45. J O Warwicker, *J. Polymer Sci.*, **A1** (1963) 3105.
46. T Morton, *Trans. Faraday Soc.*, **31** (1935) 281.
47. L B Ticknor, *J. Phys. Chem.*, **62** (1958) 1483.
48. J M Preston and P Pal, *J.S.D.C.*, **63** (1947) 430.
49. J M Preston and A H Kapadia, *ibid.*, **63** (1947) 434.
50. J Boulton, A E Delph, E Fothergill and T Morton, *J. Text. Inst.*, **24** (1933) T113.
51. J Morton, *Trans. Faraday Soc.*, **29** (1935) 262.
52. P H Till, *J. Polymer Sci.*, **24** (1957) 301.
53. A Keller, *Phil. Mag.*, **2** (1957) 1171.
54. E Fischer, *Naturforsch*, **12** (1955) 753.
55. J L Koenig and M J Hannon, *J Macromol. Sci.*, **B1** (1967) 119.
56. P F Dismore and W D Statton, *J. Polymer Sci.*, **C13** (1966) 133.
57. Y Kobayashi *et al.*, *J. Appl. Polymer Sci.*, **11** (1967) 2533.
58. W C Ingamells and K V Narasimham, *ibid.*, **20** (1976) 2271.
59. B S Sprague, *ibid.*, **C20** (1967) 159.

60. J H Dumbleton, J P Bell and T Murayama, *ibid.*, **12** (1968) 2491.
61. H Berg, *Melliand Textilber.*, **52** (1971) 448.
62. H Gullerman, *ibid.*, **53** (1972) 910.
63. F L Goodwin and S Rosenbaum, *Text. Research J.*, **35** (1965) 439.
64. Y Takagi and H Hattori, *J. Appl. Polymer Sci.*, **9** (1965) 2167, 3887.
65. G T Davis and H S Taylor, *Text. Research J.*, **35** (1965) 405.
66. N B Furwick, *J.S.D.C.*, **74** (1958) 299.
67. C H Giles, *Br. Polymer J.*, **3** (1971) 279.
68. C Renard, *Teintex*, **36** (1971) 845.
69. A R Munden and H J Palmer, *J. Text. Inst.*, **41** (1950) T609.
70. A Frey-Wyssling, *J. Polymer Sci.*, **2** (1947) 314.
71. H D Weigmann, in *Handbook of fibre science and technology*, vol. 1, part A (ed. M Lewin and S B Sello) (Marcel Dekker Inc., 1983) 24.
72. D M Marvin, *J.S.D.C.*, **70** (1954) 16.
73. E Merian, J Carbonnel, U Lerch and V Sanhuja, *J.S.D.C.*, **79** (1963) 505.
74. J O Warwicker, *J.S.D.C.*, **88** (1972) 142.
75. *Idem*, *ibid.*, **86** (1970) 303.
76. M Tsuruta and A Koshimo, *J. Appl. Polymer Sci.*, **9** (1965) 3, 81.
77. J H Rahu and U A Schwair, *Melliand Textilber*, **46** (1966) 1423.
78. R D VanVeld, G Morris and H R Billica, *J. Appl. Polymer Sci.*, **12** (1968) 2709.
79. T G Majury, *J.S.D.C.*, **70** (1954) 429.
80. K Meldrum and J S Ward, *ibid.*, **74** (1958) 140.
81. A K Doolittle, *J. Appl. Phys.*, **23** (1952) 418.
82. S Rosenbaum, *J. Polymer Sci.*, Part A, **3** (1965) 1949.
83. W C Ingamells, R H Peters and S Thornton, *J. Appl. Polymer Sci.*, **17** (1973) 3733.
84. B S Sprague, *J. Polymer Sci.*, **20** (1967) 159.
85. T Shibusaira and T Iijima, *J. Appl. Polymer Sci.*, **12** (1968) 1789.
86. A B Thompson and U W Woods, *Trans. Faraday Soc.*, **52** (1956) 1383.
87. T M A Hossain *et al.*, *J. Appl. Polymer Sci.*, **13** (1969) 541.
88. J P Bell, *ibid.*, **12** (1968) 627.
89. A Fujita, A Kishimoto and K Matsumoto, *Trans. Faraday Soc.*, **56** (1960) 424.
90. M L Williams, R F Landel and J D Ferry, *J. Amer. Chem. Soc.*, **77** (1955) 3701.
91. T Murayama, J H Dumbleton and M L Williams, *J. Macromol. Sci. (Phys.)*, (1967) 1.
92. A Koshimo and T Kakishita, *J. Appl. Polymer Sci.*, **9** (1965) 91.
93. J O Warwicker, *British Polymer J.*, **3** (1971) 69.
94. G Schreiner and W Kempter, *Textiltechnic*, (a) **33** (1983) 433; (b) **34** (1984) 324.
95. E Sada, H Kumazawa and T Ando, *J.S.D.C.*, **99** (1983) 92.
96. T Hori, Y Sato and T Shimizu, *ibid.*, **97** (1981) 6.
97. R M Rohner and H Zollinger, *Text. Research J.*, **56** (1986) 1.
98. U Meyer, R M Rohner and H Zollinger, *Melliand Textilber*, **65** (1984) 47.
99. (a) H D Weigmann, in *Handbook of fibre science and technology*, vol. 1, part A (ed. M Lewin and S B Sello) (1985) Chapter 1, 1; (b) I Rusznak, *ibid.*, Chapter 2, 51.
100. J H Hildebrand and R L Scott, *The solubility of non-electrolytes*, 3rd edn (New York: Reinhold Publishing Corporation, 1950).
101. G Scatchard, *Chem. Rev.*, **8** (1931) 321.
102. R F Blanks and J M Prausnitz, *Ind. Eng. Chem.*, **3** (1964) 11.
103. W R Moore, *J.S.D.C.*, **73** (1957) 500.
104. *Encyclopedia of polymer science and technology* (New York: Interscience, John Wiley, 1966) 833.
105. P A Small, *J. Appl. Chem.*, **3** (1953) 71.
106. A Urbanik, *AATCC Nat. Techn. Conference* (1981) 258.
107. H Gerber, *J.S.D.C.*, **94** (1978) 298.
108. E C Ibe, *J. Appl. Polymer Sci.*, **14** (1970) 837.
109. W C Ingamells and W C Thomas, *Text. Chem. Colorist*, **16** no. 3 (1984) 55.
110. Z Gur-Arieh and W C Ingamells, *J.S.D.C.*, **90** (1974) 12.
111. Z Gur-Arieh, W C Ingamells and R H Peters, *J. Appl. Polymer Sci.*, **20** (1976) 41.

112. (a) A N Derbyshire and R H Peters, *J.S.D.C.*, **71** (1955) 530; (b) W C Ingamells and A M Yabani, *ibid.*, **93** (1977) 417.
113. W R Moore and R P Sheldon, *Polymer*, **2** (1961) 315.
114. B H Knox, *J. Appl. Polymer Sci.*, **21** (1977) 225, 249, 267.
115. A S Ribnick, H D Weigmann and M G Scott, *Text. Research J.*, **43** (1976) 176.
116. W C Ingamells and N Yanumet, *British Polymer J.*, **12** (1980) 12.
117. W C Ingamells and A M Yabani, *J. Appl. Polymer Sci.*, **22** (1978) 1583.
118. K H Illers and H Breuer, *J. Colloid Sci.*, **18** (1963) 1.
119. H Jacobs and E Jenckel, *Macromol. Chem.*, **47** (1961) 72.
120. H G Zachmann and W Schermann, *Colloid Z., 2. Polym.* **241** (1970) 916.
121. E A Dimarzio and J H Gibbs, *J. Polymer Sci.*, **A1** (1963) 1417.
122. K H Illers, *Macromol. Chem.*, **38** (1960) 168.
123. G Goldbach, *Angew. Macromol. Chem.*, **32** (1975) 37.
124. R G Quynn, M A Seiminski and J L Riley, *J. Text. Inst.*, **55** (1964) T566.
125. R P Sheldon *et al.*, *J. Polymer Sci.*, **B2** (1964) 761.
126. A B Desai and G L Wilkes, *ibid.*, **C46** (1974) 291.
127. H D Weigmann *et al.*, *Text Research J.*, **46** (1976) 574.
128. P J Makerewicz and G L Wilkes, *J. Polymer Sci., Phys.*, **16** (1978) 1559.
129. *Idem.*, *Text Research J.*, **48** (1978) 136.
130. H G Zachmann, *Faserforch Textil Tech.*, **18** (1967) 95.
131. *Idem.*, *Macromol. Chem.*, **74** (1964) 29.
132. G M Venkatesh *et al.*, *J. Appl. Polymer Sci.*, **26** (1981) 223.
133. E A Gerold *et al.*, *Text. Research J.*, **49** (1979) 652.
134. (a) A S Ribnick and H D Weigmann, *ibid.*, **43** (1973) 316; (b) R D Kirby and D M Cates, *Text. Research J.*, **53** (1983) 586.
135. J H Dumbleton, *J. Polymer Sci., A2*, **7** (1969) 667.
136. A S Ribnick, *Text. Research J.*, **39** (1969) 742.
137. M V Forward and H J Palmer, *J. Text. Inst.*, **41** (1950) T267; **45** (1959) T510.
138. M I Jacobs, L Rebenfeld and H S Taylor, *J. Appl. Polymer Sci.*, **13** (1969) 427.
139. A S Ribnick and H D Weigmann, *Text. Research J.*, **43** (1973) 316.
140. R A F Moore, H D Weigmann and A Morello, *AATCC Nat. Techn. Conference* (1981) 109.
141. F L Cook and B R Kleissler, *AATCC Nat. Tech. Conference* (1983) 314.
142. S P McArthy, L Rebenfeld and H D Weigmann, *Text. Research J.*, **51** (1981) 323.
143. *Idem.*, *ibid.*, (1980) 185.
144. D Schnaitmann, D Fiebig and H Herlinger, *Melliand Textilber.*, **63** (1982) 50.
145. H L Hanna and S Shakra, *Amer. Dyestuff Rep.*, **71** (1982) 24.
146. (a) H Kellet, *J.S.D.C.*, **84** (1968) 257.
(b) N G Morton and M E Kratch, *ibid.*, **85** (1969) 639.
147. S Rosenbaum, *Text. Research J.*, **34** (1964) 159.
148. W Beckmann, *J.S.D.C.*, **77** (1961) 619.
149. O Glenz and W Beckmann, *Melliand Textilber.*, **38** (1957) 296, 783, 1152.
150. W Mayer, *Amer. Dyestuff Rep.*, **56** (1967) 869.
151. W Mayer, W Ender and A Wurtz, *Melliand Textilber.*, **47** (1966) 653, 772.
152. D G Evans and C J Bent, *J.S.D.C.*, **87** (1971) 60.
153. W Beckmann, *11th International Seminar*, Kingston, Canada (1968).
154. Society of Dyers and Colourists, *J.S.D.C.*, **89** (1973) 292.
155. V S Salvin *et al.*, *Amer. Dyestuff Rep.*, **48** (1959) 35 (14).
156. A Murray and K Mortimer, *Rev. Prog. Coloration*, **2** (1971) 67.
157. A Fern and H R Hadfield, *J.S.D.C.*, **71** (1955) 842.
158. (a) P Alexander and K A Stacey, *ibid.*, **72** (1956) 241.
(b) E Elod and A Vinh, *Textil Praxis*, **16** (1961) 718.
159. C L Zimmermann, J M Mecco and A Carlino, *Amer. Dyestuff Rep.*, **44** (1955) 296.
160. D Balmforth *et al.*, *J.S.D.C.*, **82** (1966) 405.
161. M J Schuler, *Text. Research J.*, **27** (1957) 352.
162. P Rochas and M Courmont, *Bull. Inst. Text. France*, **15** (1960) 86.

163. O Glenz, W Beckmann and W Wunder, *J.S.D.C.*, **75** (1959) 141.
164. E A Dimarzio and J H Gibbs, *J. Appl. Polymer Sci.*, **7** (1963) 1417.
165. S Rosenbaum, *ibid.*, **9** (1965) 2085.
166. H E Milson, *Amer. Dyestuff Rep.*, **44** (1955) 417.
167. H Hendrix, *Melliand Textilber.*, **43** (1962) 159.
168. T Vickerstaff, *Hexagon Digest*, **20**, 7.
169. P Rochas, *Teintex*, **24** (1959) 825.
170. *Idem.*, *Chemifasern*, **10** (1960) 666.
171. A H Brown and A T Peters, *Amer. Dyestuff Rep.*, **57** (1968) 284.
172. H Hendrix, *Melliand Textilber.*, **43** (1962) 258.
173. H Hendrix, *Z. Ges. Textilind.*, **62** (1960) 921.
174. W C Ingamells, *J.S.D.C.*, **96** (1980) 470.
175. W Waters, *J.S.D.C.*, **66** (1950) 609.
176. F M Rawicz, D M Cates and H A Rutherford, *Amer. Dyestuff Rep.*, **50** (1961) 320.
177. H Fujita and A Kishimoto, *J. Polymer Sci.*, **28** (1958) 547.
178. M K Gokale, L Peters and C B Stevens, *J.S.D.C.*, **74** (1958) 236.
179. H Fujita and A Kishimoto, *J. Polymer Sci.*, **28** (1958) 547.
180. M L Longo and M Torre, *Amer. Dyestuff Rep.*, **70** (1981) 41.
181. M G Kulkarni and R A Mashelkar, *J.S.D.C.*, **99** (1983) 131.
182. J Reiker and M Terlinden, *Melliand Textilber*, **56** (1975) 566.
183. T K Bhattacharje, *Amer. Dyestuff Rep.*, **66** (9) (1977) 68.
184. K Bredereck and E Koch, *Melliand Textilber.*, **55** (1974) 157.
185. V C Smith and F E Barwick, *Proc. AATCC Symposium Textile Solvent Technology: Update, Atlanta* (1973) 135.
186. R D Matkowsky, H D Weigmann and M G Scott, *Text. Chem. Col.*, **12** (1980) 55.
187. R A F Moore and H D Weigmann, *ibid.*, **13** (1980) 7.
188. R J Papuga, C D Liversgood and G N Mock, *ibid.*, **15** (1983) 91.
189. V N Zabashta, M Sotton and J Jacquemart, *Bull. Ci.*, **ITF3** (1974) 277.
190. L Peters and C B Stevens, *Dyer*, **115** (1956) 327.
191. *Idem.*, *J.S.D.C.*, **72** (1956) 100.
192. A Takaoka and M Aki, *Sen-i Gakkaishi*, (1965) 425, 432, 437.
193. B Koussens, *Teintex*, **38** (1973) 209.
194. W Beal and G S A Corbishley, *J.S.D.C.*, **87** (1971) 329.
195. H T Lokhande, *Colourage*, **20** (1973) 31.
196. L Peters, C B Stevens, J Budding, B C Burdett and J A W Sykes, *J.S.D.C.*, **76** (1960) 543.
197. M K Gokhale, L Peters and C B Stevens, *ibid.*, **74** (1958) 236.
198. J A Medley and D K Ramsden, *J. Text. Inst.*, **51** (1960) T1311.
199. Z Gur-Arieh and W C Ingamells, *J.S.D.C.*, **90** (1974) 8.
200. G Alberti, A Cerniani and R de Georgi, *Text. Research J.*, **55** (1985) 635.
201. W Beal and G S A Corbishley, *J.S.D.C.*, **87** (1971) 329.
202. *International critical tables of numerical data: physics, chemistry and technology*, vol. 6 (New York and London: McGraw Hill Book Co., 1926).
203. Z Gur-Arieh, W C Ingamells and R H Peters, *J.S.D.C.*, **92** (1976) 332.
204. T M A Hossain *et al.*, *J. Appl. Polymer Sci.*, **13** (1969) 541.
205. V Rossback, H Muller and D Nissen, *Textilveredlung*, **9** (1974) 339.
206. W C Ingamells, *J.S.D.C.*, **96** (1980) 466.
207. W C Ingamells, S H Lilou and R H Peters, *J. Appl. Polymer Sci.*, **26** (1981) 4095.
208. B Milicevic, *Rev. Prog. Coloration*, **1** (1969) 49.
209. A J Shipman, *ibid.*, **2** (1971) 42.
210. B Milicevic, *Text. Chem. Colorist*, **2** (1970) 18.
211. *Bayer Farben Rev., Special Edition 13* (ITMA Paris) (1971).
212. H H Hofstetter, *Melliand Textilber.*, **50** (1969) 321, 455, 845; **51** (1970) 91, 336.
213. J K Skelly, *J.S.D.C.*, **91** (1975) 177.
214. R B Love, *Rev. Prog. Coloration*, **6** (1975) 18.
215. B Milicevic, *J.S.D.C.*, **87** (1971) 503.

216. B Milicevic, *Text. Research J.*, **39** (1969) 677.
217. W Kothe, *Proc. AATCC Symposium Textile Solvent Technology: Update, Atlanta* (1973) 127.
218. P A Small, *J. Appl. Chem.*, **3** (1953) 71.
219. J A Harris and T H Guion, *Text. Research J.*, **42** (1972) 626.
220. S A Siddiqui, *Text Research J.*, **51** (1981) 527.
221. W C Ingamells and M N Thomas, *Text. Chem. Colorist*, **16** (1984) 55/11.
222. K Gerbert, *J.S.D.C.*, **87** (1971) 509.
223. STX BP 1, 312, 316 April 30th (1969).
224. G S Dodd and W C Ingamells, *J.S.D.C.*, **92** (1977) 74.
225. *Idem., ibid.*, **92** (1977) 79.
226. R B Chavan and J Vankata Rao, *Text. Research J.*, **53** (1983) 290.
227. R B Chavan, J Vankata Rao and H Ahmad, *ibid.*, **53** (1983) 347.
228. S M Betrabet, E H Daruwalla and H T Lokhande, *Text. Research J.*, **36** (1966) 684.
229. M P Kulkarni and H T Lokhande, *Text. Research J.*, **36** (1966) 684.
230. P K Chidambareswaran, *J. Appl. Polymer Sci.*, **22** (1978) 3089.
231. H T Lokhande, *ibid.*, **22** (1978) 1243.
232. *Idem., ibid.*, **22** (1978) 533.
233. ICI Technical Information D967, *Continuous thermofix dyeing of polyester cellulose blends.*
234. K V Datye *et al.*, *Proc. Symposium Contributions Chem. Synthethic Dyes and Mechanism of Dyeing, University of Bombay* (1968) 57.
235. J C Bent, T D Flynn and H H Sumner, *J.S.D.C.*, **85** (1969) 606.
236. H L Jacobs, *J. Appl. Polymer Sci.*, **13** (1969) 1191.
237. ICI Technical Information, D1055 *Classification of disperse dyes according to dyeing and heat fastness properties on polyester fibre.*
238. K Ito, *J. Appl. Polymer Sci.*, **14** (1970) 55.
239. R Casty, *Amer. Dyestuff Rep.*, **49** (1960) 952.
240. H Abel and R Casty, *Textil Praxis*, **11** (1967) 62, 809.
241. R L Elliot, N Oldfield, C B Steven and C Whewell, *J.S.D.C.*, **77** (1961) 727.
242. J C Atkinson, A Filson and J B Speakmann, *Nature*, **184** (1959) 444.

CHAPTER 4

Thermodynamics of dye sorption

HARRY H SUMNER

4.1 INTRODUCTION

When a textile fibre is immersed in a solution of dye under suitable conditions, the fibre becomes coloured, the colour of the solution decreases and dyeing has occurred. This process must not be confused with that of imbibition, in which the fibre merely becomes saturated with coloured liquid, and hence stained, whilst the concentration of dye in the solution remains unchanged. In the former case there is obviously some driving force which causes the transfer of dye from the solution to the fibre, and solutes which behave in this manner are said to be *substantive* to the fibre in question. In *Colour terms and definitions* [1] it is suggested that substantivity be defined as 'the attraction between a substrate and a dye or other substance under the conditions of test whereby the latter is selectively extracted from the application medium by the substrate'.

It became obvious from the start that different dyes show marked differences in substantivity for the same fibre, and also that the same dye shows differences on different fibres. Whilst it was (and still is) possible for the dyer to become familiar with these differences on the basis of experience and to adjust dyeing recipes accordingly, it was quickly appreciated that some numerical method of defining substantivity was required. In this way it was hoped to tabulate dyes according to their dyeing behaviour, and subsequently to use the information to select dyes of similar properties for use in mixtures or to select dyes for use under given dyeing conditions. From the point of view of the organic chemist it was thought that such information would help in the designing of new dyes by relating dyeing properties to the chemical constitution of the dye. Many attempts were made to formulate such a classification of dyes but, while useful in a limited sense, they failed to produce a completely satisfactory system because they were dependent upon arbitrary conditions of measurement, including the purity of the dye studied, and any change in these conditions could – and often did – produce a change in the classification. An example will serve to illustrate the problem. The degree of exhaustion would seem to be an obvious measure of the substantivity of a dye for the textile material concerned. This is usually quoted as a percentage, i.e.

$$\% \text{ exhaustion} = \frac{\text{amount of dye on fibre at end of dyeing}}{\text{amount of dye applied at start of dyeing}} \times 100 \quad (4.1)$$

Unfortunately the exhaustion achieved in a particular dyeing is dependent on many factors, such as the presence and concentration of other substances in the dyebath (including other dyes), the concentration of dye applied, the pH, the temperature, the time of dyeing and the liquor-to-goods ratio. The problems associated with exhaustion are somewhat alleviated by using the so-called substantivity ratio:

$$\text{substantivity ratio} = \frac{\text{concentration of dye on fibre}}{\text{concentration of dye in solution}} \quad (4.2)$$

Because concentrations are used instead of quantities the problem of the liquor-to-goods ratio is removed and, because such ratios are invariably obtained from measurements made when the system has reached an equilibrium position, so is the problem of dyeing time. All the other factors still remain, however. Of course either parameter could be used if it were feasible to spend a prodigious amount of time and effort in accumulating data for every possible set of dyeing conditions and storing this information in some large and efficient data retrieval system.

Clearly any such solution to the problem is unrealistic and any parameter or set of parameters which is to be used to define dyeing properties must be of a sufficiently fundamental nature as to be uninfluenced by variations in the dyeing conditions. Dyeing thermodynamics would appear to provide such a measure of dye and fibre properties.

4.2 BASIC CONCEPTS

4.2.1 Reversible equilibrium systems

Thermodynamic treatments of the dyeing process are almost invariably applied to truly reversible systems at equilibrium. Many dyeing processes are essentially irreversible, at least on a molecular scale, in that they can only be reversed by another drastic chemical reaction. Examples of such systems include the various forms of mordant dyeing, the oxidation of leuco vat and sulphur dyes, the azoic coupling of naphthols within the fibre and the formation of covalent bonds between dye and fibre in reactive dyeing. This chapter is not concerned with any such systems, and will consider only those which are truly reversible: disperse dyes on various fibres, direct and leuco vat dyes on cellulosic fibres, acid dyes on wool and nylon, and basic dyes on acrylic fibres.

The other essential requirement for the application of thermodynamics is that all measurements of the distribution of dye or other adsorbate must be made when the system is in equilibrium. When an undyed piece of material is placed in a suitable dyebath, uptake of dye is relatively rapid at first; this initial rate decreases, however, until eventually no further dye is adsorbed. It would be wrong to assume that the system is in a static condition at this point, and that dye molecules no longer move between the bath and the fibre. In fact interchange is still occurring, but at this point the rate of adsorption of dye by the fibre is equal to the rate of desorption from the fibre, that is, a state of equilibrium has been achieved. When the system has reached this point it is said that the concentration of dye in the fibre is *in equilibrium* with the concentration of dye in the solution. Before any measured quantities are used in the thermodynamic treatment, it is essential to ensure that true equilibrium has been reached. This is normally considered to be the stage at which dyeings for longer periods of time produce no further increase in the amount of dye taken up by the fibre. An alternative approach to that of dyeing is desorption, in which a dyed piece of material is immersed in a bath containing no dye and desorption allowed to take place until increasing time produces no change in the concentrations reached. The ideal would be to combine the two approaches for each dye–fibre system under consideration. If the system is truly reversible the two methods should give the same equilibrium position. Because of the time and effort required for such a proceeding, however, only a few such experiments have been described, most workers being content to assume reversibility. Neale [2] demonstrated that the dyeing of Cellophane sheet with Chlorazol Sky Blue FF (C.I. Direct Blue 1) was completely reversible (Table 4.1) and Boulton,

TABLE 4.1

Equilibrium of Chlorazol Sky Blue FF on Cellophane sheet [2]

Dye concentration in solution /g l^{-1}	Dye on Cellophane	
	adsorption	desorption
0.2	1.64	1.58
0.1	1.26	1.26
0.05	1.04	1.00

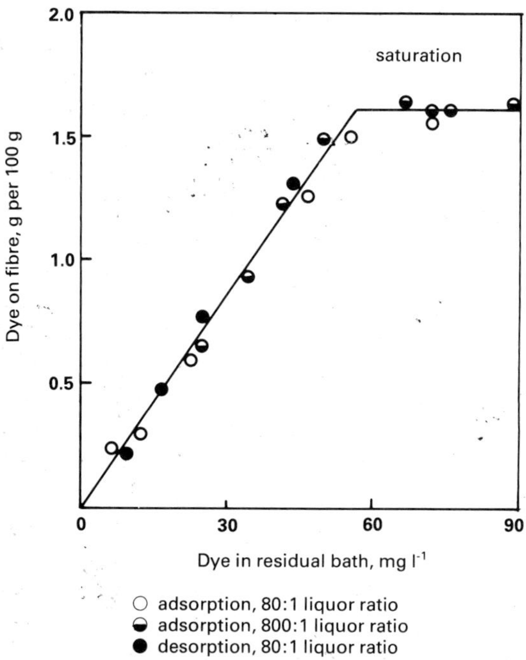

Figure 4.1 – Adsorption and desorption of C.I. Disperse Red 19 on secondary cellulose acetate at 80 °C [5]

Delph, Fothergill and Morton [3] found that the dyeing of cuprammonium rayon with Chrysophenine G (C.I Direct Yellow 12), Chlorazol Sky Blue FF and Benzopurpurine 4B (C.I. Direct Red 2) was also reversible. Studies of the dyeing of secondary cellulose acetate with a disperse dye [4, 5] and of poly(ethylene terephthalate) with 1,4-dihydroxyanthraquinone [6] demonstrated that both dyeing systems were truly reversible, in that the values obtained in both desorption and adsorption experiments lay on the same straight line (Figure 4.1).

4.2.2 Dyebath contents

In addition to ensuring that equilibrium has been achieved, other important precautions must be taken. The contents of the initial dyebath must be known precisely, and these include the adventitious additions that may have been made as a result of impurities in the dye or dyes in question and in the fibre to be dyed. Such impurities could produce significant variations in the results obtained, a fact that has not always been fully appreciated with the result that much of the earlier work on dyeing is of doubtful value. Fortunately the importance of this is now fully realised. One way of overcoming the problem is to submit the dyes and fibres to be used to a full analysis, so that the nature and amount of any impurity is known and can be included accurately in the dyebath contents. The most usual way of overcoming the problem, however, is to use pure dyes and carefully washed fibre. Such a solution is perfectly acceptable for theoretical studies, but in practice commercial dyes, with all their attendant impurities, have to be used and the method of analysis is the only one available.

Another point to be considered is that the precise contents of the final equilibrium dyebath must also be known. Whilst always of importance, such a precise knowledge was not as vital when using the earlier methods for the mathematical treatment of the results, but the modern treatments are more comprehensive and require a full knowledge of the concentrations of all the ions in the equilibrium dyebath. Again, for theoretical studies carried out in long or very long liquor-to-goods ratios concentrations in the final bath can be equated to those in the initial bath, assuming that the exhaustion of any of the ions present is experimentally insignificant. In shorter liquor ratios, such as are used in practical dyeing, this assumption can be made neither for the dye nor for any of the other ions present; either analysis of the final dyebath must be carried out or the final bath concentrations must be related in some way to those in the initial bath. The question is discussed later in the chapter.

4.2.3 Experimental methods

As described above, any experimental work must be carried out in such a way that the establishment of equilibrium is assured. There is always the danger of a false equilibrium – where the outer surfaces of the material are in equilibrium with the dye solution but the inside of the fibre is not – being mistaken for the true one. The occurrence of such ring-dyeing is well established, and if dyeing is terminated when such a state exists then any conclusions based on the result will be seriously in error. Adequate precautions must be taken to avoid such errors by carrying out preliminary experiments to establish the time required for the system to reach true equilibrium, that is, by carrying out the experiment for increasing lengths of time until increased time produces no change in the situation. Alternatively, where both adsorption and desorption methods are used, equilibrium is assured when both give the same distribution of dye.

Because of the long treatment times required with many systems to ensure that equilibrium has been reached, there is always the danger of decomposition of either the dye or the fibre, or both. This is clearly undesirable, since neither the dye–fibre system nor the application conditions are those intended, and checks are therefore essential. The simplest are to treat the dyebath, in the absence of fibre, under the conditions required and then compare the treated bath (either chromatographically or spectroscopically) with an untreated one. Similarly, the fibre is treated in a bath containing no dye and then its dyeing properties of this fibre are compared with an untreated one. Unfortunately, neither test gives a completely unambiguous result. If decomposition is observed then it is a positive finding and some action must be taken. If no decomposition is detected, however, this may be meaningful or it may not, since in some cases degradation of the fibre is 'catalysed' by dye [7, 8, 9], and also the rate of decomposition of the dye may be different when the dye is adsorbed into the fibre phase.

In view of the possibilities of degradation and decomposition it is essential to keep the dyeing time to the minimum necessary. In this context efficient agitation of the fibre in the bath or circulation of the bath over the fibre can result in a marked reduction in the time required. Even greater savings may be obtained by choosing the most suitable form of the material to be dyed, where such a choice is possible. This means selecting the material in a form which has the greatest surface area per unit mass and, in particular, avoiding closely packed systems or tightly woven materials. The surest way of reducing the time required is to increase the temperature of dyeing but this is not always possible since the results may be required at a certain temperature. In any case, increasing the temperature could be a two-edged sword: whilst increasing the rate of dyeing it could also increase the rates of decomposition of fibre and dye. Hence the effects of any changes should be checked.

4.2.4 The direct approach: adsorption

The most obvious way of carrying out the necessary dyeings is to immerse the undyed fibre in a dyebath at the chosen liquor-to-goods ratio and continue the dyeing for the required length of time, i.e. the direct approach. This essentially requires no special apparatus, except for adequate means of controlling temperature, supplying sufficient agitation and avoiding the effects of evaporation. At the end of dyeing it is then necessary to determine the concentrations of dye in the fibre and dyebath phases. Some workers have reduced the work required by determining just one concentration and obtaining the other by difference. In general they chose to measure the phase that showed the greatest change in concentration during dyeing. Thus at liquor-to-goods ratios less than *ca* 100:1 and high levels of exhaustion the bath shows the greatest change, and hence the bath is the phase to be measured. When the exhaustion is low the concentration of dye in the bath does not change very much; the percentage experimental error could be relatively large, and in turn will result in a large error when the small concentration of dye on fibre is determined by difference. In such a case it is better to measure the dye concentration on the fibre and obtain that in the bath by difference. Either method requires an accurate knowledge, not only of the total amount of dye present, but also the mass of fibre and the volume of the dyebath. At longer liquor ratios (greater than *ca* 100:1) the concentration of dye in the bath will, in general, change very little and so the fibre must be the phase to be measured. In general, however, it is undoubtedly better to measure both phases so that the mass balance can be checked to ensure that no losses have occurred either by decomposition or in the experimentation.

Measurement of the concentration of dye in the bath is relatively straightforward. The absorbance (formerly called the optical density) of the solution is measured, possibly after dilution and probably with the addition of a suitable solvent to ensure that the Beer–Lambert Law is obeyed (see Chapter 2), and converted to concentration using an appropriate calibration curve or constant. Determining the concentration of dye on the fibre is more complex, and the following is probably the best method to use. After dyeing the fibre is removed from the dyebath, squeezed to remove as much dye liquor as possible (but not rinsed) and then weighed. It is dried and then conditioned at a known ambient humidity to constant mass. The difference gives the mass of entrained liquor remaining on the fibre after squeezing and the amount of dye in this liquor can be determined from the known concentration of dye in the dyebath. Then either the whole of the fibre sample, including the dye, is dissolved in a suitable solvent, or the dye is extracted from the fibre by repeated treatments in an appropriate solvent. The absorbance of the resulting solution, diluted if required, is measured and the dye concentration obtained using a suitable calibration curve or constant. The total amount of dye on the fibre can thus be determined and corrected for the dye in the entrained liquor to give the amount of dye on a known mass of fibre. In certain cases it may be necessary to correct the dry, conditioned mass of the dyed fibre for the now known mass of dye which it contained.

4.2.5 Desorption

Since the dyeing systems under consideration are assumed to be reversible, the alternatives to the adsorption methods described are those involving desorption of dye from a previously dyed fibre. A sample of the material is first dyed with the required concentration of dye and then it is thoroughly rinsed and dried. It may then be treated in various ways, for example:

1. A quantity of the dyed fibre is immersed in a volume of blank dyebath (i.e. a bath containing all the required ingredients except the dye) at the required temperature, and dye is allowed to desorb until equilibrium is reached. Concentrations of dye in the fibre and solution phases are then determined by standard techniques.

2. The dyed material and a sample of undyed material are placed in the same blank dyebath at the required temperature and are treated until the dye concentration is the same on each. Both pieces and the solution are then analysed for dye concentration. This method not only establishes that equilibrium has been reached but also that the dyeing system is reversible.

3. For the multiple desorption method devised by Gilbert [10], a large sample of material is required which has been dyed to the required depth of shade at a temperature high enough and for a time long enough to ensure uniform distribution of the dye not only through the bulk of the material but also through the cross-section of each individual filament. (In other words, the sample is dyed to equilibrium.) The dyed material is then thoroughly rinsed, dried, conditioned and analysed for dye content. A known mass of this dyed material is introduced into bulb A of the apparatus (Figure 4.2) and a small volume of blank dyebath is added into tube B. The whole

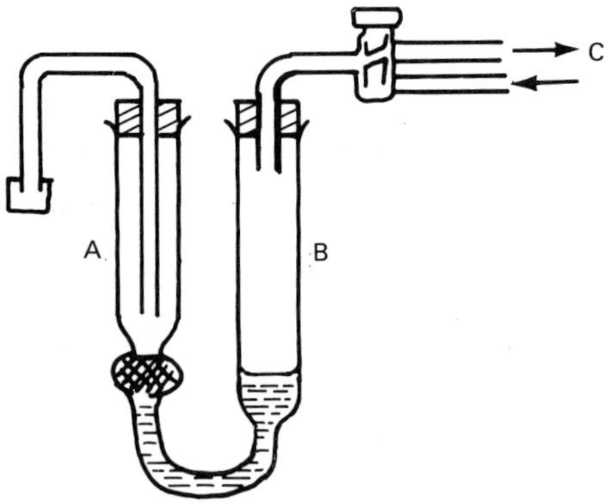

Figure 4.2 – Desorption apparatus [10]

apparatus is immersed in a thermostatted bath at the required temperature. By alternating the application and release of vacuum at C, the liquor is drawn back and forth through the dyed plug. When an adequate quantity of dye has been desorbed, as judged by the colour of the solution, the experiment is stopped and the dyed material is replaced by a fresh sample from the same original dyeing and the experiment restarted. This process is repeated using the same liquor, until the introduction of a fresh, dyed sample produces no change in the concentration of dye in the solution. The concentration of dye in the fibre is now in equilibrium with that in the solution. The concentration in solution is then measured, and that in the fibre is known from the initial analysis. Because each treatment is merely an equilibration between the solution and the outer surfaces of the fibre it requires the transfer of only very small quantities of dye, and hence the time required is quite short (15–

30 minutes per treatment) and the number of equilibrations necessary is generally small. It is claimed that the short treatment times make this a desirable method for use with substrates and dyes which are prone to degradation and decomposition. This is undoubtedly true of the equilibration part of the method, but it must be remembered that the material used had to be dyed at the start under less advantageous conditions.

4.2.6 Adsorption isotherms

The results obtained from dyeing experiments carried out to equilibrium are usually presented in the form of an adsorption isotherm (cf. section 1.5.4), in which the change in the concentration of dye on the fibre with that in the solution is followed under conditions where all the other variables, including temperature, are kept constant. The effects of changes in these other variables are described by the displacement of this concentration isotherm. Experimentally it has been found that most dyeing isotherms approximate in shape to one of three classical forms.

The partition isotherm

This is the simplest possible model. Just as a non-ionic dye dissolves in organic solvents and can be said to have affinity for them, so in dyeing the

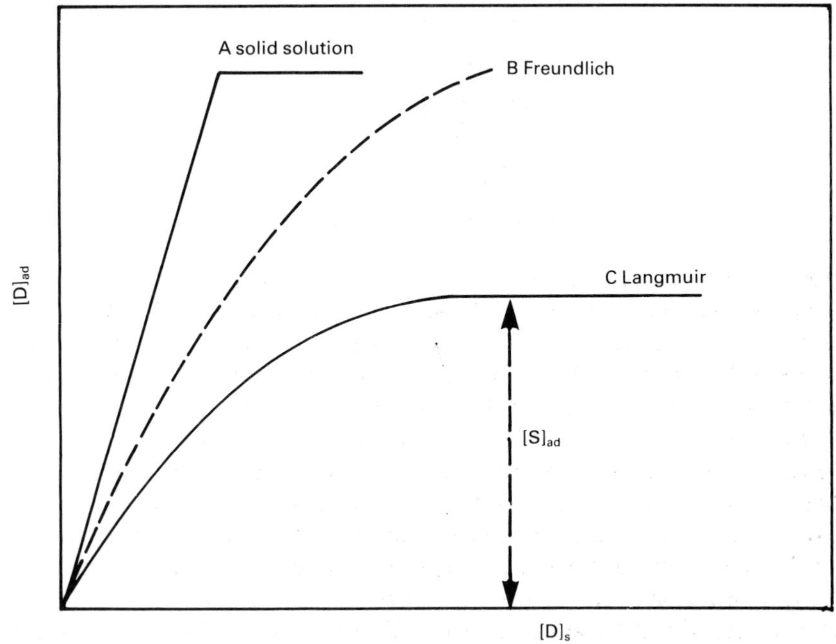

Figure 4.3 – Adsorption isotherms

dye can be considered as dissolving in the fibre because of its affinity for the latter. For this reason the process is often described as conforming to a solid solution model. When the concentration of dye in the fibre $[D]_{ad}$ (g kg^{-1} or mol kg^{-1}) is plotted against the concentration of dye in the dyebath $[D]_s$ (g l^{-1} or mol l^{-1}) a linear relationship is found which abruptly changes to a horizontal plateau (curve A in Figure 4.3). The rising, linear part is defined by the equation

$$[D]_{ad} = K[D]_s \tag{4.3}$$

and the constant K is termed the *distribution coefficient* or the *partition ratio*. The sharp break in the curve marks the saturation of the dye in the solution. Adding dye to the bath in excess of this amount does not change the concentration in solution, which remains at a value equal to the solubility of the dye under the particular experimental conditions, and hence the value of $[D]_{ad}$ remains constant (Eqn 4.3).

The Freundlich isotherm

The classical work of Freundlich on adsorption by animal charcoal led to the purely empirical conclusion that the adsorption isotherms could be defined by the equation, written in terms of dye,

$$[D]_{ad} = K[D]_s^x \tag{4.4}$$

where K is a constant and x is a fraction.

Curve B in Figure 4.3 is a typical isotherm that may be described by this relationship, and such isotherms are widely observed in practice. Eqn 4.4 is such that if logarithms of both sides are taken, a linear equation is obtained:

$$\log [D]_{ad} = \log K + x \log [D]_s \tag{4.5}$$

Thus the obedience of any set of results to the Freundlich isotherm can be confirmed by plotting log $[D]_{ad}$ against log $[D]_s$, when a linear relationship should be obtained.

The Langmuir isotherm

This was originally defined to account for the adsorption of gases on metal surfaces, and was derived by Langmuir on kinetic grounds. It has proved to be reasonably well suited to describe dye adsorption by certain textile fibres. Considered in terms of dyeing, the basic postulate is that adsorption of dye takes place on specific sites in the fibre and that when a dye molecule occupies a site that site is saturated and incapable of further adsorption. All the sites are assumed to be identical, and there is no interaction between them or between dye molecules occupying sites. Assume that a fibre contains $[D]_{ad}$ mol kg^{-1} of dye in equilibrium with a dyebath containing $[D]_s$ mol l^{-1}, and the fibre contains $[S]_{ad}$ potential dye sites per kg. Under these conditions dye will be leaving the fibre with a rate of desorption proportional to the concentration of dye in the fibre, i.e.

$$-\frac{d[D]_{ad}}{dt} = k_1[D]_{ad} \tag{4.6}$$

At the same time dye will be entering the fibre at a rate that depends on the concentration of dye in solution and the number of unoccupied sites in the fibre, which is equal to $([S]_{ad}-[D]_{ad})$. Thus the rate of adsorption will be

$$\frac{d[D]_{ad}}{dt} = k_2[D]_s([S]_{ad}-[D]_{ad}) \tag{4.7}$$

At equilibrium the rate of adsorption equals that of desorption, so that

$$k_1[D]_{ad} = k_2[D]_s([S]_{ad}-[D]_{ad}) \tag{4.8}$$

If this is rearranged and k_2/k_1 equated to k, then

$$[D]_{ad} = \frac{k[S]_{ad}[D]_s}{1 + k[D]_s} \tag{4.9}$$

This is the equation of a Langmuir isotherm, and such isotherms have the form illustrated in Figure 4.3, curve C. It is worthy of note that the horizontal portion of curve C is defined by the number of available sites in the fibre, i.e. it occurs when the fibre is saturated. This is in contrast to the horizontal part of curve A, where the limiting factor can be considered as being the solubility of the dye in the bath.

If Eqn 4.9 is inverted,

$$\frac{1}{[D]_{ad}} = \frac{1}{k[S]_{ad}[D]_s} + \frac{1}{[S]_{ad}} \tag{4.10}$$

it can be seen that if the reciprocal of the concentration of dye adsorbed $1/[D]_{ad}$ is plotted against the reciprocal of the concentration in solution $1/[D]_s$, then a straight line should be obtained with a slope of $1/k[S]_{ad}$ and an intercept on the $1/[D]_s$ axis of $1/[S]_{ad}$. This graphical method has been widely used to obtain the saturation value, that is the number of available sites $[S]_{ad}$ for use in the equation.

A further point which must be made regarding the Langmuir isotherm is that dividing Eqn 4.8 by $[D]_s$ and again equating k_2/k_1 to k gives

$$\frac{[D]_{ad}}{[D]_s} = k([S]_{ad} - [D]_{ad}) \tag{4.11}$$

which when $[D]_{ad}$ is very much smaller than $[S]_{ad}$ leads to the result that

$$\frac{[D]_{ad}}{[D]_s} \approx k[S]_{ad} \tag{4.12}$$

With any particular fibre under a given set of dyeing conditions both k and $[S]_{ad}$ should be constants, so that Eqn 4.12 then becomes a linear relationship which can easily be confused with a simple partition isotherm.

Each of the three classical isotherms described above is based on a postulated model of the adsorption process, and it has been found experimentally that most dyeing isotherms approximate in shape to one or other of them. This in turn led to the assumption that if a set of experimental dyeing results conformed to one of these classical isotherms, then the theoretical model for the particular dyeing process was defined. Whilst such an assumption (whether true or false) produced a workable model to interpret many dyeing processes, in certain cases it also resulted in problems and complications which resulted in further assumptions being necessary to make the model fit the experimental data. Care is therefore necessary when making any decision regarding the most appropriate model based on the observed shape of the experimental isotherm.

4.2.7 Activity coefficients of dyes

As indicated in Chapter 1, thermodynamic equations are based on activities of the species involved (rather than their concentrations) in the two phases, fibre and solution. The experimental methods described above produce results in terms of the concentrations of dye in the two phases, however. If the dye in both cases behaved ideally there would be no problem, because activity could then be equated to concentration. It is unlikely that dyes will behave ideally, particularly in solution, and therefore this simple equality cannot be used and some form of correction factor should be included. Lewis [11] used the relationship

$$a = cf \tag{4.13}$$

where a = the activity, c = the concentration and f = the *activity coefficient*, a factor which describes the degree of departure from ideality.

In general, as $c \to 0$, $f \to 1$ so that in very dilute solutions the situation approximates to being ideal and activity can be equated to concentration. In more concentrated solutions, however, a knowledge of the value of the activity coefficient is desirable if the thermodynamic equations are to be applied rigorously.

There is a comprehensive literature describing the available methods of measuring activity coefficients, and attempts have been made to apply certain of these techniques to the study of dye solutions. Unfortunately very little success has been achieved because of the many special problems that are encountered with dyes, principally the low concentrations of dye in solution and the presence in solutions which relate to practical dyeing of swamping concentrations of other solutes.

The alternative to direct measurement is computation. The activities of strong electrolytes in aqueous solutions have been studied by Debye and Hückel [12] who reached and demonstrated the validity of the conclusion that

$$-\ln f = Az^2 I^{\frac{1}{2}} \tag{4.14}$$

where f = the activity coefficient of the ion in question, z = the charge on the ion, A = a constant, approximately equal to 0.5, and I = the ionic strength of the solution as defined by the equation

$$I = \tfrac{1}{2}\Sigma cz^2 \tag{4.15}$$

i.e. half the sum of the concentration of each ion in the solution multiplied by the square of its valency. Although many types of dye are ionised, the theory in this simple form applies only to small spherical ions in solutions of low total ionic strength. While it has been extended to take account of the radii of the ions it is extremely doubtful whether, even then, it would be applicable to dye ions, which are far from being simple spherical ions. In addition, many dyebaths would not be considered to be of low ionic strength.

Because of the lack of experimental results and because the use of the Debye–Hückel equations is not justifiable, it is the usual practice to equate the activity coefficient of the dye ions to unity and simply to substitute concentration for activity in the thermodynamic equations. The only justification for this decision is that it seems to work in a reasonably satisfactory manner; moreover, at the moment no alternative is available.

4.2.8 Aggregation of dyes

The problems associated with the non-ideal behaviour discussed above are accompanied by other problems, which have their origin in the aggregation of dye molecules or ions in the solution and possibly in the fibre. Determinations of activity coefficients in solution for certain dyes, e.g. Chlorazol Sky Blue FF (C.I. Direct Blue 1) [13] and two acid dyes [14], gave values that would be expected from a strong electrolyte, and hence the assumption must be that they dissolve to give monomolecular solutions under the conditions of measurement. In contrast there is a considerable amount of evidence in the literature that many ionic dyes readily form aggregates in solution. In general the results show that the degree of aggregation increases as the r.m.m. of the dye increases, but that this effect can be counteracted to some extent by increasing the number of solubilising groups in the molecule. Aggregation is promoted by increasing the concentration of dye in solution and also by the presence of added electrolyte; it is reduced by increasing the temperature, however, in many cases to an extent that justifies the suggestion that the dye will not be aggregated at the usual dyeing temperatures in a normal dyebath. In other instances where the usual dyeing temperature is low and large concentrations of electrolyte are present, or where the usual temperature has been reduced for experimental reasons, such an assumption could be in error.

Many methods for studying the aggregation of dyes in solution have been used by various workers [for example, 15–20] and it would have been of considerable value if their results had provided a means of determining what proportion of dye in the solution was capable of being adsorbed by the fibre. For example if, in any particular dyeing system, the fibre is able to adsorb dye in the monomolecular form only, then the concentration of dye in solution which is in equilibrium with that on the fibre is the concentration of single molecules. Most of the methods used produce a result in the form of an aggregation number which expresses the average number of molecules in the aggregates, and this figure is generally arrived at by way of several assumptions. Such a number is useful to demonstrate that changes in the degree of aggregation do occur with changes in dyebath conditions but, except in the case of a simple monomer–dimer equilibrium, they do not permit the calculation of the concentration of the effective form in the equilibrium mixture, assuming it is already known which that form is in terms of dye adsorption.

As a result workers studying the adsorption of dyes by textile materials have had to assume that the dye solutions used are always molecularly dispersed systems and the possibility of aggregation is generally reserved to explain, in a qualitative fashion, why certain sets of results do not conform to the expected pattern. Such interpretations could possibly be correct, but quantitatively they are unsatisfactory.

4.2.9 The standard affinity

When considering the distribution of the dye between the fibre and the dyebath at equilibrium the differences in the behaviour of different dyes in a particular dyeing system, or of the same dye under different conditions of application, are generally best described using the concept of standard affinity. For thermodynamically reversible systems the basic equation has been developed in Chapter 1, namely that

$$-\Delta\mu^{\ominus} = -(\mu_{\mathrm{f}}^{\ominus}-\mu_{\mathrm{s}}^{\ominus}) = RT\ \ln\frac{a_{\mathrm{f}}}{a_{\mathrm{s}}} \qquad (4.16)$$

$-\Delta\mu^{\ominus}$ is termed the *standard affinity* and it is defined [1] as 'the difference between the chemical potential of the dye in its standard state on the fibre and the corresponding chemical potential in its standard state in the dyebath'. It is a measure of the tendency of the dye to move from the solution to the fibre when it is in its standard state in each phase. The term 'affinity' should only be used with this exact significance, despite the fact that it is frequently loosely employed to describe a variety of dyeing phenomena.

In order to calculate the standard affinity of a dye in any real dyeing system it is necessary to know the activity of the dye in each phase, and to do this requires some means of expressing the activity in terms of the measurable concentrations of dye in each phase in such a way that adequately describes the behaviour of the dye. In the solution phase there is no fundamental problem, activity being replaced by the product of concentration and activity coefficient (Eqn 4.13), but there is the real practical problem that very little, if any, information is available about activity coefficients. The only alternative is to equate activity with concentration. For non-ionic dyes, therefore, in Eqn 4.16

$$a_{\mathrm{s}} = [\mathrm{D}]_{\mathrm{s}} \qquad (4.17)$$

and the standard state is that in which the concentration is unity. For ionic dyes of the general structure Na_zD, and considering them to be strong electrolytes and hence fully dissociated in solution, it is generally assumed (again ignoring activity coefficients) that

$$a_{\mathrm{s}} = [\mathrm{Na}]_{\mathrm{s}}^{z}[\mathrm{D}]_{\mathrm{s}} \qquad (4.18)$$

and the standard state is that in which the ionic product is unity. The treatment is quite general; Na^+ may be replaced by any other cation and also, for a dye which is positively charged (i.e. a basic dye), by the appropriate anion.

The choice of a suitable function to represent the activity of dye in the fibre phase is more difficult, in that it depends on the model or mechanism of dyeing that is considered to be appropriate to the dyeing system under consideration. The different models used to describe the various common dyeing systems will be discussed later, but it must be emphasised here that the standard state will depend on the model in use. Since the calculated value of the affinity will depend on the chosen values of the standard states in both phases it is important when making comparisons between dyes, possibly in differing dyeing systems, to ensure that the standard states are the same.

4.2.10 The heat of dyeing

The second thermodynamic quantity which is of importance in describing a dyeing process is the *heat of dyeing* or the change in enthalpy of the system when adsorption takes place. It is of practical significance, because it provides a convenient description of the effect of changing the temperature on the position of the dyebath–fibre equilibrium. It may be defined as the heat absorbed (δH) when a small quantity of dye (δn moles) is transferred from a large volume of dyebath, in which the dye is in its standard state, to a large mass of dyed fibre in which the dye is also in its standard state. Theoretically the volume of the bath and the mass of fibre involved have to be so large that the transfer of dye makes a negligible change in the concentrations involved. The molar heat of dyeing ($\Delta H^\ominus$) is then given by

$$\Delta H^\ominus = \frac{\delta H}{\delta n} \tag{4.19}$$

In most dyeing systems $\Delta H^\ominus$ is negative (i.e. dyeing is an exothermic process), so that raising the temperature results in less dye being adsorbed at equilibrium, i.e. the affinity is reduced. This point is illustrated in Figure 4.4 for an azo disperse dye on secondary cellulose acetate [4, 5]. When a dye is adsorbed by a textile material the amount of heat liberated is extremely small, but it can be measured directly if a very sensitive calorimeter is used, as was demonstrated by Derbyshire with Naphthalene Orange G (C.I. Acid Orange 7) on wool [21].

The more usual way of obtaining values for the heat of dyeing is by calculation. The thermodynamic temperature T, the standard affinity $\Delta\mu^\ominus$ and the standard heat of dyeing $\Delta H^\ominus$ are related by the equation

$$\Delta H^\ominus = \frac{\delta(\Delta\mu^\ominus/T)}{\delta(1/T)} \tag{4.20}$$

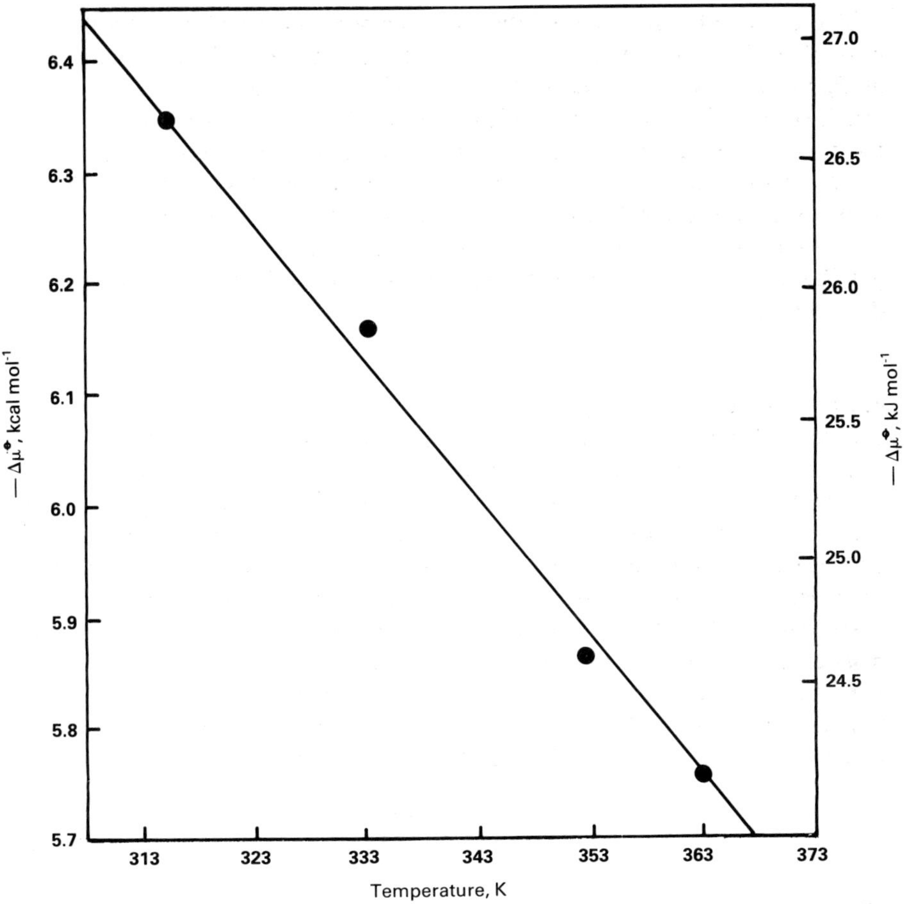

Figure 4.4 – Relation between affinity and temperature for C.I. Disperse Red 19 on secondary cellulose acetate

Hence a plot of $\Delta\mu^{\ominus}/T$ against 1/T will give a measure of $\Delta H^{\ominus}$ at any temperature by determining the slope of the curve at the selected temperature. If the plot is a straight line then it shows that $\Delta H^{\ominus}$ is a constant, independent of temperature, over the range studied. In this case Eqn 4.20 can be integrated to give

$$\frac{\Delta H^{\ominus}}{T} = \frac{\Delta\mu^{\ominus}}{T} + C \qquad (4.21)$$

where C is the integration constant. Eqn 4.21 can be used to calculate the value of $\Delta H^{\ominus}$ if the affinities $\Delta\mu_1^{\ominus}$ and $\Delta\mu_2^{\ominus}$ are known at temperatures T_1 and T_2 respectively. Substituting the appropriate values in Eqn 4.21 and eliminating C by subtraction gives

$$\Delta H^{\ominus}\left(\frac{1}{T_1}-\frac{1}{T_2}\right) = \frac{\Delta\mu_1^{\ominus}}{T_1}-\frac{\Delta\mu_2^{\ominus}}{T_2} \tag{4.22}$$

whence

$$\Delta H^{\ominus} = \frac{T_2\Delta\mu_1^{\ominus} - T_1\Delta\mu_2^{\ominus}}{T_2 - T_1} \tag{4.23}$$

Alternatively, knowing $\Delta H^{\ominus}$ and $\Delta\mu_1^{\ominus}$ at a temperature T_1 the value of $\Delta\mu_2^{\ominus}$ at temperature T_2 may be obtained from the equation in the form

$$\Delta\mu_2^{\ominus} = \frac{T_2}{T_1}(\Delta\mu_1^{\ominus}-\Delta H^{\ominus}) + \Delta H^{\ominus} \tag{4.24}$$

It must be repeated that the integration of Eqn 4.20 to give Eqn 4.21 is only possible if $\Delta H^{\ominus}$ is a constant independent of temperature over the range studied. If it is not, the use of Eqns 4.23 and 4.24 is not justified although it may be permissible if the temperature-dependence of $\Delta H^{\ominus}$ is small and (T_2-T_1) is also small.

Eqn 4.21 would indicate that if $\Delta\mu^{\ominus}$ is plotted against T and $\Delta H^{\ominus}$ is a constant, then a straight line should be obtained (cf. Figure 4.4). Extrapolation of this line to $T = 0$ would give an intercept equal to $\Delta H^{\ominus}$. Whilst possible, this method is not to be recommended since the extrapolation could cover a temperature range considerably greater than that covered by the experimental measurements.

In order to apply the above methods to obtain a value for the standard heat of dyeing it is necessary to compute a value of the standard affinity, and to do this it is necessary to have a model for the mechanism by which dyeing takes place. Generally this will be known but in some cases there may be uncertainty as to the precise mechanism. The calculation of affinities requires a satisfactory method of expressing the activities of dye in the fibre and in the dyebath, in terms of the measurable concentrations $[D]_{ad}$ (mol kg^{-1}) and $[D]_s$ (mol l^{-1}). The activity in the dyebath offers no problem, and if the activity in the fibre is assumed to be an unknown function of $[D]_{ad}$ which is independent of temperature and dye concentration ($f[D]_{ad}$), then Eqn 4.16 may be written, again ignoring activity coefficients,

$$-\Delta\mu^{\ominus} = RT \ln\frac{f[D]_{ad}}{[D]_s} \tag{4.25}$$

where $f[D]_{ad}$ = some function of dye concentration on the fibre and $[D]_s$ = dye concentration in the dyebath.

Replacing the values of $\Delta\mu_1^{\ominus}$ and $\Delta\mu_2^{\ominus}$ at temperatures T_1 and T_2 respectively, Eqn 4.22 becomes

$$\Delta H^{\ominus} \frac{(T_2 - T_1)}{RT_1T_2} = \ln \frac{f[D]_{as(1)}}{[D]_{s(1)}} \cdot \frac{[D]_{s(2)}}{f[D]_{ad(2)}} \tag{4.26}$$

If the experimental conditions are so chosen that the concentrations of dye in the fibre are the same at both temperatures, then

$$\Delta H^{\ominus} \frac{(T_2 - T_1)}{RT_1T_2} = \ln \frac{[D]_{s(2)}}{[D]_{s(1)}} \tag{4.27}$$

Thus it is possible to calculate the heat of dyeing from a knowledge of the concentration of dye in solution which is in equilibrium with the same amount of dye on the fibre at two different temperatures, i.e. the isosteric adsorption values. The desorption method due to Gilbert [10] is ideally suited to this latter method, in that a fibre containing a known amount of dye can be equilibrated with solution at different temperatures.

4.2.11 The entropy of dyeing

The third thermodynamic parameter is the standard change in the entropy of the system ($\Delta S^{\ominus}$). It is linked to the other two thermodynamic parameters by the equation

$$\Delta\mu^{\ominus} = \Delta H^{\ominus} - T\Delta S^{\ominus} \tag{4.28}$$

and because there is no direct experimental means of measuring the entropy change it is necessary to calculate its value from measurements of $\Delta\mu^{\ominus}$ and $\Delta H^{\ominus}$. This may be carried out by the direct insertion of the appropriate values in Eqn 4.28 or, alternatively, by using graphical methods. Eqn 4.21 may be rewritten:

$$\Delta\mu^{\ominus} = \Delta H^{\ominus} - TC \tag{4.29}$$

from which it follows, comparing Eqns 4.28 and 4.29, that the integration constant C is equal to the standard change in entropy $\Delta S^{\ominus}$. Thus the graph of $\Delta\mu^{\ominus}/T$ *vs* $1/T$ (Eqn 4.21), used to find the value of $\Delta H^{\ominus}$ from the slope, could also be employed to get a value of $\Delta S^{\ominus}$ from the intercept at $1/T = 0$. This involves all the problems associated with an extensive extrapolation, however, and it is clear that a more acceptable value could be obtained using

Eqn 4.28, which indicates that a graph of $\Delta\mu^{\ominus}$ against T should have a slope of $\Delta S^{\ominus}$ and that it should be a linear relationship providing $\Delta H^{\ominus}$ is a constant independent of temperature.

Whichever methods are used to obtain values of the entropy change, they will obviously be subject to a certain measure of uncertainty as a result of the experimental errors involved in the determination of $\Delta\mu^{\ominus}$ and possibly of $\Delta H^{\ominus}$, depending on the method used. The levels of uncertainty which may be produced are illustrated in Table 4.2 for a few selected examples

TABLE 4.2

The possible uncertainty in the calculated values of $\Delta H^{\ominus}$ and $\Delta S^{\ominus}$ (assuming uncertainty in $\Delta\mu^{\ominus}$ of ±1 per cent)

$\Delta\mu^{\ominus}$ /kJ (kcal) mol^{-1}		$-\Delta H^{\ominus}$ /kJ (kcal) mol^{-1}	ΔS /J (cal) K^{-1} mol^{-1}	Uncertainty in $\Delta H^{\ominus}$ ± 5.9 kJ (1.4 kcal) mol^{-1}	Uncertainty in $\Delta S^{\ominus}$ ± 16.8 J (4 cal) K^{-1} mol^{-1}
at 60 °C	at 80 °C				
25.2 (6.0)	21.8 (5.2)	81.0 (19.3)	168 (40)	±7.3%	±10%
25.2 (6.0)	22.7 (5.4)	67.2 (16.0)	126 (30)	±8.8%	±13.3%
25.2 (6.0)	23.5 (5.6)	53.3 (12.7)	84 (20)	±11%	±20%
25.2 (6.0)	24.4 (5.8)	39.1 (9.3)	42 (10)	±15.1%	±40%

and, whilst there is no suggestion that other conditions would produce the same uncertainties, it must be assumed that similarly unsatisfactory figures could result.

In addition to the uncertainties of measurement, the significance of the entropy change is not completely clear. It has generally been regarded as a measure of the degree of orientation and restraint to which the dye is subjected when it is adsorbed by the fibre as compared with its situation in solution. Thus when comparing two dyes on the same fibre, the dye that demonstrates the greatest entropy change is taken to be more rigidly attached or more fully oriented than the other. It must be remembered, however, that the measured entropy change refers to the whole system and involves not only a change in respect of the dye but also changes in respect of the water, both in the solution and on the fibre, and of the fibre itself. Hence any interpretations made or conclusions drawn must be carefully considered in the light of the possibly complicated nature of the measured entropy change.

4.3 CLASSIFICATION OF DYEING SYSTEMS

Earlier it was said that in order to use Eqn 4.16 to calculate the affinity of a dye for a particular fibre it was necessary to choose a suitable function to represent the activity of the dye in the fibre phase, and that the function chosen depended on the model or mechanism of dyeing appropriate to the system under consideration. A study of the various dyeing systems of interest to which the equation would apply, i.e. reversible systems, reveals that they can be classified as follows:

1. *Non-ionic dyes* The only major class of dyes under this heading is that of the disperse dyes, and they may be considered
   (a) on essentially non-ionic substrates, such as the hydrophobic fibres produced from polyester, secondary cellulose acetate and cellulose triacetate, and
   (b) on ionic substrates such as nylon, wool and cellulose.
2. *Ionic dyes* applied to substrates which themselves carry a charge. These may be sub-divided into two classes:
   (a) where the substrate carries the same charge as the dye, as exemplified by anionic dyes on cellulose, and
   (b) where the substrate carries an opposite charge to that on the dye, under the normal conditions of application, such as anionic dyes on wool and nylon and cationic dyes on acrylic fibres.

This sub-division of ionic dyes describes the original thinking on the mechanisms of dyeing. It has more recently been suggested that such a separation of ideas is superfluous and that there need be only one system, i.e. ionic dyes on ionic substrates. Despite this suggestion, however, the various methods of applying the theory will be discussed, initially, according to this classification.

4.3.1 Non-ionic dyes on essentially non-ionic substrates

The introduction of secondary cellulose acetate in the 1920s produced an immediate problem, in that this fibre could not be dyed with the available water-soluble dyes. It was found, however, that coloration of the fibre was achieved using weakly polar dyes of relatively small molecular size. Such dyes, because of their lack of solubilising groups, are almost insoluble in water and this meant that they had to be applied as an aqueous dispersion of essentially solid dye particles in the dyebath in order to produce acceptable strengths of dyeing. Such a method of application led naturally to the adoption of the name *disperse dyes*.

In attempting to explain the mechanism of transfer of dye molecules from the solid dye particles to the substrate, two views were expressed. One suggested that solid dye was adsorbed on the surface of the material, from where it penetrated the fibre by a process of dissolution or solid-state diffusion; in support of this it had been observed microscopically that particles of dye adhered to the surface of the fibre and then slowly disappeared as adsorption took place [22]. Support for this view came from the further observation that if secondary cellulose acetate was shaken with solid dye powder for several days at 60 °C it became dyed. When such evidence was combined with the general belief held at the time that disperse dyes were insoluble in water, it is not surprising that this view of the mechanism of dyeing was preferred to the alternative. This was that disperse dyes were transferred to the fibre from a dilute solution surrounding the particles, by the same processes as ionic dyes [23], i.e. an aqueous solution theory. Later it was found that surface adhesion of particles did not always occur, and that when it did it was not of importance. The most convincing evidence against the solid-state diffusion theory came from an experiment in which a solid dye dispersion was retained in a Cellophane bag which was, in turn, immersed in the dyebath [24]. A fibre placed in the external solution became dyed even though there was no contact with the solid dye. Such an observation demands that the dye dissolves in the dyebath, and this possibility was demonstrated unequivocally when it was shown that disperse dyes had a measurable solubility in water [25, 4, 6]. This solubility is increased significantly by the presence of surface-active agents [25].

As a result of the above observations, the currently accepted mechanism of dyeing with disperse dyes is that the dye exists in two forms in the dyebath: solid dye and dye in solution. In the presence of fibre the following equilibrium situation is set up:

$$\text{dye}_{\text{solid}} \overset{\text{solubility}}{\rightleftharpoons} \text{dye}_{\text{dissolved}} \overset{\text{affinity}}{\rightleftharpoons} \text{dye}_{\text{fibre}} \tag{4.30}$$

where the overall position of equilibrium is determined by the solubility and the affinity obtaining at the temperature of the dyeing. In the presence of surface-active agents the situation could be complicated by the presence of dye in micelles of agent, i.e.

$$\text{dye}_{\text{solid}} \rightleftharpoons \text{dye}_{\text{micelles}} \rightleftharpoons \text{dye}_{\text{dissolved}} \rightleftharpoons \text{dye}_{\text{fibre}} \tag{4.31}$$

By taking due precautions to ensure that the dyebaths contained no particulate dye, and by achieving equilibrium by both adsorption and desorption techniques, it was established that the dyeing system was truly reversible and that the results conformed to a rectilinear isotherm. Typical results are shown in Figures 4.1 and 4.5, and there are many more instances in the

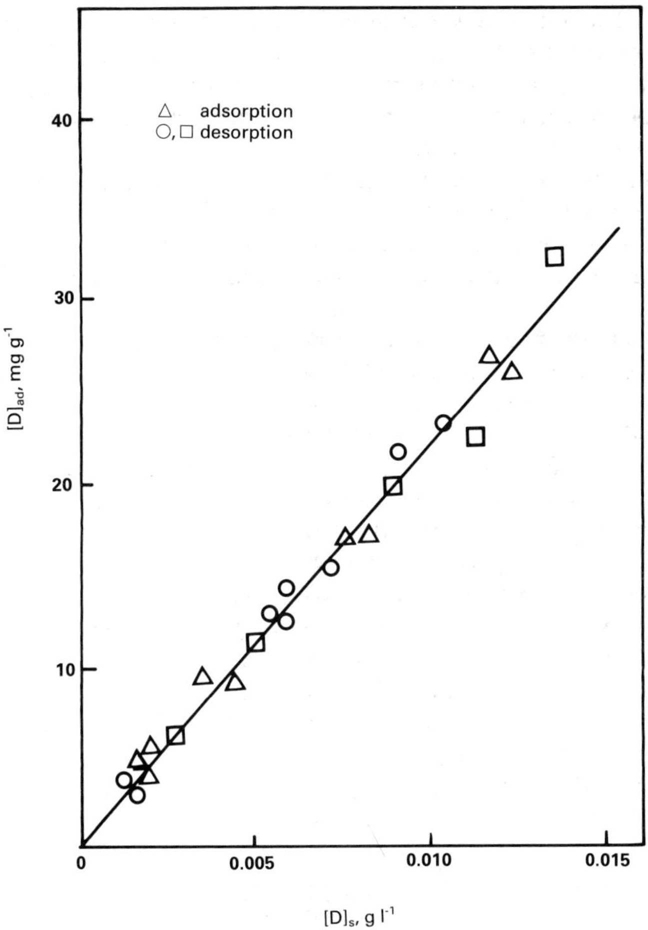

Figure 4.5 – Distribution of 1,4-dihydroxyanthraquinone between polyester fibre and water at 100 °C [6]

results of other workers using aqueous dyebaths. In all these cases the only equilibrium in the system at the end of dyeing is that between dye in true solution and dye on the fibre. Other workers using non-aqueous dyebaths have produced similar results [26, 27]. It is therefore well established that dyeing with disperse dyes is the transfer of dye molecules from a molecular dispersion (i.e. solution) into the fibre and because of the linearity of the isotherms obtained, the amount of dye adsorbed $[D]_{ad}$ relative to the concentration in the bath $[D]_s$ can be expressed by a partition coefficient K, i.e.

$$\frac{[D]_{ad}}{[D]_s} = K \qquad (4.32)$$

As more dye is introduced into the system a point will be reached at which the amount of dye in the dyebath at equilibrium exceeds the solubility of the dye. In the ideal case further additions of dye will produce no further change in the concentration of dye in solution, and hence no change in the concentration of dye on the fibre. At this point, therefore, if the abscissa denotes total dye in the bath rather than dye in solution, then the isotherm will become horizontal (Figure 4.1).

It would appear that the system of disperse dyes on hydrophobic fibres is one that is explained by the simplest of dyeing mechanisms, simple partition. There are certain complicating factors, however, which if not taken into account can lead – and indeed have led – to misleading results and consequently erroneous interpretations.

The first of these is the effects of added surface-active agent. Such agents increase the 'solubility' of the dye [25], an effect that occurs at concentrations of agent in excess of its critical micelle concentration due to adsorption of dye molecules into the micelles of the surfactant. There is no evidence available to suggest that the presence of surfactant actually increases the concentration of dye in true molecular solution. Whether or not any such change occurs, the concentration of dye in 'solution' measured in the presence of surfactant does not represent the concentration of dye in true molecular solution which is capable of taking part in the equilibrium partition between solution and fibre but some value in excess of this. Hence, even in the absence of particulate dye, the experimentally measured value of $[D]_s$ is too large for the corresponding value of $[D]_{ad}$, with the result that the calculated value of the partition coefficient K is reduced.

When the isotherm is considered it is found that the presence of surfactant does not affect the linearity but merely reduces the slope, again because the measured concentration of dye in solution is possibly in excess of that available for dyeing. As a result of such experimental observations, surface-active agents are said to reduce the affinity of a disperse dye for a particular fibre because they increase the dye's solubility in the aqueous phase. Such a statement is experimentally and for all practical purposes correct, but in mechanistic terms it is not an established truth. It is possible that it arises due to not measuring, or indeed being able to measure, the concentration of dye in true molecular solution rather than the total dye in both the solution and the micelles of surface-active agent. Some support for the view that the concentration of dye in true solution and the affinity are really unaffected by the presence of surfactant has come from the demonstration that surfactant does not change the amount of dye adsorbed at saturation [5]. This means either that the affinity and the concentration of dye available for dyeing are unaffected by surfactant, or that any reduction in the value of the affinity has been exactly compensated for by an increase in dye available for dyeing as a result of increased solubility. The latter would appear to be the more unlikely combination of effects.

The second complicating factor to be considered is that of the particle size of the dispersion being applied. In a dispersion the solubility of the particles depends on their surface energy, which in turn depends inversely on the radius of the particle, i.e. its size: hence small particles are more soluble than large ones. Thus in a heterogeneous dispersion (one in which there is a mixture of particle sizes) the solution may be in equilibrium with the smaller particles but supersaturated with respect to the larger ones. In such a case dye will crystallise on the surfaces of the larger particles, making them larger, and the smaller ones will disappear. The mean particle size will therefore increase with the result that the solubility will decrease, the concentration of dye in solution will decrease and, because of the constant partition coefficient, the amount of dye adsorbed by the fibre will decrease. This 'ageing' of a dispersion with time has been clearly demonstrated [24]. If the dispersion were homogeneous (i.e. if all the particles were the same size), this would not occur and the dispersion would be stable. Unfortunately all practical systems are heterogeneous.

Consider now the shape of the isotherm obtained, for the moment ignoring the effects of 'ageing'. On that part of the isotherm where dye is applied in excess of the saturation level, i.e. the horizontal part, more and more solid dye is being added and, at the same time, a smaller and smaller fraction is being dissolved. The mean particle size of the excess solid dye will therefore change and its solubility will change, resulting in a change in the amount of dye adsorbed. This part of the isotherm therefore need not be flat, but may rise or fall depending on how the mean particle size changes. If 'ageing' of the dispersion also occurs, then it is more than probable that the mean particle size will increase as more solid dye becomes available and the isotherm will fall.

The linear, rising part of the isotherm could also be affected by changes in particle size. In more fundamental studies carried out at long liquor-to-goods ratios, it is possible to ensure that the starting dyebath is free of particulate dye, the dye having sufficient solubility to permit the required mass of dye to be dissolved in the large volume of dyebath available. Under other more practical conditions, the demands of achieving the required depth of shade could mean that the amount of dye initially present in the aqueous phase is in excess of the solubility, and hence at the start of dyeing (and possibly for a considerable part of the dyeing time) particles of dye will be present. Complete solution will be achieved only as the dyebath becomes exhausted. The effects of crystal growth, with the resulting changes in solubility, could therefore be in evidence in this part of the isotherm.

Two extreme cases may be described. Firstly, if the rate of adsorption of dye by the fibre is greater than the rate of crystal growth, then dye in solution will be adsorbed before it can produce increased particle size. Hence particle size in the dyebath will decrease with increasing time, and the ultimate equilibrium will be achieved with a solution of dye which contains no particles. Secondly, if the rate of crystal growth is greater than the rate of dyeing, then crystals will grow resulting in a lower solubility, i.e. a lower $[D]_s$ and hence a lower $[D]_{ad}$. Such a drop in $[D]_{ad}$ will result in an even lower rate of dyeing, thereby increasing the effect. Under these circumstances, as the applied concentration is increased, the possibility of large particles forming is greater and the potential size of such particles is also increased. This will produce a decrease in $[D]_s$ and hence in $[D]_{ad}$. The normal methods of estimating the dye in the equilibrium dyebath would include the dye particles as well as the dye in molecular solution. Thus the values of $[D]_s$ used in preparing the isotherm would be too large by an amount which increased as the applied concentration increased, resulting in a curved rather than a linear isotherm.

A further effect which can occur with many disperse dyes, particularly under high-temperature dyeing conditions, is that at a certain critical temperature the crystal habit of the dye changes, generally to a more stable form which is markedly less soluble. It has been demonstrated [28] that an approximately six-fold reduction in solubility is possible by the physical modification of the particular dye studied. Such alteration would obviously produce marked changes in the concentration of dye in solution and hence in the amount on the fibre. Also, when the physical change takes place, crystals of the new modification would be surrounded by a solution of dye far in excess of their solubility and hence would grow, with all the expected consequences.

Most of the fundamental studies have been carried out using purified dyes. If commercial dyes are used a further complication is possible because, as supplied, they contain dispersing agent which is present as an aid to milling, to produce a redispersible product and to promote dispersion stability in the dyebath. Thus the greater the concentration of dye in the initial bath, the greater is the concentration of surfactant. This results in two effects: the greater the concentration of dye, the higher is the observed solubility and the lower the calculated partition coefficient, as discussed above. Both effects could markedly influence the shape of the isotherm and its linearity. Fortunately the full effects of this situation are not realised in most instances, since the initial dyebath is prepared using a swamping amount of extra surfactant.

4.3.2 The energetics of dyeing with non-ionic dyes

As described earlier and in Chapter 1, the standard affinity of a dye at temperature T K is defined by Eqn 4.16, and a value of the affinity can be calculated once the correct model for the dyeing system is decided and suitable values are ascribed to the activities of the dye in the fibre phase a_f and in the solution a_s. With non-ionic disperse dyes the linearity of the isotherms has resulted in their adsorption being regarded as a process of 'solid' solution defined by a partition coefficient (Eqn 4.32). The constancy of the partition coefficient implies that the dye is behaving ideally in both phases, so that the activities in Eqn 4.16 may be replaced by concentrations, i.e.

$$-\Delta\mu^{\ominus} = RT \ln \frac{[D]_{ad}}{[D]_s} = RT \ln K \qquad (4.33)$$

where $[D]_{ad}$ = the concentration of dye in the fibre (mol kg^{-1}), $[D]_s$ = the concentration of dye in the bath (mol l^{-1}) and the standard state in both phases is unit concentration. Using this equation the standard affinities of disperse dyes on cellulose diacetate, cellulose triacetate and polyester substrates have been calculated by many workers.

The significance of the affinities obtained is somewhat uncertain, in view of the different possible ways of expressing the concentrations of dye in either or both phases. The experimental methods give results in which $[D]_{ad}$ is in grams or moles of dye per unit mass of fibre and $[D]_s$ in grams or moles of dye per litre of solution. Whether grams or moles of dye are used as units is unimportant, as long as the same units are used for both quantities, but the measurement of fibre in terms of mass and that of solution in terms of volume result in K having the dimensions of [mass]$^{-1}$ when, in fact, it should have no dimensions. Such a situation has no serious disadvantages for many applications, e.g. the comparison of a range of dyes on a single fibre, and it might be assumed that workers who adopt this approach have equated litres to kilograms of dyebath, thereby removing the problem of dimensions with only a small resulting error. The alternative approach is to postulate that a volume of water V (l kg^{-1}) is associated with the fibre and that the dye in the fibre is contained in this volume. Hence $[D]_{ad}/V$, where $[D]_{ad}$ is in mol kg^{-1}, gives a concentration $[D]_f$ in mol l^{-1} and K_V become dimensionless. Eqn 4.33 therefore becomes

$$-\Delta\mu_V^{\ominus} = RT \ln \frac{[D]_{ad}}{V[D]_s} = RT \ln \frac{[D]_f}{[D]_s} = RT \ln K_V \qquad (4.34)$$

This concept of a volume term for fibres will be further discussed in relation to anionic dyes on cellulose. It has the advantage that the affinities of dyes on different fibres may be compared and, more particularly, that the affinities of dyes on different physical modifications of the same initial polymer may be equated by a suitable choice of V. For any given set of experimental results changing the value of V in Eqn 4.34 will alter the values of K_V and $\Delta\mu_V^\ominus$, and these values will be different from those of K and $\Delta\mu^\ominus$ obtained using Eqn 4.33 except, of course, for the case where $V = 1$, i.e. when $K = VK_V$. Obviously, therefore, it must be made absolutely clear which equation is being used.

The dyeing of cellulose diacetate, cellulose triacetate and poly(ethylene terephthalate) fibres from dyebaths containing aqueous solutions of disperse dyes has produced a lot of experimental data [6, 29, 30, 31, 32], from which affinities have been calculated. For all three fibres the isotherms were linear, reaching a saturation value as in Figure 4.3, curve A. From determinations of affinity at different temperatures values of the standard enthalpy (heat) of dyeing $\Delta H^\ominus$ have also been obtained for the different dyes on the various fibres using the methods associated with Eqns 4.20, 4.21 and 4.23. This in turn permitted calculation of the standard entropy changes involved, using Eqn 4.28. In all cases it was found that there was a decrease in entropy when the dye was adsorbed, suggesting an increase of order of the dye in the system. (Of course, the entropy of the total system – which includes fibre and water as well as dye – must increase, see section 1.2.5.)

In an effort to describe the energetics of the process, the solid dye has been used as the baseline for computation. Thus

$$\Delta H_{\mathrm{SF}}^\ominus = \Delta H_{\mathrm{SW}}^\ominus + \Delta H_{\mathrm{WF}}^\ominus \tag{4.35}$$

where the enthalpy changes concerned are $\Delta H_{\mathrm{SF}}^\ominus$ when dye moves from the crystal to the fibre (the enthalpy of solution in the fibre), $\Delta H_{\mathrm{SW}}^\ominus$ when dye moves from crystal to water (the enthalpy of solution), $\Delta H_{\mathrm{WF}}^\ominus$ when dye moves from water to fibre (the enthalpy of dyeing). The general picture that emerges from a study of a range of dyes on cellulose acetate is shown in Figure 4.6 [74]. The enthalpy of solution calculated from the effect of temperature on solubility using the equation

$$\Delta H_{\mathrm{SW}}^\ominus = R\frac{\mathrm{d}(\ln \text{solubility})}{\mathrm{d}(1/T)} \tag{4.36}$$

is positive and large in value, i.e. a rise in temperature markedly increases the solubility. The enthalpy of dyeing $\Delta H_{\mathrm{WF}}^\ominus$ is negative and numerically smaller than $\Delta H_{\mathrm{SW}}^\ominus$, so that the enthalpy of solution in the fibre $\Delta H_{\mathrm{SF}}^\ominus$ is also positive. Whilst interesting, this approach does little to clarify the situation.

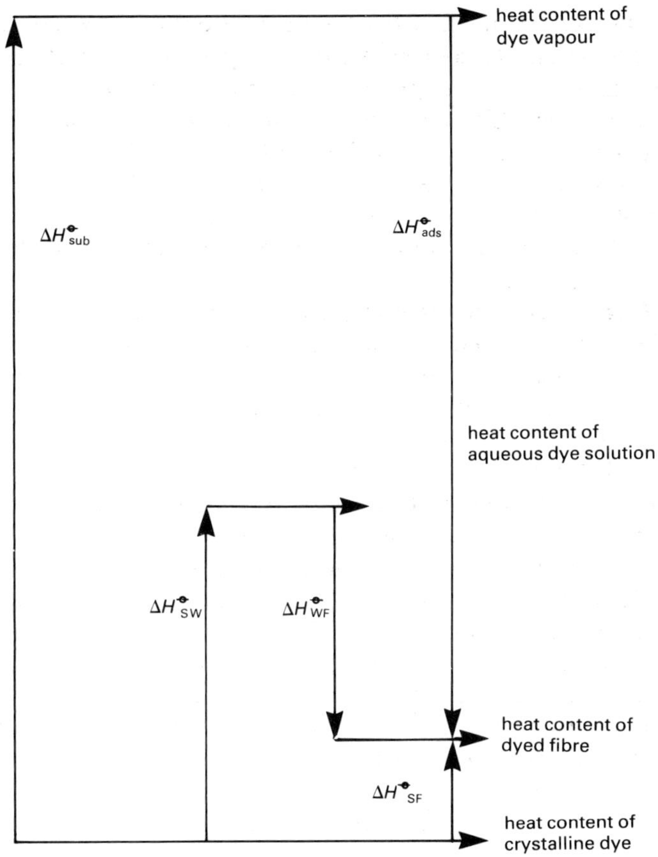

Figure 4.6 – Energy level diagram for dyeing processes

In a further attempt to resolve the energetics of the process, the vapour-phase dyeing of cellulose acetate was studied [33, 34]. Again the isotherms were linear, but the partition coefficients and hence the affinities were found to be extremely high when compared with those of the same dyes applied from aqueous solution. The enthalpies of adsorption of dye from the vapour phase are also high and found to be very close in value to the enthalpies of sublimation of the dyes when they are calculated using the equation

$$\Delta H^{\ominus}_{\text{adsorption}} = -\Delta H^{\ominus}_{\text{sublimation}} + \Delta H^{\ominus}_{\text{SW}} - \Delta H^{\ominus}_{\text{WF}} \qquad (4.37)$$

A general type of result is included in Figure 4.6. Clearly the results are very interesting and this approach is worthy of further study, particularly in view of the 'Thermosol' process, in which disperse dye is transferred from the solid or adsorbed state on one fibre to the adsorbed state on another through the vapour phase [35], and of the transfer-printing process with disperse dyes.

For many years workers have tried to establish whether the adsorption of disperse dyes by hydrophobic fibres is a process of 'solid solution' or one involving a 'site' mechanism. The idea of 'sites' being available in the fibre appears to have arisen because certain dyes on certain fibres do not produce rectilinear isotherms, and in such cases the results can be fitted using a Langmuir type of isotherm. This in turn led to the postulate that where linear isotherms are obtained they are merely the earlier part of a Langmuir isotherm, as discussed previously. Rightly or wrongly, such ideas would appear to ignore the possibility that the results are affected by such variables as particle size, solubility, the presence of dispersing agents and perhaps also modification of the fibre structure by the presence of dye. To date it must be said that the situation is not totally resolved and the understanding of disperse dye systems is still not complete, but that the weight of evidence would appear to favour the mechanism of 'solid solution', particularly the results of the vapour-phase studies [33]. This situation is discussed more fully in Chapter 2.

4.3.3 Non-ionic dyes on ionic substrates

Disperse dyes on nylon

This system is of considerable commercial importance, because of the good level-dyeing characteristics of the dyes and their ability to cover up fibre irregularities. The adsorption isotherms obtained are the same shape as those on the essentially non-ionic fibres, i.e. they are rectilinear and show a clear saturation value [36]. As a result it is again generally assumed that a 'solid solution' mechanism is operating and that affinities may be calculated using Eqns 4.33 or 4.34. The results are little affected by the pH of application unless, of course, the pH chosen results in the ionisation of the dye, in which case the system has changed to one in which both dye and fibre are ionic.

Because nylon is known to contain 'sites' on which it can be assumed that dyeing may take place, the problem of 'solid solution' *versus* 'site' adsorption has been raised with respect to this fibre. Again the situation has not been resolved, although the general tendency is to use the former model.

Disperse dyes on acrylic fibres

Although acrylic fibres are usually dyed with basic dyes, disperse dyes are used for pale to medium depths of shade. In fundamental studies linear isotherms are again obtained and affinities, enthalpies of dyeing and entropy changes have been determined for disperse dyes on this fibre [37].

Disperse dyes on cellulose and wool

These fibres are not usually dyed with disperse dyes, because of the very low affinities and saturation values which are observed. In the dyeing of blends, however, there is a problem of the staining of the wool or cellulose part of the blend by the disperse dye used for the other component. It is of interest to know the affinities of certain disperse dyes for these two fibres, and as a result adsorption studies have been carried out on cellulose [38] and on wool [31, 29]. In both cases it was found that adsorption followed the 'solid solution' model in that linear isotherms were obtained from which affinities, enthalpies of dyeing and entropy changes were obtained using the appropriate equations.

4.3.4 Affinity and partition coefficient

It is useful, in the context of disperse dyes whose adsorption on all fibres can be described by a relatively simple equation, to consider the implications of the magnitude of the affinity, and hence of the partition coefficient, on the properties of the dyes in practical applications.

The percentage exhaustion of a dyebath when dyeing is carried out to equilibrium is related to the partition coefficient (or ratio) by the equation

$$\%\text{ exhaustion} = \frac{K}{K + L} \times 100 \qquad (4.38)$$

where K = the partition coefficient as defined in Eqn 4.33 and L = the liquor-to-goods ratio.

If the partition coefficient is defined as in Eqn 4.34, then

$$\%\text{ exhaustion} = \frac{VK_V}{VK_V + L} \times 100 \qquad (4.39)$$

The results of carrying out this calculation for various values of the partition coefficient, at a selection of liquor-to-goods ratios, are given in Table 4.3.

It is also possible to illustrate how wet-fastness is related to affinity and partition coefficient by calculating how much water would be required, in a single treatment, to reduce the amount of dye on the fibre by a given percentage. If one kilogram of fibre containing D_0 g kg^{-1} of dye is treated with W litres of water and the system is allowed to come to equilibrium, then,

$$[\mathrm{D}]_0 = [\mathrm{D}]_{\mathrm{ad}} + W[\mathrm{D}]_{\mathrm{s}}$$

where $[\mathrm{D}]_{\mathrm{ad}}$ = the concentration of dye on fibre (g kg^{-1}) and $[\mathrm{D}]_{\mathrm{s}}$ = the

TABLE 4.3

The practical meaning of the partition coefficient

$-\Delta\mu^{\ominus}$ /kJ (kcal) mol^{-1} at 80 °C	at 100 °C	Partition coefficient K	% exhaustion at a liquor ratio (LR) of 5	25	100	Volume of water required to remove 5% of dye /l	Number of washes at 25:1 LR to remove 95% of dye
(0)	(0)	1	16.7	3.8	1.0	0.0526	1
6.8 (1.61)	7.2 (1.71)	10	66.7	28.6	9.1	0.526	3
13.5 (3.22)	14.3 (3.41)	100	95.2	80.0	50.0	5.26	14
20.3 (4.83)	21.5 (5.12)	1 000	99.5	97.6	90.9	52.6	122
27.0 (6.44)	28.7 (6.83)	10 000	100	99.8	99.0	526	1 183
33.8 (8.05)	35.8 (8.53)	100 000	100	100	99.9	5260	13 010

concentration of dye in solution (g l^{-1}). If, at the temperature of the treatment,

$$\frac{[\mathrm{D}]_{\mathrm{ad}}}{[\mathrm{D}]_{\mathrm{s}}} = K \tag{4.32}$$

then it can be shown that

$$K\left(\frac{[\mathrm{D}]_0}{[\mathrm{D}]_{\mathrm{ad}}} - 1\right) = W \tag{4.40}$$

If the value of W is that required for the removal of 5 per cent of the original dye present then $[\mathrm{D}]_0 = 1$ and $[\mathrm{D}]_{\mathrm{ad}} = 0.95$. The results of carrying out this calculation for various values of K are shown in Table 4.3, column 7 and illustrate the dependence of wet-fastness on affinity.

Another common problem is that of removing unwanted disperse dye from a fibre in situations where reduction clearing is not possible. The final column of Table 4.3 lists the number of washes required at a liquor-to-goods ratio of 25:1, to reduce the residual adsorbed disperse dye to less than 5 per cent of that on the fibre at the start of washing.

Because the isotherm is linear and hence the partition coefficient is a constant independent of dye concentration, each wash will remove the same fraction of adsorbed dye. If after the first wash

$$\frac{[\mathrm{D}]_{\mathrm{ad}(1)}}{[\mathrm{D}]_0} = \frac{K}{K + L} \tag{4.41}$$

then after x washes

$$\frac{[\mathrm{D}]_{\mathrm{ad}(x)}}{[\mathrm{D}]_0} = \left(\frac{K}{K + L}\right)^x \tag{4.42}$$

where $[\mathrm{D}]_{\mathrm{ad}(x)}$ = concentration of dye on fibre after x washes and $[\mathrm{D}]_0$ = concentration of dye on fibre before washing.

If $[\mathrm{D}]_{\mathrm{ad}(x)}$ is required to be at most 5 per cent of the dye that was on the fibre at the start of washing, then $[\mathrm{D}]_{\mathrm{ad}(x)}/[\mathrm{D}]_0 \leqslant 0.05$. Taking logarithms of both sides of Eqn 4.42 allows the calculation of a value of x, rounded up to the next whole number, that will fulfil this requirement for chosen values of K and L, that is, the number of washes required. The figures, although ridiculous from the point of view of practical application, do illustrate the difficulty of removing unwanted dye of medium to high affinity. Fortunately the need to remove dye of such a high degree of fastness does not arise in most applications.

It must be remembered that if the partition coefficient is defined as in Eqn 4.34, then K in Eqns 4.40 and 4.42 should be replaced by VK_V.

4.4 IONIC DYES

4.4.1 Electrical effects

In dyeing systems involving ionic dyes it is useful to consider the possible electrical effects.

When a solid containing charged groups is immersed in water it acquires an electrical potential. Even if the solid possesses no intrinsic charge of its own, the adsorption of ions from the water, or from the surrounding aqueous solution, will result in an electrical potential being generated. The presence of such a potential will result in oppositely charged ions (counterions) being attracted to the surface whilst similarly charged ions (co-ions) will be repelled. Because of this there will be an imbalance of ions near the surface under consideration, there being a layer adjacent to the surface where counterions predominate (Figure 4.7(a)). Helmholz [39], who assumed that the boundaries of this electrical double layer were sharp, treated it mathematically as a parallel-plate condenser. Since the effect of the electrical potential falls as distance from the surface increases, however, so will the concentration of counterions decrease and that of co-ions increase with increasing distance into the surrounding liquid (Figure 4.7(b)). This led Goy [39] to develop a treatment based on the concept of a diffuse electrical double layer.

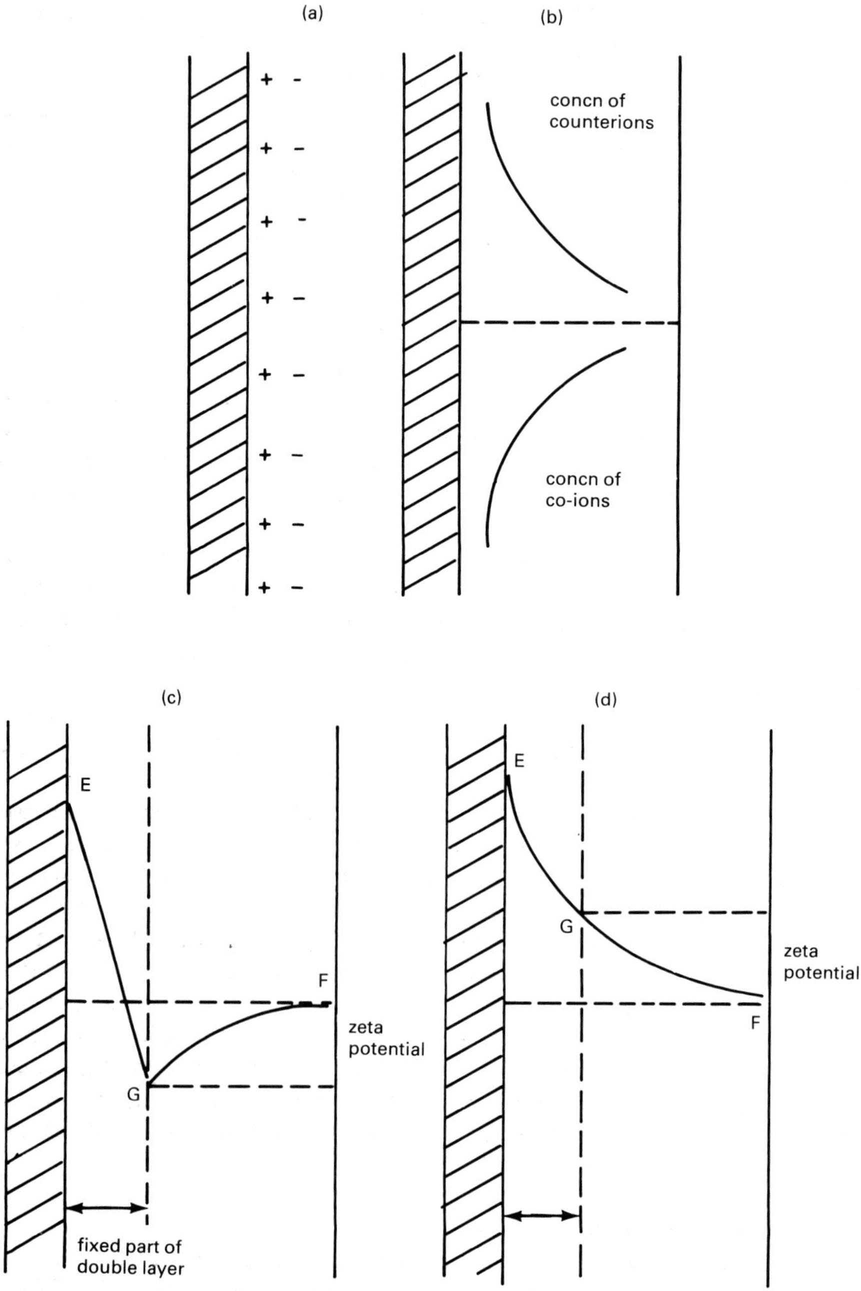

Figure 4.7 – The electrical double layer

More recently Stern [39] has shown that neither the sharp nor the diffuse theory is alone adequate and hence developed a view combining the essential features of both. Accordingly he postulated that the double layer is in two parts, one that is approximately a single solvent molecule in thickness, which remains fixed to the surface, and a second layer that extends some distance into the aqueous phase and is diffuse due to the opposing forces of thermal agitation and electrostatic attraction. The result is a sharp fall in potential in the first layer followed by a gradual fall in the second, until the bulk of the liquid is reached where the charge distribution becomes uniform. The potentials at a solid–liquid boundary may be represented in two ways, depending on the properties of the ions present in the solution. Thus Figure 4.7(c) would apply to ions with the same charge as the solid but which possess forces of attraction for the solid that exceed the electrical forces of repulsion over short but not over longer distances. Figure 4.7(d) would represent the distribution of ions with a charge opposite to that of the solid or ions for which the forces of attraction are greater than those of repulsion at all distances.

Figures 4.7(c) and (d) permit the illustration of two other quantities. The potential difference between E and F is the thermodynamic or reversible potential between the solid and the bulk of the solution, whilst that between G and F (that is, between the fixed and freely mobile portions) is the zeta or streaming potential.

The situations presented in Figures 4.7(c) and (d) would be representative of the distribution of dyes near the fibre surface. Such large ions would produce short-range forces of attraction and long-range forces of either attraction or repulsion depending on the nature of the charges on the dye and fibre. Figure 4.7(c) shows a dyeing system where the fibre and dye possess the same charge, the dye being attracted only over short distances from the fibre. This would result in a single layer of dye ions on the fibre surface, a condition which is called *unimolecular adsorption*. On the other hand Figure 4.7(d) shows a system where the attraction of the dye increases at all distances towards the fibre, resulting in the possibility of *multilayer* or *diffuse adsorption*. Both types are possible in dyeing, although in fact they are best described as extremes of the same behaviour since a dyeing system may change from one to the other during the course of a dyeing, or may be of one kind or the other depending on the presence or absence of neutral electrolytes in the dyebath.

4.4.2 The internal volume of fibres

The Donnan theory of membrane equilibrium considers the distribution of ions on the two sides of a semi-permeable membrane (see section 1.5). When this theory was applied to the treatment of the electrical double layer on a solid surface it was necessary to define a volume of liquid V l kg^{-1} adjacent to the surface, in which the potential is a constant with distance but falls to zero at the outer boundary of this layer. The potential so described is called the *Donnan potential*, ψ_{Donnan}. The thickness of this layer can be defined so that its volume equates to the volume of the diffuse layer as visualised in the double-layer theory. The outer boundary can then be regarded as a membrane separating the internal solution (i) from the external solution (s), the fixed charges on the surface being considered to be in the internal solution. Using this approach the equilibrium distribution of ions at the interface between a solution of sodium chloride and a negatively charged fibre can be described. The chemical potential of the sodium and chloride ions in the external solution can be written

$$\mu_{Na,s} = \mu^{\ominus}_{Na,s} + RT \ln [Na]_s \tag{4.43}$$

$$\mu_{Cl,s} = \mu^{\ominus}_{Cl,s} + RT \ln [Cl]_s \tag{4.44}$$

(omitting, for simplicity's sake, to write in the electrical charges) and in the internal solution

$$\mu_{Na,i} = \mu^{\ominus}_{Na,i} + RT \ln [Na]_i + F\psi_{Donnan} \tag{4.45}$$

$$\mu_{Cl,i} = \mu^{\ominus}_{Cl,i} + RT \ln [Cl]_i - F\psi_{Donnan} \tag{4.46}$$

where F is the Faraday constant and ideal behaviour is assumed, i.e. activities have been equated to concentrations. Because both solutions are ideal and of the same kind, then $\mu^{\ominus}_{Na,s} = \mu^{\ominus}_{Na,i}$ and $\mu^{\ominus}_{Cl,s} = \mu^{\ominus}_{Cl,i}$. Also, because the two phases are in equilibrium, $\mu_{Na,s} = \mu_{Na,i}$ and $\mu_{Cl,s} = \mu_{Cl,i}$, so that from Eqns 4.43 and 4.45

$$\frac{[Na]_s}{[Na]_i} = \exp(F\psi_{Donnan}/RT) = \lambda \tag{4.47}$$

and from Eqns 4.44 and 4.46

$$\frac{[Cl]_s}{[Cl]_i} = \exp(-F\psi_{Donnan}/RT) = 1/\lambda \tag{4.48}$$

where λ is the *Donnan distribution coefficient*. Hence

$$\frac{[Na]_s}{[Na]_i} = \frac{[Cl]_i}{[Cl]_s} = \exp(F\psi_{Donnan}/RT) \tag{4.49}$$

or $$[Na]_s\,[Cl]_s = [Na]_i\,[Cl]_i \tag{4.50}$$

Since many dyeing systems are influenced by ionic equilibria, the use of equations such as Eqn 4.50 provides an invaluable method of evaluating the energetics of the system. In addition, the treatment invoking an internal phase offers a means of overcoming the problem that concentrations of adsorbed or fixed ions in the fibre are measured in amounts per kilogram. By assuming that all such ions are present in the internal solution, division by V (1 kg^{-1}) gives concentrations in terms of amounts per litre of internal phase. Of course, an acceptable method for evaluating V must be available.

4.4.3 Ionic dyes applied to substrates that carry the same charge as the dye

The most common and the most important dyeing system to be considered under this heading is the application of anionic dyes to cellulose, both natural and regenerated, and this section therefore includes discussions of direct dyes, vat dyes in the reduced leuco form and reactive dyes when applied under non-reacting conditions on these substrates.

A very comprehensive study of the adsorption of the direct dye Chrysophenine G (C.I. Direct Yellow 12) on Cellophane sheet has been carried out [40], in which complete adsorption isotherms were obtained over a wide range of dye and salt concentrations at various temperatures. A typical example of the results obtained is shown by the continuous curves in Figure 4.8 for measurements at 40 °C [41], where the general shape of the curves resembles that illustrated in Figure 4.3, curve B, i.e. the dye is being adsorbed according to a Freundlich isotherm. This is apparently confirmed when the results are plotted on logarithmic scales (Figure 4.9), when it can be seen that the points lie essentially on straight lines. Such a result can be misleading, however, due to the insensitivity implicit in a log–log plot. Thus, if values of x and K in the Freundlich equation (Eqn 4.4) are calculated from the lines in Figure 4.9 (Table 4.4) and then used to calculate values of $[D]_{ad}$ for a range of values of $[D]_s$, the broken curves in Figure 4.8 result. At the higher salt concentrations, the good agreement shown in the log–log presentation is not so obvious when the experimentally obtained and the calculated isotherms are compared. Even if the fit to such an isotherm had been perfect, however, it is not easy to see how the resulting values of x and K could be used to define the properties of the dye under all conditions of application, in terms of one or two constant parameters.

Several theories have been advanced to interpret the adsorption of anionic dyes by cellulosic materials, all of them invoking the presence of an electrical

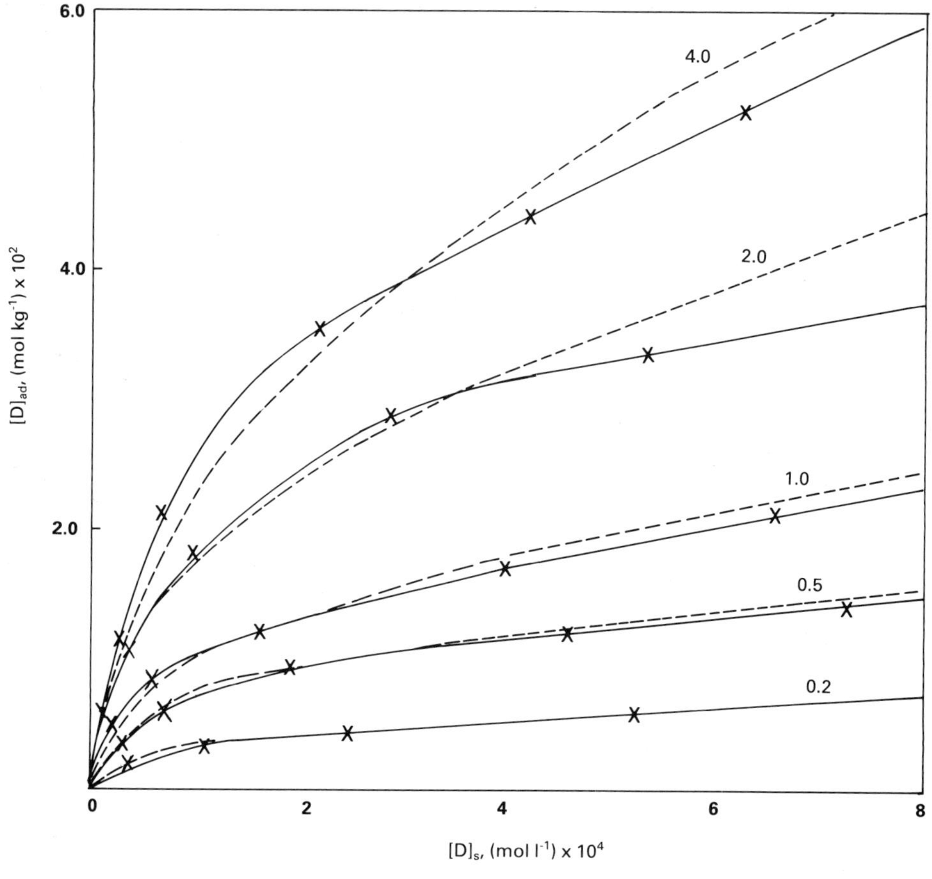

Figure 4.8 – Adsorption isotherms of Chrysophenine G on cellulose sheet at 40 °C [40]; the numbers on the curves represent concentrations (g l^{-1}) of sodium chloride in the dyebath

charge at the fibre–water interface due to the adsorption of dye ions and, indeed, assuming that it is only the dye ions that are adsorbed by the cellulose. Also inherent in these theories is that the dye is completely ionised in both phases, and that oppositely charged ions cannot move far apart so that any volume in the system must be electrically neutral. One of these theories [40], a kinetic thermodynamic approach to the problem, resulted in the development of some very complex mathematical expressions to describe the adsorption of dye. It was necessary to make a simplifying assumption to reduce the equations to experimentally manageable proportions, in that the system was reduced to a surface layer and an external solution, i.e. a system very similar to those described below. For this reason further discussion of

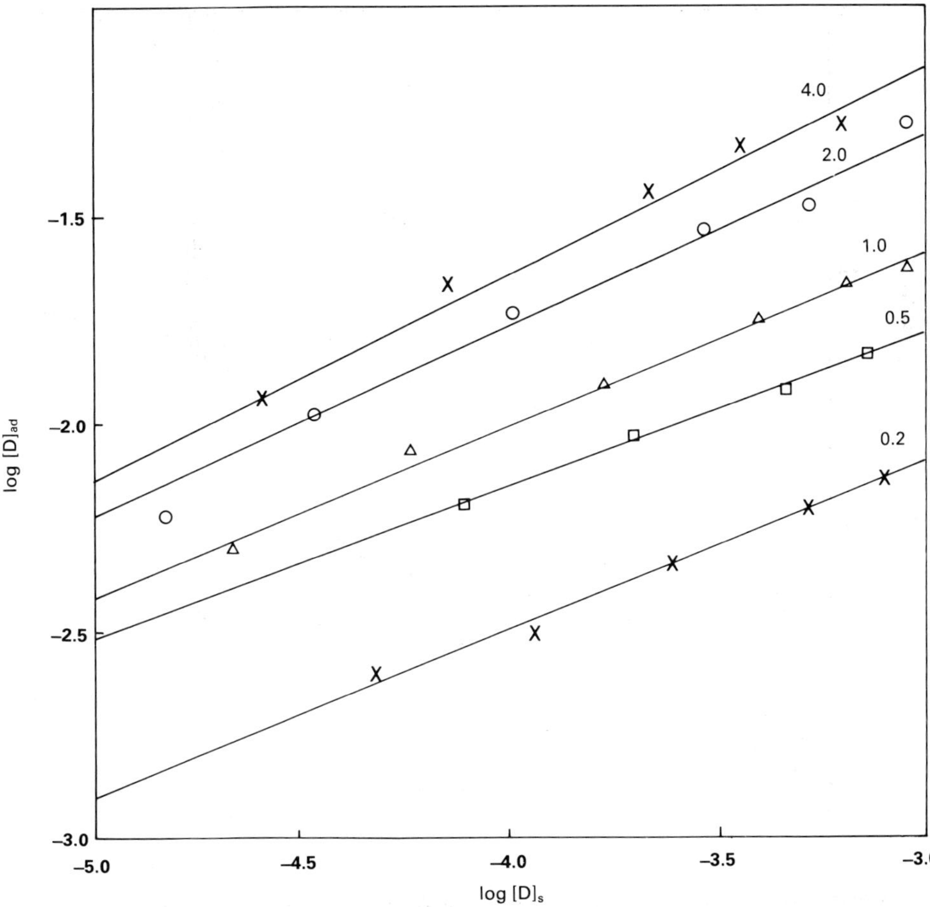

Figure 4.9 – Adsorption of Chrysophenine G on Cellophane sheet at 45 °C [40, 41]: log–log curves; the numbers on the lines represent concentrations (g l^{-1}) of sodium chloride in the dyebath

this theory is not included here; the reader is referred to the original paper for greater detail.

Probably the earliest attempt to produce an equation that would describe the situation divided the system into three parts – the fibre, a volume V of internal aqueous solution and the external solution – and applied the Donnan theory of membrane equilibrium across the boundary between the internal and external solutions [42]. Such a system involves unimolecular adsorption with the various ions distributed as shown in Figure 4.10, which assumes that the system contains only a dye (Na_zD) and sodium chloride. If it is

TABLE 4.4

Freundlich constants for the adsorption of Chrysophenine G on Cellophane sheet at 40 °C [40, 41]

Sodium chloride concentration /g l^{-1}	x	k
0.2	0.409	0.138
0.5	0.367	0.208
1.0	0.415	0.454
2.0	0.456	1.126
4.0	0.497	2.213

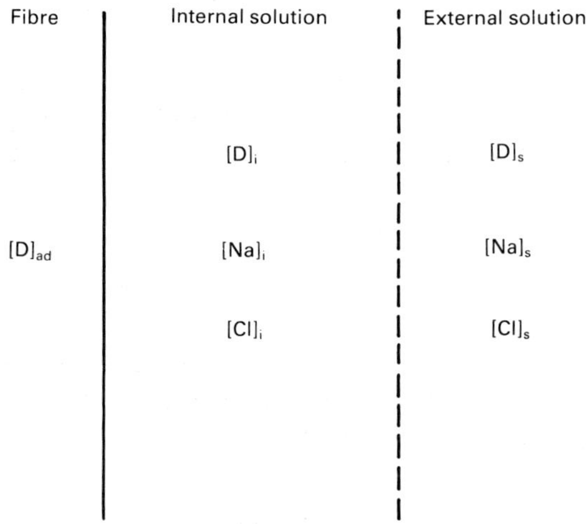

Figure 4.10 – Distribution of ions in a three-phase system

assumed that the concentration of adsorbed dye $[D]_{ad}$ (mol kg^{-1}) is in equilibrium with the concentration of dye in the internal solution $[D]_i$ (mol l^{-1}) according to a simple partition isotherm, then from Eqn 4.34

$$K_V = \frac{[D]_{ad}}{V[D]_i} = \exp\,(-\Delta\mu^{\ominus}/RT) \tag{4.51}$$

or

$$K_V = \frac{[D]_f}{[D]_i} \tag{4.52}$$

As discussed earlier, V (l kg^{-1}) must be included to make the equation dimensionally correct. Applying the Donnan equilibrium of Eqn 4.50 to the dye Na_zD,

$$\frac{[D]_i}{[D]_s} = \left(\frac{[Na]_s}{[Na]_i}\right)^z \tag{4.53}$$

Hence from Eqns 4.51 and 4.53 the affinity of the dye may be defined:

$$-\Delta\mu^{\ominus} = RT \ln K_V = RT \ln\left(\frac{[Na]_i^z[D]_{ad}}{V[Na]_s^z[D]_s}\right) \tag{4.54}$$

or, using Eqns 4.52 and 4.53,

$$-\Delta\mu^{\ominus} = RT \ln K_V = RT \ln\left(\frac{[Na]_i^z[D]_f}{[Na]_s^z[D]_s}\right) \tag{4.55}$$

All the concentrations in Eqns 4.54 and 4.55 are experimentally measurable with the exception of $[Na]_i$, and this requires to be calculated as follows. If it is assumed that the surface phase includes the surface of the fibre and that a state of electrical neutrality must exist, then

$$[Na]_i = [Cl]_i + \frac{z[D]_{ad}}{V} + z[D]_i \tag{4.56}$$

However, since the quantity of dye on the fibre which is measured at the end of dyeing will include the dye in the internal phase, then for practical purposes this equation must be written

$$[Na]_i = [Cl]_i + \frac{z[D]_{ad}}{V} \tag{4.57}$$

and substituting from Eqn 4.50 gives

$$[Na]_i = \frac{[Na]_s\,[Cl]_s}{[Na]_i} + \frac{z[D]_{ad}}{V} \tag{4.58}$$

This is a quadratic equation in $[Na]_i$, which can be solved using

$$[Na]_i = \frac{[D]_{ad}}{V}\left[\frac{z}{2} + \left(\frac{z^2}{4} + \frac{[Na]_s[Cl]_sV^2}{[D]_{ad}^2}\right)^{\frac{1}{2}}\right] \tag{4.59}$$

Again, if $[D]_{ad}/V$ is replaced by $[D]_f$, the equation becomes

$$[Na]_i = [D]_f \left[\frac{z}{2} + \left(\frac{z^2}{4} + \frac{[Na]_s[Cl]_s}{[D]_f^2} \right)^{\frac{1}{2}} \right] \tag{4.60}$$

Whilst the differences between Eqns 4.54 and 4.55 and between Eqns 4.59 and 4.60 may appear to be trivial, and indeed they are in this case, the opportunity has been taken to introduce the idea of evaluating concentrations in the fibre in the units of mol l^{-1} of internal phase, e.g. $[D]_f$, by dividing experimentally determined values in mol kg^{-1}, e.g. $[D]_{ad}$, by V at the outset. This removes the necessity of including V in the equations, which in the more complicated derivations to be discussed later will be to considerable advantage.

Another treatment based on the assumption of three phases in the system has been made [43]. This started by presuming a Langmuir distribution of dye, but when modified to the practical situation and also to include the Donnan distribution, it produced equations identical to those above.

The most widely used treatment of this dyeing system, and one which is frequently used by current workers in the field, is that of using the concept of diffuse adsorption [41, 44]. Here the system is divided into two parts, a surface layer (f) and the external solution, and any adsorbed or fixed ions on the fibre are considered to be dissolved in the surface layer. For a dye Na_zD and sodium chloride the assumed distribution is as in Figure 4.11, where all the concentrations in the internal phase have already been converted from mol kg^{-1} to mol l^{-1} of phase by dividing by V.

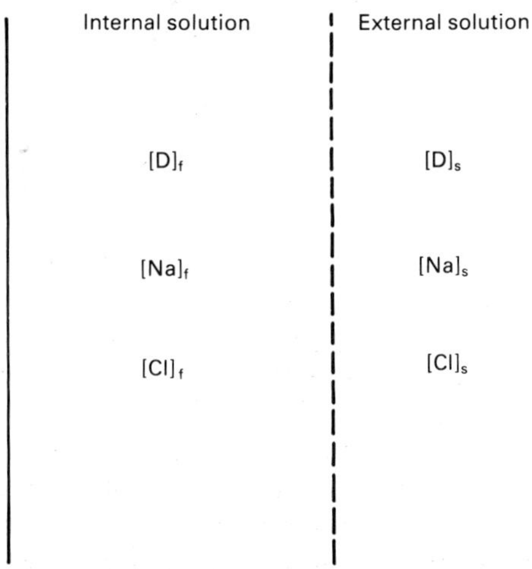

Figure 4.11 – Distribution of ions in a two-phase system

When these two phases are in equilibrium and assuming that the system is reversible, then the equation developed in Chapter 1 and discussed earlier (Eqn 4.16) may be used. If the activity of the dye is taken as $[Na]^z[D]$ in each phase, this equation becomes

$$-\Delta\mu^{\ominus} = -(\mu_f^{\ominus} - \mu_s^{\ominus}) = RT \ln \frac{[Na]_f^z[D]_f}{[Na]_s^z[D]_s} \qquad (4.61)$$

The only quantity in this equation which is not experimentally measurable is $[Na]_f$, and this can be calculated using the condition of electrical neutrality

$$[Na]_f = [Cl]_f + z[D]_f \qquad (4.62)$$

by applying the Donnan equality

$$[Cl]_f = \frac{[Na]_s[Cl]_s}{[Na]_f} \qquad (4.63)$$

to give the result

$$[Na]_f = [D]_f\left[\frac{z}{2} + \left(\frac{z^2}{4} + \frac{[Na]_s[Cl]_s}{[D]_f^2}\right)^{\frac{1}{2}}\right] \qquad (4.64)$$

The similarity between Eqns 4.55 and 4.61, and between Eqns 4.60 and 4.64, is obvious.

Throughout the above discussion it is clear that the standard state, i.e. when $a_f = 1$, is when the ionic product $[Na]_f^z[D]_f = 1$, and that $a_s = 1$ when $[Na]_s^z[D]_s = 1$.

4.4.4 The value of *V*

One problem remaining to be discussed is the choice of a value for V, the volume of the internal phase per kilogram of fibre. Inspection of Eqn 4.61 reveals that when the correct values are used in the equation not only should a constant value of $\Delta\mu^{\ominus}$ be obtained for all points on the isotherm, but a graph of log $[Na]_f^z[D]_f$ against log $[Na]_s^z[D]_s$ should be a straight line of unit slope. This method was used to establish a value of V for cotton using the results on Chlorazol Sky Blue FF (C.I. Direct Blue 1) [44]. By way of illustration a few of these results are included in Table 4.5, where values of $[D]_f$, $[Na]_f$ and $\Delta\mu^{\ominus}$ have been calculated for different assumed values of V. From these results the required log–log graphs have been drawn (Figure 4.12), and it can be seen that a value of V of 0.2 l kg^{-1} produces a line which is very close to the required slope of unity. A volume of 0.1 l kg^{-1} gives a

TABLE 4.5

Chlorazol Sky Blue FF on cotton at 90 °C [44]: calculation of affinity using different values of *V*

$[D]_s$ /mol l^{-1} $\times 10^4$	$[Na]_s$ /mol l^{-1} $\times 10^2$	$[D]_{ad}$ /mol kg^{-1} $\times 10^3$	*	$V = 0.1$	$V = 0.2$	$V = 0.22$	$V = 0.3$
0.0958	8.55	2.58	$[D]_f$	2.58	1.29	1.17	0.86
			$[Na]_f$	1.51	1.15	1.12	1.04
			$-\Delta\mu^{\ominus}$	30.7 (7.32)	25.3 (6.03)	24.7 (5.88)	22.9 (5.45)
0.252	8.56	3.79	$[D]_f$	3.79	1.90	1.72	1.26
			$[Na]_f$	1.90	1.32	1.27	1.14
			$-\Delta\mu^{\ominus}$	31.7 (7.55)	25.2 (6.01)	24.4 (5.82)	22.2 (5.29)
0.504	8.57	4.95	$[D]_f$	4.95	2.48	2.25	1.65
			$[Na]_f$	2.30	1.49	1.42	1.25
			$-\Delta\mu^{\ominus}$	32.7 (7.79)	25.4 (6.05)	24.5 (5.84)	22.0 (5.25)
1.01	8.59	6.19	$[D]_f$	6.19	3.09	2.81	2.06
			$[Na]_f$	2.74	1.66	1.59	1.36
			$-\Delta\mu^{\ominus}$	33.4 (7.95)	25.2 (6.00)	24.4 (5.82)	21.6 (5.15)
2.02	8.63	7.60	$[D]_f$	7.60	3.80	3.45	2.53
			$[Na]_f$	3.27	1.91	1.79	1.50
			$-\Delta\mu^{\ominus}$	33.9 (8.09)	25.4 (6.05)	24.3 (5.79)	21.2 (5.06)
4.04	8.71	8.84	$[D[_f$	8.84	4.42	4.02	2.95
			$[Na]_f$	3.74	2.12	1.98	1.64
			$-\Delta\mu^{\ominus}$	33.8 (8.06)	24.9 (5.93)	23.8 (5.67)	20.6 (4.91)
	Average		$-\Delta\mu^{\ominus}$	32.8 ± 1.2 (7.80 ± .28)	25.2 ± 0.17 (6.01 ± .04)	24.4 ± 0.17 (5.80 ± .04)	21.8 ± 0.59 (5.18 ± .14)

* $[D]_f$ values are given in (mol l^{-1}) × 10^2, $[Na]_f$ in (mol l^{-1}) × 10, and $-\Delta\mu^{\ominus}$ in kJ (kcal) mol^{-1}. $[NaCl]_s = 8.55 \times 10^{-2}$ mol l^{-1}.

line of greater slope and values of $\Delta\mu^{\ominus}$ that increase with increasing applied dye concentration, whereas a volume of 0.3 l kg^{-1} gives the reverse effects. The volume of 0.2 l kg^{-1} gives values of $\Delta\mu^{\ominus}$ which show no trend, merely the expected experimental variations. Previous workers have selected a *V* of 0.22 l kg^{-1}, which the figures in Table 4.5 show to be slightly inferior in achieving the right slope with no trend in $\Delta\mu^{\ominus}$. Their value was chosen using all the dyeing results available as opposed to the few considered in the table, however, and hence must be accepted as the more suitable.

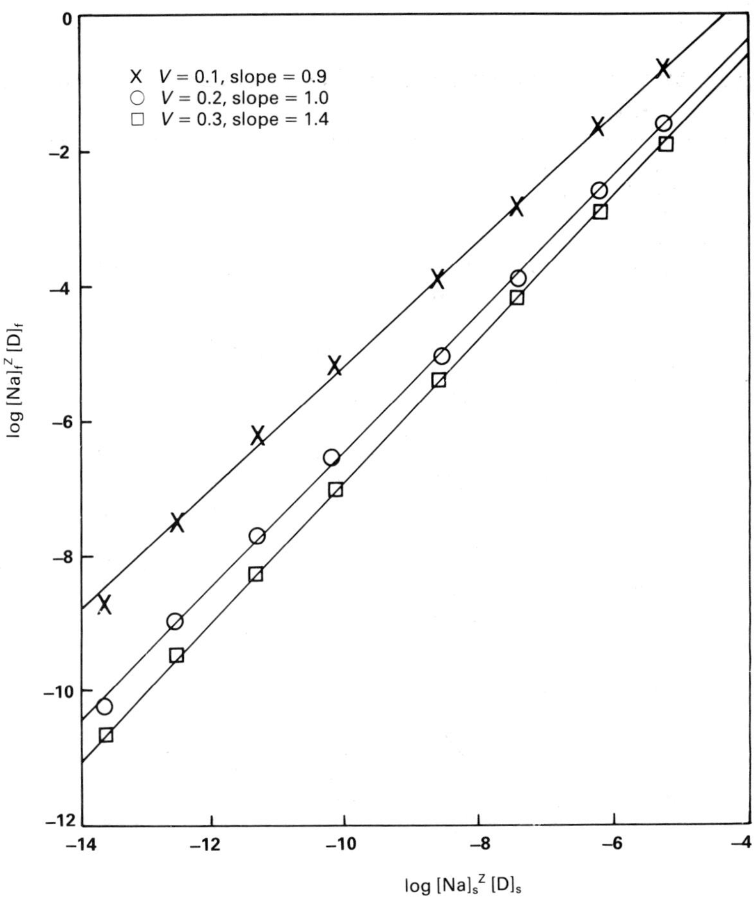

Figure 4.12 – Ionic product plot for Chlorazol Sky Blue FF on cotton at 90 °C [42]: $[D]_s = 5.04 \times 10^{-5}$ mol l^{-1}, salt variable

Another approach to determining the volume of the surface phase was based on the idea that it should correspond to the amount of water adsorbed by the fibre at 100 per cent relative humidity [42]. This was supported by the fact that the value of 0.22 l kg^{-1}, determined as above, corresponded to the value obtained for the adsorption of water by cotton at 90 °C [45]. Unfortunately, data for the other fibres were only available for 25 °C and the application of these figures (Table 4.6, column 2) to the dyeing of Chlorazol Sky Blue FF on cotton, Cellophane sheet and cuprammonium rayon gave affinities of −24.4 (−5.82), −29.5 (−7.03) and −31.4 (−7.48) kJ (k cal) mol^{-1} respectively. Such a variation is greater than might have

TABLE 4.6

Values of the volume of the internal phase in cellulosic fibres [41]

Fibre	By moisture regain /l kg^{-1}	By dye adsorption/l kg^{-1}	
		Chlorazol Sky Blue FF	Chrysophenine G
Cotton	0.22	0.22	0.30
Mercerised cotton	0.26	–	0.50
Viscose rayon	0.46	0.45	0.45
Cellophane sheet	0.33	0.45	0.45
Cuprammonium rayon	0.37	0.60	0.65

been anticipated, but is not surprising because there is no reason to assume that the surfaces available for the adsorption of the small water molecules will be the same as for the much larger dye ions [41].

If it is assumed that the volume of the internal phase is solely a function of the accessibility of the surface to the large dye molecules and that, once the correct values of V are used, the affinity of a given dye will be the same for a variety of cellulosic materials, then an alternative method of determining relative values of V for different substrates is available. This was done using the dyeing data for Chlorazol Sky Blue FF [44] on different cellulosic materials taking the volume term of 0.22 l kg^{-1} and a $\Delta\mu^{\ominus}$ of 24.4 (5.82) kJ (k cal) mol^{-1} as the base point and then calculating the necessary values of V for the other materials so as to produce the same affinity. The results are shown in Table 4.6, column 3.

Marshall and Peters made a further extensive study of the problem using measured data on Chrysophenine G (C.I. Direct Yellow 12) [46]. They used the method already described, of plotting the logarithms of the ionic product in the fibre against that in solution calculated using a series of values of V and selecting the one which gave the best straight line of unit slope. The results are given in Table 4.6, column 4. Although the log–log graphical method is not very sensitive to small changes in V (and hence the results should be regarded as approximate), the similarity between the values in columns 3 and 4 of Table 4.6 is very satisfying and would appear to confirm that, to a reasonable approximation, it is acceptable to assume that the affinity of a dye is a constant for all cellulosic fibres.

4.4.5 Determination of dye affinities

A great many affinity determinations have been made over the years and, in fact, are still being carried out, using the methods described above. The

TABLE 4.7

The effect of temperature on the affinity of direct dyes for cellulose [46, 41]

Dye	C.I. Direct	z	Fibre[a]	$-\Delta\mu^{\ominus}$ (kJ(kcal) mol^{-1}) at temperature of 60 °C	70 °C	80 °C	90 °C	100 °C
1. Those dyes which gave unit slope at 60 °C and 90 °C:								
Chrysophenine G	Yellow 12	2	B	16.0	14.3	13.4	11.8	–
				(3.8)	(3.4)	(3.2)	(2.8)	
			V	16.4			13.0	10.9
				(3.9)	–	–	(3.1)	(2.6)
			C	16.4	15.1	13.4	12.6	–
				(3.9)	(3.6)	(3.2)	(3.0)	
Durazol Fast Yellow 6G	Yellow 46	2	B	25.2	22.3	17.6	16.0	
				(6.0)	(5.3)	(4.2)	(3.8)	–
			V	24.4			13.4	
				(5.8)	–	–	(3.2)	–
Chlorazol Brown M	Brown 2	2	B	26.0	25.2	24.8	24.8	
				(6.2)	(6.0)	(5.9)	(5.9)	–
			V	27.3			25.6	23.5
				(6.5)	–	–	(6.1)	(5.6)
Durazol Grey RG	Black 61	2	B	18.9	18.0	16.8	16.0	
				(4.5)	(4.3)	(4.0)	(3.8)	–
			V	20.2			16.0	14.7
				(4.8)	–	–	(3.8)	(3.5)
Chlorazol Fast Scarlet 8B	Red 26	3	B	24.0	22.7	20.2	18.0	
				(5.7)	(5.4)	(4.8)	(4.3)	–
			V	24.0	22.3	20.6	19.3	
				(5.7)	(5.3)	(4.9)	(4.6)	–
			C	24.0	22.7	22.3	20.6	
				(5.7)	(5.4)	(5.3)	(4.9)	–
Chlorazol Sky Blue FF	Blue 1	4	B	30.2	27.7	26.5	25.2	
				(7.2)	(6.6)	(6.3)	(6.0)	–
			V	32.3			26.5	24.8
				(7.7)	–	–	(6.3)	(5.9)

continued overleaf

[a] Fibre: B = cuprammonium rayon (V = 0.65), V = viscose rayon (V = 0.45), C = bleached cotton (V = 0.30).

Dye	C.I. Direct	z	Fibre[a]	$-\Delta\mu^{\ominus}$ (kJ(kcal) mol^{-1}) at temperature of 60 °C	70 °C	80 °C	90 °C	100 °C
2. Those dyes which gave unit slope at 90 °C but not at 60 °C:								
Chlorazol Fast Orange G	Orange 23	1	B	15.5 (3.7)	–	13.9 (3.3)	12.6 (3.0)	–
			V	14.3 (3.4)	–	–	13.4 (3.2)	11.3 (2.7)
Chlorazol Violet N	Violet 3	2	B	20.6 (4.9)	19.3 (4.6)	18.0 (4.3)	18.0 (4.3)	–
			V	19.3 (4.6)	–	–	18.0 (4.3)	16.0 (3.8)
Chlorazol Fast Orange R	Orange 26	2	B	21.8 (5.2)	20.6 (4.9)	18.5 (4.4)	17.2 (4.1)	–
			V	20.2 (4.8)	–	–	17.2 (4.1)	14.7 (3.5)
Chlorazol Orange Brown X	Brown 12	2	B	20.2 (4.8)	23.1 (5.5)	–	24.0 (5.7)	–
			V	–	–	–	25.2 (6.0)	23.1 (5.5)
Chlorazol Fast Scarlet 4BA	Red 24	3	B	19.7 (4.7)	17.6 (4.2)	17.6 (4.2)	16.8 (4.0)	–
			V	18.5 (4.4)	–	–	16.0 (3.8)	12.6 (3.0)
3. Those dyes which did not give unit slope at any temperature:								
Durazol Red 2B	Red 81	2	B	16.0 (3.8)	15.1 (3.6)	14.3 (3.4)	13.4 (3.2)	–
			V	16.4 (3.9)	–	–	13.4 (3.2)	12.6 (3.0)
			C	16.0 (3.8)	14.7 (3.5)	13.9 (3.3)	13.0 (3.1)	–
Chlorazol Fast Scarlet 4B	Red 23	2	B	22.3 (5.3)	21.8 (5.2)	21.4 (5.1)	20.2 (4.8)	–
			V	21.4 (5.1)	–	–	20.6 (4.9)	17.6 (4.2)
Durazol Fast Scarlet 2G	Red 76	3	B	24.4 (5.8)	22.7 (5.4)	21.4 (5.1)	21.0 (5.0)	–
			V	–	–	–	19.7 (4.7)	–

results of a typical series of experiments are given in Table 4.7 [46] covering a number of dyes and fibres using the values of V listed in Table 4.6, column 4. The experiments were carried out in the presence of 0.01M or 0.02M sodium chloride over a range of applied concentrations for each dye. The affinities quoted in the table are the averages of the results so obtained. When the log–log plots of the ionic products were made it was found that the theoretical unit slope was not always obtained, but that the deviation from unity decreases as the temperature increased. A slope less than unity suggested that all the dye in solution was not available for dyeing, assuming the correct

V term was being applied, which in turn the authors interpreted as being possibly due to the aggregation of dye in solution, an effect which would be greatest at the lower temperatures. In Table 4.7 the dyes are classified relative to this behaviour, according to the sub-headings in the table. The consistency of the affinities on the different fibres, extending as it does over a range of dyes varying in basicity from one to four, led the authors to conclude that the measured affinities are indeed those of the dyes for 'cellulose' in whatever form it is presented, and that the volume term is a useful means of specifying the difference in dyeing behaviour of the different fibres.

4.4.6 The effect of temperature on affinity

As described earlier, the effect of temperature on dye uptake can be expressed using the standard enthalpy (heat) of dyeing $\Delta H^{\ominus}$, which can be determined by measuring the affinity at different temperatures and plotting $\Delta\mu^{\ominus}/T$ against $1/T$. The slope of the line obtained gives $\Delta H^{\ominus}$. This has been done [46] for the dyes listed in Table 4.7, when it was found that the dyes in group 1 produced straight lines from which $\Delta H^{\ominus}$ could be readily obtained giving the results shown in Table 4.8. It can be seen immediately that the

TABLE 4.8

Heats of dyeing of direct dyes on cellulosic fibres/kJ (kcal) mol^{-1} [46]

Dye	Cuprammonium rayon	Viscose rayon	Cotton	Mean
Chrysophenine G	−59(−14)	−59(−14)	−59(−14)	−59(−14)
Durazol Fast Yellow 6G	−130(−31)	–	–	−130(−31)
Chlorazol Brown M	−42(−10)	−50(−12)	–	−46(−11)
Durazol Grey RG	−59(−14)	−67(−16)	–	−63(−15)
Chlorazol Fast Scarlet 8B	−88(−21)	−80(−19)	−80(−19)	−84(−20)
Chlorazol Sky Blue FF	−84(−20)	−97(−23)	–	−92(−22)

enthalpy of dyeing for a given dye on different fibres is the same to a reasonable approximation, thereby confirming that the energetics of the system are those between dye and cellulose irrespective of the form in which the fibre is presented. As a piece of corroborative evidence this is even more valuable, since in the calculation of $\Delta H^{\ominus}$ by subtracting affinities at two temperatures the volume term is eliminated from the equation, so that the constancy of the results does not rely on a suitable choice for the value of V.

When the remaining dyes in Table 4.7 were considered, it was found that the graph of $\Delta\mu^{\ominus}/T$ against $1/T$ did not always produce straight lines thereby suggesting that $\Delta H^{\ominus}$ was changing with temperature, possibly due to aggregation effects; sufficient dyes did, however, produce a linear relationship on one or other of the fibres studied to discourage too much speculation relating to their positions in groups 2 and 3.

4.4.7 The Langmuir approach

As a result of some of the problems associated with the Donnan approach of diffuse adsorption, several workers have attempted to apply the Langmuir isotherm to the dyeing of cellulose with direct dyes. To do this requires a knowledge of the number of sites in cellulose, whatever they may be – that is, figures for the saturation value of cellulose are required. Attempts to establish these by direct measurement [47] using high concentrations of both dye and sodium chloride in the dyebath have resulted in large uptakes of dye but produced results that are not very meaningful, in that calculation of the surface area of cellulose required to accommodate so much dye led to values much greater than those obtained by other methods, justifying the conclusion that the dye was in a highly aggregated state both in the fibre and in solution. Work at very low concentrations of dye gives clear evidence of monolayer formation, however [48].

An alternative method of determining a saturation value is by extrapolation of the reciprocal plot, either in the form of $1/[D]_{ad}$ *vs* $1/[D]_s$, which tended to give curved lines at the lower salt concentrations, or better $1/[D]_{ad}$ *vs* $1/[D]_i$, i.e. the reciprocal of the dye concentration in the internal phase where

$$[D]_i = \lambda^2[D]_s \qquad (4.65)$$

and λ is the Donnan coefficient. This latter plot did give straight lines and hence saturation values by extrapolation, but the results showed so great a spread between dyes as to render them of little significance.

Despite the failure to determine saturation value for anionic dyes on cellulose, which is a fundamental requirement to applying a Langmuir isotherm, work has still been carried out in attempts to achieve this end [for example, 49, 50 and 51] resulting in a mathematical treatment of increasing complexity and some interesting experimental observations. In summary, however, this work seems to provide more questions than answers, resulting in the Donnan approach, as exemplified by Eqns 4.61–4.64, being the one most widely used in the study of this dyeing system. Calculation of the affinities of anionic dyes on cellulosic substrates has produced results that have been of considerable use in the interpretation of dyeing phenomena with these dyes. Unfortunately the results are not perfect, in that affinity – which by definition should be a constant at any one temperature with changing application conditions – is frequently found to change in systematic ways, for example, with increasing dye or electrolyte concentration. It could be that the fault lies with the assumed ideality of the system, i.e. that concentrations cannot be truly equated to activities. Alternatively, or in addition, it could be that the equations used are not complete and attempts to correct this have been considered as follows.

4.4.8 The partial oxidation of cellulose

The first attempt to correct the Donnan equations came with the realisation that cellulose itself is not electrically neutral. Due to the effects of partial oxidation cellulose contains carboxylic acid groups which can be ionised, thereby producing a negative charge on the fibre, the magnitude of which will depend on the number of carboxylic acid groups present. The effect of the number of such groups was clearly demonstrated [52] when a series of 'so-called' oxycelluloses was prepared from cotton in such a way that the various samples obtained contained various amounts of carboxylic acid group. Subsequent dyeing of these samples with Chlorazol Sky Blue FF in the presence of 5.0 g l^{-1} of sodium chloride at pH 6.2 revealed that the greater the amount of acid group in the cellulose, the smaller was the amount of dye taken up. Obviously some modification was required to the equations used to calculate affinity, and the first step in this direction was to modify the electrical neutrality equation (Eqn 4.62) by including carboxylic acid groups, assuming that they were all ionised, i.e.

$$[Na]_f = [Cl]_f + z[D]_f + [COO^-]_f \tag{4.66}$$

This was still only an approximation to the truth, however; it was found, for any one cellulosic fibre containing a fixed amount of acid group, that the effect was dependent on the pH of the bath [53] and also upon other changes in the bath conditions [54, 55], demonstrating that the carboxylic acid groups are not always completely ionised. A method of dealing with this situation and the reasons for it will be discussed later. It is interesting, however, that despite the conclusive findings described above many subsequent workers seem to have chosen to ignore the need for some such correction.

4.4.9 Leuco vat dyes on cellulose

A further modification to Eqn 4.62 arose as a result of quantitative studies into the dyeing of leuco vat dyes on cellulose. The equilibrium thermodynamic approach used with direct dyes can be successfully applied to vat dyes when they are in the soluble, reduced form. As was pointed out [56], however, the study of the equilibrium state is practically difficult with vat dyes because some of them are not stable to prolonged treatment in the strongly alkaline solutions needed to keep the reduced form soluble, and their ready re-oxidation requires the exclusion of air during dyeing. Furthermore the complex nature of the dyebath, which contains sodium, dye, hydroxide, dithionite and often chloride ions, as well as the oxidation products of the reducing agent, makes exact quantitative treatment difficult if not impossible.

In order to apply the method of calculation used with direct dyes (Eqns 4.61–4.64) it was again assumed that only dye ions had affinity for the fibre and the simplifying assumption was made at first that all the other anions could be replaced by an equivalent quantity of chloride ions, that is, a concentration c_y of a particular ion of basicity y was equated to a concentration yc_y of chloride ions. The electrical neutrality Eqn 4.62 therefore became

$$[\mathrm{Na}]_f = [\mathrm{eqCl}]_f + z[\mathrm{D}]_f \tag{4.67}$$

where

$$[\mathrm{eqCl}]_f = [\mathrm{Cl}]_f + [\mathrm{OH}]_f + \Sigma y[c_y]_f \tag{4.68}$$

When the value of $[\mathrm{Na}]_f$ obtained from Eqn 4.67 was used in the standard affinity equation (Eqn 4.61), assuming $[\mathrm{Na}]_s = [\mathrm{eqCl}]_s$, $z = 2$ and $V = 0.25\ \mathrm{l\ kg^{-1}}$, the affinities in Table 4.9 were obtained for five purified

TABLE 4.9

Affinities of leuco vat dyes for cotton at 40 °C in the presence of 0.15M electrolyte [56]

Dye	$-\Delta\mu^\ominus$ /kJ (kcal) mol^{-1}
1-Benzoylaminoanthraquinone	7.1 (1.68)
1,4-Dibenzoylaminoanthraquinone	13.7 (3.26)
1,5-Dibenzoylaminoanthraquinone	10.5 (2.49)
Flavanthrone	24.1 (5.73)
Pyranthrone	23.4 (5.56)

vat dyes [56]. These results confirmed that the approach could be applied to the adsorption of the alkali-soluble leuco forms of vat dyes, and demonstrated that the high exhaustion of vat dyes is not the result of intrinsically high affinity but of the large electrolyte concentrations used in their application.

In order to remove the uncertainties associated with the use of dithionite as a reducing agent, the adsorption of two purified vat dyes was studied on cuprammonium rayon in which reduction was achieved catalytically using hydrogen in the presence of palladium-impregnated asbestos, so that the dyebaths contained only sodium, vat dye, hydroxide and chloride ions [57]. At the same time the small but significant amount of hydroxide ions adsorbed by the cellulose should be taken into account. To this end an isotherm for the adsorption of sodium hydroxide by cuprammonium rayon was measured at the temperature used for the subsequent dyeings. This produced a straight line of slope *ca* 3.14, and since

$$K_{\mathrm{NaOH}} = \exp\,(-\Delta\mu^{\ominus}/RT) = \frac{[\mathrm{Na}]_{\mathrm{f}}[\mathrm{OH}]_{\mathrm{f}}}{[\mathrm{Na}]_{\mathrm{s}}[\mathrm{OH}]_{\mathrm{s}}} = \frac{[\mathrm{NaOH}]_{\mathrm{f}}^{2}}{[\mathrm{NaOH}]_{\mathrm{s}}^{2}} \qquad (4.69)$$

the value of K is the square of the slope, i.e. *ca* 10, or alternatively $\Delta\mu^{\ominus}_{\mathrm{NaOH}}$ has a value of about -5.96 kJ (-4.12 kcal) mol^{-1}. This figure, although small, means that the hydroxide ions are adsorbed by the fibre, thereby increasing the charge on the fibre, and hence this adsorption must be considered in the electrical neutrality equation. This becomes

$$[\mathrm{Na}]_{\mathrm{f}} = [\mathrm{Cl}]_{\mathrm{f}} + [\mathrm{OH}]_{\mathrm{f}} + z[\mathrm{D}]_{\mathrm{f}} \qquad (4.70)$$

Applying Donnan theory and the value of $[\mathrm{OH}]_{\mathrm{f}}$ from Eqn 4.69 gives

$$[\mathrm{Na}]_{\mathrm{f}} = \frac{[\mathrm{Na}]_{\mathrm{s}}[\mathrm{Cl}]_{\mathrm{s}}}{[\mathrm{Na}]_{\mathrm{f}}} + \frac{K[\mathrm{NaOH}]_{\mathrm{s}}}{[\mathrm{Na}]_{\mathrm{f}}} + z[\mathrm{D}]_{\mathrm{f}} \qquad (4.71)$$

from which $[\mathrm{Na}]_{\mathrm{f}}$ can be obtained in the usual way by solving the quadratic equation.

Whilst it is possible to obtain affinities without correcting for hydroxide ion adsorption, the affinities so found were inconsistent and the shapes of the log $[\mathrm{Na}]_{\mathrm{f}}^{z}[\mathrm{D}]_{\mathrm{f}}$ *vs* log $[\mathrm{Na}]_{\mathrm{s}}^{z}[\mathrm{D}]_{\mathrm{s}}$ plots were not linear. When corrected using Eqn 4.71 the plots produced straight lines of unit slope as predicted by theory. It was therefore proved that some allowance must be made for the adsorption of hydroxide ions if reasonable values of the affinity are to be derived [57].

In this same reference it was also suggested that a better approach to that of using equivalent chloride ion concentration, as described by Eqn 4.68 [56], was to apply the Donnan equilibrium to both mono- and divalent anions separately. Thus for a divalent anion An^{2-} Donnan theory states

$$[\mathrm{Na}]_{\mathrm{f}}^{2}[\mathrm{An}^{2-}]_{\mathrm{f}} = [\mathrm{Na}]_{\mathrm{s}}^{2}[\mathrm{An}^{2-}]_{\mathrm{s}} \qquad (4.72)$$

and the equation for electrical neutrality in the fibre phase becomes

$$[\mathrm{Na}]_f = [\mathrm{Cl}]_f + [\mathrm{OH}]_f + 2[\mathrm{An}^{2-}]_f + z[\mathrm{D}]_f \tag{4.73}$$

which by applying the appropriate Donnan equations and including the correction for adsorbed hydroxide ions gives

$$[\mathrm{Na}]_f = \frac{[\mathrm{Na}]_s[\mathrm{Cl}]_s}{[\mathrm{Na}]_f} + \frac{K[\mathrm{NaOH}]_s}{[\mathrm{Na}]_f} + \frac{2[\mathrm{Na}]_s^2[\mathrm{An}^{2-}]_s}{[\mathrm{Na}]_f^2} + z[\mathrm{D}]_f \tag{4.74}$$

This is a cubic equation in $[\mathrm{Na}]_f$, which can be solved since all the other concentrations are known.

No results of applying Eqn 4.74 are available, but two other studies have applied the correction necessary to account for the adsorption of hydroxide ions to results obtained in the presence of sodium dithionite in the dyebath, using Eqn 4.71 but with [Cl] equated to [eqCl] (Eqn 4.68). The first of these papers [58] studied the adsorption of approximately 60 purified anthraquinone derivatives and a series of condensed-ring quinones by cuprammonium rayon in an attempt to establish a relationship between chemical structure and affinity. The second [59] reports the results of dyeing cotton with a range of commercial samples of vat dyes under conditions of varying electrolyte concentration and temperature, and confirms that some correction for the adsorption of hydroxide ions is necessary if meaningful results are to be obtained.

4.4.10 Reactive dyes on cellulosic materials

In this dyeing system dye is applied from an alkaline solution, and from the outset it became clear that a complete picture could only be obtained if the effects, both chemical and physical, of the action of alkali on the fibre were known. For many years a great deal of investigative work had been carried out on the system cellulose–sodium hydroxide–water in attempts to explain the mechanism of mercerisation [60]. Although each piece of work gave self-consistent results, very little mutual agreement was achieved and the mechanism was not explained.

Eventually the suggestion was made [61, 62] that the adsorption of sodium hydroxide by cellulose was ionic, the hydroxyl groups in the cellulose acting as a weak acid which is neutralised by hydroxide ions to form sodium cellulosate. On this assumption, the dissociation constant K_{OH} of these groups was obtained by applying Donnan theory to the adsorption of sodium hydroxide to give the value of 1.84×10^{-14} at 25 °C [63], a figure which compares favourably with the dissociation constants of other carbohydrates determined by other workers.

This work was of particular interest to dyeing with reactive dyes, in that the fixation of these dyes appeared to involve the ionised hydroxyl groups of the cellulose. In order to substantiate this hypothesis it was necessary to calculate the number of such ionised groups present at any alkaline pH. This should be possible using the value of K_{CellOH} previously obtained, except that this value was determined using very high concentrations of alkali when the cellulose was very highly swollen. At the much lower concentrations of alkali used in reactive dyeing it was necessary to correct for the fact that not all the hydroxyl groups would be available [64]. This was done by introducing an 'availability' term into the calculation. The necessary equations were derived as follows [65].

If cellulose dissociates according to the equation

$$\mathrm{CellOH} \rightleftharpoons \mathrm{CellO^-} + \mathrm{H^+} \tag{4.75}$$

(where [CellO] = the concentration of ionised hydroxyl groups in the fibre) then, assuming ideal behaviour and ignoring activity effects,

$$K_{\mathrm{OH}} = \frac{[\mathrm{CellO}][\mathrm{H}]}{[\mathrm{CellOH}]} \tag{4.76}$$

If Q_{OH} is the total concentration of available cellulose hydroxyl groups, then $[\mathrm{CellOH}] = Q_{\mathrm{OH}} - [\mathrm{CellO}]$, so that

$$K_{\mathrm{OH}} = \frac{[\mathrm{CellO}][\mathrm{H}]}{Q_{\mathrm{OH}} - [\mathrm{CellO}]} \tag{4.77}$$

or, rearranging,

$$[\mathrm{CellO}] = \frac{K_{\mathrm{OH}} Q_{\mathrm{OH}}}{K_{\mathrm{OH}} + [\mathrm{H}]} \tag{4.78}$$

Since $[\mathrm{H}][\mathrm{OH}] = K_{\mathrm{w}}$, the ionic product of water, this can be written in terms of [OH]:

$$[\mathrm{CellO}] = \frac{K_{\mathrm{OH}} Q_{\mathrm{OH}} [\mathrm{OH}]}{K_{\mathrm{w}} + K_{\mathrm{OH}}[\mathrm{OH}]} \tag{4.79}$$

Considering the system of cellulose in equilibrium with an aqueous solution of sodium hydroxide and sodium chloride, for electrical neutrality in the fibre phase

$$[\mathrm{Na}]_{\mathrm{f}} = [\mathrm{Cl}]_{\mathrm{f}} + [\mathrm{OH}]_{\mathrm{f}} + [\mathrm{CellCOO}] + [\mathrm{CellO}] \tag{4.80}$$

where [CellCOO] = the concentration of ionised carboxylic acid groups in the fibre, which in the initial treatment were assumed to be fully ionised under the conditions of the experiments.

All the concentrations in the fibre phase have been previously corrected for V, the internal volume, to give concentrations in mol l^{-1} of internal phase. Applying the Donnan distribution for $[Cl]_f$ given by Eqn 4.63 and the corresponding value for $[OH]_f$ given by Eqn 4.81,

$$[OH]_f = \frac{[Na]_s[OH]_s}{[Na]_f} \tag{4.81}$$

together with Eqn 4.79, in which the concentration of hydroxide ions is that in the internal phase, then Eqn 4.80 becomes

$$[Na]_f = \frac{[Na]_s[Cl]_s}{[Na]_f} + \frac{[Na]_s[OH]_s}{[Na]_f} + [CellCOO] + \frac{K_{OH}Q_{OH}[Na]_s[OH]_s/[Na]_f}{K_w + K_{OH}[Na]_s[OH]_s/[Na]_f} \tag{4.82}$$

The only unknown in this equation is $[Na]_F$ and, using the values given in Table 4.10, it can be solved for any selected dyebath conditions to give the appropriate value of $[Na]_f$.

If at the start of the experiment the fibre is completely in the acid form, i.e. carboxylic acid groups are present as —COOH and hydroxyl groups

TABLE 4.10

Constants used in the cellulosic calculations [65]

Quantity	Viscose rayon	Cotton
K_w	6.87×10^{-15}	6.87×10^{-15}
K_{OH}[a]	1.84×10^{-14}	1.84×10^{-14}
Concentration of ionised carboxylic acid [CellCOO⁻]/mol l^{-1}	4.6×10^{-2}	9.1×10^{-3}
Availability/%	74	41
Number of glucose residues/mol kg^{-1}	6.17	6.17
Available glucose residues/mol kg^{-1}	4.57	2.47
Available glucose residues (Q_{OH})/mol l^{-1}	10.15	11.22
V/l kg^{-1}	0.45	0.22

[a] This value is that calculated for 25 °C by Neale [63]. Since no information is available for the heat of ionisation, it has been used here in work at 20 °C.

as —OH, then the amount of hydroxide ion removed from the aqueous solution under any particular experimental conditions will be given by

$$[\mathrm{OH}]_{ad} = [\mathrm{OH}]_f + [\mathrm{CellCOO}] + [\mathrm{CellO}] \tag{4.83}$$

Values of $[\mathrm{OH}]_{ad}$ can be readily obtained from the values of $[\mathrm{Na}]_f$ calculated, using Eqn 4.81 to get $[\mathrm{OH}]_f$ and then Eqn 4.79 for [CellO], [CellCOO] being assumed constant under the alkaline conditions used since the acid groups will be fully ionised. A comparison of the results obtained with experimental measurements of hydroxide ion uptake showed good agreement [65], and it would appear that the adsorption of hydroxide ion by cellulose can be successfully predicted from this theoretical treatment. A further demonstration of the validity of this approach concerns the application of dyes under alkaline conditions and emphasises the need to include the ionised fibre groups in subsequent calculation of affinity [65].

When dye is present in the system it is necessary, as in the simpler treatment (Eqns 4.62 and 4.66), to allow for the fact that adsorbed dye will affect the value of $[\mathrm{Na}]_f$. Thus Eqn 4.80 becomes

$$[\mathrm{Na}]_f = [\mathrm{Cl}]_f + [\mathrm{OH}]_f + [\mathrm{CellCOO}] + [\mathrm{CellO}] + z[\mathrm{D}]_f \tag{4.84}$$

which by making the same substitutions gives an equation which is the same as Eqn 4.82 but with the added term $z[\mathrm{D}]_f$. Again this can be solved to give a value of $[\mathrm{Na}]_f$ for the particular dyeing conditions, and this can be used in Eqn 4.61 to calculate the affinity.

The principal effect observed when applying water-soluble anionic dyes to cellulose from alkaline dyebaths is that there is a marked fall in substantivity as the pH of the dyebath increases. This is illustrated in Table 4.11, column 3, where the amount of dye adsorbed by viscose rayon at room temperature $[\mathrm{D}]_f$ is listed against the pH of the baths, which were of long liquor-to-goods ratio and contained the same concentrations of dye and electrolyte. Using these results in Eqn 4.61 permits the calculation of the affinity, which by definition should be a constant for all dyebath conditions at any one temperature. Since $[\mathrm{D}]_s$, $[\mathrm{D}]_f$ and $[\mathrm{Na}]_s$ are known from measurement of the composition of the dyebath, the degree of constancy achieved must be a direct indication of the validity of the method of calculating $[\mathrm{Na}]_f$. The results in column 4 of the table were obtained using Eqn 4.67 [56], which ignores the adsorption of hydroxide ions and the presence of ionised carboxylic acid groups, and it can be seen that the calculated affinity falls as the alkalinity increases. The results in column 5 were obtained using Eqns 4.70 and 4.71 [57], i.e. taking account of hydroxide ion adsorption. They show a smaller dependence of affinity on pH, but there is still a fall at higher pH. Eqn 4.84

TABLE 4.11

Effect of pH on substantivity and results of calculating affinity by different methods [65]

Dye used:

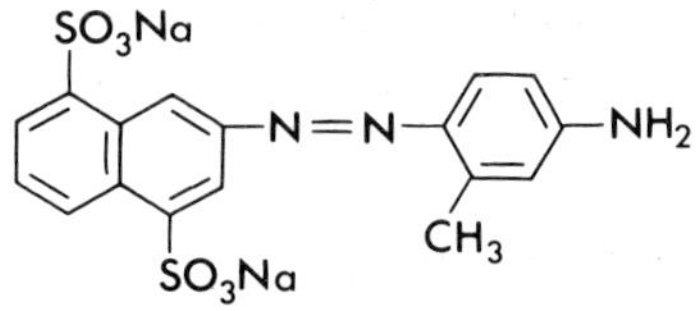

pH	$[D]_s$ /mol l^{-1} $\times 10^4$	$[D]_f$ /mol l^{-1} $\times 10^3$	Affinity (kJ (kcal) mol^{-1}) calculated by method of		
			Fowler *et al.* [58]	Peters and Simons [57]	Sumner [65]
7.0	1.20	14.1	11.7 (2.78)	11.7 (2.78)	11.9 (2.83)
8.0	1.20	13.8	11.8 (2.81)	11.8 (2.81)	12.0 (2.86)
9.5	1.19	13.9	11.8 (2.82)	11.8 (2.82)	12.1 (2.87)
11.0	1.19	12.9	11.6 (2.77)	11.7 (2.78)	12.0 (2.85)
11.9	1.27	8.90	10.5 (2.51)	11.0 (2.61)	11.7 (2.79)
12.6	1.31	4.55	8.8 (2.09)	10.3 (2.45)	11.9 (2.83)
12.7	1.30	4.06	8.5 (2.02)	10.3 (2.45)	12.0 (2.85)

[65] was used to calculate the affinities in column 6 and they exhibit perfect constancy, taking into account the expected experimental error.

The success of this approach in accounting for the experimentally observed adsorption of hydroxide ions by both cotton and viscose rayon, together with its ability to account for the marked loss of substantivity of an anionic dye with increasing alkalinity of application, is very strong evidence that the adsorption of hydroxide ions is a neutralisation effect producing ionised hydroxyl groups in the fibre which, because of their negative charge, reduce the adsorption of anionic dye. Moreover, the achievement of agreement over a wide range of alkalinity suggests that the 'availability' and the V term are not significantly changed by increasing the bath pH.

4.4.11 Further considerations

The treatment considered above was for dyebath conditions on the alkaline side of neutral in the presence of fairly high concentrations of electrolyte, that is, conditions in which it was reasonable to assume that the carboxylic acid groups would be fully ionised. Such an assumption would not be valid in all cases, and for more general application the concentration of ionised carboxyl groups must be replaced by a pH-dependent term. Such a term can be derived in exactly the same way as that for hydroxyl groups (Eqns 4.75–4.79), starting from

$$\text{CellCOOH} \rightleftharpoons \text{CellCOO}^- + \text{H}^+ \tag{4.85}$$

to give

$$[\text{CellCOO}] = \frac{K_{\text{COOH}} Q_{\text{COOH}}}{K_{\text{COOH}} + [\text{H}]} \tag{4.86}$$

or the equivalent equation in terms of [OH]:

$$[\text{CellCOO}] = \frac{K_{\text{COOH}} Q_{\text{COOH}}[\text{OH}]}{K_\text{w} + K_{\text{COOH}}[\text{OH}]} \tag{4.87}$$

Eqns 4.80 and 4.84 ignore the presence of a further positive ion, namely the hydrogen ion. For the purposes for which these equations were used this was completely justified, because of the overwhelming concentrations of sodium ion present. For other conditions this could be a serious omission, and for general application the term $[\text{H}]_\text{f}$ should be included.

Taking into account that there may be multivalent anions in addition to dye ions present [57], that there may be more than one dye in the system, that the fibre may contain more than one type of carboxylic acid group or hydroxyl group and that it may also contain one or more fixed dyes, then the equation for electrical neutrality in the fibre phase becomes, in general terms,

$$[\text{Cat}^+]_\text{f} + [\text{H}]_\text{f} = [\text{OH}]_\text{f} + [\text{An}^-]_\text{f} + 2[\text{An}^{2-}]_\text{f} + 3[\text{An}^{3-}]_\text{f} + \Sigma z_i[\text{D}]_{\text{f},i} + \Sigma y_j[\text{D}]_{\text{fixed},j} + \Sigma[\text{CellCOO}] + \Sigma[\text{CellO}] \tag{4.88}$$

where $[\text{Cat}^+]_\text{f}$ = the sum of the concentrations of all monovalent cations in the fibre (excluding H^+), $[\text{An}^-]_\text{f}$ = the same for all monovalent anions (excluding OH^-), and $[\text{An}^{2-}]_\text{f}$ and $[\text{An}^{3-}]_\text{f}$ = respectively the sums of the divalent and trivalent anions (other than dye ions). $\Sigma z_i[\text{D}]_{\text{f},i}$ is the ionic sum of the concentrations of all the unfixed anionic dyes, and $\Sigma y_j[\text{D}]_{\text{fixed},j}$ the same for all fixed anionic dyes. $\Sigma[\text{CellCOO}]$ and $\Sigma[\text{CellO}]$ = the sums of the concentrations of all the various ionised carboxylic acid and hydroxyl groups respectively.

In the same way that the electrical neutrality equation 4.80 could be rewritten using the various Donnan and other derived equalities to give Eqn 4.82 in which the only unknown was $[Na]_f$, so can Eqn 4.88. At this point, however, it is more convenient to define certain of the equalities in the general sense rather than the specific. Thus, by Donnan theory and using the nomenclature defined above,

$$[An^-]_f = \frac{[Cat^+]_s[An^-]_s}{[Cat^+]_f} \tag{4.89}$$

$$[An^{2-}]_f = \frac{[Cat^+]_s^2[An^{2-}]_s}{[Cat^+]_f^2} \tag{4.90}$$

$$[An^{3-}]_f = \frac{[Cat^+]_s^3[An^{3-}]_s}{[Cat^+]_f^3} \tag{4.91}$$

and

$$[An^-]_f = \frac{[H]_s[An^-]_s}{[H]_f} \tag{4.92}$$

Equating the expressions for $[An^-]_f$ from Eqns 4.89 and 4.92 shows that

$$[Cat^+]_f = \frac{[H]_f[Cat^+]_s}{[H]_s} \tag{4.93}$$

Substituting in Eqn 4.88 gives an equation in which the only term which is not experimentally measurable is $[Cat^+]_f$, and hence a value of this can be obtained by solving the equation. This value can then be used in the equation

$$-\Delta\mu^\ominus = RT \ln\frac{[Cat^+]_f^{z,i}[D]_{f,i}}{[Cat^+]_s^{z,i}[D]_{s,i}} \tag{4.94}$$

which is the generalised form of Eqn 4.61, to calculate the affinity of a single dye or those of the component dyes of a mixture.

Alternatively, writing the Donnan equations (Eqns 4.89–4.91) in terms of hydrogen ion as exemplified by Eqn 4.92, and using Eqns 4.78 and 4.86, together with the fact that K_w = [H][OH], permits the formulation of an equation in which the only quantity which is not experimentally measurable is $[H]_f$. Thus

$$\frac{[\mathrm{H}]_{\mathrm{f}}[\mathrm{Cat}^+]_{\mathrm{s}}}{[\mathrm{H}]_{\mathrm{s}}} + [\mathrm{H}]_{\mathrm{f}} = \frac{K_{\mathrm{w}}}{[\mathrm{H}]_{\mathrm{f}}} + \frac{[\mathrm{H}]_{\mathrm{s}}[\mathrm{An}^-]_{\mathrm{s}}}{[\mathrm{H}]_{\mathrm{f}}} + \frac{2[\mathrm{H}]_{\mathrm{s}}^2[\mathrm{An}^{2-}]_{\mathrm{s}}}{[\mathrm{H}]_{\mathrm{f}}^2} + \frac{3[\mathrm{H}]_{\mathrm{s}}^3[\mathrm{An}^{3-}]_{\mathrm{s}}}{[\mathrm{H}]_{\mathrm{f}}^3}$$

$$+\sum z_i[\mathrm{D}]_{\mathrm{f},i} + \sum y_j[\mathrm{D}]_{\mathrm{fixed},j} + \sum \frac{K_{\mathrm{COOH}}Q_{\mathrm{COOH}}}{K_{\mathrm{COOH}}+[\mathrm{H}]_{\mathrm{f}}} + \sum \frac{K_{\mathrm{OH}}Q_{\mathrm{OH}}}{K_{\mathrm{OH}}+[\mathrm{H}]_{\mathrm{f}}} \quad (4.95)$$

and this equation can be solved for $[\mathrm{H}]_{\mathrm{f}}$. The affinity equation (Eqn 4.94) can be modified using Eqn 4.93, to give Eqn 4.96, which includes $[\mathrm{H}]_{\mathrm{f}}$:

$$-\Delta\mu^{\ominus} = RT \ln \frac{[\mathrm{H}]_{\mathrm{f}}^{z,i}[\mathrm{D}]_{\mathrm{f},i}}{[\mathrm{H}]_{\mathrm{s}}^{z,i}[\mathrm{D}]_{\mathrm{s},i}} \quad (4.96)$$

This latter approach, using hydrogen ion concentrations, is probably the more convenient and informative method for reasons which will be discussed later.

In all of the above one possibility has been ignored: that the system may contain multivalent cations. Considering a monovalent dye anion, then the affinity equation for a monovalent cation, using Eqn 4.93, would be:

$$-\Delta\mu^{\ominus} = RT \ln \frac{[\mathrm{Cat}^+]_{\mathrm{f}}[\mathrm{D}]_{\mathrm{f}}}{[\mathrm{Cat}^+]_{\mathrm{s}}[\mathrm{D}]_{\mathrm{s}}} = RT \ln \frac{[\mathrm{H}]_{\mathrm{f}}[\mathrm{D}]_{\mathrm{f}}}{[\mathrm{H}]_{\mathrm{s}}[\mathrm{D}]_{\mathrm{s}}} \quad (4.97)$$

and for a divalent cation, using the Donnan equalities, Eqn 4.92 and $[\mathrm{Cat}^{2+}]_{\mathrm{f}}[\mathrm{An}^-]_{\mathrm{f}}^2 = [\mathrm{Cat}^{2+}]_{\mathrm{s}}[\mathrm{An}^-]^2_{\mathrm{s}}$,

$$-\Delta\mu^{\ominus} = RT \ln \frac{[\mathrm{Cat}^{2+}]_{\mathrm{f}}[\mathrm{D}]_{\mathrm{f}}^2}{[\mathrm{Cat}^{2+}]_{\mathrm{s}}[\mathrm{D}]_{\mathrm{s}}^2} = RT \ln \frac{[\mathrm{H}]_{\mathrm{f}}^2[\mathrm{D}]_{\mathrm{f}}^2}{[\mathrm{H}]_{\mathrm{s}}^2[\mathrm{D}]_{\mathrm{s}}^2} = 2RT \ln \frac{[\mathrm{H}]_{\mathrm{f}}[\mathrm{D}]_{\mathrm{f}}}{[\mathrm{H}]_{\mathrm{s}}[\mathrm{D}]_{\mathrm{s}}} \quad (4.98)$$

If care is taken with the experimental procedure to ensure that only one type of cation is present then there is no problem; a value of the affinity can be obtained using the appropriate equation. If a mixture of mono- and divalent cations is present, however, there is the problem of which equation should be used. All the values $[\mathrm{H}]_{\mathrm{f}}$, $[\mathrm{D}]_{\mathrm{f}}$, $[\mathrm{H}]_{\mathrm{s}}$ and $[\mathrm{D}]_{\mathrm{s}}$ will be the same for both ions, since they are present in the same dyebath, but using them in Eqn 4.98 will give an affinity twice that derived from Eqn 4.97. Fortunately, in dyeing and in particular in practical dyeing, almost all the electrolytes in common use are sodium or potassium salts and produce only monovalent cations; for this reason this problem can generally be ignored.

4.5 IONIC DYES APPLIED TO OPPOSITELY CHARGED SUBSTRATES

The most important dyeing systems to be considered under this heading are the application of

(a) anionic dyes, i.e. acid dyes, to wool and polyamide fibres;
(b) cationic dyes, i.e. basic dyes, to acrylic fibres.

4.5.1 Anionic dyes on wool

As a result of its chemical structure wool contains a considerable number of both acidic and basic groups and hence, in contrast to cellulose, it was natural to consider that the adsorption of ions by wool was directly concerned with the presence of these groups; this in turn led to the idea that ions are adsorbed by wool on specific sites in the fibre. It is certain that the ionised groups constituting these sites are the side-chain —NH_3^+ and side-chain —COO^- groups of the wool protein. There will be some contribution from other ionised groups, e.g. hydroxyl, but this will be only small. The ionisation of the —NH_2 groups to —NH_3^+ and of —COOH groups to —COO^- is pH-dependent and hence so are also the respective concentrations of these groups and the overall charge on the protein. This in turn will alter the receptivity of the protein to other ions, including dye anions. This dependence on pH of the charge on the protein is best illustrated by considering a model, water-soluble protein containing (per molecule)

100 aspartic acid residues (Asp) with $pK_{COOH} = 3.86 = pK_{Asp}$
30 lysine residues (Lys) with $pK_{NH^2} = 8.95 = pK_{LysH^+}$
50 arginine residues (Arg) with $pK_{NH_2} = 13.20 = pK_{ArgH^+}$

Assuming that the pK values in the protein are the same as for the free amino acids then the net charge on the protein can be calculated from

$$\begin{aligned}\text{net charge} &= [\text{LysH}^+] + [\text{ArgH}^+] - [\text{Asp}^-] \\ &= \frac{[\text{H}][\text{lysine}]}{K_{\text{LysH}^+} + [\text{H}]} + \frac{[\text{H}][\text{arginine}]}{K_{\text{ArgH}^+}\,[\text{H}]} - \frac{K_{\text{Asp}}[\text{aspartic}]}{K_{\text{Asp}} + [\text{H}]}\end{aligned} \tag{4.99}$$

where [lysine], [arginine] and [aspartic] represent *total* concentrations, charged and uncharged. The amounts of the various ions and the resultant net charge are shown in Table 4.12 over a range of pH values, and the net charge is shown as a titration curve in Figure 4.13.

Starting at low pH, all the aspartic acid residues will be un-ionised (—COOH), whereas lysine and arginine will be protonated (—NH_3^+). The total charge is +80. It can be seen that as the pH increases the protein begins to lose hydrogen ions, due to the ionisation of the carboxylic acid groups,

TABLE 4.12

Titration of a hypothetical, water-soluble protein (for details see text)

	pH:	2.0	3.86	4.0	5.0	7.0	8.95	11.0	13.2	14.0
Aspartic acid	—COOH	93.6	50	42	6.8	0.1	0	0	0	0
	$—COO^-$	1.4	50	58	93.2	99.9	100	100	100	100
Lysine	$—NH_3^+$	30	30	30	30	29.7	15	0.26	0	0
	$—NH_2$	0	0	0	0	0.3	15	29.7	30	30
Arginine	$—NH_3^+$	50	50	50	50	50	50	49.7	25	6.8
	$—NH_2$	0	0	0	0	0	0	0.3	25	43.2
Net charge		+78.6	+30	+22	−13.2	−20.2	−35	−50	−75	−93.2

and the negatively charged groups so formed result in a decrease in the net positive charge on the protein. At pH 3.86 the carboxyl groups will be half-ionised, and at a pH between 4 and 5 the number of negative groups will equal the number of positive groups, so that the protein will carry a net charge of zero. This is the *isoelectric point* of the protein. Further increase in pH results in further loss of hydrogen ion until, at about pH 6, the change becomes negligibly small and all the carboxyl groups may be assumed to be ionised. Nothing further happens until approximately pH 7, when loss of hydrogen ion from the positively charged amino groups of lysine becomes significant. This change is half-complete at pH 8.95, and for all practical purposes is complete at approximately pH 11. Eventually the much more basic arginine begins to lose its hydrogen ions.

Several points are worth noting about the titration curve.

(a) The number of molecules of acid or base bound is, at all pH values, numerically equivalent to the net charge on the protein.
(b) The maximum acid-binding capacity is determined by the number of basic groups present in the protein, although the process of acid-binding is in fact the back-titration of ionised carboxylic acid groups.
(c) The maximum base-binding capacity is fixed by the number of carboxylic acid groups and the process is one of titration of basic groups.
(d) The pH ranges over which the various titrations occur is determined by the pK values of the groups concerned.

Included in Figure 4.13 are approximate curves for the titration of gelatine and of wool. The curve for gelatine, a soluble protein, resembles that of the model quite closely, whereas that of wool, an insoluble protein, is rather different in that it shows a region between pH 4 and pH 12 in which only

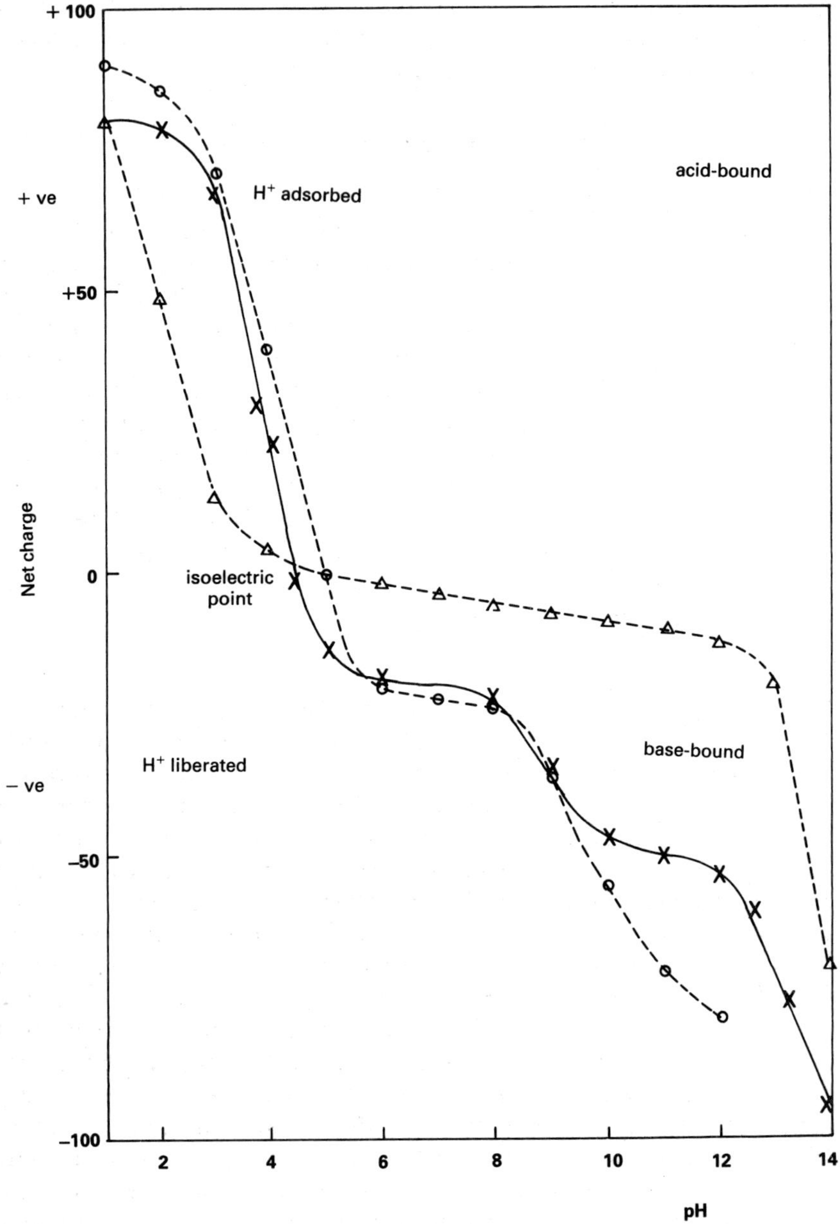

Figure 4.13 – Titration curve of model soluble protein (—×—×—×—); also included are approximate curves for gelatine (- - - ○ - - - ○ - - - ○ - - -) and wool (- - - △ - - - △ - - - △ - - -)

small amounts of acid or base are bound. This has been interpreted as being due to the electrical charge carried by the fibre and to the need when dealing with water-swollen, solid proteins to consider the system as being

$$\text{adsorbate in dyebath} \rightleftharpoons \text{adsorbate in internal solution} \rightleftharpoons \text{adsorbate in fibre}$$

with the proviso that each part of the system must be electrically neutral. If a protein is considered at its isoelectric point (i.e. no charge on the fibre) then the adsorption of a hydrogen ion will render the fibre positively charged, making it more difficult for a second proton to be adsorbed. The positive charge on the fibre will attract into the internal phase a counterion, however, so that the adsorption becomes a simultaneous uptake of positive and negative ions. This extra requirement makes it more difficult for the fibre to adsorb protons and results in the titration curve being displaced to lower pH. The converse is true on the other side of the isoelectric point, where the curve is moved to higher pH values. The result is the long, rather flat portion of the curve which is not due to changes in the dissociation constants of the constituent amino acids, but to the charge on the fibre. In spite of this the process of adsorption of acids and bases, outlined above for the model protein, still applies.

4.5.2 The reactivity of wool

Before moving to a theoretical study of the adsorption of acids by wool it is useful to consider the reactivity of wool, in particular its reactions with water. Wool fibres, and indeed all protein fibres, are very susceptible to hydrolysis, particularly at elevated temperatures. Many of the acid side-chains are in the form of the amide and these readily lose ammonia to give the carboxylic acid. The peptide links are also susceptible to hydrolysis, particularly under acid or alkaline conditions, producing new carboxylic acid and amino end-groups. In addition the protein can contain some water-soluble, low-r.m.m. material. Hence any aqueous dyeing experiment involving wool may well be carried out in a dyebath that contains a greater or lesser quantity of unknown material as a result of fibre hydrolysis and the dissolution of water-soluble fractions. Thus there is a situation in which both the fibre and the dyebath are changing in some undefined manner which is a function of time, temperature and dyeing conditions. In addition wool has such a relatively large capacity for adsorbing ions that great care is required, and this is not always obvious from experimental reports, to ensure that adventitious ions in unknown amounts are not introduced by the fibre into the dyeing system.

Clearly all of the above makes it very difficult to carry out meaningful experiments to determine the thermodynamics of wool dyeing, and this in turn means that much of the data in the literature must be viewed with a certain amount of suspicion unless the experimental processes are adequately selected, described and carried out. Such a situation is illustrated in Table 4.16 on p. 331, which gives the results of applying a pure free dye acid to wool. Since the only known addition to the dyebath is the dye acid, and since this is a strong acid and hence will be fully dissociated, the equivalent concentrations of hydrogen ion and dye ion in the dyebath should be the same. The difference, shown in column 3 of the table, is marked and indicates that the dyebath contained other ionic material of unknown composition. It is obvious that this will not be an isolated case and clearly, despite the considerable amount of work that has been carried out to establish a theory which will describe the titration of wool, none of the proposed models has been thoroughly tested because of the difficulties of obtaining accurate experimental data.

4.5.3 The theoretical analysis of the titration curve

Taking the general picture of a water-swollen fibre forming an equipotential volume, surrounded by an external bath, then it is possible to define, for each mobile ion in the system, that

$$\mu_f = \mu_f^{\ominus} + RT \ln a_f \tag{4.100}$$

where f indicates the fibre,

$$\mu_i = \mu_i^{\ominus} + RT \ln a_i \tag{4.101}$$

where i indicates the internal solution and

$$\mu_s = \mu_s^{\ominus} + RT \ln a_s \tag{4.102}$$

where s indicates external solution.

At equilibrium, within the equipotential volume, $\mu_i = \mu_f$ and hence

$$-\Delta\mu_{i,f}^{\ominus} = -(\mu_f^{\ominus} - \mu_i^{\ominus}) = RT \ln\frac{a_f}{a_i} \tag{4.103}$$

There will be a difference in potential between this equipotential volume and the external solution and this must be allowed for, so that at equilibrium $\mu_i = \mu_s - \psi F$ and hence

$$-\Delta\mu_{si}^{\ominus} = -(\mu_i^{\ominus} - \mu_s^{\ominus}) = RT \ln\frac{a_i}{a_s} + \psi F \tag{4.104}$$

where ψF is the electrical work required to move one mole of cation from the external to the internal solution against this potential. Combining Eqns 4.103 and 4.104, and considering only cations, gives:

$$-\Delta\mu^{\ominus}_{s,f} = -(\Delta\mu^{\ominus}_{s,i} - \Delta\mu^{\ominus}_{i,f}) = RT\ \ln\frac{a^{+}_{f}}{a^{+}_{s}} + \psi F \qquad (4.105)$$

and a similar equation can be written for anions:

$$-\Delta\mu^{\ominus}_{s,f} = RT\ \ln\frac{a^{-}_{f}}{a^{-}_{s}} - \psi F \qquad (4.106)$$

By adding Eqns 4.105 and 4.106 the unknown electrical work term is eliminated:

$$-\Delta\mu^{\ominus}_{\pm} = RT\ \ln a^{+}_{f}a^{-}_{f} - RT\ \ln a^{+}_{s}a^{-}_{s} \qquad (4.107)$$

where $\Delta\mu^{\ominus}_{\pm}$ is the combined affinities of the cation and anion.

It was at this point, where values had to be assigned to the activities of the ions in the fibre, that a difference of opinion arose and this led to two different approaches being formulated, namely the Langmuir or Gilbert–Rideal method [67] and the Donnan method [68]. The differences between the two are described pictorially in Figure 4.14 for a simple, monobasic strong acid HX.

4.5.4 The Langmuir (Gilbert–Rideal) method

In this approach it is assumed that the fibre contains sites for both cations and anions (Figure 4.14(a)). The activities of both in the fibre are therefore equated, in the Langmuir sense, to $\theta/(1-\theta)$ where

$$\theta_H = \frac{[H]_f}{[S]_H} \quad \text{and} \quad \theta_X = \frac{[X]_f}{[S]_X} \qquad (4.108)$$

and $[H]_f$ and $[X]_f$ are the amounts of the two ions adsorbed and $[S]_H$ and $[S]_X$ are the maximum amounts possible, i.e. the saturation values. Eqn 4.107 therefore becomes, for the monobasic acid,

$$-\Delta\mu^{\ominus}_{\pm} = RT\ \ln\left(\frac{\theta_H}{1-\theta_H}\cdot\frac{\theta_X}{1-\theta_X}\right) - RT\ \ln\ ([H]_s[X]_s) \qquad (4.109)$$

and at the standard state $\theta/(1-\theta)$ equals unity, i.e. $\theta = 0.5$, and the situation is one of half-saturation of the available sites.

In the case of wool adsorbing acid from a solution that contains no other

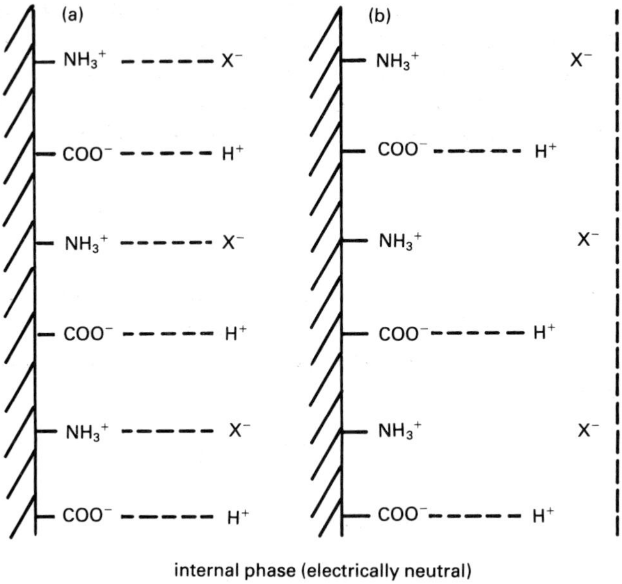

Figure 4.14 – Pictorial representation of two approaches to wool dyeing theory: (a) Gilbert–Rideal, (b) Donnan

electrolytes then, for electrical neutrality in the solution, $[H]_s = [X]_s$ and since in wool the numbers of positive and negative sites are very nearly equal, i.e. $[S]_H \approx [S]_X$, then $[H]_f = [X]_f$ and Eqn 4.109 becomes

$$-\Delta\mu^{\ominus}_{\pm} = 2RT \ln\left(\frac{\theta_H}{1-\theta_H}\right) - 2RT \ln [H]_s \qquad (4.110)$$

or

$$\frac{-\Delta\mu^{\ominus}_{\pm}}{4.606RT} = \log\left(\frac{\theta_H}{1-\theta_H}\right) + pH_s = -\log\left[\left(\frac{1}{\theta_H}-1\right)[H]_s\right] \qquad (4.111)$$

This equation makes two points.

(a) If $\log [\theta_H/(1-\theta_H)]$ is plotted against pH_s then a straight line of negative unit slope should be obtained.
(b) The pH at which half-saturation occurs, i.e. at which $\theta_H/(1-\theta_H) = 1$ and $\log [\theta_H/(1-\theta_H)] = 0$, is a measure of the affinity.

4.5.5 The Donnan method

In the above treatment the sites for hydrogen ion adsorption are clearly defined, leading to un-ionised —COOH groups, but those concerned with anion adsorption are not. The Donnan method avoids this difficulty by assuming that, while hydrogen ions are adsorbed on specific sites in the fibre, the anions are merely present in the internal solution in accordance with the Donnan theory of membrane equilibrium at a concentration of $[X]_i$ mol l^{-1} (Figure 4.14(b)). Thus Eqn 4.107 becomes, for an acid HX,

$$-\Delta\mu_+^\ominus = RT \ln\left(\frac{\theta_H}{1-\theta_H}.[X]_i\right) - RT \ln([H]_s[X]_s) \qquad (4.112)$$

It must be noted that this equation now defines the affinity of the cation alone, since the anion affinity has been equated to zero, i.e. no combination takes place. If no other electrolyte is present then, for electrical neutrality in the internal phase,

$$V[X]_i = V[H]_i + [H]_f \qquad (4.113)$$

and assuming that $V[H]_i$ is negligible compared to $[H]_f$

$$[X]_i \approx \frac{[H]_f}{V} \qquad (4.114)$$

so that

$$-\Delta\mu_+^\ominus = RT \ln\left(\frac{\theta_H}{1-\theta_H}\cdot\frac{[H]_f}{V}\right) - RT \ln([H]_s[X]_s) \qquad (4.115)$$

Since $[H]_f = \theta_H[S]_H$,

$$-\Delta\mu_+^\ominus - RT \ln [S]_H + RT \ln V =$$

$$RT \ln\left(\frac{\theta_H^2}{1-\theta_H}\right) - RT \ln([H]_s[X]_s) \qquad (4.116)$$

But in the absence of other electrolyte $[H]_s = [X]_s$, so that

$$\frac{-\Delta\mu_+^\ominus}{2.303RT} - \log\left(\frac{[S]_H}{V}\right) = \log\left(\frac{\theta_H^2}{1-\theta_H}\right) + 2\text{pH} \qquad (4.117)$$

Thus by this treatment a plot of log $[\theta_H^2/(1-\theta_H)]$ against pH should give a straight line of slope -2.

4.5.6 Comparison of the two approaches

An analysis [41] of the results of titrating wool with hydrochloric acid [69] revealed that both methods gave a linear relationship, but that the Gilbert–Rideal equation gave a slope of 0.88 instead of 1.0 whilst the Donnan method gave a slope of 1.4 instead of 2.0. Whether these discrepancies are due to assuming that the experimental results are accurate and uninfluenced by fibre decomposition and its attendant problems, or whether they have their source in the theoretical treatment, is impossible to decide. For example, both approaches take the measured adsorption as $[H]_f$ whereas it is in fact $[H]_f + V[H]_i$. The assumption that $V[H]_i$ is negligible is possibly not always valid. It may also be that the chosen values of $[S]_H$ and V are not the best possible. Whatever the reason, the results of this analysis do not indicate that a preference should be given to either approach.

When the effect of added electrolyte (potassium chloride) on the titration curve of hydrochloric acid on wool is considered, the Gilbert–Rideal equation (Eqn 4.110) becomes

$$-\frac{\Delta\mu_{\pm}^{\ominus}}{2.303RT} = 2\log\left(\frac{\theta_H}{1-\theta_H}\right) + \text{pH} - \log\,[\text{Cl}]_s \qquad (4.118)$$

and a plot of $\log\,[\theta_H/(1-\theta_H)]$ against pH, at constant chloride ion concentration, should give a straight line of slope -0.5 which is displaced to higher values of pH with increasing chloride ion by the difference in the value of $\log\,[\text{Cl}]_s$. Analysis of a set of results [41, 69] did indeed give straight lines with the correct displacement of pH with electrolyte concentration, but the slopes varied from approximately 0.4 to 0.5 as chloride ion concentration increased from 0.005 to 0.2 mol l^{-1}.

When the effect of added potassium chloride was studied using the Donnan approach [68] attention was concentrated on the internal rather than the external pH. By Donnan membrane theory,

$$[\text{H}]_i[\text{Cl}]_i = [\text{H}]_s[\text{Cl}]_s \qquad (4.119)$$

For electrical neutrality

$$[\text{H}]_f + V[\text{H}]_i = V[\text{Cl}]_i = a \qquad (4.120)$$

where a = the total concentration of hydrogen ions taken up by the fibre. Substituting $[\text{Cl}]_i = a/V$ from Eqn 4.120 into Eqn 4.119, followed by taking logarithms and substituting pH_i for $-\log\,[\text{H}]_i$ and pH_s for $-\log\,[\text{H}]_s$, leads to

$$\text{pH}_i = \text{pH}_s + \log\frac{a}{V} - \log\,[\text{Cl}]_s \qquad (4.121)$$

This equation applies when $[\mathrm{H}]_s$ and $[\mathrm{Cl}]_s$ are not equal (e.g. in the presence of potassium chloride). In the special case where $[\mathrm{H}]_s = [\mathrm{Cl}]_s$ it simplifies to

$$\mathrm{pH_i} = 2\mathrm{pH_s} + \log\frac{a}{V} \tag{4.122}$$

When, from the experimental data, the amount of acid adsorbed was plotted against $\mathrm{pH_s}$ (the external pH), different curves were obtained for each concentration of added chloride ion present but when plotted against $\mathrm{pH_i}$ (the internal pH) the plots resolved to a single curve. These are the results expected from Eqn 4.122.

Thus, bearing in mind the provisos listed above with regard to fibre decomposition and possible wrong assumptions, it would appear that both approaches give a satisfactory interpretation of the effect of electrolyte on the titration curve. Further critical studies have been carried out [70, 71, 72, 73] without establishing which of these approaches and others is to be preferred, but there was some indication [70] that the Gilbert–Rideal (Langmuir) method gave a truer interpretation of the adsorption of acids by wool. For this reason this approach will be adopted for the remainder of this discussion. The application of the Donnan theory is described elsewhere [74].

4.5.7 The saturation uptake of wool

Before the Gilbert–Rideal theory can be used to calculate the affinity of a monobasic acid for wool, using Eqns 4.109 or 4.111, it is necessary to determine values of $[\mathrm{S}]_\mathrm{H}$ and $[\mathrm{S}]_\mathrm{X}$, the saturation values for cations and anions respectively. The values chosen greatly affect the value of θ obtained, particularly as θ approaches unity.

By inspection of the titration curve for hydrochloric acid it is clear that $[\mathrm{S}]_\mathrm{H}$ must lie between 0.8 and 0.9 mol kg^{-1} of dry fibre, but a better estimate may be obtained if Eqn 4.111, in which it has already been assumed that $[\mathrm{S}]_\mathrm{H} \approx [\mathrm{S}]_\mathrm{X}$ for wool, is rearranged to give Eqn 4.111a, since $\theta_\mathrm{H} = [\mathrm{H}]_\mathrm{f}/[\mathrm{S}]_\mathrm{H}$:

$$\exp\left(\frac{\Delta\mu^\ominus_\pm}{2RT}\right) = \left(\frac{1}{\theta_\mathrm{H}} - 1\right)[\mathrm{H}]_\mathrm{s} = \left(\frac{[\mathrm{S}]_\mathrm{H}}{[\mathrm{H}]_\mathrm{f}} - 1\right)[\mathrm{H}]_\mathrm{s} = K \tag{4.111a}$$

Rearranging Eqn 4.111a leads to:

$$\frac{1}{[\mathrm{H}]_\mathrm{f}} = \frac{K}{[\mathrm{S}]_\mathrm{H}} \cdot \frac{1}{[\mathrm{H}]_\mathrm{s}} + \frac{1}{[\mathrm{S}]_\mathrm{H}} \tag{4.123}$$

Thus a graph of $1/[\mathrm{H}]_\mathrm{f}$ against $1/[\mathrm{H}]_\mathrm{s}$ should give a straight line of slope

$K/[S]_H$ and intercept $1/[S]_H$. Such a result is found, and the value of $[S]_H$ thus obtained is probably more satisfactory than that given by inspection of the titration curve, although there are problems due to extrapolating lines of poor linearity. By a similar analysis it can be shown that a graph of $1/[X]_f$ against $1/[X]_s$ should also give a straight line with an intercept of $1/[S]_X$. Using this method with the hydrochloric acid titration results gave a saturation value $[S]_H$ of 0.82 mol kg^{-1} [74].

TABLE 4.13

Affinity of hydrochloric acid for wool at 0°C [41, 69]

Acid alone			Constant ionic concentration (0.1M)		
pH	θ	$-\Delta\mu^{\ominus}_{\pm}$ /kJ (kcal) mol^{-1}	pH	θ	$-\Delta\mu^{\ominus}_{\pm}$ /kJ (kcal) mol^{-1}
1.09	0.927	22.8 (5.43)	1.09	0.929	22.6 (5.37)
1.47	0.940	27.8 (6.62)	1.38	0.935	24.5 (5.83)
1.66	0.830	24.5 (5.83)	1.68	0.908	24.3 (5.78)
1.71	0.793	23.9 (5.68)	1.99	0.858	23.7 (5.64)
1.74	0.770	23.6 (5.63)	2.27	0.810	23.6 (5.62)
1.93	0.695	23.9 (5.68)	2.54	0.735	23.0 (5.48)
2.06	0.622	23.7 (5.65)	2.81	0.655	22.76 (5.42)
2.28	0.549	24.7 (5.88)	3.21	0.544	22.8 (5.42)
2.34	0.475	23.9 (5.70)	3.65	0.435	23.1 (5.50)
2.45	0.427	24.2 (5.75)	4.21	0.305	23.4 (5.57)
2.73	0.305	24.6 (5.86)	4.62	0.214	23.4 (5.56)
3.04	0.207	25.6 (6.10)	5.03	0.149	23.5 (5.60)
3.27	0.134	25.6 (6.10)	5.47	0.092	23.3 (5.55)
3.62	0.082	26.8 (6.38)	–	–	–
3.84	0.055	27.1 (6.45)	–	–	–

An alternative method has been proposed [8] which in certain cases produces improved linearity and hence more certain extrapolation. Again it is assumed that for a monobasic acid $[S]_H \approx [S]_X$ and $\theta_H = \theta_X$, so that Eqn 4.109 may be written

$$-\Delta\mu^\ominus_\pm = RT \ln\left(\frac{\theta_X}{1-\theta_X}\right)^2 - RT \ln [H]_s[X]_s \qquad (4.109a)$$

from which it follows that

$$\left(\frac{\theta_X}{1-\theta_X}\right)^2 = [H]_s[X]_s \exp\left(-\frac{\Delta\mu^\ominus_\pm}{RT}\right) \qquad (4.124)$$

Substituting K from Eqn 4.111a and rearranging then gives

$$\frac{1}{[X]_f} = \frac{1}{K[S]_X}\left(\frac{1}{([H]_s[X]_s)^{\frac{1}{2}}}\right) + \frac{1}{[S]_X} \qquad (4.125)$$

A graph of $1/[X]_f$ against $[H]_s[X]_s^{-\frac{1}{2}}$ should give a straight line of slope $1/K[S]_X$ and intercept $1/[S]_X$. A value of 0.90 mol kg^{-1} of dry wool was found from measurements of the adsorption of the free acid of Metanil Yellow YK (C.I. Acid Yellow 36) on wool.

Using a saturation value of 0.82 mol kg^{-1} of dry wool the standard affinity of hydrochloric acid for wool has been calculated from the experimental results obtained [41, 69] in the absence and presence of potassium chloride, using Eqns 4.111 and 4.118 respectively. The results are shown in Table 4.13, which shows that the consistency is reasonably satisfactory, although the results with acid alone tend to rise when θ is less than 0.3. The mean affinities at various ionic concentrations (Table 4.14, column 2) show a drift to lower values as the ionic concentration increases, however. The last column in the table shows that the drift can be eliminated almost completely by using activities for the ions in solution instead of concentrations, that is, by multiplying the latter by the mean activity coefficient $f_\pm$. Eqn 4.118 therefore becomes

$$\frac{-\Delta\mu^\ominus_\pm}{2.303RT} = 2 \log\frac{\theta_H}{1-\theta_H} - RT \log [H]_s - RT \log [Cl]_s - 2RT \log f_\pm \qquad (4.126)$$

To extend this correction to other acids is only rarely possible, because the required values of the activity coefficients are seldom available. Hence it is necessary to revert to the usual procedure and equate them to unity.

TABLE 4.14

Mean affinity of hydrochloric acid for wool at 0 °C at different ionic concentrations [41]

Ionic concentration/M	$-\Delta\mu^{\ominus}_{\pm}$ (mean) /kJ (kcal) mol^{-1}	$2RT \ln f_{\pm}$ /kJ (kcal) mol^{-1}	Corrected $-\Delta\mu^{\ominus}_{\pm}$ /kJ (kcal) mol^{-1}
Acid alone	24.6 (5.86)	0.46 (0.11)	25.1 (5.97)
0.01	24.4 (5.81)	0.46 (0.11)	24.9 (5.92)
0.02	23.9 (5.69)	0.59 (0.14)	24.5 (5.83)
0.04	23.7 (5.64)	0.80 (0.19)	24.5 (5.83)
0.10	23.6 (5.61)	1.13 (0.27)	24.7 (5.88)
0.20	23.1 (5.51)	1.42 (0.34)	24.5 (5.85)
0.50	22.3 (5.32)	1.97 (0.47)	24.3 (5.79)
1.00	22.0 (5.25)	2.40 (0.57)	24.6 (5.82)
Mean	23.43 ±0.84 (5.58 ± 0.20)		24.61 ± 0.25 (5.86 ± 0.06)

In order to compare the affinities of a range of acids it is preferable to use a standard set of experimental conditions for all the acids, and the most obvious choice is when the fibre is half-saturated since then the acid in the fibre is in its standard state. As described earlier, under such conditions Eqn 4.111 becomes

$$-\Delta\mu^{\ominus}_{\pm} = 4.606\ RT.\mathrm{pH}_{\mathrm{mid}} \qquad (4.127)$$

where $\mathrm{pH}_{\mathrm{mid}}$ is the pH in the solution at which the fibre is half-saturated. This treatment can be applied to any titration curve, but again the principal problem remains of assigning a value to $[S]_H$, the maximum acid-binding power of wool. The figures in Table 4.15 have been calculated using a value of 0.90 mol kg^{-1} of dry wool. It must be noted that the affinities obtained in this way are the sums of the affinities of the cation and the anion, i.e. $\Delta\mu^{\ominus}_{\pm}$. Although thermodynamically incorrect, it is possible to separate the affinities of the two ions if $\Delta\mu^{\ominus}_{\mathrm{HCl}}$ is taken as a starting point and $\Delta\mu^{\ominus}_{\mathrm{Cl}}$ is equated to zero, arbitrarily. The value of $\Delta\mu^{\ominus}_{\mathrm{HCl}}$ has been determined by several workers

TABLE 4.15

Affinity of acids for wool at 0 °C [41]

Acid	pH_{mid}	Total affinity /kJ (kcal) mol^{-1}	Anion affinity /kJ (kcal) mol^{-1}	Source ref.
Hydrochloric	2.12	−22.3 (−5.30)	–	
Sulphuric acid monoethyl ester	2.21	−23.1 (−5.50)	−0.80 (−0.20)	74
Sulphuric[a]	2.96	−47.7 (−11.36)	−3.2 (−0.76)	74
Sulphuric[a] (at 22 °C)	2.90	−50.8 (−12.10)	−3.4 (−0.80)	75
Trichloracetic	2.62	−27.3 (−6.50)	−5.0 (−1.20)	74
Benzenesulphonic	2.50	−26.0 (−6.20)	−3.8 (−0.90)	74
4-Toluenesulphonic	2.53	−26.5 (−6.30)	−4.2 (−1.00)	74
4-Toluenesulphonic[b]	2.63	−27.3 (−6.50)	−5.0 (−1.20)	76
o-Xylenesulphonic	2.61	−27.3 (−6.50)	−5.0 (−1.20)	74
Naphthalene-β-sulphonic	3.11	−32.6 (−7.75)	−10.3 (−2.45)	74
Naphthalene-β-sulphonic[b]	3.18	−33.2 (−7.90)	−10.9 (−2.60)	76
Anthraquinone-β-sulphonic[b]	3.63	−38.0 (−9.05)	−15.8 (−3.75)	76
2-Nitrobenzenesulphonic	2.72	−28.4 (−6.75)	−6.1 (−1.45)	74
2,4-Dinitrobenzenesulphonic	3.03	−31.7 (−7.55)	−9.5 (−2.25)	74
2,5-Dichlorobenzenesulphonic	2.98	−31.1 (−7.40)	−8.8 (−2.10)	74
1-chloro-4-nitrobenzene-2-sulphonic	2.98	−31.1 (−7.40)	−8.8 (−2.10)	74
Chromic[c]	3.30	−34.4 (−8.20)	−12.2 (−2.90)	77
Picric	3.72	−39.9 (−9.25)	−16.6 (−3.95)	74
1-Naphthol-2,4-dinitro-7-sulphonic	4.06	−42.4 (−10.10)	−20.2 (−4.80)	74
2,4,6-Trinitroresorcinol	3.48	−36.3 (−8.65)	−14.1 (−3.35)	74

[a] Dibasic acid. [b] Corrected from 22 °C to 0 °C assuming $\Delta H = 0$.
[c] Assumed to be acting as a monobasic acid.

[69, 77], whose results give an average figure of -22.3 kJ (-5.3 kcal) mol^{-1}, a value which is only slightly affected by temperature. This may be taken as the affinity of the hydrogen ion and subtracted from the total affinities of the monobasic acids to obtain the affinities of the various anions (Table 4.15, column 4).

It is obvious that there is a clear tendency for the affinity of the anion to increase with increasing r.m.m. and complexity or with the introduction of polar groups. Hence the affinity cannot be due only to electrostatic effects – otherwise all monobasic acids would have the same value as hydrochloric acid. Clearly there must be a considerable contribution from non-electrostatic forces of attraction.

With regard to the acids listed, three points must be made. Sulphuric acid is treated as dibasic, so that the value of the $\Delta\mu^{\ominus}$ term has been obtained by subtracting $2\Delta\mu^{\ominus}_{H}$ from the total affinity. Chromic acid is considered monobasic under the conditions used (since its second dissociation has a pK of 6.4), but complications arise because of changes in the valency state during the experiment giving trivalent chromium ions [78]. The adsorption of weak acids by wool is quite different from that of strong acids, and their behaviour suggests that they are essentially adsorbed in the undissociated state [75].

4.5.8 The adsorption of free dye acids: affinity determination

In this method wool is dyed to equilibrium from solutions containing increasing amounts of pure, free dye acid. The dyebaths therefore contain varying concentrations of hydrogen and dye ions but, because no other ions are present, these two concentrations must be equivalent. Similarly, because wool has approximately equal amounts of negatively and positively charged fibre groups, the concentrations of hydrogen and dye ions in the fibre must also be equivalent. This would be so in the ideal case, but in practice fibre decomposition, the presence of water-soluble fibre fractions and the possibility that adventitious electrolyte may be added with the wool result in this equivalence being destroyed. For the dyebath the presence in the final equilibrium solution of other electrolytes is demonstrated by the lack of agreement between $[H]_s$ and $[D]_s$ in Table 4.16, where the difference between the two is given under the heading $[U]_s$. The fact that part of this effect is due to wool decomposition, and is therefore time-dependent, is demonstrated in Table 4.17 and it has been observed that this effect becomes worse as the affinity of the dye increases. In similar experiments [79] the dyebath was found to contain a sparingly soluble, crystalline deposit which was similar to those found with certain amino acids.

Nevertheless, despite these problems, the method gives consistent results and titration curves have been obtained for several free dye acids [8]. These are shown in Figure 4.15, together with curves for hydrochloric and

TABLE 4.16

The adsorption of Metanil Yellow free acid by wool [41]

Assuming the wool contains: 0.6 mol kg^{-1} of a basic group with $pK_A = 12.5$
0.18 mol kg^{-1} of a basic group with $pK_A = 10.3$
0.05 mol kg^{-1} of a basic group with $pK_A = 6.1$
0.79 mol kg^{-1} of an acidic group with $pK_A = 4.1$
0.027 mol kg^{-1} of an acidic group with $pK_A = 10.3$
and using: $V = 0.3$ l kg^{-1}, $z = 1$, $[Na]_s = 0$, $[Cl]_s = 0$.

pH	$[H]_s$ /mol l^{-1} $\times 10^6$	$[U]_s$ /mol l^{-1} $\times 10^4$	$[D]_s$ /mol l^{-1} $\times 10^4$	$[D]_f$ /mol l^{-1} $\times 10$	$-\Delta\mu^{\ominus}$/kJ (kcal) mol^{-1} Langmuir	$-\Delta\mu^{\ominus}$/kJ (kcal) mol^{-1} New treatment (Eqn 4.167)
5.77	1.70	0.491	0.508	1.23	46.6 (11.1)	21.0 (5.00)
5.83	1.48	1.39	1.40	2.23	47.9 (11.4)	22.9 (5.45)
5.77	1.70	2.34	2.36	3.10	47.9 (11.4)	23.4 (5.58)
5.58	2.63	3.34	3.37	4.27	47.9 (11.4)	23.5 (5.60)
5.48	3.31	4.02	4.05	5.63	48.3 (11.5)	24.3 (5.79)
5.27	5.37	4.36	4.41	6.50	47.9 (11.4)	23.8 (5.66)
5.30	5.01	5.49	5.54	7.97	49.1 (11.7)	24.8 (5.90)
4.92	12.0	5.43	5.55	9.33	47.9 (11.4)	23.5 (5.59)
4.63	23.4	6.11	6.34	13.2	48.7 (11.6)	24.0 (5.71)
3.98	105	11.3	12.4	18.4	46.6 (11.1)	21.1 (5.03)
				Mean	47.9 ± 0.9 (11.4 ± 0.2)	23.2 ± 1.2 (5.53 ± 0.28)

naphthalene-β-sulphonic acids for comparison. It can be seen that the main effect of increasing affinity is to move the curve to higher pH values, just as was found for the simpler acids by the change in pH_{mid} in Table 4.15. On any curve, too high a pH results in minimal adsorption whilst too low a pH produces high exhaustion and a consequently low concentration of dye in the bath, which in turn minimises levelling and migration possibilities. Thus for each dye there is an optimum pH range which produces good exhaustion combined with adequate levelling. This range moves to higher values as the

TABLE 4.17

pH and dye concentration of a solution of Naphthalene Orange G free acid in contact with wool at 60 °C for varying times [41]

Time /days	Dye concentration /(mol l^{-1}) × 10^3	pH	$[H]_s$ /(mol l^{-1}) × 10^3	$[U]_s$ /(mol l^{-1}) × 10^3
0	21.0	1.70	20.0	1.0
1	9.80	2.13	9.40	0.40
2	9.66	2.25	8.75	0.91
3	9.62	2.37	7.99	1.62
5	9.62	2.55	6.53	3.09
7	9.62	2.72	4.47	5.15

dye anion affinity increases, i.e. in practical terms, from levelling → milling → supermilling dyes.

The total dye affinity $\Delta\mu^\ominus_\pm$ may be calculated for each point on the titration curve by substituting the measured values in Eqn 4.109 with [X] equated to [D]. The derived Eqn 4.111 cannot be used, because of the lack of equivalence between [H] and [D] in solution. Because of possible fibre decomposition during the prolonged treatment required to reach equilibrium, the method is limited to dyes of low affinity (i.e. fast-diffusing dyes) which can be isolated as the free acid. The method does yield consistent results, however, as shown in Table 4.16, column 6.

With polybasic dyes a further problem arises as to whether the dye is assumed to be occupying only one site or as many sites as it has charges. In the former case, for a dye carrying a charge of z then $z\theta_D = \theta_H$ and Eqn 4.109 becomes

$$-\Delta\mu^\ominus_\pm = -(z\Delta\mu^\ominus_H + \Delta\mu^\ominus_D) = RT \ln \left(\frac{\theta_H}{1-\theta_H}\right)^z \left(\frac{\theta}{z-\theta_H}\right) - RT \ln [H]^z_s[D]_s \qquad (4.128)$$

Alternatively, if the dye occupies z sites then $\theta_D = \theta_H$, and

$$-\Delta\mu^\ominus_\pm = -(z\Delta\mu^\ominus_H + \Delta\mu^\ominus_D) = (z+1)\, RT \ln \frac{\theta_H}{1-\theta_H} - RT \ln [H]^z_s[D]_s \qquad (4.129)$$

The relative merits of the two approaches are difficult to assess. The steric problem of z sites being in the required position favours the former, but the

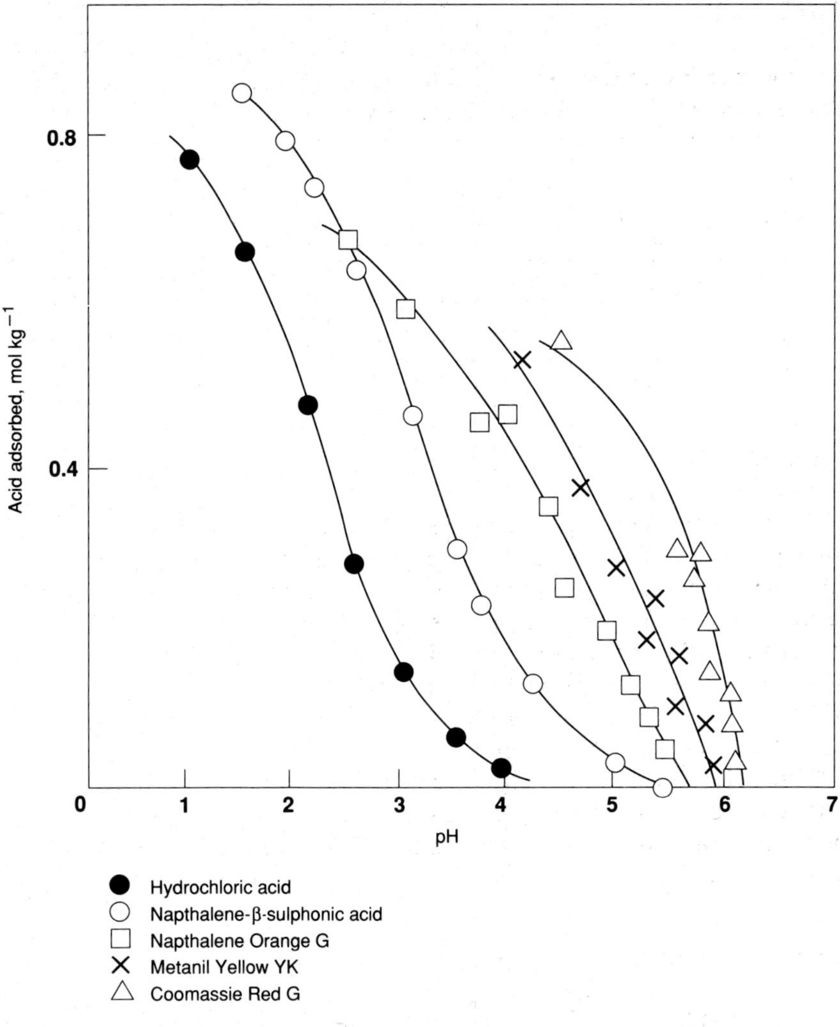

Figure 4.15 – Titration curves of wool with free acids [8]

fact that the fibre is saturated when an equivalent amount of dye is adsorbed favours the latter. However, measurements on the trisulphonated dye Naphthalene Scarlet 4R (C.I. Acid Red 18) [41] gave affinities which, in consistency, seemed marginally to favour the idea that three sites are occupied by one dye ion (Table 4.18).

TABLE 4.18

Affinity of Naphthalene Scarlet 4R (free acid) for wool at 60 °C [41]

θ_H	$-\Delta\mu^\ominus_\pm$ kJ (kcal) mol^{-1} for each dye anion occupying one site	three sites
0.080	110.0 (26.2)	113.0 (26.9)
0.166	106.3 (25.3)	108.8 (25.9)
0.523	99.1 (23.6)	102.5 (24.4)
0.855	95.3 (22.7)	102.9 (24.5)
0.945	95.3 (22.7)	105.4 (25.1)
Mean	101.2 ± 5.9 (24.1 ± 1.4)	106.7 ± 3.9 (25.4 ± 0.9)

4.5.9 Determination of affinity by displacement

It was known empirically that when applying low-affinity, levelling dyes to wool, the addition of neutral electrolyte to an acid dyebath reduces the amount of dye adsorbed, the effect increasing with increasing electrolyte concentration. Because of its overall effect on the quality of the dyeing the electrolyte was said to be acting as a *levelling agent*. This observation indicates that, when more than one type of anion is present in the dyebath, there is competition for sites in the wool, the final distribution of dye and electrolyte anions depending not only on the affinity of each but on their relative concentrations. The effect is observed only with dyes of low affinity, and only in acid solution. In neutral or alkaline solution the addition of electrolyte promotes adsorption in the same way as in the dyeing of cellulose. The pH of the changeover is determined by the isoelectric point of the dyed fibre.

This competition for sites led to the development of a method of comparing the affinities of dye anions with that of an inorganic anion by determining how much dye was displaced by given concentrations of inorganic electrolyte Na_yX [10]. It was assumed that under the experimental conditions the amount of sodium ions taken up by the fibre was negligible in comparison with that of hydrogen ions, in other words that the dyebath contained an acid H_yX and the dye acid H_zD. Again assuming that the number of basic and acidic sites in wool are the same, then

$$\theta_H = \theta_X + \theta_D \tag{4.130}$$

where θ_X is the fraction of positive sites occupied by the non-dye ion and assuming, if both ions are polybasic, that they occupy a number of sites equivalent to their charge. The affinity of the dye anion is given therefore by

$$-\Delta\mu_D^{\ominus} = RT \ln\frac{\theta_D}{1 - \theta_D - \theta_X} - RT \ln [D]_s + z\psi F \tag{4.131}$$

and that of the non-dye anion by

$$-\Delta\mu_X^{\ominus} = RT \ln\frac{\theta_X}{1 - \theta_D - \theta_X} - RT \ln [X]_s + y\psi F \tag{4.132}$$

Multiplying by y and z respectively and subtracting removes the electrical work term, to give

$$-(y\Delta\mu_D^{\ominus} - z\Delta\mu_X^{\ominus}) = RT \ln\left(\frac{\theta_D}{1 - \theta_D - \theta_X}\right)^y \left(\frac{1 - \theta_D - \theta_X}{\theta_X}\right)^z - RT \ln\frac{[D]_s^y}{[X]_s^z} \tag{4.133}$$

If the desorbing electrolyte is chosen so that the basicity of its anion is the same as that of the dye, then $y = z$ and Eqn 4.133 becomes

$$-(\Delta\mu_D^{\ominus} - \Delta\mu_X^{\ominus}) = RT \ln\frac{\theta_D}{\theta_X} \cdot \frac{[X]_s}{[D]_s} \tag{4.134}$$

The problems of estimating $[X]_f$ in order to obtain a value of θ_X may be overcome by carrying out the experiment at a sufficiently low pH that $\theta_H \rightarrow 1$, so that (from Eqn 4.130) $\theta_X = 1 - \theta_D$. Under these conditions

$$-(\Delta\mu_D^{\ominus} - \Delta\mu_X^{\ominus}) = RT \ln\frac{\theta_D}{1 - \theta_D} \cdot \frac{[X]_s}{[D]_s} \tag{4.135}$$

The results of desorbing Naphthalene Orange G (C.I. Acid Orange 7) from wool using sodium chloride at 60 °C are shown in Table 4.19 [41], where it has been assumed that the affinity of the chloride ion is so small that $\Delta\mu_X^{\ominus} = 0$. It can be seen that results are reasonably consistent over a wide range of salt concentrations.

The displacement method is a quite elegant means of determining affinities,

TABLE 4.19

Affinity of Naphthalene Orange G by desorption with sodium chloride at 60 °C

θ_D	Dye in solution /(mol l^{-1}) × 10^5	Chloride ion in solution /mol l^{-1}	$-\Delta\mu^\ominus$ /kJ (kcal) mol^{-1}
0.271	0.805	0.01	17.0 (4.05)
0.271	1.02	0.02	18.0 (4.28)
0.271	1.52	0.05	19.7 (4.69)
0.271	3.18	0.10	19.6 (4.66)
0.271	4.18	0.15	20.0 (4.75)
0.271	5.66	0.20	19.9 (4.73)
0.271	6.51	0.25	18.1 (4.32)
0.271	11.9	0.50	20.4 (4.85)
0.403	0.964	0.01	18.1 (4.32)
0.403	1.16	0.02	19.6 (4.66)
0.403	3.01	0.05	19.4 (4.63)
0.403	6.65	0.10	19.2 (4.56)
0.403	8.33	0.15	23.3 (5.54)
0.403	11.1	0.20	19.7 (4.68)
0.403	14.4	0.25	19.6 (4.67)
0.403	29.6	0.50	19.5 (4.64)
0.403	33.0	1.00	21.1 (5.02)
		Mean	19.3 ± 0.8 (4.6 ± 0.2)

particularly if the apparatus shown in Figure 4.2 is used. There is an economy of experimental effort, and the short times of treatment involved are said to minimise decomposition of the wool although this could be offset by the use of very acid solutions to ensure that $\theta_H \rightarrow 1$. The limitation that the method applies only to low-affinity dyes because of the low affinity of the displacing inorganic ions can be overcome by using, as a displacing anion, a dye of contrasting colour. By this method it is possible to measure the amounts of each dye in the fibre and in the equilibrium solution by spectrophotometric means, thereby eliminating the need to keep the fibre saturated with ions, i.e. there is no need to use strongly acid solutions. From the results, the affinities of the two dyes may be calculated; it has however been found empirically that determining the affinity of one dye using a dye of known affinity as displacing agent rather than an inorganic ion gives better results, in that they are more closely comparable with those from other methods.

4.5.10 The adsorption of acid dyes by wool from neutral solution

The adsorption on to wool of the monobasic dye Naphthalene Orange G from neutral solutions has been studied at 60 °C in the presence of 0.1 mol l^{-1} of sodium chloride [9]. In order to express the behaviour of the dye quantitatively Eqn 4.109 was modified by the replacement of [H] by [Na] throughout to give, for a dye of basicity z,

$$-\Delta\mu^{\ominus}_{\pm} = RT \ln \left(\frac{\theta_{Na}}{1 - \theta_{Na}}\right)^{z} \cdot \frac{\theta_D}{1 - \theta_D} - RT \ln [Na]^{z}_{s}[D]_{s} \qquad (4.136)$$

or, assuming that $\theta_{Na} = \theta_D$,

$$-\Delta\mu^{\ominus}_{\pm} = (z + 1)RT \ln \frac{\theta_D}{1 - \theta_D} - RT \ln [Na]^{z}_{s}[D]_{s} \qquad (4.137)$$

When this equation was used to evaluate the results they were found to be inconsistent, in that affinity decreased markedly with increasing concentration of dye applied and that on average it was much lower than values obtained by other methods, i.e. titration or desorption under acid conditions. In a further study [41] the saturation values of various level-dyeing acid dyes were determined from the adsorption isotherms using the method of reciprocal plotting; this revealed that the saturation value of wool dyed from neutral solution was very much lower in every case than that dyed from acid solution, i.e. *ca* 300 instead of 900 mmol kg^{-1}. When dibasic dyes were studied the discrepancy was even greater, with a saturation value of *ca* 100 mmol kg^{-1}. In addition, when the sodium chloride concentration was increased it was found that whilst at high applied dye concentrations the isotherm was displaced to higher values of dye adsorbed, as would be expected, at low concentrations the adsorption was reduced.

It must be concluded that the reversal in the effect of added electrolyte occurs at a pH which is determined by the quantity and basicity of the adsorbed dye. Equally it must be concluded that the understanding of the neutral dyeing of wool is far from satisfactory.

4.5.11 The enthalpy of dyeing and the entropy of dyeing of acid dyes on wool

Studies of the effect of temperature on the affinity of acid dyes for wool have produced results that are very difficult to interpret. Few, if any, demonstrate a linear relationship when $-\Delta\mu^{\ominus}/T$ is plotted against $1/T$ as required by Eqn 4.21 [80], and indeed some results produce a parabolic relationship indicating that the enthalpy of dyeing goes from negative, through zero, to positive as the temperature increases [81]. Such results have tended to be over-emphasised – perhaps correctly, but possibly not, in view of the difficulties associated with measuring affinities on wool due to factors such as fibre degradation and dyebath contamination, and to the problems of calculating entropy values discussed previously. It would appear, however, that large values of the enthalpy of dyeing can be observed.

4.5.12 The maximum dye-combining power of wool

The theories described for the adsorption of simple acids and dyes by wool demand that the maximum acid-binding power of wool should be an amount equivalent to the number of basic groups in the fibre, although there is some doubt of the precise meaning of this in terms of polybasic anions. Most of the experimental evidence, particularly with low-affinity anions, indicates that an effect which may be considered as fibre saturation occurs when wool has adsorbed 0.8–0.9 moles of monobasic anion per kilogram of dry wool. Certain high-affinity dyes have been studied which show adsorptions in excess of this amount, however [82, 83]. Experiments in which wool was first deaminated with nitrous acid and then acetylated showed a decrease in the maximum amount of hydrochloric acid taken up to a value of 0.008 mol kg^{-1}, but this wool still adsorbed dye in excess of this (0.07 mol kg^{-1}) indicating that forces other than coulombic interaction between positive sites and negative dye were operating [84]. Thus, although there is much evidence that the dyeing of wool can be explained by the presence in the fibre of positive sites, certain discrepancies must be noted. These are

(a) the residual adsorption of dyes on completely deaminated wool;
(b) the excessive uptake of certain dyes;

(c) the upward trend of the titration curves of strong acids at low pH values;
(d) the fact that weak acids are adsorbed in the undissociated state in amounts far in excess of the number of basic groups in the fibre [85].

Although several suggestions have been advanced to account for the excessive adsorption, e.g. the ionisation of the weakly basic peptide groups to produce new sites at acid pH values, the situation is far from clear.

4.5.13 Anionic dyes on nylon

In the dyeing of wool problems arise because of fibre decomposition and dyebath contamination. Although less well realised this also occurs with polyamides, considerable quantities of ammonia being liberated into the dyebath from the decomposition of acid amide groups [86, 66]. That some discrepancy does occur is illustrated in Table 4.21, column $[U]_s$ (p. 344), which is the difference between $[H]_s$ and $[D]_s$. Using a pure, dye-free acid with no other addition these should be the same, and hence the difference must indicate the presence of some unknown cation or anion. Also the imido groups may hydrolyse, in particular under acid or alkaline conditions [87, 66]. Both effects are dependent on the internal pH of the fibre rather than the external pH in the bath, the hydrolysis of imido groups being particularly observed under conditions of 'overdyeing' when very severe acidic conditions can develop inside the fibre phase due to the excess of negative ions present. Measurements of the uptake of HCl by nylon 6, in which the experimental technique minimised the amount of adventitious impurity in the dyebath [88, 66], gave results that were quite different from those obtained by the conventional method. Clearly, as with wool, the problems of obtaining meaningful results are considerable but nevertheless progress has been made in establishing a mechanism of dyeing.

Nylon 6.6 is made by condensing stoichiometric proportions of hexamethylenediamine with adipic (hexane-1,6-dioic) acid. A product thus made should contain equal numbers of amino and carboxyl groups at the ends of the polymer chains but, because the chain length has to be controlled, small amounts of a monocarboxylic acid are added which reacts with some of the amino groups thereby terminating the growing chains with acyl groups. The finished product therefore contains fewer basic than acidic end-groups with, in addition, some terminal acyl groups. A typical nylon 6.6 has, by analysis, the constitution shown in Table 4.20.

The presence of basic and acidic groups in nylon suggested a similarity to wool and indeed, both experimentally and theoretically, the problem of the dyeing of nylon was attacked along similar lines. As with wool the adsorption of simple acids and alkalis was studied as a preliminary to the study of more complex dye ions, and the results were very similar. Typical acid titration

TABLE 4.20

Constitution of a typical nylon yarn

Group	mol kg^{-1} of dry yarn
Amine	0.036
Carboxyl	0.090
Amide	8.85
Terminal acyl	0.063

curves produced for hydrochloric acid [41, 74, 89] show evidence of an inflection in the curve at acid pH (*ca* 2–3), at which point the amounts of acid adsorbed corresponded reasonably well with the number of basic end-groups in the fibre. At lower pH further acid is adsorbed, an effect which has been attributed to the ionisation of the amido groups producing more basic sites [90, 91, 92]. Curves taken to pH values higher than 7 show a shallow inflection at pH 6–7 which corresponds to the pH at which the remaining ionised carboxyl groups are equal in amount to the basic groups present, i.e. the isoelectric point of fibre. (See Figure 4.13 for a soluble protein.) The dependence of the amount of acid adsorbed on the number of basic end-groups has been demonstrated by deamination and by acetylation [74, 89] when the amount of acid adsorbed was reduced to a very small value. Additions of neutral electrolyte resulted in the displacement of the hydrochloric acid titration curve to higher pH values [74, 89]. When nylon is titrated with other acids the titration curve moves to higher pH values as the size and complexity of the anion increases [89].

Thus, in all respects related to the adsorption of simple acids, nylon behaves qualitatively in a manner exactly similar to wool and hence it is reasonable to apply the same basic equations. Certain modifications are necessary, however, to take into account the fact that, unlike wool, in most nylons the numbers of basic and acidic groups are not the same, acidic groups being in excess.

Starting from Eqn 4.109, if $[H]_f$ is the concentration of hydrogen ions bound to the nylon and C is the saturation value (which, for normal dyeing conditions, is equal to the total number of carboxyl groups),

$$\theta_H = \frac{[H]_f}{C}$$

Also, if the content of amino groups is B, then for adsorption of the acid HX,

$$\theta_X = \frac{[X]_f}{B}$$

and the equation becomes

$$-\Delta\mu^\ominus_\pm = RT \ln \frac{[H]_f}{C - [H]_f} \cdot \frac{[X]_f}{B - [X]_f} - RT \ln [H]_s[X]_s \quad (4.138)$$

$[H]_f$ is the total concentration of hydrogen ions in the fibre, i.e. the amount adsorbed at any particular point on the titration curve $[H]_{ad}$ plus the amount present at the start of the titration which, if the titration was started at the isoelectric point of the fibre, is equal to the difference between the numbers of acidic and basic end-groups in the fibre:

$$[H]_f = [H]_{ad} + (C - B) \quad (4.139)$$

Clearly if the titration is started at some other point Eqn 4.139 no longer applies and, in this case, when $[H]_{ad}$ is plotted against pH_s the shape of the titration curve will be different.

For a titration that starts at the isoelectric point, substituting in Eqn 4.138 using Eqn 4.139 gives

$$-\Delta\mu^\ominus_\pm = RT \ln \frac{[H]_{ad} + (C - B)}{B - [H]_{ad}} \cdot \frac{[X]_f}{B - [X]_f} - RT \ln [H]_s[X]_s \quad (4.140)$$

At any point in the titration the fibre must be electrically neutral, and hence the number of unfilled positive sites must equal that of unfilled negative sites. Thus, if the only electrolyte present is the acid HX, then

$$C - [H]_f = B - [X]_f \quad (4.141)$$

Combining Eqns 4.139 and 4.141 $[H]_{ad} = [X]_f$, and also, for electrical neutrality in solution, $[H]_s = [X]_s$. Substituting these equalities in Eqn 4.140 and rearranging gives

$$\frac{-\Delta\mu^\ominus_\pm}{2.303RT} = \log k \quad (4.142)$$

where

$$k = \frac{[X]_f(C - B + [X]_f)}{(B - [X]_f)^2[H]_s^2)} \quad (4.143)$$

and $[X]_f$ is the amount adsorbed as observed in the titration experiment.

Rearranging Eqn 4.143,

$$\frac{[[X]_f([X]_f + C - B)]^{\frac{1}{2}}}{[H]_s} = Bk^{\frac{1}{2}} - [X]_f k^{\frac{1}{2}} = r \qquad (4.144)$$

and a graph of r (calculated from experimental values of $[X]_f$) against $[X]_f$ should be linear with a slope of $-k^{\frac{1}{2}}$ and an intercept of $Bk^{\frac{1}{2}}$ from which the affinity and the maximum dye uptake, i.e. the amino group content, can be calculated.

Using data obtained from the titration of nylon 6.6 with hydrochloric acid [74, 93] straight-line relationships are indeed obtained when r is plotted against $[Cl]_f$, giving values for $-\Delta\mu^{\ominus}_{\pm}$ of 43.7 (10.4), 45 (10.7) and 45.8 (10.9) kJ (kcal) mol^{-1} at 18, 34.8 and 55.5 °C respectively. These must be compared with others obtained by calculating the affinity at the various points along the titration curve using Eqn 4.142 [41]. The average so obtained was 44.5 kJ (10.6 kcal) mol^{-1} at room temperature although there was some variation in the results (±1.7 kJ) with a trend to higher values of $-\Delta\mu^{\ominus}_{\pm}$ as the pH increased.

The alternative approach of applying the Donnan method, as described for wool, is available when affinities may be calculated using the equivalent to Eqn 4.116. Variations in the values obtained are greater than in those using the Langmuir equations, however, which would therefore appear to be preferable.

4.5.14 The affinity of dyes for nylon

The simplest practical method would appear to be the direct measurement of the titration curve of the pure dye free acid, but even this is difficult as the experiment must be carried out under conditions in which the fibre is not saturated with dye, i.e. in the pH range 4–7. This in turn requires the application of very dilute, unbuffered solutions of dye and a very precise experimental technique. Work is reported on two dibasic dyes, Naphthalene Red EA (C.I. Acid Red 13) and Solway Blue BN (C.I. Acid Blue 45), using nylon 6.6 [41]. Two approaches have been used for calculating the dye affinity from these results, both assuming a nylon with an amine content of 0.037 and a carboxylic acid content of 0.090 mol kg^{-1}. The first uses Eqn 4.109 in the form for a dye of basicity z:

$$-\Delta\mu^{\ominus}_{\pm} = RT \ln\left(\frac{\theta_H}{1-\theta_H}\right)^z \cdot \frac{\theta_D}{1-\theta_D} - RT \ln [H]_s^z[D]_s \qquad (4.145)$$

which becomes, assuming that the maximum number of sites available to both hydrogen and dye ions is determined by the number of basic groups in the fibre and that $[H]_f = z[D]_f$,

$$-\Delta\mu^{\ominus}_{\pm} = (z + 1)RT \ln \frac{\theta_D}{1 - \theta_D} - RT \ln [H]^z_s[D]_s \qquad (4.146)$$

Alternatively Eqn 4.140 may be used in the form for a dye of basicity z [74] which gives, when $[H]_{ad}$ is equated to $z[D]_f$,

$$-\Delta\mu^{\ominus}_{\pm} = RT \ln \left(\frac{C - B + [D]_f}{B - [D]_f}\right)^z \cdot \frac{z[D]_f}{B - z[D]_f} - RT \ln [H]^z_s[D]_s - RT \ln z \qquad (4.147)$$

The affinities obtained for Naphthalene Red EA using the two equations are shown in Table 4.21. Neither set shows a definite trend but there is some variation, most probably due to random experimental error. Unfortunately the results are not the same and this is repeated when other methods of calculation are used, indicating that a problem still remains in deciding on the correct form of equation to use. Providing that the same equation is used, however, then the affinities so obtained may be compared.

In order to avoid the practical problems of measuring acid titration curves, affinities have been determined using a desorption method. Unfortunately, whilst some of the practical problems are overcome, the essential question remains of the form of equation that should be applied to the results. Using the Gilbert–Rideal equations to deduce a formula for calculating the relative affinity of a dye anion by desorption with sodium chloride [94] gave, for a monobasic dye,

$$-(\Delta\mu^{\ominus}_{D} - \Delta\mu^{\ominus}_{Cl}) = RT \ln \frac{\theta_D}{1 - \theta_D} - RT \ln [D]_s + RT \ln [Cl]_s \qquad (4.148)$$

and, for a dibasic dye,

$$-(\Delta\mu^{\ominus}_{D} - 2\Delta\mu^{\ominus}_{Cl}) = RT \ln \left(\frac{\theta_D}{1 - \theta_D}\right)^2 \cdot \frac{\theta_D}{2 - \theta_D} - RT \ln [D]_s + 2RT \ln [Cl]_s \qquad (4.149)$$

These were applied to the results of desorptions from nylon during which the fibre was kept as saturated as possible with hydrogen ions, and in this way the affinities of six acid dyes were obtained (Table 4.22). Other workers [95] used Eqn 4.142 in the form for a monobasic dye, which gives Eqn 4.150 as the equivalent of Eqn 4.143:

TABLE 4.21

The titration of nylon 6.6 with Naphthalene Red EA free acid [41]

Fibre end-groups: —NH_2 = 0.037 mol kg^{-1} $K_{NH_2} = 2.6 \times 10^{-11}$
—COOH = 0.090 mol kg^{-1} $K_{COOH} = 1.6 \times 10^{-5}$
Temperature = 60 °C, V = 0.1 l kg^{-1}, z = 2, $[Na]_s$ = 0, $[Cl]_s$ = 0.

					$-\Delta\mu^{\ominus}$/kJ (kcal) mol^{-1}			
					Langmuir		New treatment (Eqn 4.167) NH_2 (mol l^{-1}) =	
pH	$[H]_s$ /(mol l^{-1}) × 10^7	$[U]_s$ /(mol l^{-1}) × 10^5	$[D]_s$ /(mol l^{-1}) × 10^5	$[D]_f$ /(mol l^{-1}) × 10	Eqn 4.146	Eqn 4.147	0.37	0.36
4.71	195.0	52.8	27.4	1.79	115.1 (27.4)	121.0 (28.8)	40.7 (9.7)	50.8 (12.1)
5.08	83.2	10.6	5.35	1.76	114.6 (27.3)	118.4 (28.2)	47.5 (11.3)	52.1 (12.4)
5.15	70.8	3.83	2.27	1.72	114.6 (27.3)	119.3 (28.4)	48.7 (11.6)	51.7 (12.3)
5.39	40.7	4.06	2.24	1.66	115.1 (27.4)	117.6 (28.0)	49.6 (11.8)	51.3 (12.2)
5.79	16.2	3.88	2.02	1.53	115.1 (27.4)	120.1 (28.6)	51.7 (12.3)	52.5 (12.5)
6.10	7.94	3.61	1.85	1.40	115.5 (27.5)	119.7 (28.5)	53.3 (12.7)	54.2 (12.9)
6.17	6.76	2.63	1.35	1.22	113.4 (27.0)	118.0 (28.1)	53.3 (12.7)	53.3 (12.7)
6.30	5.01	2.37	1.21	1.13	113.8 (27.1)	118.4 (28.2)	53.8 (12.8)	54.2 (12.9)
6.37	4.27	2.03	1.04	1.07	113.0 (26.9)	118.9 (28.3)	54.2 (12.9)	54.6 (13.0)
6.28	5.25	1.66	0.855	1.05	115.5 (27.5)	118.0 (28.1)	53.3 (12.7)	53.8 (12.8)
6.72	1.91	1.68	0.850	0.890	116.0 (27.6)	121.4 (28.9)	57.5 (13.7)	57.9 (13.8)
6.54	2.88	1.45	0.740	0.780	113.8 (27.1)	118.4 (28.2)	54.2 (12.9)	54.6 (13.0)
6.70	2.00	1.20	0.610	0.745	113.8 (27.1)	119.7 (28.5)	56.7 (13.5)	56.7 (13.5)
6.94	1.15	0.552	0.282	0.570	114.2 (27.2)	122.6 (29.2)	60.1 (14.3)	60.1 (14.3)
6.83	1.48	0.669	0.342	0.460	110.9 (26.4)	119.7 (28.5)	56.7 (13.5)	57.9 (13.8)
6.90	1.26	0.211	0.112	0.415	113.4 (27.0)	123.1 (29.3)	60.1 (14.3)	60.5 (14.4)
				Means	114.2 ± 1.26 (27.2 ± 0.3)	119.7 ± 1.7 (28.5 ± 0.4)	53.3 ± 4.6 (12.7 ± 1.1)	54.6 ± 2.9 (13.0 ± 0.7)

TABLE 4.22

Affinity of acid dyes for nylon at 98 °C and pH 3.2 [94]

Dye	C.I. generic name	z	$-(\Delta\mu_{D}^{\ominus} - z\Delta\mu_{Cl}^{\ominus})$ /kJ (kcal) mol^{-1}
Metanil Yellow YK	Acid Yellow 36	1	26.0 (6.2)
Anthraquinone Blue RXO		1	25.2 (6.0)
Orange RO		1	24.4 (5.8)
Orange 11	Acid Orange 7	1	21.0 (5.0)
Anthraquinone Green G		2	20.6 (4.9)
Anthraquinone Blue B		2	9.7 (2.3)

$$k = \frac{[D]_f(C - B + [D]_f)}{(B - [D]_f)^2.[H]_s[D]_s} \tag{4.150}$$

Rearranging Eqn 4.150 gives the equivalent to Eqn 4.144 but in terms of $[D]_f$:

$$\left[\frac{[D]_f(C - B + [D]_f)}{[H]_s[D]_s}\right]^{\frac{1}{2}} = Bk^{\frac{1}{2}} - [D]_f k^{\frac{1}{2}} = r \tag{4.151}$$

TABLE 4.23

Affinity of acid dyes for nylon [95]

Dye	Temp /°C	Saturation value B /mmol kg^{-1}	$k \times 10^{11}$	$-\Delta\mu_{HD}^{\ominus}$ /kJ (kcal) mol^{-1}	$-\Delta H^{\ominus}$ /kJ (kcal) mol^{-1}	$\Delta S^{\ominus}$ /J (cal) K^{-1} mol^{-1}
Metanil Yellow YK (C.I. Acid Yellow 36)	21	30.3	40.0	71.2 (16.96)	38.2 (9.1)	113.4 (27)
	80	29.2	3.0	78.0 (18.56)		
Quinizarine Blue (C.I. Acid Violet 43)	21	28.0	31.0	70.6 (16.81)	38.2 (9.1)	109.2 (26)
	80	28.3	2.3	77.1 (18.36)		

Plotting r against $[D]_f$ for results from both adsorption and desorption experiments gave the same straight line, from which the affinity and the maximum dye uptake could be calculated (Table 4.23).

Another, more comprehensive investigation into the dyeing of nylon with acid dyes [96] used a statistical, thermodynamic argument to derive Eqn 4.152:

$$-(\Delta\mu_D^{\ominus} - \Delta\mu_Y^{\ominus}) = RT \ln \frac{\theta_D}{1 - \theta_D} \cdot \frac{[Y]_s}{[D]_s} - RT \ln \frac{z}{y} \qquad (4.152)$$

where Y is an anion of charge y, and applied this to the results from the desorption of three dibasic acid dyes with chloride ions. The results for one of the dyes are shown in Table 4.24, where it can be seen that a reasonable consistency in the value of the affinity is obtained using this equation.

An alternative equation was derived using the principles of the Donnan approach:

$$-\Delta\mu_D^{\ominus} = RT \ln \frac{\theta_D}{(1 - \theta_D)^z} \cdot \frac{[Cl]_s}{[D]_s} - RT \ln zV \qquad (4.153)$$

where V = the volume of the internal phase in $1\ kg^{-1}$ of dry fibre; but this gave results that were far from consistent and must be considered unsatisfactory. In this work some 14 dyes were studied using chloride ion desorption, with the results shown in Table 4.25. In addition certain of the dyes were desorbed using sulphate ions, when it was found that the affinities obtained were very much higher than with chloride. To correct this it was considered that, at the pH of the experiments (3.20), the sulphate would not be present as SO_4^{2-} but as HSO_4^- (the pK for the second dissociation of sulphuric acid being 1.92) and hence an amount of 10.79 kJ (2.87 kcal) mol^{-1} was subtracted from the measured affinities. This tended to overcorrect but did give results which were closer to those from chloride desorption.

The use of desorption techniques does not seem to be the answer to the problems associated with the measurement of titration curves since, once again, the availability of different equations, all purporting to give affinity but in fact giving different values from the same experimental data, leads to uncertainty and consequent lack of confidence that an acceptable equation has been derived.

4.5.15 The overdyeing of nylon

Adsorption of anionic dye in excess of the number of basic groups available in the fibre is much more important for nylon than it is for wool, because of the smaller number of basic groups involved, i.e. *ca* 30–50 mmol kg^{-1} for

TABLE 4.24

Affinity of Solway Blue A (C.I. Acid Blue 69) by desorption from nylon 6.6 using sodium chloride [74, 96]

Fibre end-groups: $—NH_2$ = 0.030 mol kg^{-1} $K_{NH_2} = 2.6 \times 10^{-11}$
$—COOH$ = 0.090 mol kg^{-1} $K_{COOH} = 1.6 \times 10^{-5}$
Temperature = 75 °C, pH = 3.52, $[H]_s = 3.02 \times 10^{-4}$, $z = 2$.

		$-\Delta\mu^{\ominus}$/kJ (kcal) mol^{-1}				
		Sites		New treatment (Eqn 4.167)		
θ_D	[NaCl] /mol l^{-1}	Donnan (Eqn 4.153)	Gilbert–Rideal (Eqn 4.152)	$V = 0.05$	$V = 0.1$	$V = 0.2$
0.346	0.051	18.0 (4.29)	18.4 (4.38)	17.6 (4.19)	17.2 (4.10)	16.8 (4.01)
	0.101	20.2 (4.82)	18.6 (4.44)		17.1 (4.07)	
	0.201	23.1 (5.50)	19.5 (4.64)		17.3 (4.13)	
0.447	0.051	19.4 (4.61)	19.3 (4.59)		18.5 (4.41)	
	0.101	21.6 (5.15)	19.4 (4.62)		18.1 (4.32)	
	0.201	24.3 (5.79)	20.1 (4.79)		17.9 (4.27)	
0.460	0.051	19.4 (4.62)	19.4 (4.62)		18.6 (4.42)	
	0.101	21.2 (5.05)	19.0 (4.52)		17.6 (4.18)	
	0.201	24.3 (5.78)	20.2 (4.80)		17.9 (4.25)	
0.549	0.051	20.5 (4.88)	19.8 (4.72)	20.0 (4.76)	19.2 (4.58)	18.4 (4.39)
	0.101	24.4 (5.81)	20.0 (4.76)		18.6 (4.42)	
	0.201	25.7 (6.12)	21.1 (5.03)		18.5 (4.40)	
	Means	21.8 ± 2.3 (5.20 ± 0.56)	19.6 ± 0.7 (4.66 ± 0.17)		18.1 ± 0.6 (4.30 ± 0.15)	

nylon compared with 800–900 mmol kg^{-1} for wool. Typical titration curves for three acid dyes on nylon 6.6 are shown in Figure 4.16 and it can be seen that three regions are distinguishable [91]. Region A is typical of anionic titration; its position on the pH axis is defined by the affinity of the anion

TABLE 4.25

The affinities of acid dyes for nylon by desorption at 75 °C [96]

Dye	z	$-(\Delta\mu_D^\ominus - \Delta\mu_Y^\ominus)$/kJ (kcal) mol^{-1} by desorption with		
		chloride	sulphate	hydrogen-sulphate
Naphthalene Red J	1	32.2 (7.66)	–	–
Naphthalene Orange 1	1	25.9 (6.17)	32.8 (7.80)	20.7 (4.93)
Naphthalene Orange RO	1	25.3 (6.03)	32.8 (7.80)	20.7 (4.93)
Naphthalene Orange G	1	24.4 (5.82)	30.8 (7.33)	18.7 (4.46)
Lissamine Fast Red 7BP	2	27.8 (6.62)	35.9 (8.54)	(5.67)
Naphthalene Red EA	2	27.4 (6.52)	–	–
Lissamine Fast Red 6B	2	22.6 (5.37)	31.0 (7.39)	19.0 (4.52)
Lissamine Fast Yellow 2G	2	22.3 (5.31)	30.7 (7.31)	18.7 (4.44)
Azo Geranine 2G	2	21.5 (5.11)	–	–
Solway Blue BN	2	21.2 (5.05)	30.2 (7.18)	18.1 (4.31)
Naphthalene Fast Orange 2G	2	21.0 (5.00)	26.9 (6.41)	14.9 (3.54)
Solway Blue A	2	19.6 (4.67)	28.6 (6.80)	16.5 (3.93)
Naphthalene Scarlet 4R	3	24.7 (5.88)	–	–
Tartrazine N	3	21.0 (5.01)	27.5 (6.54)	15.4 (3.67)

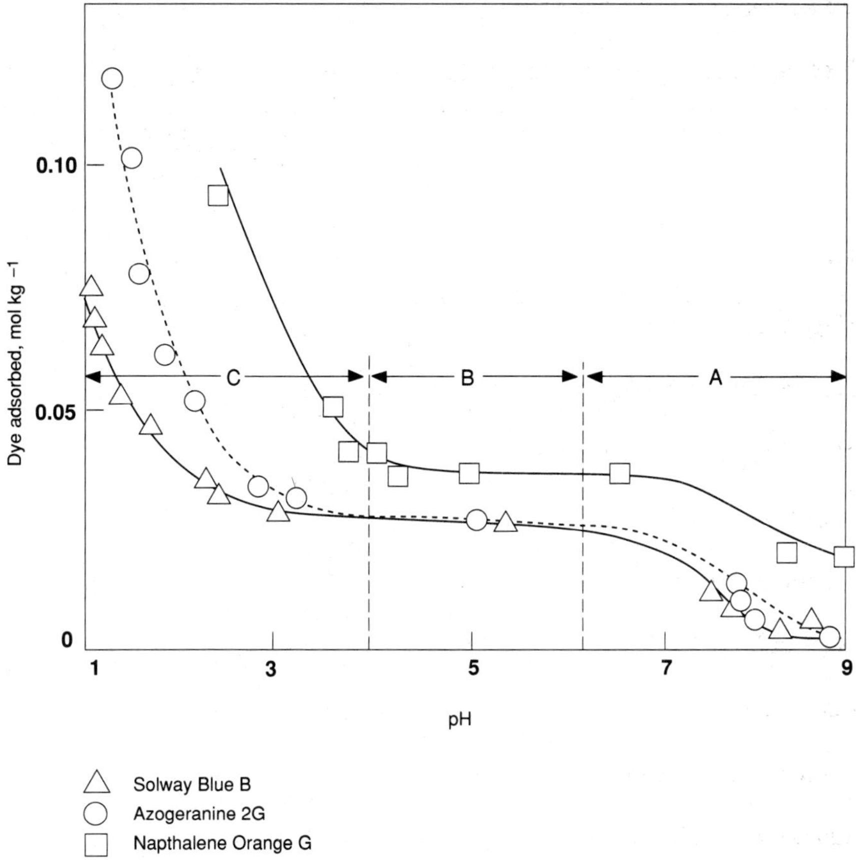

Figure 4.16 – Adsorption of acid dyes on nylon yarn at different pH values at 60 °C [91]

used. As the pH decreases this changes to a plateau region B, which extends over several pH units at values again determined by the affinity of the anion. This region is not always well defined but corresponds essentially to the number of amino end-groups present in the nylon, i.e. it is a region of electrostatic saturation; this has been clearly demonstrated with nylons containing different amounts of end-group [41], when the amount of dye adsorbed was in linear relationship to the number of end-groups present

[97]. At lower pH values is region C, where dye is adsorbed in amounts which exceed the normal (so-called) 'saturation value' and adsorption continues to increase with decreasing pH apparently without limit. The change from region B to region C is clearly also dye-dependent and with certain dyes can occur at relatively high pH values. This dyeing in excess of the amount of available amino end-group is referred to as *overdyeing*. It can also occur with certain dyes as the concentration applied is increased at constant pH [96].

Several explanations have been offered to explain this adsorption of dye in excess of the number of available basic sites. One is that in this region of low pH the amido groups of the nylon chain are hydrolysed, thereby producing amino groups for further dye adsorption [95]; the process is favoured by the need for the large amounts of dye taken up to be accompanied by hydrogen ions, thereby making the internal pH of the fibre even lower than that in the external bath. That such hydrolysis occurs is demonstrated by the fall in viscosity of solutions of overdyed nylon. In addition, if overdyed nylon is stripped of dye and then redyed at higher pH values, then in regions A and B the dye uptake is greater than with the untreated fibre showing that more basic sites are now available. An alternative explanation of overdyeing is that, at the high levels of acidity in the fibre, the main-chain amido groups become ionised to give —$OCNH_2^+$—, which then provide extra sites for dye adsorption. This of course would be only a temporary feature, and would not result in increased uptake after stripping and redyeing unless such ionisation was a precursor to hydrolysis.

Such explanations of overdyeing revolve around the presence in the fibre of an increased number of dyeing sites, and as such they are the result of the almost universal adoption of a site theory to explain the dyeing of nylon. The possibility that overdyeing and indeed all dyeing of nylon occurs as a result of attractive forces which are non-electrostatic in nature has only recently been considered [98, 99], an approach described in section 4.6. With such a mechanism, of course, overdyeing would be achieved without the formation of new sites but when it occurs it results in fibre degradation due to the development of highly acidic conditions in the internal fibre phase. The changing electrical environment in the fibre, as a result of degradation, will favour the adsorption of even more dye.

4.5.16 Cationic dyes on acrylic fibres

The polymerisation of acrylonitrile and its copolymers is carried out using redox initiators and the residues of these, attached to the chain ends, result in most acrylic fibres containing sulphate or sulphonic acid end-groups. Because of these anionic groups acrylic fibres can be dyed with cationic (basic) dyes as discussed in Chapter 3. The capacity of these fibres to take up basic dye is increased if the co-monomer also carries an acidic group, for example sulphonic, phosphoric or carboxylic acid. However, whilst the dye uptake of fibres containing only strongly acidic groups is unaffected by pH changes over a wide range, because the anionic groups remain ionised, those containing weakly acidic groups (such as carboxylic acid) show a decrease in dye uptake as the pH is reduced due to the group changing to the un-ionised form.

For acrylic fibres not containing acid co-monomers, the uptake of basic dye is limited by the number of acidic end-groups in the fibre. This was established [100, 101, 102] when reciprocal plots of $1/[D]_f$ against $1/[D]_s$ were found to give saturation values of 31–35 mmol kg^{-1}, which agreed very closely to the number of end-groups measured by titration or by sulphur analysis (Figure 4.17).

The fibre carries acidic groups and it has been considered that it should

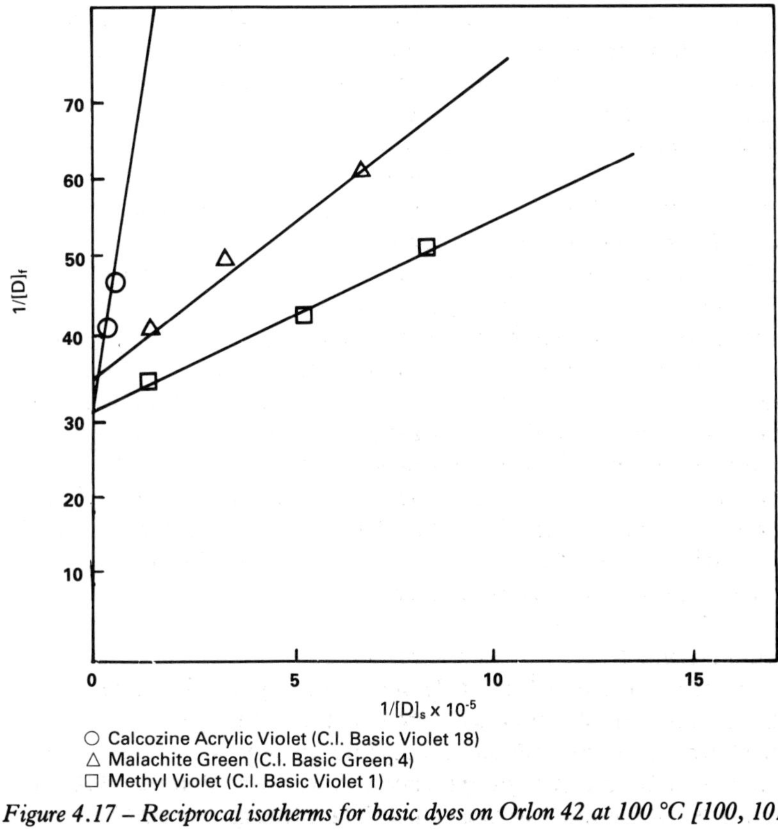

Figure 4.17 – Reciprocal isotherms for basic dyes on Orlon 42 at 100 °C [100, 101]

act as an ion-exchange medium in which hydrogen ions associated with the acid groups are exchanged for dye cations [103, 104], thus:

$$\text{Fibre—SO}_3\text{H}_f + \text{D}_s^+ \rightleftharpoons \text{Fibre–SO}_3\text{D}_f + \text{H}_s^+$$

which would have an equilibrium constant

$$K_H^D = \frac{[H]_s[D]_f}{[D]_s[H]_f} \tag{4.154}$$

Although the convention adopted throughout is that fibre concentrations are in terms of mol l^{-1} of fibre phase, there is no need to define a V term in the above since it appears at top and bottom and hence cancels out. Since the fibre must be considered to be electrically neutral, then

$$[S]_f = [H]_f + [D]_f$$

where $[S]_f$ = the number of available acidic sites, and hence

$$K_H^D = \frac{[H]_s}{[D]_s} \cdot \frac{[D]_f}{[S]_f - [D]_f} \tag{4.155}$$

Assuming the pH of the bath to remain constant, i.e. $[H]_s$ = constant, then Eqn 4.155 may be rearranged to give

$$\frac{1}{[D]_f} = \frac{1}{[S]_f} + \frac{1}{K'[S]_f} \cdot \frac{1}{[D]_s} \tag{4.156}$$

where

$$K' = K_H^D/[H]_s$$

Thus a plot of $1/[D]_f$ against $1/[D]_s$, measured at constant pH, should give a straight line with an intercept of $1/[S]_f$. This has been done [101] using a variety of fibres, and the values of $[S]_f$ obtained agreed well with the results from the titration of the dissolved fibres. It was also found, however, that this saturation value increased as the pH of application increased, a result that was interpreted as being due to the presence of weakly acidic end-groups in the fibre which only ionised at higher pH.

The action of added sodium ions in reducing the equilibrium uptake of dye confirms the ion-exchange mechanism, demonstrating that ions such as sodium can exchange with either hydrogen ions or dye ions. In order to account for the action of sodium ions two ion-exchange constants have been defined [103]:

$$K_D^H = \frac{[H]_f}{[D]_f} \cdot \frac{[D]_s}{[H]_s} \quad (4.157)$$

$$K_D^{Na} = \frac{[Na]_f}{[D]_f} \cdot \frac{[D]_s}{[Na]_s} \quad (4.158)$$

(It must be noted that these equations are in the reciprocal form compared to Eqn 4.154.) If no other ions exert any significant influence then, for electrical neutrality,

$$[S]_f = [D]_f + [H]_f + [Na]_f \quad (4.159)$$

Replacing $[H]_f$ and $[Na]_f$, using Eqns 4.157 and 4.158 respectively, gives

$$[S]_f - [D]_f = \frac{[D]_f}{[D]_s}(K_D^H[H]_s + K_D^{Na}[Na]_s) \quad (4.160)$$

If a measured constant K_M is defined such that

$$\frac{1}{K_M} = ([S]_f - [D]_f) \cdot \frac{[D]_s}{[D]_f} = K_D^H[H]_s + K_D^{Na}[Na]_s \quad (4.161)$$

then this can be rearranged to give Eqn 4.161a:

$$\frac{1}{[D]_f} = \frac{1}{[S]_f K_M} \cdot \frac{1}{[D]_s} + \frac{1}{[S]_f} \quad (4.161a)$$

and, at constant $[H]_s$ and $[Na]_s$, a plot of $1/[D]_f$ against $1/[D]_s$ should again be linear but having a slope that varies with pH and salt concentration. This was found to be so. In addition, if $1/K_M$ is plotted against $[H]_s$ with $[Na]_s$ constant or against $[Na]_s$ with $[H]_s$ constant, then a linear relationship should be obtained in both cases. Again this was found [103].

Using various modifications of these equations [103] it was possible to describe how the Langmuir adsorption constant could be affected by sodium and hydrogen ion concentrations and, in addition, how the separate ion-exchange 'constants' were influenced by the amount of dye present.

If Eqn 4.155 is rearranged then, in the absence of sodium ions, it becomes

$$\frac{[D]_f}{[D]_s} = \frac{K_H^D}{[H]_s}([S]_f - [D]_f) \quad (4.162)$$

and hence if $[D]_f/[D]_s$ is plotted against $[D]_f$ a straight line of slope $K_H^D/[H]_s$ should result, with an intercept when $[D]_f/[D]_s = 0$ of $[S]_f$. From the results of experiments using five representative basic dyes on films made from a copolymer of acrylonitrile with *ca* 7 per cent of vinyl acetate [105] it was indeed found that linear relationships were obtained but that with some dyes this obtained only over 80 per cent of the concentration range; at higher values of $[D]_f$ a 'tail' was shown, indicating that under these conditions dye was being adsorbed in excess of the number of acidic end-groups (Figure 4.18(a)). The results obtained from the linear plots and from extrapolation

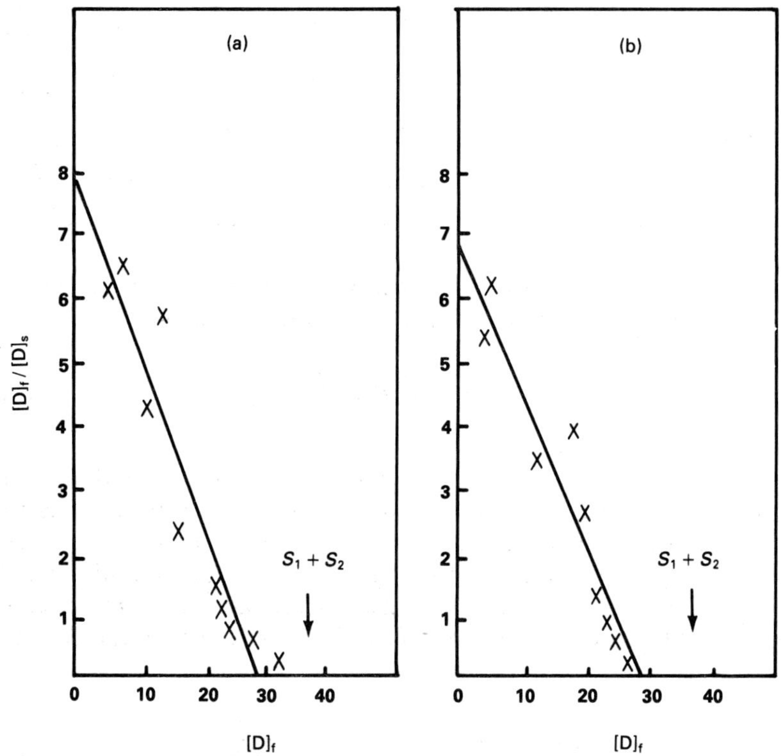

Figure 4.18 – Langmuir plots of adsorption isotherms ($[D]_f$ in mmol kg^{-1}) [104]: (a) C.I. Basic Blue 22, (b) C.I. Basic Red 14

of the linear part of the curves produced consistent values for the intercept of 26–28 mmol kg^{-1} (S_1) shown in Table 4.26. However, the total amount of end-group from titration measurements was 35.1 mmol kg^{-1} ($S_1 + S_2$). Since with certain dyes the amount of dye adsorbed reached values approximat-

TABLE 4.26

Adsorption of basic dyes by an acrylic copolymer film [105]

pH = 3.4, $[H]_s = 3.98 \times 10^{-4}$ mol l^{-1}, $K_1 = K_H^D/[H]_s$.

Dye	Temp. /°C	$K_1 \times 10^{-6}$	$K_1S_1 \times 10^{-4}$	S_1 /mmol kg^{-1}	K_H^D	$-\Delta\mu^{\ominus}_{\mathrm{B}}$ /kJ (kcal) mol^{-1}
C.I. Basic Blue 22	99.8	0.29	0.81	28	115	14.7 (3.50)
C.I. Basic Red 14	99.1	2.50	6.90	28	995	21.4 (5.09)
C.I. Basic Red 24	94.6	1.30	3.40	27	518	19.1 (4.55)
C.I. Basic Violet 19	99.3	3.30	8.50	26	1312	22.2 (5.29)
Oxazine dye	99.1	3.70	9.50	26	1472	22.6 (5.37)

ing to $(S_1 + S_2)$ it was considered that two kinds of fibre end-group were operating, which led to the defining of an isotherm based on

$$[D]_f = \frac{K_1[S]_1[D]_s}{1 + K_1[D]_s} + \frac{K_2[S]_2[D]_s}{1 + K_2[D]_s} \qquad (4.163)$$

which was shown to fit the data for those dyes which showed this 'tail'. Unfortunately not all dyes conform with this equation, which represents a weakness in the proposition. It has also been suggested [102] that, because at a certain level the amount of dye adsorbed increases linearly with the concentration of dye applied, a constant-partition 'solution' process is superimposed on the Langmuir isotherm. This led to the formulation of the equation

$$[D]_f = \frac{K_1[D]_s}{1 + K_1[D]_s} \cdot [S]_f + K_3[D]_s \qquad (4.164)$$

which also gave a good fit to the data.

Just as for nylon, the apparent 'overdyeing' of acrylic fibres is the consequence of the adoption of a site mechanism of dyeing, and again the possibility must be considered that dyeing is the result of attractive forces which are non-electrostatic in nature [98, 99]. This will be dealt with later. However, since Eqn 4.155, based on an ion-exchange mechanism, appears to give a satisfactory interpretation of the greater part of the dyeing process, the constant $K_{\mathrm{H}}^{\mathrm{D}}$ can be regarded as a measure of affinity, so that $-\Delta\mu_{\mathrm{D}}^{\ominus} = RT \ln K_{\mathrm{H}}^{\mathrm{D}}$.

Some values of $-\Delta\mu_{\mathrm{D}}^{\ominus}$ are given in Table 4.26.

4.6 A GENERAL EQUATION FOR ALL IONIC DYEING SYSTEMS

The electrical neutrality equation developed for use in the calculation of the affinity of anionic dyes for cellulose (Eqn 4.88) and its expanded version after applying Donnan theory (Eqn 4.95) have been described. The improvements associated with the application of the various parts of these equations in isolation, i.e. the allowance for multivalent anions in the bath [57] and the prediction of the adsorption of alkali by cellulose [65], together with the determination of dye affinities over a range of alkaline pH [65], suggests most strongly that the equations are meaningful and that their use is satisfactory for anionic dyes on cellulosic materials.

In the case of anionic dyes on protein and polyamide substrates problems arise in the application of either the Gilbert–Rideal [67] or the Donnan [68] treatments to practical systems that involve the interpretation of the adsorption of either simple acids or anionic dyes. Thus, for example,

(a) they do not predict or take account of the fact that either acid or dye can be adsorbed in excess of the so-called 'saturation' value;
(b) there is uncertainty when dealing with polybasic ions in that it is difficult to decide whether they occupy a stoichiometric number of sites or only one site.

These difficulties can be removed if the Donnan approach is fully developed [99]. Eqn 4.88 is the electrical neutrality equation for a fibre containing only ionisable acidic groups, i.e. potentially negative groups. If this equation is extended to include ionisable basic groups, i.e. potentially positive groups, then it could be applied to proteins and polyamides. Thus

$$\Sigma[\mathrm{NH_3}^+]_\mathrm{f} + [\mathrm{H}]_\mathrm{f} + [\mathrm{Cat}^+]_\mathrm{f} = [\mathrm{OH}]_\mathrm{f} + [\mathrm{An}^-]_\mathrm{f} + 2[\mathrm{An}^{2-}]_\mathrm{f} + 3[\mathrm{An}^{3-}]_\mathrm{f} + \Sigma z_i[\mathrm{D}]_{\mathrm{f},i} + \Sigma y_j[\mathrm{D}]_{\mathrm{fixed},j} + \Sigma[\mathrm{COO}^-]_\mathrm{f} + \Sigma[\mathrm{O}^-]_\mathrm{f} \qquad (4.165)$$

Using the various Donnan equalities and

$$[\mathrm{NH_3}^+] = \frac{Q_{\mathrm{NH_2}}[\mathrm{H}]_\mathrm{f}}{K_{\mathrm{NH_2}} + [\mathrm{H}]_\mathrm{f}} = \frac{Q_\mathrm{B}[\mathrm{H}]_\mathrm{f}}{K_\mathrm{B} + [\mathrm{H}]_\mathrm{f}} \qquad (4.166)$$

where $Q_{\mathrm{NH_2}} = Q_\mathrm{B}$ = the amount of each available basic group in the fibre with a dissociation constant of K_B, then

$$\sum \frac{Q_B[H]_f}{K_B + [H]_f} + [H]_f + \frac{[H]_f[Cat^+]_s}{[H]_s} =$$
$$\frac{K_w}{[H]_f} + \frac{[H]_s[An^-]_s}{[H]_f} + 2\frac{[H]_s^2[An^{2-}]_s}{[H]_f^2} + 3\frac{[H]_s^3[An^{3-}]_s}{[H]_f^3}$$
$$+ \sum z_j[D]_{f,i} + \sum y_j[D]_{fixed,j} + \sum \frac{K_A Q_A}{K_A + [H]_f} \quad (4.167)$$

where the two terms for acidic groups in Eqn 4.165 have been grouped together using the general nomenclature K_A and Q_A. Knowing the equilibrium dyebath conditions this equation can be solved for the only unknown, $[H]_f$, and the value obtained used in Eqn 4.96 to calculate the affinity.

An equation is thus produced which is applicable to all fibres and dyeing systems. The idea of having such an equation has been described [98] but, whilst depending on the same basic concepts, there are differences in nomenclature and in the method of approach. For continuity it was felt desirable, therefore, to proceed from the equations and the nomenclature already developed in this chapter.

Ignoring the problems associated with fibre decomposition and dyebath contamination, the results of using Eqns 4.167 and 4.96 to calculate the affinity of Metanil Yellow YK (C.I. Acid Yellow 36) for wool are presented in the last column of Table 4.16. The various constants used are defined in the table. Also included are the previously published values obtained using the Langmuir equations [41], and it can be seen that the latter are much larger because they include the 'affinity' of the hydrogen ion, whereas those from the current calculation do not, since it gives the affinity of the dye anion alone. Despite this the parallel behaviour of the two sets of results is clear. Similarly the results obtained from the experimental titration of nylon 6.6 with the pure free acid of the dibasic dye Naphthalene Red EA (C.I. Acid Red 13) are set out in Table 4.21. Again the various constants used are included in the table, that of V being chosen quite arbitrarily, without confirmation, from a general study of the results obtained from numerous dyeings on nylon 6.6. Also included are the results of the Langmuir calculation [41], which are again much larger due to the inclusion of the 'affinities' of two hydrogen ions. Nevertheless, the parallel behaviour of the two sets of results must again be recognised. Comparison of the last two columns of the table reveals how sensitive the calculated results are, particularly at the higher concentrations of dye on fibre, to the value used for the amino end-group content.

The alternative method of determining affinities is the desorption method [10], and Table 4.24 shows the affinities obtained by applying Eqns 4.166, 4.167 and 4.96 to the results of a series of desorption experiments carried out on nylon 6.6 dyed with Solway Blue A (C.I. Acid Blue 69) using sodium chloride as desorbent [96]. The constants are included again in the table, the value for the amino end-group content being that deduced by the original authors. The affinities obtained using the general equations again represent a perfectly acceptable set of results which closely parallel those obtained using either of the approaches using 'sites'. Also included are the effects of the rather drastic changes in the value of the V term used, and it can be seen that these are relatively small and in no way change the relationship of the results to one another.

Clearly this new treatment can be used to calculate affinities from the results of equilibrium dyeings on basic fibres which are as consistent as might be expected in view of the experimental problems associated with the dyeing of these fibres. By a simple modification to Eqn 4.167 this method can be used to predict the amount of dye adsorbed from dyebaths of varying dye concentration with dyes of different affinities. If the logarithmic term in Eqn 4.96 is put equal to A, rearrangement leads to

$$[\mathrm{D}]_{\mathrm{f},i} = \frac{A[\mathrm{H}]_{\mathrm{s}}^{z,i}[\mathrm{D}]_{\mathrm{s},i}}{[\mathrm{H}]_{\mathrm{f}}^{z,i}} \tag{4.168}$$

If this is written for a number of dyes then the term $\Sigma z_i[\mathrm{D}]_{\mathrm{f},i}$ in Eqn 4.167 can be replaced by

$$\sum z_i \frac{A_i[\mathrm{H}]_{\mathrm{s}}^{z,i}[\mathrm{D}]_{\mathrm{s},i}}{[\mathrm{H}]_{\mathrm{f}}^{z,i}} \tag{4.169}$$

After introducing the required values of A and $[\mathrm{D}]_{\mathrm{s}}$, together with all the other variables, the equation can be solved for $[\mathrm{H}]_{\mathrm{f}}$ which is then used in Eqn 4.169 to obtain $[\mathrm{D}]_{\mathrm{f}}$ for each of the individual dyes applied.

Figure 4.19 presents curves, calculated as above, for a series of monobasic dyes of increasing values of A assuming that they have been applied to nylon 6.6 as pure free acid in the absence of other electrolyte. The dyes are taken to be strong acids, so that dye concentration in the bath can be derived from solution pH, assuming ideal behaviour. It has also been assumed that there is no restriction as a result of dye solubility. The shapes of the curves, in terms of concentration of dye applied and hence solution pH, are typical of those obtained by experiment in that all the curves have a point of inflection at which the amount of dye adsorbed is equivalent to the number of basic end-groups in the fibre, and that further dye adsorption in excess of this amount (i.e. 'overdyeing') is predicted. The curves also show the expected displacement to higher values of pH as the affinity of the dye increases.

The equations used therefore predict that 'overdyeing' is possible without invoking the formation of new 'sites' in the fibre, either from the ionisation

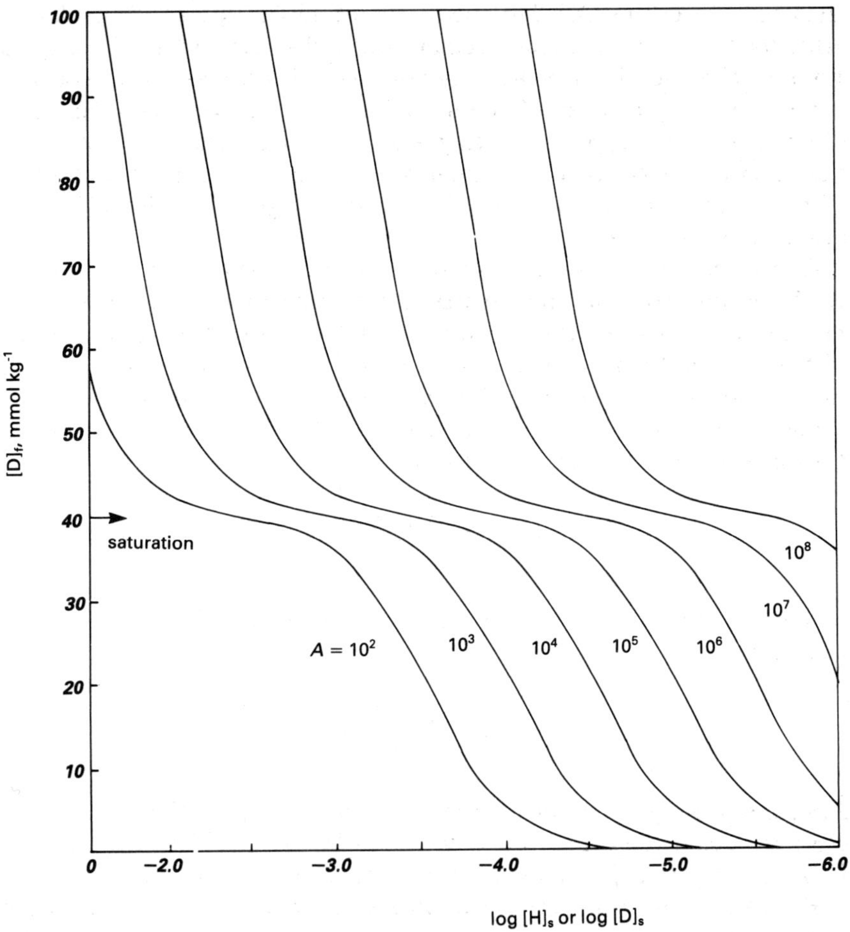

Figure 4.19 – Calculated isotherms at different affinities for pure dye free acid applied to nylon at 60 °C

Fibre contains 0.04 mol kg^{-1} of —NH_2 group $K_A = 2.6 \times 10^{-11}$

0.08 mol kg^{-1} of —COOH group $K_A = 1.6 \times 10^{-5}$

$V = 0.1\ l\ kg^{-1}$

Dye monobasic, i.e. $z = 1$; *no other addition, i.e.* $[H]_s = [D]_s$

of other groups such as amido linkages, or as a result of fibre degradation. The production of fibre degradation by 'overdyeing' has been demonstrated [106], and this is not surprising when the calculated values of the internal pH of the fibre are considered. These show that, for dyes of all affinities under these dyeing conditions, the internal pH of the fibre at 'saturation' is

2.45 in all cases and that further dye adsorption results in increased acidity to values which can be exceptionally high [99]. Even though such ideal behaviour would not be encountered in practice, the internal pH must be sufficiently low under practical conditions to produce fibre degradation, resulting in more basic end-groups and a further practical enhancement of the results in Figure 4.19. The conclusion must be drawn, however, that 'overdyeing' produces fibre degradation rather than the often postulated reverse that degradation results in 'overdyeing'.

It is obvious from a study of Eqns 4.165 and 4.167 that the possibility of different basicities of either dye or electrolyte have been taken into account and hence none of the problems associated with equations based on 'sites' are encountered in this treatment.

The same equations may be used to do similar calculations for the dyeing system of cationic dyes on acrylic fibres. They do have to be modified, however, in order to accommodate the fact that the dye ion has an opposite charge. Thus when applying Eqn 4.167 to obtain $[\mathrm{H}]_\mathrm{f}$, the term $\Sigma z_i[\mathrm{D}]_{\mathrm{f},i}$ appears on the left-hand side of the equation, because the charge on the dye is now positive. Also Eqn 4.96 cannot be used to calculate the affinity since this is defined for a cationic dye of basicity z, with a univalent anion X, by

$$-\Delta\mu^\ominus = RT\ln\frac{[\mathrm{D}]_\mathrm{f}[\mathrm{X}]_\mathrm{f}^z}{[\mathrm{D}]_\mathrm{s}[\mathrm{X}]_\mathrm{s}^z} \qquad (4.170)$$

which, using the Donnan equation $[\mathrm{H}]_\mathrm{f}[\mathrm{X}]_\mathrm{f} = [\mathrm{H}]_\mathrm{s}[\mathrm{X}]_\mathrm{s}$, becomes

$$-\Delta\mu^\ominus = RT\ln\frac{[\mathrm{D}]_\mathrm{f}[\mathrm{H}]_\mathrm{s}^z}{[\mathrm{D}]_\mathrm{s}[\mathrm{H}]_\mathrm{f}^z} \qquad (4.171)$$

When a value of $[\mathrm{D}]_\mathrm{f}$ is required from a known affinity the procedure is the same as for anionic dyes except that $\Sigma z_i[\mathrm{D}]_{\mathrm{f},i}$ is replaced by $\Sigma z_i A_i[\mathrm{D}]_{\mathrm{s},i}[\mathrm{H}]_\mathrm{f}^{z_i}/[\mathrm{H}]_\mathrm{s}^{z,i}$ and used on the left-hand side of Eqn 4.167.

4.7 THE APPLICATION OF THEORETICAL EQUATIONS TO PRACTICAL DYEING [99]

The various treatments described so far have relied on the three basic assumptions that

(a) dyeing has proceeded to equilibrium,
(b) the concentrations of all the ions in the final dyebath are known and
(c) the system is ideal in that concentrations may be used for activities.

In studies designed to determine the amount of dye taken up, and hence the affinity of the dye, achieving conditions (a) and (b) is no problem and experience indicates that the ideality of the system is adequate. Ample time can be given in most cases and very long liquor-to-goods ratios can be used, so that it is reasonable to assume that all concentrations, with the possible exception of dye, will be unchanged from the start of the experiment. Even if the studies are carried out at short liquor ratios, when differences between start and end are probable, the final dyebath is always available for complete analysis, even though this could be a very time-consuming procedure.

When the exercise is one of prediction, however – for example, of dye uptake from a known affinity – no analysis is possible because no final dyebath is available. The only ionic concentrations that are known with any certainty are those which it is proposed to put in the initial dyebath and these will change to a greater or lesser degree, depending on the liquor ratio, in order to fulfil the requirements of electrical neutrality in both the fibre and dyebath phases and those of the Donnan membrane equilibrium. This effect could be particularly important for any ion applied in low concentration. It is clear, therefore, that any system of prediction must relate final dyebath concentrations of dye and all other ions to the initial, known concentrations applied and that provision must be made for the effects of different liquor-to-goods ratios.

Regarding the problem of dyeing time, the only thing to be done in this respect is to assume that a situation is reached in the time available, which is defined by commercial considerations, that will not be significantly different in terms of practical tolerances from the equilibrium position. Also any errors could be minimised by carrying out the necessary calibration dyeings (required to establish the affinities of the dyes under consideration) using the same practical conditions as those for which the prediction is being made. Such a procedure would also have the added advantage of equalising, between calibration and prediction, other effects such as departures from ideal behaviour and possible aggregation effects.

The first step in any prediction procedure is therefore to determine the concentrations of all the ions in the initial system, not only those due to deliberate additions but all contributions from other sources as well. Thus, for example, the fibre itself may contain ions carried over from any pretreatment it may have had, depending on the efficiency of any intermediate washing process. In the case of a redyeing it may contain fixed dye (if the previous stage had been a reactive dyeing) but it will certainly contain unfixed dye, which must be taken into account if the exercise is one of 'shading' or 'topping-up'. The dyes being applied will be of commercial quality and will therefore contain ions, other than those of dye, which have been either required in manufacture or added during standardisation. In addition there must be sufficient acid or alkali introduced into the bath to ensure that the pH at the end of dyeing will be as required. This will not only add hydrogen or hydroxide ions but also the relevant co-ions, which will be anions or cations respectively. If a weak acid or a buffer system is used then a prior calculation of the various ionic concentrations at the final, required bath pH would be necessary.

Having obtained the amounts of all the various ions in the system at the start of dyeing it is then necessary to determine how they will be distributed between the fibre and the dyebath phases at the end of dyeing. When 1 kg of dry fibre is introduced into a dyebath of L litres, where L is the liquor-to-goods ratio, then the original bath will divide into two parts, V litres of internal surface phase and $(L - V)$ litres of dyebath phase. The amounts of the various ionic entities in the two phases must add up to the original amounts put in the dyebath. Thus if the amount of the monovalent cation (Cat^+) in the original dyebath is x_0 moles, then

$$x_0 = V[\mathrm{Cat}^+]_\mathrm{f} + (L - V)[\mathrm{Cat}^+]_\mathrm{s} \tag{4.172}$$

using the nomenclature defined for Eqn 4.88. Substituting the Donnan distribution $[\mathrm{Cat}^+]_\mathrm{s} = [\mathrm{Cat}^+]_\mathrm{f}[\mathrm{H}]_\mathrm{s}/[\mathrm{H}]_\mathrm{f}$ and rearranging gives

$$[\mathrm{Cat}^+]_\mathrm{f} = \frac{x_0[\mathrm{H}]_\mathrm{f}}{V[\mathrm{H}]_\mathrm{f} + (L - V)[\mathrm{H}]_\mathrm{s}} \tag{4.173}$$

The concentration in the bath at the end of dyeing $[\mathrm{Cat}^+]_\mathrm{f}$ is thus described in terms of the amount initially added, x_0. Similar derivations give expressions for $[\mathrm{An}^-]_\mathrm{f}$, $[\mathrm{An}^{2-}]_\mathrm{f}$ and $[\mathrm{An}^{3-}]_\mathrm{f}$ and these can all be substituted in Eqn 4.165, together with the terms for basic and acidic groups and that for $[\mathrm{OH}]_\mathrm{f}$, to give an extended equation in which all concentrations are in terms of initial amounts present with the exception of those related to dye. The term referring to fixed dye can be computed from known experimental values and the remaining term, namely $\Sigma z_i[\mathrm{D}]_{\mathrm{f},i}$, will vary depending on the information available and the objective of the calculation in the following way.

If the dyeing has been carried out and the concentrations of each dye in the fibre, $[\mathrm{D}]_{\mathrm{f},i}$ mol l^{-1}, and in solution, $[\mathrm{D}]_{\mathrm{s},i}$ mol l^{-1}, are known by measurement then the value of $\Sigma z_i[\mathrm{D}]_{\mathrm{f},i}$ is known and can be used directly in the extended form of Eqn 4.165 which is then solved for $[\mathrm{H}]_\mathrm{f}$. The affinity of each dye may then be obtained using Eqn 4.96.

If the affinities at the dyeing temperature A_i and the masses of the dyes applied initially are known and the object is to predict the concentrations of the dyes on the fibre at the end of dyeing $[\mathrm{D}]_{\mathrm{f},i}$, then, for each dye, the dye applied at the start of dyeing, d_i moles, will have been divided between the fibre and the bath so that

$$d_i = V[\mathrm{D}]_{\mathrm{f},i} + (L - V)[\mathrm{D}]_{\mathrm{s},i} \tag{4.174}$$

Combining Eqn 4.174 with Eqn 4.168 gives

$$[\mathrm{D}]_{\mathrm{f},i} = \frac{A_i[\mathrm{H}]_{\mathrm{s}}^{z,i}d_i}{[\mathrm{H}]_{\mathrm{f}}^{z,i}(L - V) + A_i[\mathrm{H}]_{\mathrm{s}}^{z,i}V} \tag{4.175}$$

which relates the dye on fibre at the end of dyeing to the amount originally applied. Using this relationship to replace the term $\sum z_i[\mathrm{D}]_{\mathrm{f},i}$ in Eqn 4.165 for all dyes, then the extended equation can be solved for $[\mathrm{H}]_\mathrm{f}$ and the value obtained used in Eqn 4.175 to calculate $[\mathrm{D}]_{\mathrm{f},i}$ for each dye present. In the above, amounts of dye have been considered in moles and dye concentrations in mol l^{-1}. In practice it is more usual to quote dye in grams or g l^{-1} of commercial product, i.e. impure dye, but the equations used will accept this provided that the quantities are suitably corrected for dye purity and r.m.m.

Probably the most interesting application of the developed equations is that, if the amount of a dye of known affinity required on the fibre is stipulated, then it is possible to calculate how much dye is required in the initial bath to produce this amount on the fibre under specified conditions of application. If the dyes used are pure then there is no problem, the extended Eqn 4.165 in the form using $\sum z_i[\mathrm{D}]_{\mathrm{f},i}$ is solved for $[\mathrm{H}]_\mathrm{f}$ and then the amount of dye required, d_i moles, is obtained from a rearranged Eqn 4.175. If the dyes are of commercial quality and contain ionic impurities, however, then not only is the quantity of dye required unknown but other ionic concentrations in the bath are also unknown, because the contribution from any electrolyte in the dye is not available. This problem can be resolved by a process of successive approximation [99] using Eqns 4.165 and a modified Eqn 4.175 described above, making due correction of dye concentrations to mol l^{-1} using the purity and r.m.m. of each dye, until the results from successive calculations are within the predetermined limits of accuracy required.

Thus by carrying out the necessary calibration experiments it is possible to predict the results of a dyeing from known initial conditions, or to calculate the amount of dye that must be applied to produce a desired result. To do this with commercial brands of dye requires a knowledge of the dye's r.m.m. and basicity, the purity of the brand and the nature of the ionisable impurities it contains, because the treatment requires all concentrations to be known and in mol l^{-1}. Also, because the treatment assumes ideal behaviour (which may or may not be justified in all cases), it would obviously be advantageous if the calibration experiments were carried out under the same dyeing conditions as those applying in the subsequent prediction stage, or at least conditions that are closely similar.

It may be considered that the treatment developed presents an 'overkill' situation, particularly in terms of obtaining the concentrations of all the ionic species present in the system. For many dyeing systems this is indeed true, but in others such care is warranted. The solution of the equations is clearly a computer operation and once the complete program is written it can be used in totality or suitably abbreviated, using the appropriate data input.

4.8 THE INTERNAL pH OF FIBRES

In any dyeing system for which the theoretical model requires the division of the aqueous phase into two parts, the external dyebath and the internal fibre phase of V l kg^{-1}, the concentration of any ion – even those which are non-substantive – can be quite different in the two phases, because of the need to satisfy the requirements of electrical neutrality and the Donnan theory of membrane equilibrium. The hydrogen ion is no exception, so that the pH of the internal fibre phase (pH_i or pH_f) can be quite different from that in the external bath (pH_s). This possibility has been recognised; for example, the results from measurements of the membrane potentials of cellulose sheets [54] were more effectively rationalised when values of the internal pH were calculated and used to determine the number of ionised carboxylic acid groups in the cellulose. Similarly, an analysis of the adsorption of hydrochloric acid by wool in the presence of added neutral electrolyte benefited from the use of internal rather than external pH [68]. More recently, studies into the mechanism of fixation of reactive dyes to cellulose required a knowledge of the degree of ionisation of the cellulose hydroxyl groups, a quantity which could only be sensibly obtained from a knowledge of the internal pH [65, 107].

The importance in all ionic dyeing systems of internal pH, and in particular the way in which it differs from the external dyebath pH, has been highlighted by the recent tendency to solve the electrical neutrality equation (e.g. Eqns 4.95 and 4.167) for $[H]_f$ rather than $[Na]_f$ as was done originally, and to use $[H]_f$ in the affinity equation (Eqn 4.96) and derivatives thereof. Numerous examples are available, but only two will be considered here.

Figure 4.20 shows the results of calculating the internal pH over a range of external pH for an undyed, cellulose-type fibre in the presence of various concentrations of added neutral electrolyte [108]. These calculations reveal that the internal pH never exceeds that in the outer bath. In high concentrations of added electrolyte the difference is only small, but at low concentrations it can be considerable. Thus, for example, if a piece of this cellulosic material is perfectly washed in water at pH 5–7 then an internal pH of 2.6 would be obtained, and storage of the material under these conditions could result in fibre degradation due to acid hydrolysis. The presence of a low concentration of electrolyte in the rinsing water (0.001 mol l^{-1} of sodium chloride $\approx$ 60 parts per million) results in a much higher internal pH and consequently less tendering of the material. Fortunately the latter situation is in practice

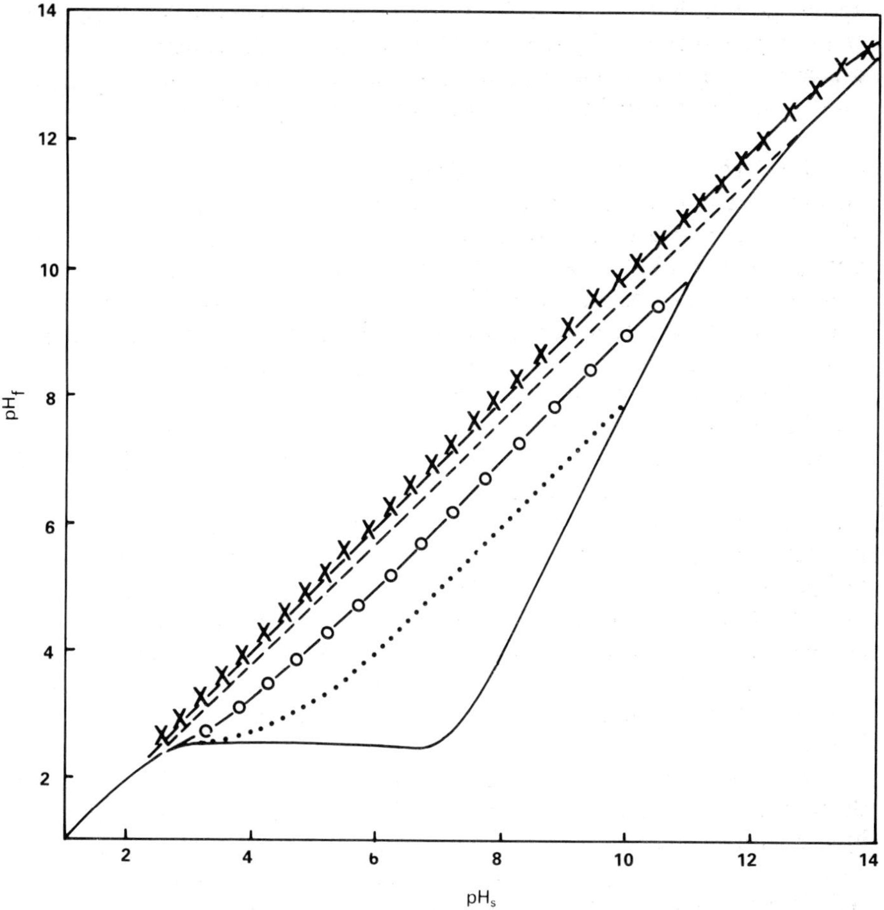

Figure 4.20 – Cellulosic fibre, internal versus external pH [108]
Fibre contains 1×10^{-2} mol l^{-1} of —COOH group *pK = 1×10^{-3}*
11.22 mol l^{-1} of —OH group *pK = 1.84×10^{-13}*
Added electrolyte (mol l^{-1}):

————————	*nil*
....	*1×10^{-4} NaCl*
-O—O—O—O-	*1×10^{-3} NaCl*
— — — —	*1×10^{-2} NaCl*
—X—X—X—X	*1 NaCl*

the more likely although over-enthusiastic rinsing in water from an ion-exchange column or from molten snow is not entirely impossible or unknown.

The presence in the fibre of anionic dye ions can aggravate the effect. Figure 4.21 shows the results of calculating the internal pH of a cellulosic-

Fibre and added electrolyte as in Figure 4.20
External pH = 6
Equivalent % shade for dye of r.m.m. 1000 and 50% purity

with z = 1	*0.2%*	*2%*	*20%*
z = 2	*0.1%*	*1%*	*10%*
z = 4	*0.05%*	*0.05%*	*5%*

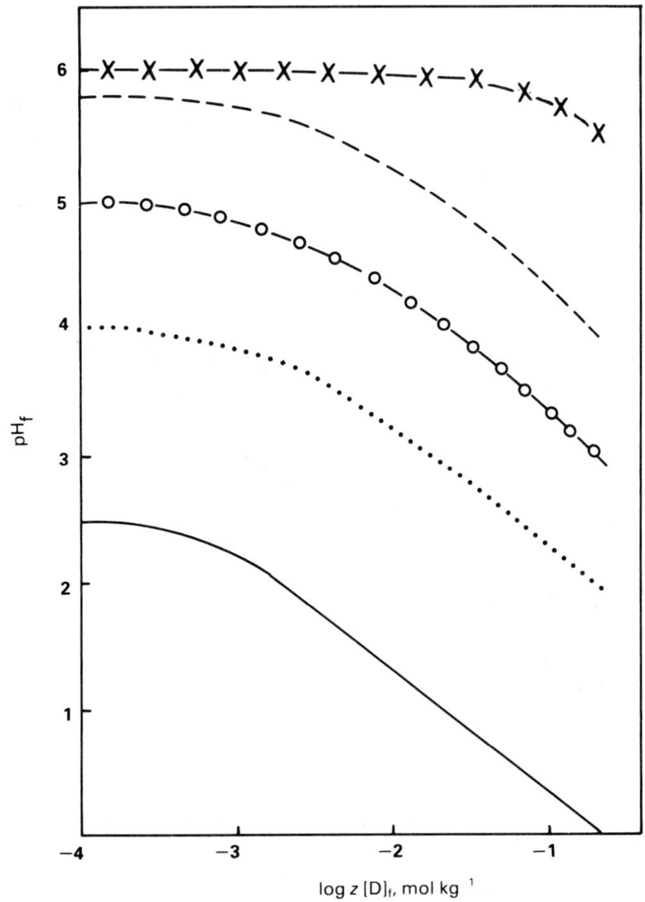

Figure 4.21 – Effect of anionic dye on internal pH

type fibre containing anionic dye when it is immersed in water at pH 6 containing various concentrations of sodium chloride. It can be seen that extremely high internal acidities are predicted for quite average depths of shade and not unreasonably low electrolyte concentrations in the bath. (It must be noted that these calculations presume ideal behaviour; in practice, activity effects would result in some modification of these results.) This situation is not observed with direct dyes, since before such low levels of electrolyte can be achieved most if not all of the dye has been desorbed from the fibre. With fixed reactive dyes, however, the situation is perfectly feasible when relatively pure water is used, resulting in hydrolysis of the dye–fibre

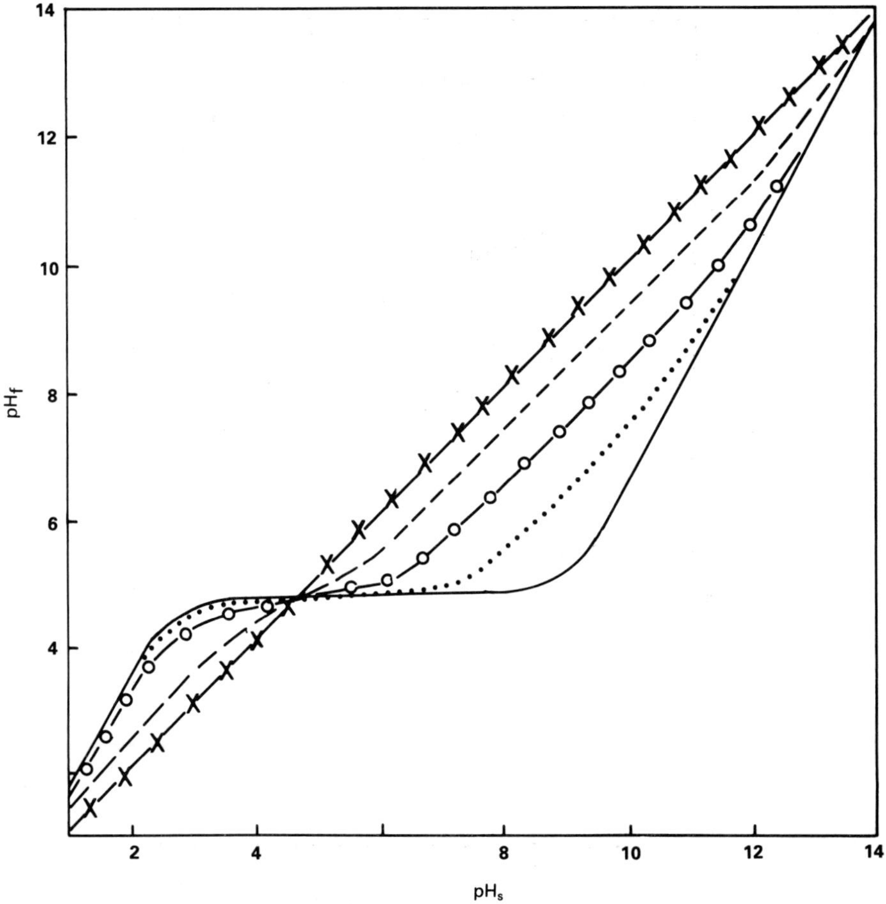

Figure 4.22 – Nylon-type fibre, internal versus external pH

Fibre contains 4×10^{-1} mol l^{-1} of —NH_2 group *pK = 2.6×10^{-11}*

8×10^{-1} mol l^{-1} of —COOH group *pK = 1.6×10^{-5}*

Added electrolyte (mol l^{-1}):

———————	*nil*
………………	*1×10^{-3} NaCl*
-O—O—O—O-	*1×10^{-2} NaCl*
— — — —	*1×10^{-1} NaCl*
—X—X—X—X	*1 NaCl*

bond and possible fibre degradation, both during the later stages of the washing-off process and also when heat is applied in any subsequent drying process. It is possible that attempts to remove the last traces of unfixed reactive dye fail because, if the conditions are right, loose dye is being produced by hydrolysis at a rate equal to or greater than the rate at which it is being washed off. In addition if a perfectly washed material is achieved it could develop further loose dye during drying under these conditions.

When the same calculations are carried out for a nylon-type fibre, the result is quite different (Figure 4.22); the curves for the various electrolyte concentrations, in the absence of dye, intersect at a point, the isoelectric point of the fibre. At external pH less than at this point, i.e. where the fibre carries a positive charge, the internal pH is always higher than that in the bath and the presence of added electrolyte serves to lower it, but never below that in the external bath. On the other side of the isoelectric point, i.e. where the fibre carries a negative charge, the situation is reversed and is similar to that for the cellulosic-type fibre (Figure 4.20). Again, the effects of the presence of anionic dye in the fibre are of considerable interest. If the nylon-type fibre under consideration is immersed in 0.0001 mol l^{-1} of hydrochloric acid, i.e. a theoretical bath pH of 4, the internal pH falls as the amount of dye on the fibre is increased (Figure 4.23). The change is comparatively slight until the isoelectric point of the dyed fibre is reached, a situation indicated by the crossover point of the curves, but from then on the most dramatic changes are predicted, the internal pH rapidly reaching a value equivalent to molar acid. (Again, it must be remembered that activity effects have been ignored.) The magnitude of the effect is progressively reduced by the presence of increasing concentrations of neutral electrolyte. It would appear that the tendering of nylon which has been 'overdyed' with a water-soluble anionic dye is due to this effect. It is clear from both sets of curves in Figures 4.22 and 4.23 that at the isoelectric point the internal pH is the same as that in the external bath, but Figure 4.23 also demonstrates that the isoelectric point occurs where the total negative charge, i.e. that from both the dye and the ionised carboxylic acid end-groups, is equal to the number of positively charged end-groups. In other words, the fibre is not 'saturated' when the amount of dye alone equates to the amount of positively charged end-group but when the total negative charge on the fibre reaches this value.

A consideration of the results depicted in Figures 4.20 and 4.22 leads to an appreciation of the effects of added neutral electrolyte in the two dyeing systems concerned. From Eqn 4.96,

$$\left(\frac{[\mathrm{H}]_\mathrm{f}}{[\mathrm{H}]_\mathrm{s}}\right)^z \cdot \frac{[\mathrm{D}]_\mathrm{f}}{[\mathrm{D}]_\mathrm{s}} = \text{constant} \qquad (4.176)$$

Fibre and added electrolyte as in Figure 4.22
External pH = 4
Equivalent % shade for dye of r.m.m. 1000 and 50% purity

with z = 1	*0.2%*	*2%*	*20%*
z = 2	*0.1%*	*1%*	*10%*
z = 4	*0.05%*	*0.5%*	*5%*

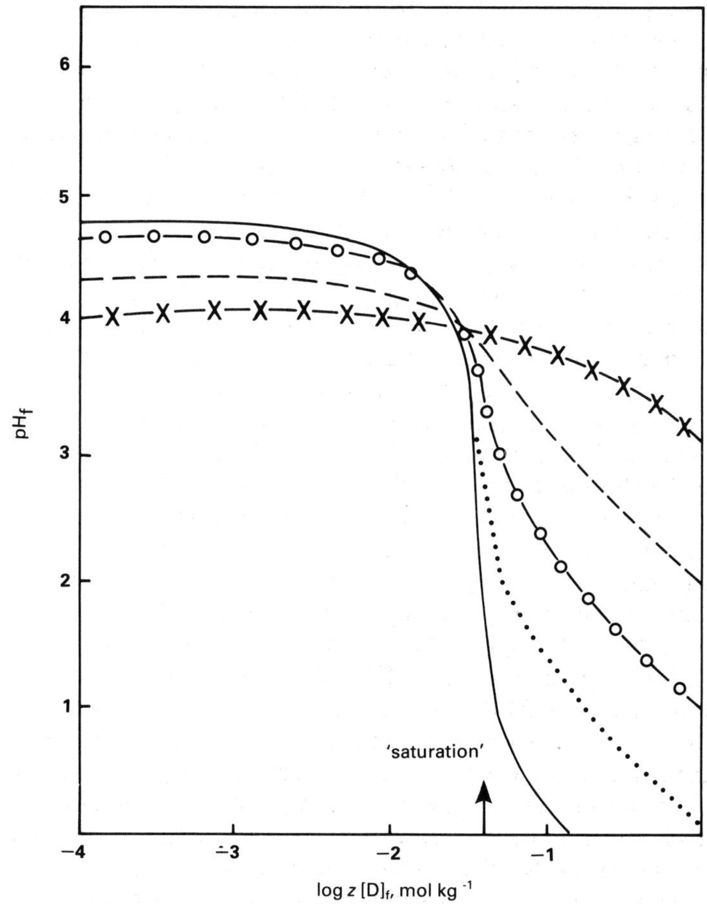

Figure 4.23 – Effect of anionic dye on internal pH [107]

For the cellulosic-type fibre (Figure 4.20), as the concentration of electrolyte increases at any given value of $[H]_s$, then $[H]_f$ decreases. Thus $[H]_f/[H]_s$ decreases and in order to maintain a constant value $[D]_f/[D]_s$ must increase, that is, at constant $[D]_s$ the amount of dye on the fibre $[D]_f$ must increase, and hence neutral electrolyte increases dye uptake. With the nylon-type fibre

(Figure 4.22) the effect at pH below the isoelectric point is the opposite. At constant $[H]_s$, increasing electrolyte increases $[H]_f$ and hence also the value of $[H]_f/[H]_s$. To maintain constancy $[D]_f/[D]_s$ must decrease, so that at constant $[D]_s$ the amount on the fibre $[D]_f$ must decrease, i.e. neutral electrolyte reduces dye uptake. At pH above the isoelectic point the nylon-type fibre behaves like the cellulosic-type fibre.

REFERENCES

1. The Society of Dyers and Colourists, *J.S.D.C.*, **89** (1973) 411.
2. S M Neale, see T Vickerstaff, *The physical chemistry of dyeing*, 2nd edn (London: Oliver and Boyd, 1954) 191.
3. J Boulton, A E Delph, F Fothergill and T H Morton, *J. Textile Inst.*, **24** (1933) 113.
4. C L Bird, F Manchester and P Harris, *Faraday Soc. Disc.*, No. 16 (1954) 85.
5. C L Bird and F Manchester, *J.S.D.C.*, **71** (1955) 604.
6. M J Schuler and W R Remington, *Faraday Soc. Disc.*, No. 16 (1954) 201.
7. J Steinhardt and C H Fugitt, *J. Res. Nat. Bur. Standards*, **29** (1942) 215.
8. D R Lemin and T Vickerstaff, *J.S.D.C.*, **63** (1947) 405.
9. A B Meggy, *Trans. Faraday Soc.*, **43** (1947) 502.
10. G A Gilbert, *Proc. Roy. Soc.*, **A183** (1944) 167.
11. G N Lewis, any textbook on physical chemistry.
12. P Debye and E Hückel, any textbook on physical chemistry.
13. C S Chadwick and S M Neale, *J. Polymer Sci.* **28** (1958) 355.
14. B Milićević and G Eigenmann, *Helv. Chim. Acta*, **47** (1964) 1039.
15. C Robinson and J L Moilliet, *Proc. Roy. Soc.*, **A143** (1934) 630.
16. E Valkó, *Rev. of Progr. in Color.*, **3** (1972) 50.
17. E Coates, *J.S.D.C.*, **85** (1969) 355.
18. P J Hillson and R B McKay, *Trans. Faraday Soc.*, **61** (1965) 374.
19. D R Lemin and T Vickerstaff, *Trans. Faraday Soc.*, **43** (1947) 491.
20. A N Derbyshire, *Trans. Faraday Soc.*, **51** (1955) 909.
21. A N Derbyshire, *Hexagon Digest*, **21** (1955) 12.
22. V Kartaschoff, *Helv. Chim. Acta*, **8** (1925) 928.
23. R. Clavel and T Stanisz, *Rev. Gen. Matières Col.*, **28** (1923) 145.
24. T Vickerstaff and E Waters, *J.S.D.C.*, **58** (1942) 116.
25. C L Bird, *J.S.D.C.*, **70** (1954) 69.
26. H J White, *Text. Research J.*, **30** (1960) 329.
27. K V Datye, S C Pitkar and U M Parao, *Textilveredlung*, **6** (1971) 593.
28. W Biedermann, *J.S.D.C.*, **87** (1971) 105.
29. C L Bird and P Harris, *J.S.D.C.*, **73** (1957) 199.
30. W McDowell and R Weingarten, *J.S.D.C.*, **85** (1969) 595.
31. C L Bird and J Firth, *J. Textile Inst.*, **51** (1960) T1342.
32. M Kusunose and K Odajima, *J. Soc. Text. Cell. Ind. Japan*, **14** (1958) 49.
33. F Jones and R Seddon, *Text. Research J.*, **34** (1964) 373; **35** (1965) 334.
34. F Jones and J Kraska, *J.S.D.C.*, **82** (1966) 333.
35. C J Bent, T D Flynn and H H Sumner, *J.S.D.C.*, **85** (1969) 606.
36. A Johnson, R H Peters and A S Ramadan, *J.S.D.C.*, **80** (1964) 129.
37. A Takaoka, A Katayama, N Kuioki and N Konishi, *Sen-i Gakkaishi*, **21** (1965) 158.
38. J R Aspland and C L Bird, *J.S.D.C.*, **77** (1961) 9.
39. (a) K J Laidler and J M Meiser, *Physical chemistry* (New York: Benjamin Cummings, 1982).
    (b) S Glasstone and D Lewis, *Elements of physical chemistry* (London: Macmillan Press Ltd, 1962).
40. H F Willis, J O Warwicker, H A Standing and A R Urquhart, *Trans. Faraday Soc.*, **41** (1945) 506.

41. T Vickerstaff, *The physical chemistry of dyeing*, 2nd edn (London: Oliver and Boyd, 1954).
42. J Hanson, S M Neale and W A Stringfellow, *Trans. Faraday Soc.*, **31** (1935) 1718.
43. J Crank, *J.S.D.C.*, **63** (1947) 293.
44. W J Marshall and R H Peters, *J.S.D.C.*, **63** (1949) 17.
45. A R Urquhart and A M Williams, *J. Textile Inst.*, **15** (1924) T559.
46. W J Marshall and R H Peters, *J.S.D.C.*, **65** (1949) 17.
47. S M Neale and W A Stringfellow, *J.S.D.C.*, **56** (1940) 19.
48. A Johnson, K C Maheshwari and L W C Miles, *SIRTEC*, Ap 1969, 557.
49. E H Daruwalla and A P D'Silva, *Text. Research J.*, **33** (1963) 40.
50. S R Iyer, G Srinivasan, N T Baddi and M R Ravikrishnan, *Text. Research J.*, **34** (1964) 807.
51. S R Iyer, G Srinivasan and N T Baddi, *Text. Research J.*, **38** (1968) 693.
52. J Hanson, S M Neale and W A Stringfellow, *J.S.D.C.*, **56** (1940) 17.
53. M M Bilsbury, J T Martin and H A Standing, *Shirley Inst. Mem.*, **26** (1953) 217.
54. S M Neale and P T Standring, *Proc. Roy. Soc.*, **A213** (1952) 530.
55. E H Daruwalla, P J Kangle and G M Nabar, *Text. Research J.*, **31** (1961) 712.
56. J A Fowler, A G H Michie and T Vickerstaff, *Melliand Textilber.*, **32** (1951) 296.
57. R H Peters and J Simons, *J.S.D.C.*, **70** (1954) 557.
58. R H Peters and H H Sumner, *J.S.D.C.*, **71** (1955) 130.
59. R H Peters and H H Sumner, *J.S.D.C.*, **73** (1957) 12.
60. G W Blanco, *Ind. Eng. Chem.*, **20** (1928) 926.
61. P N Pavlov, *Kolloid Zhur.*, **44** (1928) 44.
62. S Lipatov and N Sokolova, *Z. Anorg. Allgem. Chem.*, **192** (1930) 383.
63. S M Neale, *J. Textile Inst.*, **20** (1929) T373.
64. S P Saric and R K Schofield, *Proc. Roy. Soc.*, **A185** (1946) 431.
65. H H Sumner, *J.S.D.C.*, **76** (1960) 672.
66. I D Rattee and M M Breuer, *The physical chemistry of dye adsorption* (London: Academic Press, 1974).
67. G A Gilbert and E K Rideal, *Proc. Roy. Soc.*, **A182** (1944) 335.
68. L Peters and J B Speakman, *J.S.D.C.*, **65** (1949) 63.
69. J Steinhardt and M Harris, *J. Res. Nat. Bur. Stands.*, **24** (1940) 335.
70. B Oloffsen, *J.S.D.C.*, **67** (1951) 57; **68** (1952) 506.
71. L Peters, *J. Textile Inst.*, **51** (1960) T1290.
72. M M Breuer, *Trans. Faraday Soc.*, **60** (1964) 1006.
73. A R Mathieson and C S Whewell, *J. Appl. Polymer Sci.*, **8** (1964) 2029.
74. R H Peters, *Textile chemistry*, vol. 3, *The physical chemistry of dyeing* (Amsterdam: Elsevier, 1975).
75. J Steinhardt, C H Fugitt and M Harris, *J. Res. Nat. Bur. Stands.*, **26** (1941) 293.
76. J B Speakman and E Stott, *Trans. Faraday Soc.*, **31** (1935) 1425; **34** (1938) 1203.
77. J B Speakman and G H Elliot, *Symposium on fibrous proteins* (S.D.C. 1946) 116.
78. P Carlene, F Rowe and J B Speakman, *J.S.D.C.*, **62** (1946) 329.
79. B Benson and P Larose, *Can. J. Res.*, **F28** (1950) 238.
80. B G Ferrini, Y Kimura and H Zollinger, *Proc. Intern. Wool Text. Res. Conf., Paris (CIRTEL)*, **3** (1965) 291.
81. L Peters and G Lister, *Faraday Soc. Disc.*, No. 16 (1954) 24.
82. B Skinner and T Vickerstaff, *J.S.D.C.*, **61** (1945) 193.
83. F L Goodall and C Hobday, *J.S.D.C.*, **55** (1939) 529.
84. J B Speakman and E Stott, *J.S.D.C.*, **50** (1934) 341.
85. J Steinhardt, C H Fugitt and M Harris, *J. Res. Nat. Bur. Stands.*, **30** (1943) 123.
86. D Marfell, PhD thesis (Leeds University, 1971).
87. M A White, PhD thesis (Leeds University, 1968).
88. G Chantry, PhD thesis (Leeds University, 1971).
89. A R Mathieson, C S Whewell and P E Williams, *J. Appl. Polymer Sci.*, **8** (1964) 2009.
90. M Harris and A M Sookne, *J. Res. Nat. Bur. Stands.*, **26** (1941) 289.
91. R H Peters, *J.S.D.C.*, **61** (1945) 95.
92. J Boulton, *J.S.D.C.*, **62** (1946) 65.

93. F T Wall and A Beresniewicz, *J. Phys. Chem.*, **60** (1956) 692.
94. F C McGrew and A K Schneider, *J. Amer. Chem. Soc.*, **72** (1950) 2547.
95. W R Remington and E K Gladding, *J. Amer. Chem. Soc.*, **72** (1950) 2553.
96. E Atherton, D A Downey and R H Peters, *Text. Research J.*, **25** (1955) 988.
97. H J Palmer, *J.S.D.C.*, **67** (1951) 609.
98. R McGregor and P W Harris, *J. Appl. Polymer Sci.*, **14** (1970) 513.
99. H H Sumner, *J.S.D.C.*, **102** (1986) 301, 341, 392.
100. T Vogel, J M A De Bruyne and C L Zimmerman, *Amer. Dyes Rept.*, **47** (1958) 881.
101. D Balmforth, C A Bowers and T H Guion, *J.S.D.C.*, **80** (1964) 577.
102. (a) O Glenz and W Beckmann, *Melliand Textilber.*, **38** (1957) 296; (b) W Beckmann, *J.S.D.C.*, **77** (1961) 616.
103. S Rosenbaum, *Text. Research J.*, **33** (1963) 899.
104. S Rosenbaum, *J. Appl. Polymer Sci.*, **7** (1963) 1225.
105. R J Harwood, R McGregor and R H Peters, *J.S.D.C.*, **88** (1972) 216.
106. C D O'Briain and R H Peters, *J.S.D.C.*, **69** (1953) 435.
107. H H Sumner and C D Weston, *Amer. Dyes Rept.*, **52** (1963) 442.
108. H H Sumner, *J.S.D.C.*, **81** (1965) 193.

CHAPTER 5

Diffusion and rates of dyeing

FRANCIS JONES

5.1 DIFFUSION

Diffusion is the process by which matter is transported from one part of a system to another as a result of random molecular motion. This eventually leads to an equalisation of chemical potential and concentration throughout the system. Where there is more than one phase, the chemical potential of the diffusant is constant throughout the system only when true dynamic equilibrium has been achieved, even though the concentrations in each phase may differ considerably. In dyeing processes one phase consists initially of the fibre or filament and the other is usually a dye solution in which the undyed fibre is immersed. Although apparently a simple transfer process of dye passing from solution to the interior of the fibre, the mechanism of dyeing in its entirety is more complex. It is necessary to study at least three stages, any of which may control or influence the rate at which dye molecules are transferred. These stages are

(a) the diffusion of dye through the bulk solution of dyebath to the surface of the polymer,
(b) the possible adsorption of dye molecules on this surface and
(c) the diffusion of dye from the surface to the interior of the fibre.

Thus, three factors at least must be examined before the dyeing process can be elucidated.

In view of the complexity of the diffusion process, real dyeing systems are studied on the basis of simpler idealised situations, but these simpler models have imposed on them considerable restrictions or conditions to make them amenable to mathematical treatment. It is necessary to study diffusion in terms of simple models before any advances can be made in dealing with real or experimental investigations of actual dyeing processes.

The student's first acquaintance with diffusion probably comes with Graham's Law, which states that the rate of diffusion of a gas is inversely proportional to the square root of its density. The rate at which a gas as a whole is able to diffuse is much smaller than the velocities of the individual molecules, however, owing to the large number of molecular collisions in the gas. It is interesting to consider the effect on the rate of diffusion of a gas when the gas is confined to a capillary. De Boer [1] has visualised this effect by scaling-up the size of each molecule to that of a bee, i.e. by using a magnification factor of approximately 30 million with respect to both length and time. In turn, the 'capillary' tube will be 100 ft wide and 20 miles long. An individual 'bee' could fly the length of the tube directly in five minutes. Entering the tube in an arbitrary direction, however, the 'bee' will collide with the wall. After each collision with the wall the 'bee' may remain there for one hour (dwell time) before flying off in an arbitrary direction. It can be deduced that, on average, the 'bee' takes 50 years to traverse the tube, a distance which it could fly in five minutes. When the example is that of a dye molecule moving down a capillary in a substrate such as cellulose, with many solute–dye molecule collisions in addition to collisions with capillary walls, it is surprising that diffusion occurs at all. Fortunately rates of diffusion are strongly dependent on temperature which affects the dwell time.

If we now consider an iodine solution, for example, on top of which clear water is poured without disturbing the lower layer, then a concentration gradient develops in the water. Consider any horizontal section and two thin equal elements of volume separated by the section. In a given time, on average, a definite fraction of the molecules in the lower element will cross the section to the higher element by random collision processes. The same fraction will also cross from the higher to the lower element. This fraction of concentration is constant, but since the lower element initially had more iodine molecules per unit volume of element the *number* moving across the section in an upward direction is greater than the number of molecules moving downwards. There is therefore a mass transfer due to differences in concentration, as a result of random motion. Further, with time, the concentration difference becomes less, until equilibrium is established. This equilibrium is dynamic in that once it is attained molecules will continue to cross the section, but at the same rate in each direction.

5.1.1 Fick's laws of diffusion

The fundamental concepts of mass transfer are comparable with those of heat conduction. Fick [2] (1855) was the first to adapt the latter concepts, developed earlier by Fourier [3], to cover quantitative diffusion in an isotropic medium. He used the hypothesis that the rate of transfer of a diffusant through a section of unit area is proportional to the concentration gradient measured normal to the section. Expressed in mathematical form this statement is

$$F = -D\frac{\partial C}{\partial x} \tag{5.1}$$

where F is the rate of transfer per unit area of section, C is the concentration of the diffusing substance, x is the space coordinate measured normal to the section and D is the diffusion coefficient. The term $\partial C/\partial x$ is therefore the concentration gradient. If F and C are both expressed in terms of the same unit of quantity, e.g. grams or moles, then D is independent of this unit and has the dimensions of length^2 time^{-1}, e.g. $m^2\ s^{-1}$. The negative sign in this equation arises because diffusion occurs in a direction opposite to that of an increasing concentration.

Eqn 5.1 is referred to as Fick's first law and describes the rate of transfer in a steady state of flow across the section, i.e. when the concentration gradient is constant. It applies only to an isotropic medium in which the diffusion properties at any point are the same relative to all directions. In an anisotropic medium, such as most crystals and fibres, the diffusion properties depend on the directions in which the diffusant is moving.

Fick's second law of diffusion is of more fundamental consequence and refers to non-steady-state conditions, in that it describes the accumulation of matter in a medium at a given point as a function of time. It is derived from Eqn 5.1 as follows.

ABCD and $A^1B^1C^1D^1$ are faces, each of area $2dy \times 2dz$, perpendicular to

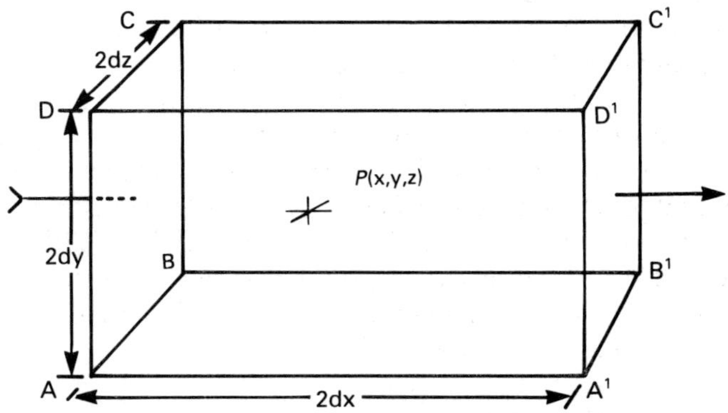

Figure 5.1 – Diffusion through an element of volume

the x-axis as shown in Figure 5.1. The rate at which diffusant enters the element through the face ABCD in the plane $(x - dx)$ is given by

$$\text{rate of entering} = 4dy\ dz\ (F_x - \frac{\partial F_x}{\partial x}dx) \tag{5.2}$$

where F_x is the rate of transfer through unit area of the corresponding plane at P. Similarly, the rate of loss of diffusant through the plane $A^1B^1C^1D^1$ is given by

$$\text{rate of loss} = 4\mathrm{d}y\,\mathrm{d}z\,(F_x + \frac{\partial F_x}{\partial x}\mathrm{d}x) \tag{5.3}$$

Subtracting the rate of loss from the entering rate gives the contribution to the rate of increase in concentration of diffusant in the element. Therefore,

$$\text{rate of increase} = -8\mathrm{d}x\,\mathrm{d}y\,\mathrm{d}z\frac{\partial F_x}{\partial x} \tag{5.4}$$

Similarly for the other faces, the contributions to the rate of increase or accumulation are $-8\mathrm{d}x\,\mathrm{d}y\,\mathrm{d}z\,(\partial F_y/\partial_y)$ and $-8\mathrm{d}x\,\mathrm{d}y\,\mathrm{d}z\,(\partial F_z/\partial z)$. However, since the rate at which the diffusant accumulates in the element is also given by $8\mathrm{d}x\,\mathrm{d}y\,\mathrm{d}z\,(\partial C/\partial t)$, it follows that

$$\frac{\partial F_x}{\partial x} + \frac{\partial F_y}{\partial y} + \frac{\partial F_z}{\partial z} + \frac{\partial C}{\partial t} = 0 \tag{5.5}$$

If the diffusion coefficient D is constant and independent of x, y and z directions and F_x, F_y and F_z are given by Eqn 5.1, then Eqn 5.5 becomes

$$\frac{\partial C}{\partial t} = D\left(\frac{\partial^2 C}{\partial x^2} + \frac{\partial^2 C}{\partial y^2} + \frac{\partial^2 C}{\partial z^2}\right) \tag{5.6}$$

For one-dimensional diffusion, say in the x-direction, Eqn 5.6 becomes

$$\frac{\partial C}{\partial t} = D\frac{\partial^2 C}{\partial x^2} \tag{5.7}$$

Eqn 5.7 is then an expression of Fick's second law of diffusion.

5.1.2 Diffusion in anisotropic media

Eqns 5.1 and 5.7, representing Fick's two laws, are based on the assumption that the diffusion coefficient is constant. In many real systems the value of D depends on the concentration of the diffusant in the film or fibre. Also, when the medium is not homogeneous, D may be a function of the coordinates x, y and z. In such circumstances it follows from Eqn 5.6 that

$$\frac{\partial C}{\partial t} = \frac{\partial}{\partial x}\left(D\frac{\partial C}{\partial x}\right) + \frac{\partial}{\partial y}\left(D\frac{\partial C}{\partial y}\right) + \frac{\partial}{\partial z}\left(D\frac{\partial C}{\partial z}\right) \tag{5.8}$$

or, considering the x-direction only, Fick's second law may be generalised as

$$\frac{\partial C}{\partial t} = \frac{\partial}{\partial x}\left(D\frac{\partial C}{\partial x}\right) \tag{5.9}$$

For anisotropic media such as non-cubic crystals, fibres and films, it is not always true that diffusion proceeds in a direction normal to the surface which is maintained at constant concentration. Under steady-state conditions, the rate of transfer of diffusant as described by Eqn 5.1 must then be modified, since it now depends not only on the concentration gradient in the x-direction but also on the y- and z-concentration gradients, i.e. Eqn 5.1 becomes

$$-F_x = D_{11}\frac{\partial C}{\partial x} + D_{12}\frac{\partial C}{\partial x} + D_{13}\frac{\partial C}{\partial z} \tag{5.10}$$

$$-F_y = D_{21}\frac{\partial C}{\partial x} + D_{22}\frac{\partial C}{\partial y} + D_{23}\frac{\partial C}{\partial z} \tag{5.11}$$

$$-F_z = D_{31}\frac{\partial C}{\partial x} + D_{32}\frac{\partial C}{\partial y} + D_{33}\frac{\partial C}{\partial z} \tag{5.12}$$

The diffusion coefficients have the significance that, e.g., F_x is made up not only from the contribution of diffusion in the x-direction but also from components of concentration gradients in both the y- and the z-directions, these being controlled by diffusion coefficients D_{12} and D_{13} respectively. By transforming these 'Cartesian' coordinates to regular coordinates the system may be 'brought into line' so that

$$\frac{\partial C}{\partial t} = D_1\frac{\partial^2 C}{\partial \xi^2} + D_2\frac{\partial^2 C}{\partial \eta^2} + D_3\frac{\partial^2 C}{\partial \zeta^2} \tag{5.13}$$

The new axes ξ, η and ζ may be called the ***principal axes of diffusion***, and D_1, D_2 and D_3 the principal diffusion coefficients. Thus the principal axis of diffusion may not coincide with the geometric axes of a film, cylinder or sphere substrate. In most experiments. conditions are so arranged that mass transfer occurs in one direction only, in the sense that the concentration gradient lies only along the x-axis. By substituting the F_x, F_y and F_z values given by Eqns 5.10–5.12 in Eqn 5.5 the rate of accumulation of diffusant is given by

$$\frac{\partial C}{\partial t} = D_{11}\frac{\partial^2 C}{\partial x^2} + D_{22}\frac{\partial^2 C}{\partial y^2} + D_{33}\frac{\partial^2 C}{\partial z^2} + (D_{23} + D_{32})\frac{\partial^2 C}{\partial y\,\partial z} + (D_{31} + D_{13})\frac{\partial^2 C}{\partial x\,\partial z} + (D_{12} + D_{21})\frac{\partial^2 C}{\partial x\,\partial y} \tag{5.14}$$

It is clear that, since C and $\partial C/\partial x$ are everywhere independent of y and z, the diffusion equation is governed by

$$\frac{\partial C}{\partial t} = D_{11}\frac{\partial^2 C}{\partial x^2} \tag{5.15}$$

and D_{11} in these experiments is the measured diffusion coefficient. If the direction of diffusion is chosen to be that of a principal axis, then D_{11} is equal to one or other of D_1, D_2 or D_3, the principal diffusion coefficients. If a sheet cut from a crystal or a film is used so that its normal has direction cosines (l, m, n) relative to the principal axes of diffusion, then the measured diffusion coefficient is related to the principal diffusion coefficients. Even if a concentration gradient exists in one direction only, it is clear from Eqns 5.10–5.12 that diffusional flow is along this direction if the direction coincides with a prinicpal axis of diffusion.

5.1.3 Diffusion in the steady state

Consider diffusion through a plane sheet of thickness l, whose surfaces $x = 0$ and $x = l$ are maintained at constant concentrations C_1 and C_2 respectively. Initially the sheet contains no diffusant, but diffusant accumulates and after a time a steady state is reached, as shown in Figure 5.2, in which the concentration remains constant at all points x within the sheet. Provided that the diffusion coefficient is constant, Eqn 5.7 representing Fick's second law reduces to

$$\frac{d^2 C}{dx^2} = 0 \tag{5.16}$$

and on integration twice with respect to x and introducing the boundary concentration conditions at $x = 0$ and $x = l$ we obtain

$$\frac{C - C_1}{C_2 - C_1} = \frac{x}{l} \tag{5.17}$$

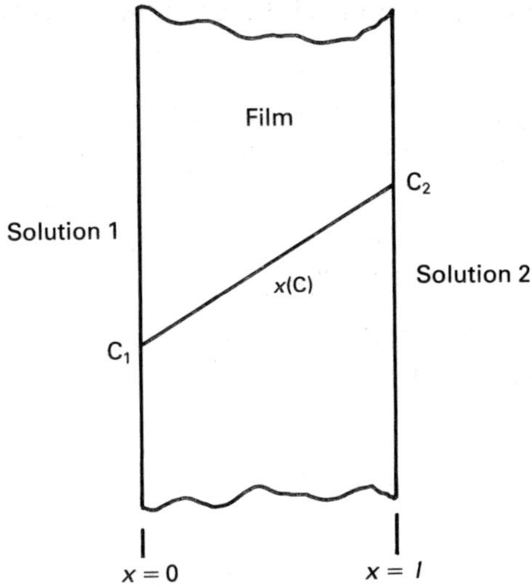

Figure 5.2 – Steady-state diffusion with constant surface concentrations

The concentration changes linearly through the sheet and the rate of transfer across one plane in the sheet is the same as the rate of transfer across any other plane. The rate of transfer F is thus given by

$$F = -D\frac{\partial C}{\partial x} = D\frac{(C_1 - C_2)}{l} \tag{5.18}$$

Since the gradient shown in Figure 5.2 is positive, there is a change of sign. If the thickness and the *surface* concentrations C_1 and C_2 are known, then D can be calculated from a single observation of the flow rate.

In simple experiments surface concentrations are not usually known. Provided, however, there is a linear relationship between the concentration in solution C_s and the concentration of the diffusing substance in the film C_f, i.e. when $C_f/C_s = K$, where K is a partition coefficient, Eqn 5.18 still holds. In this case C_1 and C_2 are the concentrations of the solutions on either side of the sheet.

5.2 MEASUREMENT OF DIFFUSION

5.2.1 Daynes' time-lag method [4]

The above expressions describing diffusion in the steady state enable the diffusion coefficient to be experimentally determined provided it is assumed that no dimensional changes occur in the film during diffusion, that the diffusion coefficient is constant and that the diffusant concentration on the receiving side of the film approaches zero. Initially, diffusant begins to build up in the film and, until steady-state conditions (i.e. a linear concentration gradient) have been achieved, both the rate of flow and the concentration of diffusant at any point within the film will vary with time. During this period the amount of diffusant Q_t passing through the sheet in time t is given [5] by

$$\frac{Q_t}{C_1}=\frac{Dt}{l^2}-\frac{1}{6}-\frac{2}{\pi^2}\sum_1^{\infty}\frac{(-1)^n}{n^2}\exp\left(\frac{-Dn^2\pi^2t}{l^2}\right) \quad (5.19)$$

where C_1 is the concentration of diffusant on the entry side of the film, l is the film thickness and n is an integer. As t becomes large, the steady state is approached and the exponential term becomes negligibly small so that a graph of Q_t against t tends to the line

$$Q_t = \frac{DC_1}{l}\left(t-\frac{l^2}{6D}\right) \quad (5.20)$$

The intercept on the t-axis L is given by

$$L = \frac{l^2}{6D} \quad (5.21)$$

and the rate of transfer, once a steady state has been reached, is given by the slope of the linear portion of the curve. Figure 5.3, taken from the work of Bird [6], shows a typical curve obtained by this method. Using cellulose acetate film of thickness 6.38×10^{-6} m and a dispersion of the dye 1-methylamino-4-(β-hydroxy)ethylaminoanthraquinone, which maintains a constant concentration of dye molecules in solution on the entrant side of the film, the intercept L on the time axis is 49.5 min. The diffusion coefficient, using Eqn 5.21, is therefore

$$D=\frac{l^2}{6L}=\frac{(6.38\times 10^{-6})^2}{6\times 49.5\times 60}=2.28\times 10^{-15}\ \mathrm{m^2\ s^{-1}} \quad (5.22)$$

when time is expressed in seconds. When the solution concentration on the entry side is lowered by the diffusion process, the surface concentration may fall. It is necessary, therefore, to use a relatively large volume of diffusant solution, so that this decrease may be minimised. The converse situation must apply to the exit side, since the concentration conditions must approach zero and in practice be consistent with the lower limit of the experimental

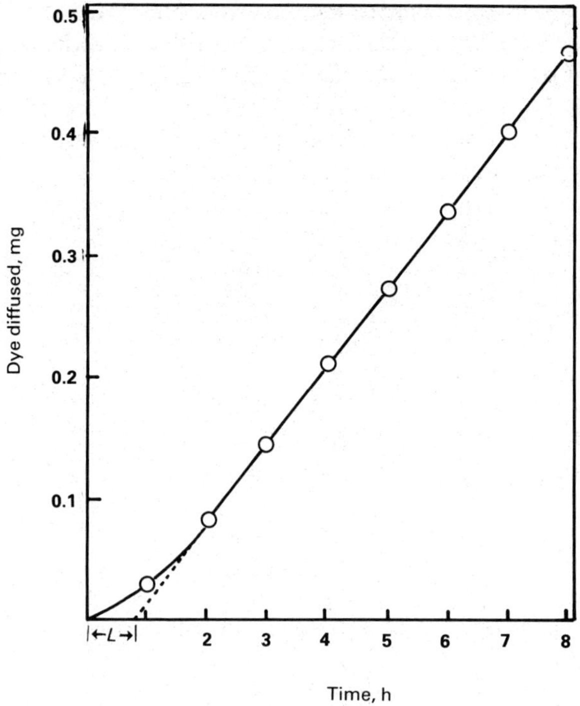

Figure 5.3 – Diffusion through a film under steady-state conditions

determination of the amount of diffusate. Sumner and Taylor [7] have discussed the practical and theoretical difficulties which arise when this method is used to determine diffusion coefficients of water-soluble dyes. By varying the dye concentration on the solution side of the film the dependence of the diffusion coefficient on the concentration of dye within the film may be determined, and this method has been used by Sada *et al.* [8] in studying the two-stage diffusion process of acid dye in nylon film.

5.2.2 Diffusion in the non-steady state

During the initial period before the Q_t against t plot becomes linear, diffusant accumulates in the film under a non-linear concentration gradient. This condition is known as the non-steady state, since the amount of diffusant in the film is continuously increasing. The same situation arises in normal dyeing processes where a filament or fibre is immersed in a dyebath, whether or not the dyebath concentration is maintained constant. If the contribution to mass transfer through the ends of a filament or through the edges of a film is considered negligible, the dye penetrates the empty filament or film along a non-linear concentration gradient from both sides of the film or from the surface of the filament. When the dye molecules reach the centre of the material the overall concentration gradient begins to decrease owing to accumulation from both sides, until finally no measurable gradient exists. The situation thus differs from steady-state diffusion, and the total amount of diffusant taken up in time t can be determined by integrating Fick's second equation (Eqn 5.7) over the period $t = 0$ to $t = t$. For small times, a solution given by Crank [9] for diffusion into a film is

$$\frac{C_t}{C_\infty} = 4\left(\frac{Dt}{l^2}\right)^{\frac{1}{2}}\left(\frac{1}{\pi} + 2\sum_{n=0}^{\infty}(-1)^n \operatorname{ierfc}\frac{nl}{2(Dt)^{\frac{1}{2}}}\right) \tag{5.23}$$

where C_t is the amount of diffusant accumulated after time t and C_∞ is the amount accumulated at equilibrium, i.e. the saturation equilibrium concentration within the film, relevant to the constant concentration of the solution; l again is the film thickness and n is a positive integer. Eqn 5.23 describes a diffusion process controlled by a constant diffusion coefficient, which is a necessary condition for applying Eqn 5.7. The value of D can be deduced from the initial gradient of the graph of C_t/C_∞ as a function of $(t/l^2)^{\frac{1}{2}}$. Within the limits of experimental error this graph is linear up to a value of $C_t/C_\infty = 0.5$, or to larger values when the diffusion coefficient is a function of concentration which increases with increasing concentration.

An approximation to Crank's equation, described by Vickerstaff [10], which can sometimes be used with satisfactory results in the early stages of diffusion, relies on the fact that until the diffusant penetrates to the centre of the substrate the latter can be regarded as an infinitely thick slab of material into which diffusion is occurring from one face. This applies to both film and fibre, and Eqn 5.23 reduces to

$$\frac{Ct}{C_\infty} = 2\left(\frac{Dt}{\pi}\right)^{\frac{1}{2}} \tag{5.24}$$

in the initial stage of uptake. A plot of C_t/C_∞ against $t^{\frac{1}{2}}$ should therefore be linear, with a slope proportional to the square root of the diffusion coefficient. A solution by Crank [11] of Fick's second equation which is more suitable to moderate and large times is given by Eqn 5.25:

$$\frac{C_t}{C_\infty} = 1 - \frac{8}{\pi^2}\sum_{m=0}^{\infty}\frac{1}{(2m+1)^2}\exp\left[-D\frac{(2m+1)\pi^2 t}{l^2}\right] \tag{5.25}$$

where the symbols have the same meaning as in Eqn 5.23, except that m is now the integer. Again, the use of this expression requires the assumption of a diffusion coefficient that does not vary with concentration. The time $t_{\frac{1}{2}}$ when $C_t/C_\infty = 0.5$, is referred to as the *time of half-sorption* or *time of half-dyeing*, and at this instant Eqn 5.25 can be simplified so that

$$\frac{t_{\frac{1}{2}}}{l^2} = \frac{1}{\pi^2 D} \ln\left[\frac{\pi^2}{16} - \frac{1}{9}\left(\frac{\pi^2}{16}\right)^9\right] \tag{5.26}$$

to within an error of ±0.001 per cent. Since in the right-hand side of Eqn 5.26 there is only one constant, π, which is known, the value of D is given by

$$D = 0.049\,19 \times \frac{l^2}{t_{\frac{1}{2}}} \tag{5.27}$$

If, therefore, the time of half-dyeing and the thickness or diameter of the film or fibre respectively are known, it is possible to calculate a diffusion coefficient at a diffusant concentration within the material corresponding to that at the time of half-dyeing.

Alternatively, during the later stages of diffusion, only the first term in the series expressed in Eqn 5.25, i.e. when $m = 0$, need be considered. By taking logarithms of the reciprocal values of both sides and plotting $\ln(C - C_t)$ against t a curve is obtained which approaches a straight line. The gradient of the line is given by

$$\frac{\mathrm{d}[\ln(C_\infty - C_t)]}{\mathrm{d}t} = \frac{-D\pi^2}{l} \tag{5.28}$$

from which the value of D at a diffusant concentration C_t can be obtained. The extent of the linear portion of the graph has been investigated by Frensdorff [12].

Earlier expressions have been given by McBain [13] for diffusion into a film where edgewise diffusion is negligible, as follows:

$$\frac{C_t}{C_\infty} = 1 - \frac{8}{\pi^2}\left\{\exp[-(\pi/l)^2 Dt] + \frac{1}{9}\exp[-(3\pi/l)^2 Dt] + \frac{1}{25}\exp[-(5\pi/l)^2 Dt] + \ldots\right\} \tag{5.29}$$

and by Hill [14] for diffusion into an infinitely long cylinder or filament of radius r,

$$\frac{C_t}{C_\infty} = 1 - 0.692\ \{\exp(-5.785Dt/r^2) + 0.190\ \exp(-30.5Dt/r^2) + 0.0775\ \exp(-74.9Dt/r^2)\ldots\} \quad (5.30)$$

The reader may usefully compare Eqns 5.29 and 5.30 with Eqn 5.25 for different values of *m*. It has been pointed out by Vickerstaff [14] that both these solutions are of the form

$$\frac{C_t}{C_\infty} = 1 - A\exp(-B\beta) - C\exp(-E\beta) - \ldots \quad (5.31)$$

where $\beta = Dt/r^2$ or Dt/l^2 and $A, B, C, E \ldots$ are known constants. If an arbitrary value is assigned to β then the ratio C_t/C_∞ can be calculated. For

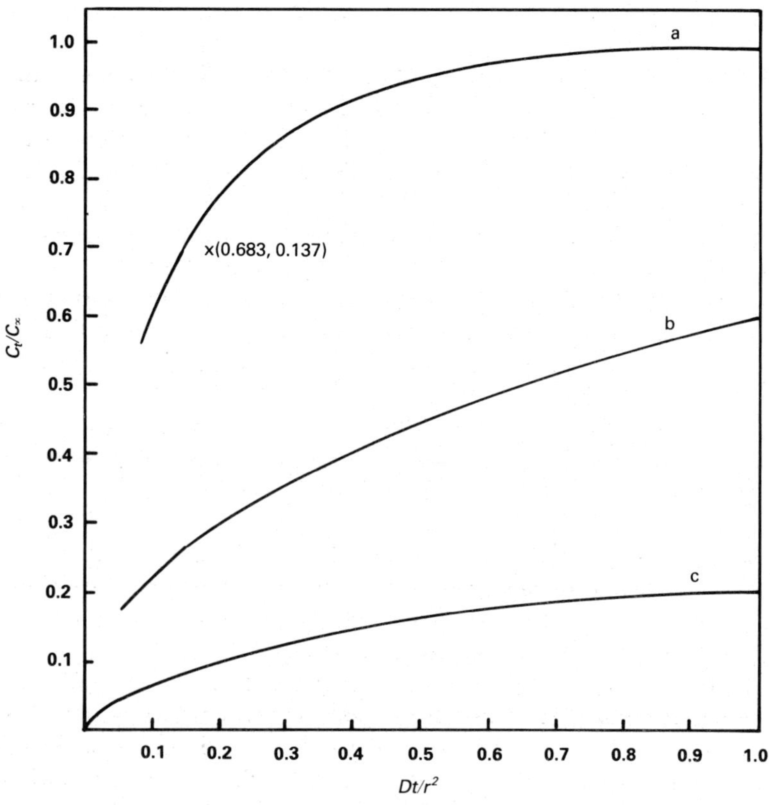

Figure 5.4 – Diffusion into a cylinder from a solution of constant concentration: (a) $DT/r^2 = 0.0$–1.0; *(b)* $DT/r^2 = 0.0$–0.1; *(c)* $DT/r^2 = 0.0$–0.01

each value of Dt/r^2 in the case of a cylinder, or Dt/l^2 for a film there is a specific value of C_t/C_∞. Since t and l or r are known and assuming no change in the latter dimensions, it is simply necessary to know C_t at time t and C_∞, the saturation equilibrium value, to determine the diffusion coefficient.

The variations of C_t/C_∞ for different values of Dt/r^2 in the case of a cylinder are shown in Figure 5.4, plotted from data supplied by Peters [15] based on Eqn 5.25. The values are given on as large a scale as possible so that the readers may use the graph for their own experimental results. Consider the following example: using nylon filament of radius $r = 1.407 \times 10^{-5}$ m, the amount of pure C.I. Disperse Red 11 absorbed by the filament from a dyebath dispersion at 80 °C after 30 min was 0.701 g per 100 g nylon. The dispersion acts as an infinite dyebath. On dyeing to equilibrium for 24 h the amount of dye absorbed was 1.026 g per 100 g nylon. The value of C_t/C_∞ is therefore 0.701/1.026 = 0.683. This value corresponds to the position of x in Figure 5.4, the value of Dt/r^2 being 0.137. At the time of sampling $t = 1800$ s, so $D = 0.137r^2/t = 0.137 \times (1.407 \times 10^{-5})^2/1800 = 1.507 \times 10^{-14}\ m^2 s^{-1}$.

It may be noted that the method utilises the term r^2 in obtaining a diffusion coefficient. Since film thickness or fibre radius is necessarily small in order that a saturation equilibrium value C_∞ may be obtained in a realistic time, the errors in determining l or r will be reflected in considerably less accurate values of D.

In most instances it is necessary to prolong the dyeing time to as much as 24 h or more to determine a value for C_∞, and during this time the structure of the substrate may alter. One way of eliminating this source of error is to take the ratio of dye concentrations in the fibre at two different times t_1 and t_2. Applying these values, C_{t1} and C_{t2} separately to Eqn 5.31 and then dividing it is seen that

$$\frac{C_{t1}}{C_{t2}} = \frac{1 - A \exp(-B\beta_1) - C \exp(-E\beta_1) - \ldots}{1 - A \exp(-B\beta_2) - C \exp(-E\beta_2) - \ldots} \tag{5.32}$$

The values of D_{t1}/r^2 and D_{t2}/r^2 for C_{t1}/C_∞ and C_{t2}/C_∞ respectively can be found by reference to Figure 5.4. For a fixed ratio of the times chosen, a graph of C_{t1}/C_{t2} can be plotted against D_{t1}/r^2 and, knowing C_{t1}/C_{t2}, the average diffusion coefficient can be calculated.

Numerical solutions of the equation for diffusion into a plane sheet immersed in a solution of constant concentration have been given by Newman [16] and Crank [17]. Vickerstaff [18] has presented such data in tabular form. Figure 5.5 shows the variation of C_t/C_∞ with $(Dt/l^2)^{\frac{1}{2}}$ for diffusion into a plane sheet of half-thickness l under these conditions.

The figures on the curves refer to a third dimensionless variable, which is

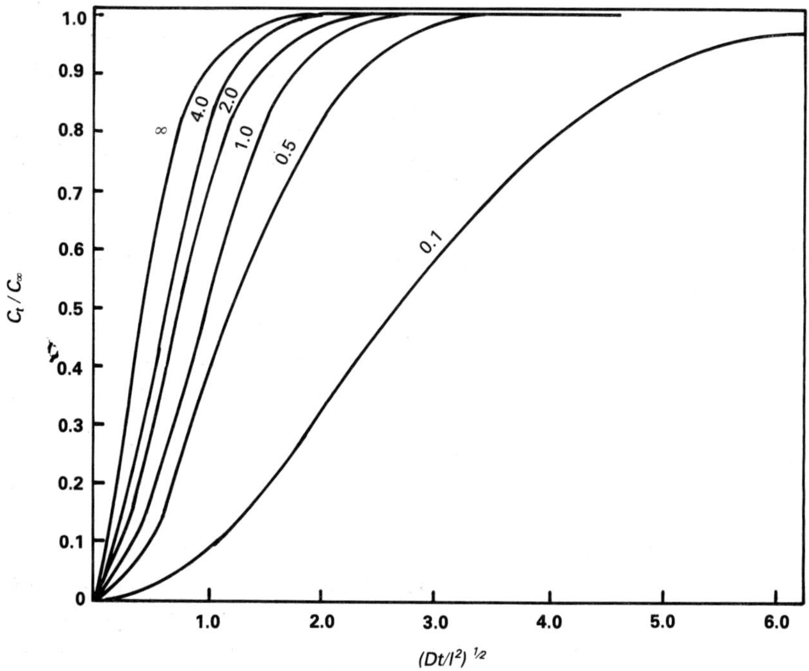

Figure 5.5 – Diffusion into a plane sheet from solution of constant concentration

related to the existence of a diffusional boundary layer extending from the substrate surface into the solution and which will be described later. With efficient stirring this layer is minimised and, assuming it has no effect on the rate of transfer, the curve marked ∞ should be referred to in determining diffusion coefficients. The sigmoidal shape of the rate curves shown in Figure 5.5 during the initial stages of diffusion, particularly for lower values of the variable, is probably due to the limitation of the equation used by Newman at low values of t.

5.2.3 Boundary layers in diffusion

As stated above, the rate of stirring of the solution can affect the rate of diffusion into a substrate by influencing the rate at which the diffusant molecules in the solution are brought to the surface of the solid phase. Even when stirring and convection are at their most efficient, the presence of a more or less static region around the fibre or film is still possible. This region may be idealised as a diffusional boundary layer and has been defined [19] as that region in which the diffusant concentration changes its value from that at the substrate surface to 99 per cent of the concentration in the bulk liquid phase.

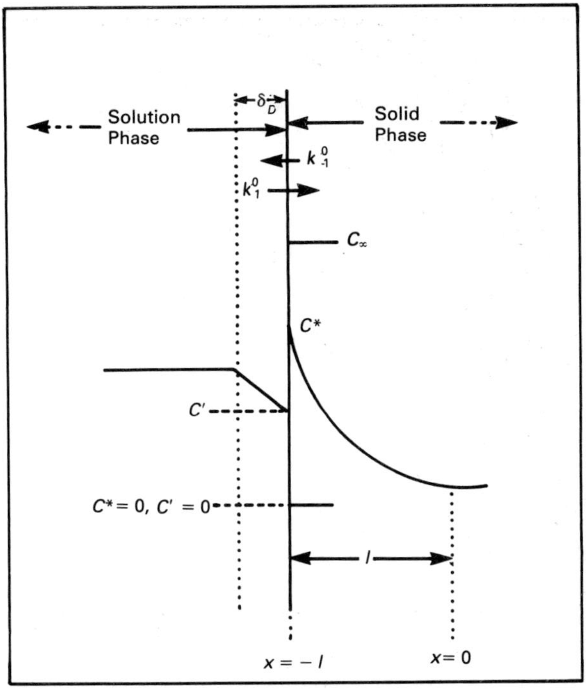

Figure 5.6 – Diffusion into a substrate through a diffusional boundary layer

The formation of this boundary layer may occur under both non-steady- and steady-state conditions, and an analysis by using a simple model system indicates that under certain conditions restriction of the diffusion process may occur within this layer. Figure 5.6 shows diagrammatically the model on which this concept is based for non-steady-state diffusion [20]. Diffusion occurs from a solution at concentration C across a diffusional boundary layer of thickness δ_D to the liquid side of the substrate surface where the concentration is C'. It is assumed that in this layer steady-state conditions apply so that the rate of transfer, governed by Fick's first equation (Eqn 5.1), is given in this case by

$$F = -D_s \frac{(C' - C)}{\delta_D} \tag{5.33}$$

where D_s is the diffusion coefficient of the solute within the layer. The substrate is a plane sheet of thickness $2l$ such that the interfaces x are situated at $x = \pm l$. On the substrate side of the interface, where the concentration is C^*, diffusion occurs along a non-linear gradient and the rate of transfer at

$x = -l$ is given by

$$F = -D_f\left(\frac{\partial C}{\partial x}\right)_{x=-l} \tag{5.34}$$

where D_f is the diffusion coefficient, assumed constant, in the substrate. To solve the equation, two conditions are necessary. Firstly, the sheet initially must contain no diffusant and, secondly, the rate of supply of molecules through the boundary layer δ_D must be equal to or greater than the rate of removal of molecules from the surface to the interior of the substrate. When the two rates are equal, expressions 5.33 and 5.34 may be equated to give

$$-D_s\frac{(C' - C)}{\delta_D} = -D_f\left(\frac{\partial C}{\partial x}\right)_{x=-l} \tag{5.35}$$

This second condition implies that there is no restriction on diffusion at the interface, whereas there may be restriction within the boundary layer.

By assuming that equilibrium is immediately established between concentrations on either side of the interface, a constant ratio or partition coefficient K can be defined which is equal to k_1^0/k_{-1}^0, where the rate constants k_1^0 and k_{-1}^0 are those representing the passage of diffusant into and out of the interface respectively. Therefore

$$KC' = C^*_{x=-l} \tag{5.36}$$

When the whole system has reached equilibrium C^* becomes $C\infty$ and C' becomes C and, from Eqn 5.36,

$$KC = C_\infty \tag{5.37}$$

If C is the *saturation* solubility of the diffusant in the bulk liquid phase, the partition coefficient K is the thermodynamic equilibrium partition coefficient found, for example, with disperse dyes. Eqn 5.37, as given here, is not however limited to saturation concentrations in the liquid phase.

Substituting Eqns 5.36 and 5.37 into Eqn 5.35 gives for the boundary condition at $x = -l$,

$$\frac{D_s}{D_f\delta_D}\left(\frac{C^*_{x=-l}}{K} - \frac{C_\infty}{K}\right) = \left(\frac{\partial C}{\partial x}\right)_{x=-l} \tag{5.38}$$

or

$$\left(\frac{\partial C}{\partial x}\right)_{x=-l} + h(C - C^*_{x=-l}) = 0 \tag{5.39}$$

where
$$h = \frac{D_s}{D_f \delta_D K} \tag{5.40}$$

This boundary condition also applies to the surface $x = +l$, and it can be seen that the model applies equally well to diffusion in a cylinder or filament of radius a during non-steady-state conditions.

The significance of h now becomes apparent when we refer back to the dimensionless variable given on the curves of Figure 5.5, since the value of this variable L is ha or hl. Since

$$L = hl = \frac{D_s l}{D_f \delta_D K} \tag{5.41}$$

the rate of diffusion into the substrate will depend on the value assigned to L, which in turn will depend not only upon the ratio D_s/D_f and δ_D, the boundary layer thickness, but also on K, the partition coefficient of the diffusant. If the two rate constants k_1^0 and k_{-1}^0 are equal and opposite, then the whole diffusion process can be described in terms of simple physical absorption and desorption. *It is important to recognise that this is not a characteristic of dyeing systems, which require a partition coefficient greater than unity or strongly in favour of the diffusant in the polymer phase.* Any study of diffusion coefficient in dyeing systems therefore requires not only a knowledge of the diffusion coefficient, but also quantitative information on the partition coefficient and interaction forces between dye and polymer which can influence partition.

The value of L is also inversely proportional to the thickness of the boundary layer. Consequently L can be increased by decreasing the thickness of the layer, which may be accomplished by increasing the rate of flow of the solution. An upper limit to the rate of diffusion, as $L \to \infty$, can be achieved and above this further increases in rate of flow have no effect, the observed rate of diffusion being controlled solely by the rate of diffusion within the substrate. McGregor and Peters [20] have extended this fundamental treatment to include practical systems in the dyeing of acid dyes on wool and nylon, and disperse dyes on cellulose acetate.

5.2.4 Diffusion in finite baths

So far we have considered only diffusion from a solution in which the concentration remains constant. If there is only a limited amount of solution, however, the solute concentration in the bath will decrease as solute diffuses into the substrate. The sole condition for diffusion evaluation in this case is that the *total* amount of the diffusant in the system remains constant. This situation is more readily comparable with actual dyeing processes. In some cases it is experimentally more convenient to determine the rate of sorption of the solute by the substrate by measuring the changes in concentration in the solution with time, rather than the rate of increase in solute concentration in the substrate. In view of the above condition it becomes necessary in this method to eliminate both loss of diffusant by adsorption on the vessel surfaces and loss by volatilisation of either the solute or the solvent.

A method for determining the diffusion coefficient from finite baths under the restriction of the above conditions has been given in the form of mathematical solutions to Fick's second equation by Carman and Haul [21]. For a plane sheet one solution [22] is of the form

$$\frac{C_t}{C_\infty} = 1 - \sum_{n=1}^{\infty} \frac{2\alpha(1+\alpha)}{1+\alpha+\alpha^2 q_n^2} \exp\left[\frac{4Dq_n^2 t}{l^2}\right] \qquad (5.42)$$

where the q_n values are the non-zero positive roots of

$$\tan q_n = -\alpha q_n \qquad (5.43)$$

and $\alpha = a/l$, the ratio of the volume of the solution to that of the sheet. Where there is a partition coefficient K, such that the concentration of diffusant just within the sheet surface is some constant multiple of the concentration in the solution, then $\alpha = a/Kl$. Under the same conditions the expression for diffusion uptake in an infinite cylinder is given by Eqn 5.44:

$$\frac{C_t}{C_\infty} = 1 - \sum_{n=1}^{\infty} \frac{4\alpha(1+\alpha)}{4+4\alpha+\alpha^2 q_n^2} \exp\left[\frac{q_n^2 Dt}{r^2}\right] \qquad (5.44)$$

where the q_n terms are the non-zero roots of the equation

$$\alpha q_n J_0(q_n) + 2J_1(q_n) = 0 \qquad (5.45)$$

where J_0 and J_1 are Bessel functions of order zero and one respectively. The solutions given in Eqns 5.42 and 5.44 are suitable for numerical computation, and Figure 5.7 shows the family of curves obtained for five values of final percentage uptake when C_t/C_∞ is plotted against $(Dt/r^2)^{\frac{1}{2}}$.

Similarly shaped curves can be obtained for diffusion into a plane sheet and alternative solutions [23] have been given for conditions when Dt/r^2 and Dr/l^2 take small or intermediate values, since the smaller these values are, the more terms in Eqns 5.44 and 5.42 are needed for a given accuracy. More recently Barrett and Cook [24] have produced data by computer iterative methods which relate C_t/C_∞ and Dt/r^2 to a greater degree of precision than

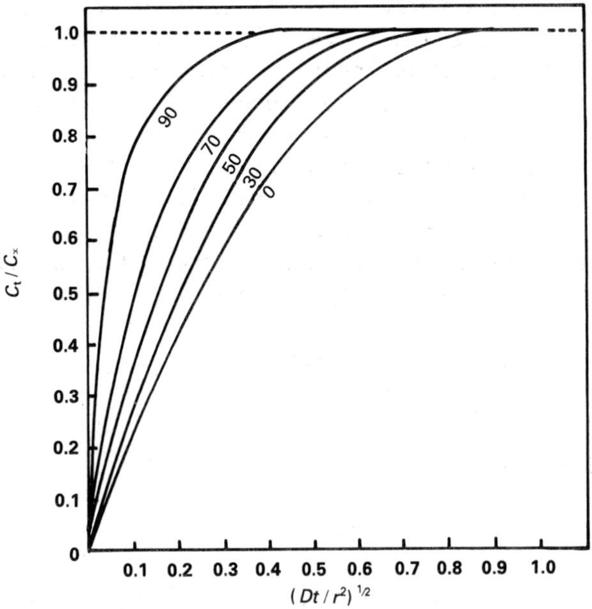

Figure 5.7 – Sorption by an infinite cylinder from a stirred finite bath

those shown in Figure 5.7 for a range of fractional exhaustions from 30 to 99.5 per cent.

5.2.5 Approximation to diffusion equations

Solutions to the fundamental diffusion equations such as Eqns 5.42 and 5.44 involve the tedious calculation of q_n values, and although these equations are often quoted it is common practice to resort to empirical equations which approximate more or less accurately to diffusion processes and which take into account the final degree of exhaustion of the diffusant solution. Nevertheless, Etters and Urbanik [25] have detailed simplified methods for obtaining q_n, and by using iterative techniques with a programmable calculator have determined $(Dt/r^2)^{\frac{1}{2}}$ values applicable to the Wilson equation (Eqn 5.42) and the Carman and Hall equation (Eqn 5.44) for geometrically spaced equilibrium exhaustion values. Their results, given to five decimal places, show that the higher the fractional equilibrium value the more closely they approach that of Wilson in accuracy, with the Carman and Hall equation more accurate than that of Crank (eqn 5.23) for higher values of C_t/C_∞.

To overcome the complexities in using such equations several workers [26, 27, 28, 29] have proposed empirical equations, usually obtained by curve-fitting techniques to rates of uptake, and generally corresponding to the expression

$$\frac{C_t}{C_\infty} = 1 - \exp\left[-k(Dt/r^2)^x\right] \tag{5.46}$$

Eqn 5.46 and similar proposed modifications do not, however, coincide completely with the Wilson and Hall equations over the whole range of fractional uptakes. In Eqn 5.46 values of x and k can be obtained from the slope and intercept of plots of log $[-\ln(1 - C_t/C_\infty)]$ against log (Dt/r^2), the latter being obtained from graphical or tabular data derived from Eqn 5.44 for different fractional uptakes. Constant values of x and k do not necessarily fit over the full range of fractional uptake, and Eqn 5.46 fits the theoretical points best over the first and last 25 per cent of fractional uptake [26].

On the other hand Shibusawa [28] has found good agreement by subdividing C_t/C_∞ into three intervals, corresponding to three equations similar to Eqn 5.46 which agree with the Carman and Hall equation. Additionally, Etters [29] gives equations used to determine the constants in Eqn 5.46 provided the range of fractional uptake, i.e., C_t/C_∞, lies within the region 30.0 to 99.5 per cent, and also for infinite bath conditions where the fractional exhaustion approaches zero or a very small number. One kinetic expression claimed to cover the full range of exhaustion values is the modified Cegarra–Puente [30] equation, i.e.

$$-\ln\left[1 - (C_t/C_\infty)^2\right] = \exp(kt)^a \tag{5.47}$$

which takes into account the exponent a over the time t.

One must perforce begin to speculate whether the agreement of the results of empirical equations with the Carman and Hall and other fundamental diffusion equations to within three to five decimal places over the whole range of exhaustions is not self-defeating, since diffusional models must eventually be compared with real systems and practical diffusion coefficients can be subject to considerable experimental error.

All the above methods contain within them a factor that we can call α, which is related to the fractional uptake and which predicts that the rates of sorption or dyeing from finite baths are strongly dependent on the volume ratio of bath to substrate. The final fractional uptake of the diffusant E is related to this ratio by the expression

$$\alpha = \frac{1 - E}{E} \tag{5.48}$$

The rate of dyeing therefore increases as the degree of exhaustion increases.

If we consider a dye of high affinity applied from a limited solution, then it is possible that the solution is rapidly depleted of dye, but the diffusion flux within the fibre is restricted by the high affinity, resulting in an annular ring of dye just within the fibre surface. If the same dye is applied under conditions where exhaustion is lower, e.g. in low electrolyte concentrations, then the dye concentration at the fibre surface will be relatively low and diffusion is attained to a much greater penetration distance than in the former case. This has been observed by Boulton and Morton [31] for both a direct dye and a vat dye. As exhaustion is decreased the rate of dyeing becomes lower, even though the diffusing properties of the dye within the fibre remain unchanged [32]. Any comparison of diffusants or dyes on this basis, therefore, must be carried out by arranging experimental conditions so that the degree of exhaustion is the same for each dye. Boulton [33] was able to compare the times of half-dyeing for a large number of direct dyes on viscose rayon yarn under conditions which give rise to a 50 per cent exhaustion, namely by the addition of a predetermined amount of salt. The times of half-dyeing are then comparable with the rate of diffusion and hence with the diffusion coefficient. Table 5.1 gives some of the results obtained by Boulton, which

TABLE 5.1

Diffusion coefficients and times of half-dyeing of direct dyes

Dye	Diffusion coeff. (Neale [34]) /(m^2 min^{-1}) × 10^{12}	Time of half-dyeing (Boulton [33])/min
Icyl Orange R	180	0.11
Chlorazol Fast Orange G	156	0.07
Chrysophenine G	90	0.26
Diazo Brilliant Orange G	50	0.22
Benzo Fast Bordeaux 6B	40	1.6
Durazol Red 2B	30	0.8
Chlorazol Fast Helio BK	27	1.2
Chlorazol Black SD	23	1.5
Chlorazol Fast Black BK	20	0.3
Chloramine Rose B	15.2	1.7
Chlorazol Violet R	10.2	0.5
Chlorazol Sky Blue FF	8.0	15.9
Benzopurpurine 10B	5.2	27.7
Chlorazol Fast Blue 2BN	5.1	24.1
Benzopurpurine 4B	4.6	8.9
Chlorantine Green BL	1.73	55
Chlorantine Blue 3GLL	1.25	125
Chlorazol Drab RH	1.24	115
Chlorazol Fast Orange AG	0.71	155

may be compared with the diffusion coefficients of the same dyes, but diffusing into cellulose films, given earlier by Neale [34]. The correlation is not perfect, but is adequate to show the trend that a low time of half-dyeing is related to a high diffusion coefficient. Neale's results show that the values of the diffusion coefficients are spread widely over a 3000-fold range from 1.80×10^{-10} m^2 min^{-1} to 7.1×10^{-13} m^2 min^{-1}.

This relation between the diffusion coefficients and times of half-dyeing has been further amplified by Militky and Rais [35]. At half-dyeing times the diffusion coefficient as represented by $D_{t_{\frac{1}{2}}}/r^2$ can be determined from the expression

$$\frac{D_{t_{\frac{1}{2}}}}{r^2} = \left[1.36 + \frac{2.64}{1 - E_s}\right]^{-2} \tag{5.49}$$

where E_s is the fractional equilibrium exhaustion. Use of Eqn 5.49 in comparison with other methods for determining D at times of half-dyeing gave agreement to within ±3.0 per cent over a full range of practical exhaustion values.

5.3 REAL SYSTEMS

The above procedures for determining diffusion coefficients all assume that the diffusion coefficient is constant and independent of the concentration of dye in the substrate. In actual dyeing processes several factors contribute to a variable diffusion coefficient, in addition to those of concentration-dependence and affinity of the dye for the substrate. In non-steady-state diffusion, concentration varies in a non-linear manner with the penetration distance. It is easy to see therefore that the total amount of dye absorbed at any instant may be represented by a concentration gradient which can be either steep or shallow, indicating respectively the presence of strong or weak absorption forces between dye and substrate molecules. Also, in the initial stages of diffusion and at the diffusion boundary, molecules enter a substrate that is devoid of dye, whereas later molecules have to diffuse through a substrate to which some dye molecules are already attached, thus influencing the diffusion process.

Certain other parameters, which may be variable and interrelated, which also influence the diffusion processes and rates of dyeing at the microscopic level have been assessed [36] and summarised as:

(a) variations in the nature and concentration of ionisable groups both in the dye and in the polymer substrate;
(b) interactions between all species present in both phases;
(c) changes in physical structure of the substrate and the release and

formation of stresses together with changes in amorphous or 'available' regions;

(d) changes in structure and molecular orientation of the solvent in both main phases and in boundary layers.

Since these parameters are interrelated it is extremely difficult to assess their individual effects on the rate of dyeing or on the diffusion coefficient. Because of factors such as these, diffusion coefficients determined by measuring the total amount absorbed at intervals of time up to an equilibrium value – as required, for example, when using Eqns 5.29 and 5.30 – are considered to be average values and are referred to as apparent diffusion coefficients D_{app}. They depend on the concentration range over which they are measured. The most likely average, $\overline{D}$, of these apparent diffusion coefficients is given approximately by the expression

$$\overline{D} \approx \frac{1}{C_0} \int_0^{C_0} D(C)\,dC \qquad (5.50)$$

where $D(C)$ is the relation between the diffusion coefficient and the dye concentration in the substrate over the concentration range 0 to C_0. The integral between the concentration limits can be obtained graphically from the area under the curve. Alternatively, $\overline{D}$ may be defined over the whole concentration range 0 to C_∞ or over any concentration range interval C_1 to C_2. The assumption that $\overline{D}$ and D_{app} are the same has been tested by Crank and Henry [37], who chose several possible relations for $D(C)$, including those in which the diffusion coefficient increases or decreases linearly with concentration and where the relation exhibits a maximum in the diffusion coefficient. For each function the rate-of-dyeing curve was calculated from the solution of Fick's second equation (Eqn 5.9) for a diffusion coefficient which is a function of concentration, viz.

$$\frac{\partial C}{\partial t} = \frac{\partial}{\partial x}\left(D(C) \frac{\partial C}{\partial x} \right) \qquad (5.51)$$

A range of rate-of-dyeing curves is thus obtained, and from the time of half-dyeing obtained from each of these curves a value of the apparent diffusion coefficient can be calculated. A comparison of each of these values of D_{app} was then made with the corresponding most likely average $\overline{D}$ obtained from Eqn 5.50 above. $\overline{D}$ values were found to be approximately equal to D_{app}, but only to within 20 per cent for absorption processes in some cases, showing that the original assumption may not be valid. A practical example of this procedure, showing in detail the method by which $\overline{D}$ may be calculated for a family of rate curves at different saturation equilibrium values, has been given by Crank and Park [38].

The above-mentioned rate-of-dyeing curves calculated from concentration-dependent diffusion coefficients do not vary significantly in shape and are similar to those calculated for a constant diffusion coefficient. This insensitivity to a varying diffusion coefficient is illustrated in Figure 5.8, where

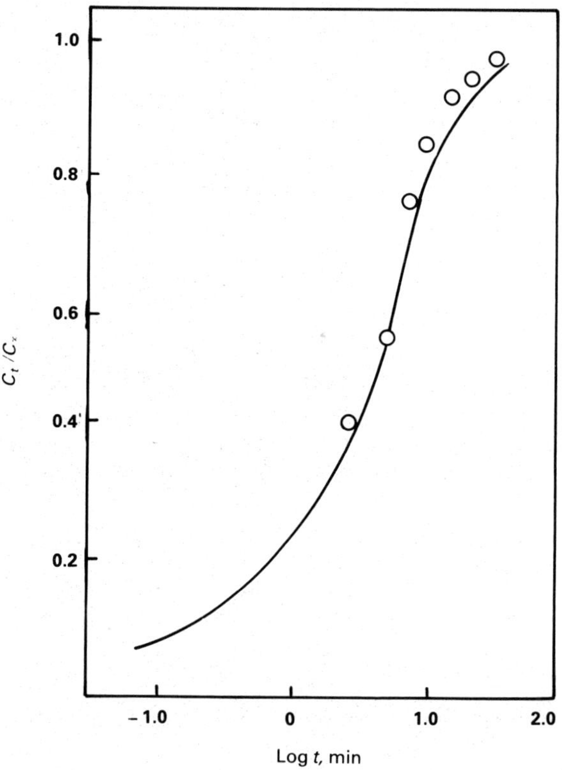

Figure 5.8 – Insensitivity of rate of dyeing to variations in diffusion coefficients

experimental points obtained by Garvie and Neale [39] for diffusion of Chlorazol Sky Blue FF in viscose sheet are compared with the continuous line which represents the best-fitting curve, assuming a constant diffusion coefficient. If the diffusion coefficient is assumed to vary linearly, e.g. where the variation can be expressed by the equation

$$D(C) = D_0(1 + \beta C) \quad (5.52)$$

where D_0 is the initial diffusion coefficient and β is a constant, the calculated values are virtually coincident with the line calculated for a constant diffusion coefficient. The reason for this insensitivity is not far to seek and has already been mentioned. Although the distribution of dye throughout the substrate at any instant is affected by the relation between $D(C)$ and C, the total amounts of dye taken up by the substrate in different cases, represented by the areas under the distribution profiles, may be the same, even though the shapes of the profiles themselves may be considerably different. The integral curve of the total fractional uptake C_t/C_∞ when plotted against D_t/r^2 (Figure 5.4), used to determine the diffusion coefficient, will therefore be insensitive to the manner in which the diffusion coefficient varies with concentration. It follows that further progress from a practical standpoint can be made only if the diffusion process is 'frozen' at intervals and the actual dye distribution observed experimentally. This has been achieved by the use of multiple membranes and cross-sectioning of fibres and films.

5.3.1 Concentration-distribution curves

The multiple-membrane experiment of Neale [39] was the first attempt to apply the technique to dyeing studies. Chlorazol Sky Blue FF was allowed to diffuse into a composite membrane consisting of sheets of Cellophane clamped together, which was immersed in a well-stirred dye solution. After a predetermined time the composite was removed and each cellulose layer analysed for dye content. The resulting data enabled a concentration–distance relation shown in Figure 5.9 to be constructed. The process was repeated for specific periods of time t, to give a set of curves as shown in Figure 5.9.

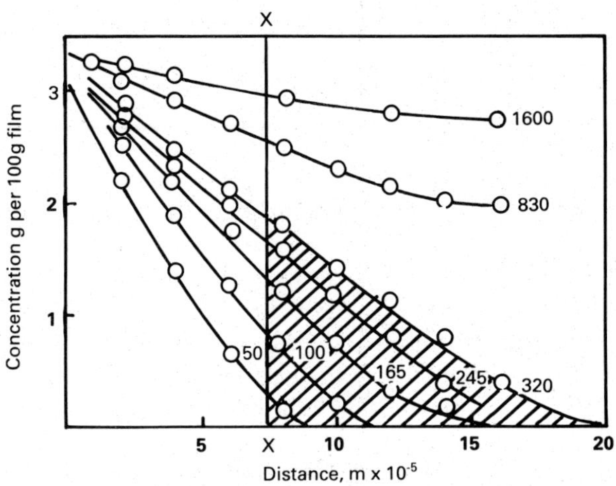

Figure 5.9 – Distribution curves for Chlorazol Sky Blue FF through a composite cellulose membrane [30] (numbers on curves are dyeing times in min)

At any point X within the composite it is possible to measure the slope $\partial C/\partial x$ and by measuring the areas to the right of X, under the curves obtained at different times (e.g. the area shown shaded for a time of 320 min in Figure 5.9) the total dye flux F having passed the plane of X at these times, can be determined. Referring to Fick's first law, i.e.

$$F = -DA\frac{\partial C}{\partial x} \tag{5.53}$$

and knowing the area of the film, we thus have all the factors required for calculating D. By selecting X for different layers within the composite film the variation of D with C can be ascertained.

An alternative experimental approach [40] is to sandwich a film of material between two metal plates. Edgewise diffusion of dye into the film occurs on immersion in a dyebath. After removal, the concentration of dye along the length of the film is determined. As noted by the authors, the rate of penetration of dye is dependent on its direction relative to the extrusion axis of the film, so this method in conjunction with a multiple-membrane method would be useful in elucidating the structural anisotropy of films in those cases where diffusion is relatively rapid. The edge-diffusion method has since been used to study the diffusion of Chrysophenine G in regenerated-cellulose sheet [41], disperse dyes in nylon [42, 43] and basic dyes in acrylic fibres [44]. A modification of the multiple-membrane method, consisting of the dyeing of a roll of polyester film wound tightly around a 1-cm diameter glass former, has been employed by Sekido [45] and more recently in other studies (e.g. [60]).

Both edgewise diffusion and the dyeing of multiple membranes require longer dyeing times than those which apply in practice, and any abnormalities occurring in the early stages of dyeing are difficult to detect. Moreover, the interfaces between individual films in the composites may influence the rate of diffusion, and any penetration of dye solution between the film and the metal plate in edgewise diffusion could affect the results. A general method which overcomes some of these difficulties has been developed by Peters and Petropoulos [46] and continued by McGregor [47]. A film of material, supported in a frame, is dyed over a period with rapid agitation of the solution. After dyeing, surplus dye solution is removed from the film, which is then freeze-dried to immobilise the dye, and a cross-section is prepared and mounted on a microscope slide. The changes in absorbance across the section are determined by scanning with a microdensitometer. As the instrument slit crosses the edge of the section the absorbance rises steeply. It is assumed that the actual position of the edge is located at the point where the absorbance reaches half its maximum value. Diffusion profiles [46], as

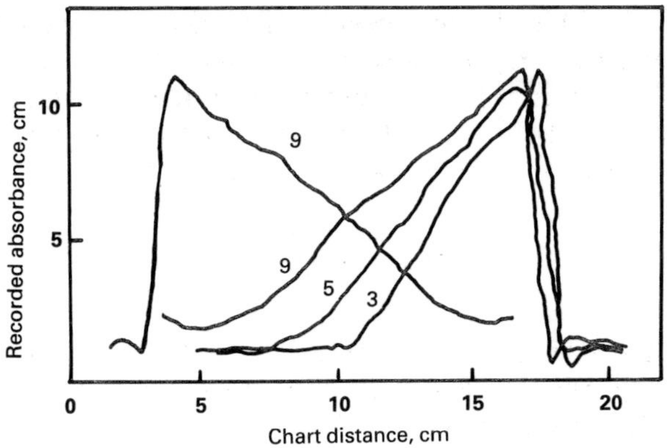

Figure 5.10 – Recorded diffusion profiles for Chlorazol Sky Blue FF in Cellophane (numbers on curves are dyeing times in min)

recorded by the microdensitometer for Chlorazol Sky Blue FF in Cellophane film (thickness 8 μm), are shown in Figure 5.10.

It is necessary to correct these curves for deviations from the Lambert–Beer law arising from thickness variation, and to localise the position of the edge of the section. The corrected or normalised curves obtained in this manner for the same dyeing times of Chlorazol Sky Blue FF are shown in Figure 5.11. Extrapolation of the curves to the edge of the section shows that a constant surface concentration applies over these dyeing times and that agitation of the dye solution is adequate. For a constant surface concentration and assuming no overall change in volume of the substrate on

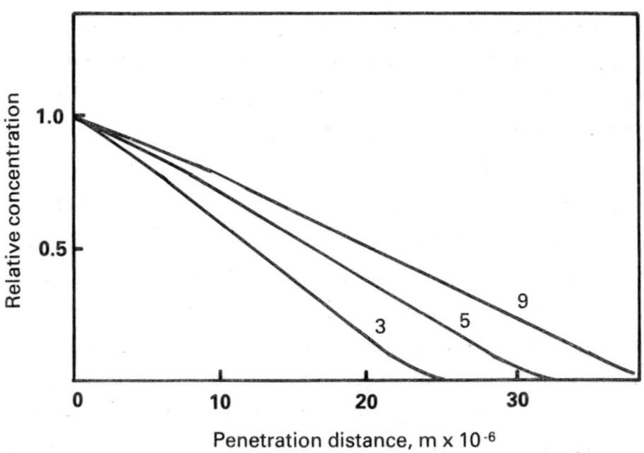

Figure 5.11 – Corrected diffusion profiles for Chlorazol Sky Blue FF (substrate and dyeing times as in Figure 5.10)

diffusion, the data from corrected diffusion profiles are sufficient to determine the diffusion coefficient and its variation with concentration as described by Garvie and Neale [39]. A more accurate method of analysis is that due to Matano [48] and described by Crank [49] for unidimensional diffusion according to Eqn 5.54:

$$\frac{\partial C}{\partial t} = \frac{\partial}{\partial x}\left(D\frac{\partial C}{\partial x}\right) \tag{5.54}$$

Provided that D is a function of C only, the dye concentration may be expressed by a single variable $x/2t^{\frac{1}{2}}$, and Eqn 5.54 may be reduced to an ordinary differential equation by its use.

When this variable ($\eta = x/2t^{\frac{1}{2}}$) is applied to Eqn 5.54, then, after integrating between the limits $C = 0$ and $C = C_1$ and reintroducing x and t, Eqn 5.55 is obtained:

$$D(C) = -\frac{1}{2t}\frac{\partial x}{\partial C}\int_0^{C_1} x\, dC \tag{5.55}$$

$D(C)$ is calculated from the graph of C against x at given time t. The variation

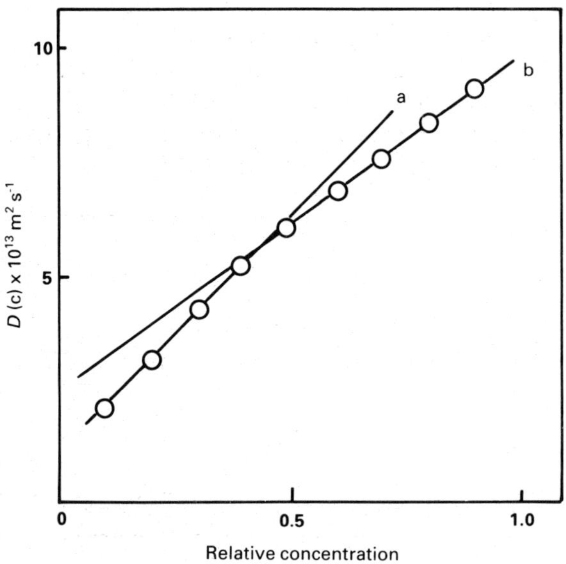

Figure 5.12 – Relation between diffusion coefficient and relative concentration for Chlorazol Sky Blue FF (the abscissa represents the concentration of dye relative to its equilibrium saturation value)

of the diffusion coefficient with relative concentration c of Chlorazol Sky Blue FF is shown in Figure 5.12. The experimental results [50] obtained from dyeing at 90 °C in the presence of 5 g l^{-1} salt may be expressed by two linear equations of the form $D(c) = D_0(1 + \beta c)$, where β is the slope of the lines a or b in Figure 5.12. When relative concentrations are replaced by actual concentrations, the intercept D_0 is found to agree approximately with the value of 0.80×10^{-13} m^2 s^{-1}, given by a reassessment [51] of Neale's result [39]. The discrepancy can be accounted for by possible differences in the physical nature of the cellulose substrates used.

Information about the relation between diffusion coefficients and concentration, derived from diffusion profiles obtained from dyeings carried out under the different conditions prevailing in dyeing processes, is useful in elucidating mechanisms of dyeing. The inclusion of thermodynamic sorption data is essential, however, since dyeing is not only a process of mass transfer but also one in which interaction in different degree between dye and substrate occurs. Peters [52] has classified dyeing behaviour into four main groups according to whether or not the dye is ionised in solution and whether the substrate does or does not contain accessible sites with which the dye moiety may interact. Within each of these groups the variation of the diffusion $D(C)$ with concentration appears to follow a distinct pattern. These variations and their causes are discussed in more detail below.

5.3.2 Disperse dyes on hydrophobic fibres

The dyeing of non-ionised disperse dyes on hydrophobic substrates such as secondary cellulose acetate and triacetate, polyesters, nylon and Courtelle, which do not contain specific sites for dye–polymer interaction, is typified by the dyeing of Duranol Red 2B (1-amino-4-hydroxyanthraquinone) on

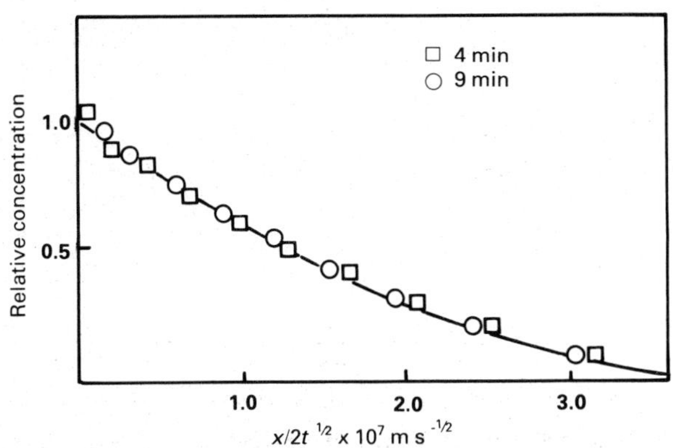

Figure 5.13 – Corrected profile for nylon 6.6 dyed with an aqueous dispersion of Duranol Red 2B at 55 °C

nylon 6.6 film. The normalised diffusion profiles for dyeing times of 4 min and 9 min are shown as experimental points in Figure 5.13, from which it is seen that a constant surface concentration is achieved. The coincidence of the experimental results for the two dyeing times shows that the diffusion coefficient is not time-dependent and that changes in polymer structure do not occur during the dyeing process. The continuous line is that calculated from Fick's law for a constant diffusion coefficient. The agreement between the experimental points and the calculated curve shows that diffusion of this type is governed by a concentration-independent coefficient and that the system exhibits ideal behaviour. The ideal behaviour is attributed to a low degree of interaction between dye and substrate, which allows free movement of the dye molecules through the polymer matrix (in contrast with ionised dyes on this substrate – see below).

Influence of carriers in dyeing with disperse dyes (see also Chapter 3)
Owing to the compact nature of the arrangement of polymer chains within the fibre, the rate of dyeing of polyesters at temperatures up to 100 °C is markedly low. It follows that the values of apparent diffusion coefficients will also be low. The value for the apparent diffusion coefficient of 1-amino-4-hydroxyanthraquinone applied to Terylene (ICI) staple fibre [53] at 90 °C is $6.95 \times 10^{-17}\ m^2\ s^{-1}$. Its value [54] in nylon 6.6 film at temperatures much less than this (55 °C) is approximately 1000 times greater. To improve the rate of dyeing of polyester fibre to a realistic value, it is necessary to employ temperatures above 100 °C, or alternatively to add organic compounds, referred to as 'carriers', to the dyebath. The term 'carrier' is a misnomer, still extensively used, originating from the idea [55] that the compound and the dye formed a complex which could 'carry' the dye into the fibre, thus causing more rapid dyeing. It is now known that the carrier is adsorbed by the fibre and operates by modifying the structure of the amorphous regions [56]. Its effectiveness is dependent on affinity [57] and on its plasticising action on the fibre, which allows the chain segments in the polymer to have a greater freedom of movement, thus facilitating dye penetration. In some cases [58] the structural changes caused by carrier action may be retained if the carrier is subsequently removed. Dyeing after such pretreatment can then proceed more rapidly. Recorded distribution–distance profiles [54], shown in Figure 5.14, illustrate this effect.

Curve a shows the dye distribution after dyeing cellulose triacetate film for 3 min with an aqueous dispersion of Duranol Red 2B. If the film to be dyed is first pretreated with a carrier and then dyed under the same conditions, curve b is obtained. When the carrier is completely extracted before dyeing, the distribution curve (c) shows that diffusion is more rapid than is diffusion into untreated film. These recorded profiles, if corrected as described

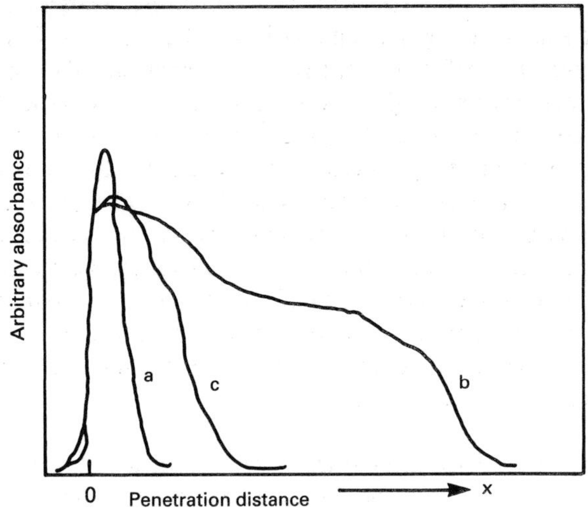

Figure 5.14 – Recorded diffusion profiles (Duranol Red 2B on cellulose triacetate films)

previously, show that diffusion is governed by a constant diffusion coefficient. When dye and carrier are applied simultaneously the diffusion profile becomes sigmoidal in shape, suggesting that diffusion of dye is controlled by the rate at which the carrier 'opens up' the structure to allow dye penetration. The extent to which the dye diffusion rate is increased depends on the concentration of carrier and not on its molecular structure [59]. Similar effects have been shown [60] in vapour-phase dyeing for the simultaneous diffusion of a model dye, 1-aminoanthraquinone, and salicylic acid. This model dye shows a concentration-independent diffusion coefficient in the absence of salicylic acid, and the increase of diffusion coefficient in co-diffusion is attributed to increases in polyester chain separation. On the other hand, in aqueous dyeing with these reagents, the same author earlier attributes the magnitude and type of change in D_{app} to competition for dye adsorption 'sites' in the polymer, although the latter remain undefined [61].

In anhydrous heat-fixation methods (e.g. Thermosol) Eqn 5.24 has been applied [62] to some 34 disperse dyes on polyester to determine their apparent diffusion coefficients. These range from 5×10^{-12} to $6 \times 10^{-13}\ m^2\ s^{-1}$ at 200 °C under conditions in which a constant surface concentration of dye can be maintained.

$$C_t/C_\infty = 2(Dt)^{\frac{1}{2}} \qquad (5.24)$$

Application of Eqn 5.24 to integral rate-of-dyeing curves is justified only when constant surface conditions can be maintained. Because of possible experimental error, however, a constant surface concentration may not always be assumed in the diffusion profiles of disperse dyes determined by the Petropoulos and Peters [46] method. In fact, the more recent studies of Cometto [63] and Patterson [64] show evidence of a time-dependent surface concentration. The method employed by the latter authors differs from that already described in that a dyed filament of circular cross-section is scanned across its diameter, using a microspectrophotometer whose slit-width is narrow in comparison with the diameter of the filament. The relative concentration distribution across the diameter is determined by means of a computer program and cross-sectioning is unnecessary. Figure 5.15 shows a

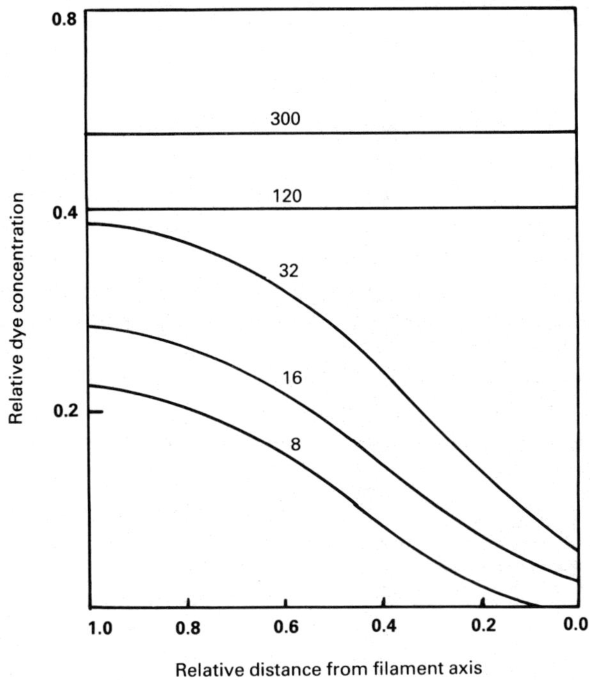

Figure 5.15 – Concentration–distance profiles on nylon 6.6 filament of C.I. Disperse Red 1 dyed at 85 °C (numbers on curves are dyeing times in min)

typical result for the diffusion profile C.I. Disperse Red 1 on a nylon 6.6 filament, where the sigmoidal shape of the curves in the initial stages of dyeing can be seen. The linear profiles at higher relative dye concentration are comparable with the hypothetical case [65] of dyeing in a discontinuous system which is controlled solely by the dye concentration, the rate of accumulation of dye at high relative concentration being given by the equation

$$\frac{dC}{dt} = B(1 - C) \tag{5.56}$$

where B is a constant. For a real dyeing system in which the dye concentration is independent of position in the fibre phase, i.e. for the dyeing times of 120 and 300 min in Figure 5.15, the diffusion process must be controlled not by the rate of penetration of dye within the fibre, but by some external factor such as a surface barrier or the rate of supply of dye molecules to the fibre surface.

More recently doubt has been cast on the idea that dye diffusion into polyester can be interpreted solely by the free volume or chain segment separation model; account must also be taken of the pore model of diffusion in which dye can migrate into fibres along capillary channels of molecular dimensions, particularly when carriers are present. Hori *et al.* [66] postulate this more generalised diffusion model, which they represent by Eqn 5.57:

$$\ln D_m = \frac{\ln \phi_0 \delta^2}{6} + \frac{\ln \Theta}{\tau} - \ln K - \frac{\Delta H_H}{T} - \frac{AB}{B + (T - T_g)} \tag{5.57}$$

The first three terms relate to the pore theory and, together with the enthalpy of dyeing ΔH_H, can be determined from experimental thermodynamic data; K is the partition coefficient, Θ the fractional volume and τ the tortuosity factor. The last two terms relate to the free volume model (as described in Chapter 3), the last term being obtained from swelling measurements on the fibre in presence of carrier and the lowering of the glass transition temperature T_g. In the diffusion of 1,4-diaminoanthraquinone into polyester from solvents which swell the fibre to different extents, good agreement is found between the diffusion coefficient found experimentally and the value of D_m calculated from Eqn 5.57. The predominance of the free volume mechanism is shown by the fact that the major contribution to D_m comes from the last term in this equation.

5.3.3 Dyeing with ionised dyes on substrates without sites

Under normal dyeing conditions with direct dyes cellulosic fibres remain largely un-ionised, with no formally charged sites at which dye interaction can occur, although there may possibly be a small number of ionised carboxylic acid groups. The substrate is water-swollen and contains a system of pores and channels, the dye being able to enter via the dye solution contained in this system. The dyes generally contain two or more ionised sulphonic acid groups and, when absorbed on the pore surfaces, they are still accompanied by attendant (usually Na^+) gegenions to preserve electrical neutrality. When sodium chloride is added to the dye solution, the diffusion of sodium cations is facilitated by mass action and the amount of dye anion adsorbed by the fibre is increased. The diffusion coefficient of the dye is therefore a function of both dye and electrolyte concentration. We have already seen (Figure 5.12) that the *D*(*c*) value of Chlorazol Sky Blue FF is a linear function of concentration within the cellulose in the presence of a fixed amount of added electrolyte. When different amounts of sodium chloride are added this linearity is maintained [47], particularly at lower dye concentrations within the cellulose, as can be seen from Figure 5.16, although the slope decreases with increasing salt concentration. It is apparent that *D*(*c*), during the initial stages of dyeing where $c \rightarrow 0$, is depressed by an increasing salt concentration. Also, at higher concentrations of dye and at high electrolyte concentration the slopes of the graph decrease, reflecting possible aggregation of the linear dye molecules. Figure 5.17, which is derived from the data presented in Figure 5.16, illustrates the effect of salt concentration on *D*(*c*) at different stages in the diffusion process. It is considered that the initial rapid rise is due to the response of *D*(*c*) to the lowering of the electrical potential barrier to diffusion as the concentration of electrolyte increases. A similar behaviour is observed when acid dyes are applied to cellulose as the

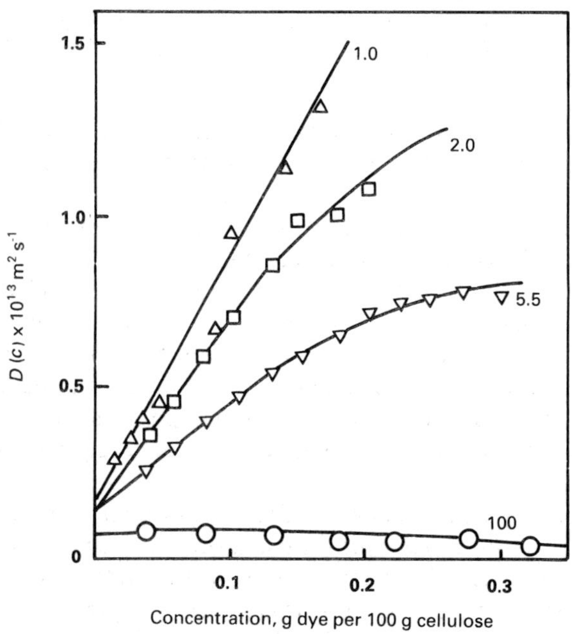

Figure 5.16 – Relation between D(c) and C for Chlorazol Sky Blue FF at 90 °C at different salt concentrations (g l^{-1})

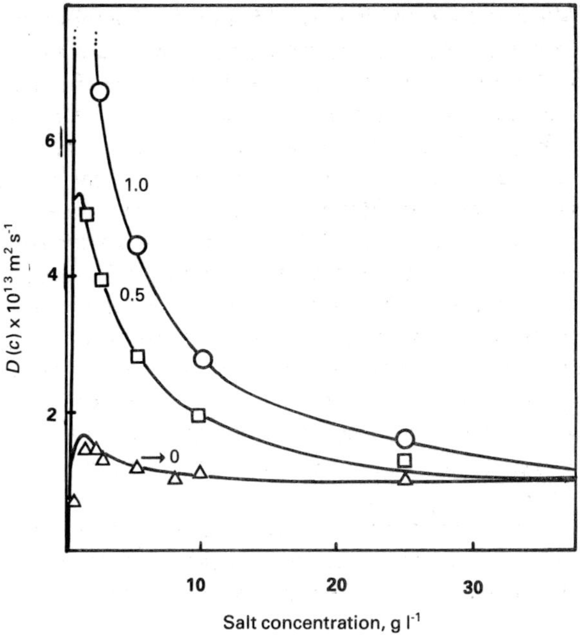

Figure 5.17 – Variation of D(c) with salt concentration at fixed values of adsorbed Chlorazol Sky Blue FF (numbers on curves are amounts (g) of adsorbed dye per 100 g cellulose)

dye or electrolyte concentration is increased, and Warwicker [41] has obtained similar results for the dye Chrysophenine G from edgewise diffusion methods.

A cautionary note may be inserted at this point. The above conclusions are derived from diffusion-profile data. The average diffusion coefficient at any specific salt concentration may be obtained from the results shown in Figure 5.17. This average diffusion coefficient $\overline{D}$ varies with salt concentration in a manner similar to that observed by Neale and Stringfellow [67], in that it gradually increases, reaches a maximum, and then falls as the amount of salt added is increased. The maximum value occurs at a much higher salt concentration than that pertaining to $D(c)$ from profile studies and illustrates the limitation of any conclusions derived from studying the variation of $\overline{D}$ with C.

The generally accepted model for diffusion of direct dyes into cellulosic fibres is based on a heterogeneous system in which dye anions diffuse through the liquid in the pores of the substrate and are simultaneously adsorbed on pore walls [68]. When the pore diameter is much larger than the size of the dye anion, the latter may be considered to be diffusing in the internal liquid, under the same conditions as in the bulk liquid phase and as described by

Fick's first equation. Contribution to the total flux by diffusion of adsorbed dye along the pore surface may take place, but this is likely to be much slower than diffusion through the internal liquid phase. In rate-of-dyeing measurements the amount of dye C absorbed at any instant is the total of the dye present in the pore liquid at a concentration C_p and the amount adsorbed on the pore walls C_A. The volume fraction α occupied by the pores of the substrate must be taken into account, so that

$$C = \alpha C_p + C_A \tag{5.58}$$

In the absence of an electrical diffusion potential and electrolyte, the rate of transfer may be written

$$F = \alpha D_0 \frac{\partial C_p}{\partial x} + D_A \frac{\partial C_A}{\partial x} \tag{5.59}$$

where D_0 is the diffusion coefficient of the dye in the internal liquid phase and D_A is that of the adsorbed dye. Eqns 5.58 and 5.59 give for the concentration-dependent diffusion coefficient the expression

$$D(C) = \alpha(D_0 - D_A)\frac{dC_p}{dC_A} + D_A \tag{5.60}$$

Where surface migration along the pore plays little or no part, $D_A \to 0$ and consequently

$$D(C) = \alpha D_0 \frac{dC_p}{dC_A} \tag{5.61}$$

Since $C_A \gg C_p$, the difference between C_A and C_p is negligible, so Eqn 5.61 can be written in terms of C instead of C_A and the slope dC_p/dC_A is taken to be the slope of the internal adsorption isotherm dC_0/dC_A, where C_0 is the dye concentration in the bulk liquid phase.

Eqn 5.61 has elements similar to that developed more recently by Weisz, Hicks and Zollinger [69], who used a rigid inorganic substrate of constant pore size to obtain

$$D_{app} = \gamma n D \left(\frac{P}{b}\right)\left(\frac{C_0}{C_f}\right) \tag{5.62}$$

Here P is the fraction of solid accessible to diffusion and b is a tortuosity factor having a value of less than $3^{\frac{1}{2}}$. C_0/C_f is the inverse ratio of the total dye concentration per unit volume of fibre to its concentration in the external phase and with which it is in equilibrium, and n is the partition coefficient between the mobile internal dye in the capillary liquid and the dye in the dyebath. γ is a constant for any particular dyeing system and depends on the form of the sorption isotherm; for 'weak' sorption its value approaches unity and for 'strong' sorption it is approximately 1.6. Eqn 5.62 applies equally to diffusion accompanied by immobilisation of dye ions on specific sites and to systems in which the concentration of mobile dye in the pores is not equal to that in the dyebath. By taking these factors into account, Weisz has shown that the values of D considered to be the 'true' diffusion coefficients are very much larger than the apparent diffusion coefficients determined by Neale (see Table 5.1) and have a much narrower range. Some quoted values taken from the extremes of this range are given in Table 5.2.

TABLE 5.2

Apparent and 'true' diffusivities of some direct dyes

Dye	Apparent diffusion coeff. [70] /(m^2 min^{-1}) × 10^{12}	'True' diffusion coeff. [69] /(m^2 min^{-1}) × 10^{8}
Icyl Orange R	180	2.52
Chlorazol Violet R	10.2	0.66
Chlorazol Sky Blue FF	8.0	0.66
Chlorazol Fast Orange AG	0.71	0.036

A comparison of all these recalculated diffusion coefficients shows that they approach in magnitude the value of the diffusion coefficient of the respective dye in water, some being smaller but none larger. When the ratio of pore to dye molecular dimensions approaches unity there are considerable deviations from the aqueous diffusivity values. The model is still of only limited value, however, since it takes no account of the electrical potential barriers, particularly in presence of electrolyte, which may alter or reduce the effective pore size. Under these circumstances the speed of diffusion is, in large measure, determined by the diffusion of the inorganic ions that accompany the dye.

5.3.4 Dyeing with ionised dyes on substrates containing charged sites

The dyeing of polyamide, e.g. nylon 6 or nylon 6.6, and wool fibres with dyes that contain ionised sulphonic acid groups is an example of dyeing on ionised sites. These fibres contain carboxyl and amino groups which, under acid conditions, are protonated according to

$$H_3N^+—F—COO^- \overset{H^+}{\rightleftharpoons} \underset{(I)}{H_3N^+—F—COOH} \overset{X^-}{\rightleftharpoons} X^-H_3N^+—F—COOH$$

$$\updownarrow D^-$$

$$\underset{(II)}{D^-H_3N^+—F—COOH} + X^- \qquad (5.63)$$

where —F— represents the polymer chain and D^- the ionised dye. The residual positive charge on I is neutralised by the anion X^-, usually acetate or formate, which is always present as the gegenion. Consequently, assuming that electrical neutrality is preserved, the absorption of dye anion to give II is accompanied by the replacement of the relatively small anion X^- originating from the acid. In this last step the reaction from left to right is strongly favoured, since the ionised dye sulphonic acid is a strong electrolyte and its attachment, on account of its molecular size and shape, occurs also through hydrogen-bonding groups and van der Waals forces operating between dye and polymer chains [71]. From the stoichiometric view of the above scheme the amount of dye adsorbed will normally tend to a maximum which is determined by the number of amino groups in the fibre, provided that these are all accessible to dye anions [72], and by the basicity or number of negatively ionised groups in the dye. However, dye in excess of the amino group content can be absorbed at near neutral pH values, e.g. in 'overdyeing' nylon. Dye can also be absorbed in excess of the number of amino groups at lower pH values. By virtue of their strong affinity, dyes take with them a certain number of protons. Alternatively, at lower pH values, overdyeing may be due to protonation of amide groups, i.e. protons enter first and dye anions follow. The adsorption isotherm under these conditions of low pH is linear and the diffusion coefficient is not so dependent on concentration as is the case at lower concentrations of acid.

The adsorption of positively charged basic dye ions by acrylic fibres is also an example of dyeing with ionised dye on ionised sites. In the dyeing of fibres such as Acrilan or Orlon the charged sites are ionised sulphato and sulphonic acid [73], and possibly carboxylic acid [74], groups. The interaction of dye cation and negative site can be formally represented as

$$—F—SO_3H + D^+ \rightleftharpoons —F—SO_3^-D^+ + H^+ \qquad (5.64)$$

Over a large concentration range the equilibrium dye saturation corresponds to the total number of available ionised groups within the fibre. As in the dyeing of nylon fibres with acid dyes, it is possible to assume an exchange mechanism operating, although distribution of dye cations between the fibre and solution phases cannot be discounted.

For both nylon and acrylic fibre the diffusion mechanism is usually interpreted in terms of a substrate consisting of a system of pores and channels filled with dye solution through which the dye ions can diffuse with simultaneous adsorption on, and migration along, the pore surface. Distribution concentration–distance profiles for two acid dyes of different basicities, viz. trisulphonated C.I. Acid Red 18 and monosulphated C.I. Acid Red 8, in nylon 6.6 film have been determined by Hopper *et al.* [75]. The normalised curves for the two dyes, shown in Figure 5.18, fall into two main types (A and B) depending on the acid conditions during application.

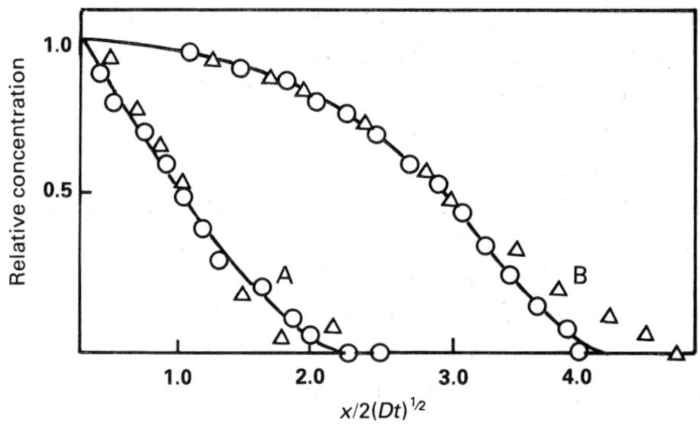

Figure 5.18 – Distribution of C.I. Acid Red 18 (○) and C.I. Acid Red 8 (△) in nylon 6.6 film

The normalised curves A were obtained from the results of examination of cross-sections of films dyed at pH 6.5 and 8.4 for the trisulphonated and monosulphonated dyes respectively. Under these conditions the instantaneous surface dye concentration, which is equivalent to C_∞, the dye saturation value, is such that it is only 50 per cent of that required to saturate the total amino end-groups and no overdyeing is possible. At pH 3.2 and 4.8 for the two dyes respectively, curves of type B are obtained. Conditions show that complete surface saturation is achieved and the sigmoidal shape of this curve indicates a marked dependence of the diffusion coefficient on concentration. Changes in surface concentration controlled by changes in pH thus modify the profile shape. Similarly, at a constant pH, changes in electrolyte (salt) concentration cause shape changes from type B to type A and these changes are reflected in changes in the value of $D(c)$, determined according to Matano's method (Eqn 5.55), when plotted against relative dye concentration within the film. The reduced dependence of $D(c)$ on dye concentration on adding sodium chloride is seen in Figure 5.19.

To account for this lower dependence on concentration and the change in shape of the distribution–distance profiles, Hopper *et al.* [75] introduced a factor to account for the fact that all the ionised sites are not accessible to

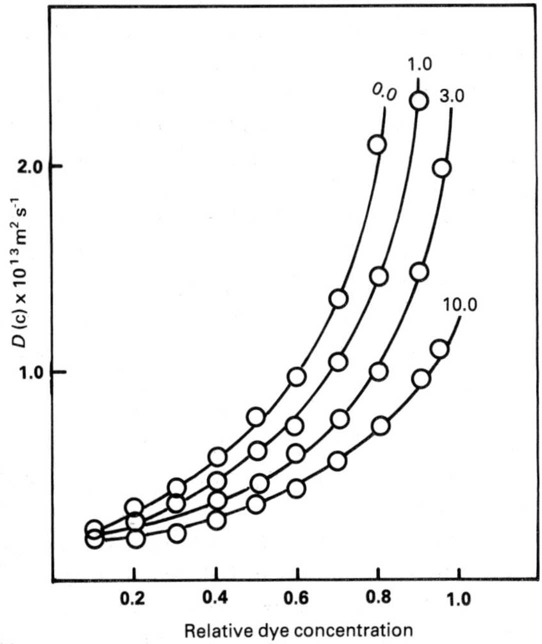

Figure 5.19 – Effect of salt concentration on D(c) in dyeing of C.I. Acid Red 18 on nylon 6.6 at 60 °C and pH 3.2 (numbers on curves are salt concentrations (g l^{-1}) used)

dye. The behaviour is most satisfactorily explained by the use of Eqn 5.65:

$$D(C_A) = D_{0,i}\left[\frac{1 + (z + 1)\theta_A}{1 - \alpha\theta_A}\right] \tag{5.65}$$

where $D(C_A)$ is the concentration-dependent diffusion coefficient of the dye adsorbed on the polymer, z is the basicity or number of negative charges on the dye anion, θ_A is the fractional site saturation of the dye and $D_{0,i}$ is the diffusion coefficient of the dye anion in the internal liquid phase. The accessibility factor α is related to the coefficient K, describing the partition between dye adsorbed on the polymer and dye in the internal liquid phase, in accordance with Eqn 5.66:

$$\alpha = \left[1 - \frac{zD_{0,i}K}{D_{0,x}}\right] \tag{5.66}$$

$D_{0,x}$ is the diffusion coefficient of the displaced acid anion (reaction 5.63) in the internal phase. The concentration-dependent $D(C_A)$ should therefore vary according to the number of charges carried by the dye anion and the nature and rate of outward diffusion of the displaced anion X^-, as well as being dependent on the partition coefficient.

Petropoulos and Japanese co-workers [76] have pointed out that, in the case of polybasic acid dyes on nylon, contributions to the total diffusional flux should also include a contribution from the non-ionised fraction of the dye by superimposing the Henry's Law sorption curve on the curved Langmuir isotherm. The Langmuir type of adsorption was considered [77] to be instantaneous on pore sites throughout the nylon, and the resultant concentration profile could also be expressed by the single independent variable $x/2t^{\frac{1}{2}}$ as, for example, shown in Figure 5.13. In reaching this conclusion it was pointed out that the diffusion coefficient should be arrived at by a method involving pore ratios and tortuosity factors as developed by Zollinger [69], thus illustrating that the dyeing of nylon with acid dyes is a heterogeneous process. More recently [78], the same authors have found poor agreement between permeation rates determined by steady- and non-steady-state diffusion under the same dyebath conditions, and it is now considered that the diffusion mechanism is a two-stage process controlled by a region within the nylon available for rapid, but not instantaneous, penetration, followed by dye permeation into a region formed by slow relaxation of polymer chains.

Observations have been made with respect to the ion-exchange mechanism of dyeing acrylic films with basic dyes [79] and of the dyeing of anionically modified polyester with cationic dyes [80]. The distribution–distance profiles were again sigmoidal in shape and the variation of $D(c)$ with relative concentration of dye was of the shape illustrated in Figure 5.19, in the absence of salt. If the desorbed cation (reaction 5.64) is increased in size by pretreating the acrylic film to replace the hydrogen of the sulphonic acid group, e.g. with a cetylpyridinium cation, $D(c)$ becomes less concentration-dependent since this cation, now displaced by dye, has a lower mobility in the internal liquid. Also, the distribution–distance profiles change from sigmoidal to type A curves (Figure 5.18) obtained by calculation assuming a concentration-independent diffusion coefficient [79].

When the difference in mobility between the entering and leaving ions is large the diffusion process will be modified, since an electrical potential gradient will be set up. The rate of transfer F_D for dye ions into a semi-infinite plane slab under these conditions may be expressed by

$$F_D = -D_0\left[\frac{\partial C_D}{\partial x} + \frac{C_D F}{RT}\frac{\partial \psi}{\partial x}\right] \tag{5.67}$$

where $\partial\psi/\partial x$ is the electrical potential gradient and F is the Faraday constant. A similar expression holds for the rate of transfer F_X of the desorbing ion, and the electrical potential gradient can be eliminated only if the two rates of transfer are considered equal and opposite, i.e. if $F_D + F_X = 0$. At present this assumption of electrical neutrality must be adopted, since no actual measurements of the developed potential gradient have as yet been made.

There is, however, one important difference in comparing the diffusion characteristics of acrylic fibres with those of nylon in dyeing with basic dyes. This is that the glass transition temperature T_g of acrylics lies within the practical dyeing temperature range, and the diffusion kinetics will therefore be strongly temperature-dependent. Above T_g, Coulombic dye interaction as expressed by Eqn 5.64 will tend to a maximum as ionic sites become available. In the presence of a carrier such as 4-nitrophenol [81] or benzyl alcohol [82], increased segmental chain mobility causes a considerable reduction in T_g so that maximum site saturation can occur at temperatures as low as 54 °C. In addition the lower permittivity (dielectric constant) of benzyl alcohol will cause suppression of ionisation of both sites and dye; interaction between the two will be suppressed but chain mobility will be increased.

5.3.5 Reactive dyes

In dyeing cellulose with reactive dyes, the diffusion mechanism does not differ theoretically from that described for direct dyes on this substrate. Under alkaline conditions, however, the progress of the dye is hindered not only by physical adsorption on the pore walls of the substrate, but also by chemical reaction between the reactive residues in the dye and the available ionised sites (normally hydroxyl groups) in the substrate. If this reaction is a first-order reaction or, where there is an excess of ionised sites, a pseudo-first-order reaction, the rate of fixation is proportional to the concentration of dye C_s in the solution. The diffusion equation then becomes [83]

$$\frac{\partial C}{\partial t} = D\frac{\partial^2 C}{\partial x^2} - kC_s \tag{5.68}$$

where k is the first-order rate constant. The system is analogous with the conduction of heat along a wire, which loses heat from its surface at a rate proportional to its temperature. For diffusion into and simultaneous reaction with an infinite filament it can be shown that initially or, more strictly, for small values of kt,

$$\frac{\mathrm{d}C}{\mathrm{d}t} \approx C^*\left(\frac{\mathrm{D}}{\pi t}\right)^{\frac{1}{2}}(1 + kt) \tag{5.69}$$

and for larger values of kt,

$$\frac{dC}{dt} \approx C^*(Dk)^{\frac{1}{2}} \tag{5.70}$$

where C^* is the surface saturation concentration of the dye. Under actual dyeing conditions the reactive dye is removed from the system not only by reaction with the substrate, but also by hydrolysis with the aqueous solvent. The rate of loss by hydrolysis dh/dt is a second-order reaction given by

$$\frac{dh}{dt} = k'C_s[OH^-]_s \tag{5.71}$$

for an infinite dyebath, which does not effectively change in concentration. The efficiency of fixation E, which is the ratio of the rate of reaction with the substrate to that of hydrolysis, is governed according to Eqn 5.72, obtained by combining Eqns 5.70 and 5.71:

$$E = \frac{dC}{dh} = \frac{C^*}{C_s}\frac{(Dk)^{\frac{1}{2}}}{k'[OH^-]_s} \tag{5.72}$$

It can be seen [84] that the primary factor determining efficiency is the substantivity ratio C^*/C_s, whereas the diffusion coefficient D and the reactivity ratio k/k' play a comparatively minor role. To obtain a more accurate insight into the overall mechanism of reaction it would be necessary to consider the reaction and hydrolysis rates of the dye within the substrate and their variation with pH.

One approach has been to study the diffusion of dyes which, in structure, are comparable with reactive dyes, but do not contain reactive residues [85]. By using hydrolysed reactive dye to distinguish between the diffusion of reactive and non-reactive species and based on the rate equation 5.68, Rusznak, Zobor and Levai [86] have derived an equation describing the ratio of the amounts of diffused non-bonded and covalently bonded reactive dye in cellulosic substrates. Experimental and calculated dyeing rates were found to agree only in the initial stages of dyeing. Divergencies in later stages were considerable and attributed in different systems to possible aggregation, pH changes and the use of different alkaline buffers. Thus it appears that the heterogeneous diffusion mechanisms associated with reactive dyes coupled with simultaneous or subsequent chemical reaction still present a problem seeking an answer.

5.3.6 Diffusion of ionised dyes in wool

As a fibre, wool is heterogeneous, both chemically and physically. Its diameter varies considerably from fibre to fibre, even within samples of the same origin. Each fibre is also surrounded by a thin hydrophobic epicuticle layer that controls the penetrability of the fibre with respect to dyes and other solutes. In addition, the permeability of the epicuticle may differ in different parts of the fibre, and this factor alone complicates any mathematical treatment of diffusion data [67]. Removal of the epicuticle, either in processing or by accidental damage, increases the initial rate of dyeing. Further, as a result of its hydrophobic properties, the penetrability of the undamaged fibre by dyes is dependent on the hydrophilic character of the dye anion. Thus a dye anion containing three ionised sulphonic acid groups was shown [88] to be incapable of penetrating the epicuticle in the early stages of dyeing, whereas a monosulphonated dye could diffuse more readily. Unsulphonated water-soluble 1:2 metal-complex dyes, although of relatively large molecular size, are least affected by these variations in surface characteristics [89].

Earlier determinations of diffusion coefficients did not take into account the retarding effect of the epicuticle. Diffusion coefficients for Orange II obtained by Alexander and Hudson [90] gave mean values of 3×10^{-13} and $5 \times 10^{-13}\ m^2\ s^{-1}$ at 25 and 35 °C respectively. The value of $5.35 \times 10^{-10}\ m^2\ s^{-1}$ for the diffusion coefficient of this dye in water is very much higher [91]. The temperature-dependence of the diffusion coefficient in the fibre is much greater at high dye concentrations, and Alexander and Hudson concluded that under these conditions diffusion occurred by displacement of keratin chains and not through discrete pores in the fibre.

On the other hand, Medley and Andrews [92] have taken the epicuticle layer into account by setting up a model of a plane sheet with a diffusion barrier covering the surface. The model applies equally well to fibres possessing a barrier layer, since the theoretical considerations are limited to diffusion in the early stages in a semi-infinite medium. In the absence of a barrier the amount of dye absorbed with time is described by Eqn 5.24, with the consequence that a plot of fractional uptake against $t^{\frac{1}{2}}$ is linear. This linearity is shown in the calculated curve B in Figure 5.20, the coincident experimental points being obtained from the sorption of the free acid of Orange II by wool from which the epicuticle had been physically removed. Chemical removal of the epicuticle was avoided since this process may affect the diffusional properties of the underlying bulk phase.

In the presence of a barrier such as an epicuticle, which absorbs a negligible quantity of dye, a new boundary condition describing the continuity of flow across the inside surface of the barrier must be imposed. If it is assumed that the concentration gradient across the barrier, of thickness δ, is linear, the rate of transfer F can be written

$$F = D_b \frac{(C_b - SC_p)}{\delta} \qquad (5.73)$$

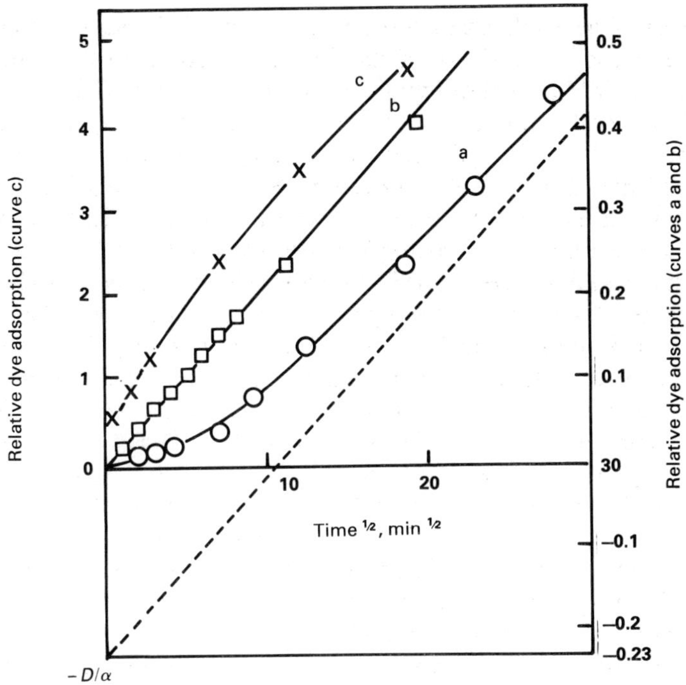

Figure 5.20 – Relative sorption of Orange II free acid by virgin wool (curve a), abraded wool (curve b) and horn-keratin membrane (curve c)

where D_b is the diffusion coefficient in the barrier, C_b is the equilibrium concentration of dye in the barrier and S is the equilibrium partition coefficient between the concentration of dye in the barrier and the dye in the bulk interior phase C_∞, i.e. $S = C_b/C_\infty$. C_p is the dye concentration in the fibre just beyond the inner barrier surface at a point P. The diffusional flow of dye at this point is therefore governed by the rate of supply through the barrier surface, i.e.

$$-D\frac{\partial C}{\partial x} = D_b\frac{(C_b - SC_p)}{\delta} = \alpha(C_\infty - C_p) \qquad (5.74)$$

According to Andrews and Medley, α represents the 'admittance' of the barrier and $\alpha = D_bC_b/\delta C_\infty$. D is the diffusion coefficient in the bulk fibre phase.

If α/D is constant, the solution for the diffusion equation is one formally dealt with by Crank [93] in connection with diffusion and surface evaporation. This solution may be expanded in two forms for describing initial and ultimate behaviour. The latter case is given by

$$\frac{C_t}{C_\infty} = 2\left(\frac{Dt}{\pi}\right)^{\frac{1}{2}} - \frac{D}{\alpha} + \frac{1}{\alpha^2}\left(\frac{D^3}{\pi}\right)^{\frac{1}{2}} t^{-\frac{1}{2}} + \ldots . \tag{5.75}$$

Eqn 5.75 shows that, once the initial lag due to the barrier effect has been overcome, a plot of C_t/C_∞ against $t^{\frac{1}{2}}$ approaches a line of slope $2(D/\pi)^{\frac{1}{2}}$ with an intercept on the sorption axis of $-D/\alpha$. Such a result is shown as the calculated curve A in Figure 5.20, the experimental points being obtained with dyed virgin Merino wool. The initial 'time-lag' can readily be observed and the asymptote of this curve has an intercept at $-D/\alpha$. In order to compare the results of curves A and B, it was assumed that the diffusion coefficient of the dyes was the same in abraded and virgin wool and that the process of abrasion caused a 2.5-fold increase in the surface of the abraded fibres. The results suggest that the barrier controls the initial rate of strike.

Curve C represents a similar curve for horn-keratin sheet. This material does not possess an epicuticle layer and has been previously used [94] as a model substrate for wool. The behaviour is anomalous, however, in that the curvature in the initial stages of dyeing is in the reverse direction. The effect is possibly due to the rough or porous nature of this type of substrate.

From experiments similar to those carried out by Medley, Rattee [95] concludes that the barrier to diffusion is centred in the exocuticle region located just below the outer surface of the epicuticle, since complete chemical reduction of the exocuticle corresponds to the point where the value of $-D/\alpha$ is very close to zero. This conclusion was verified [96] by the use of dye labelled with the radioactive tracer ^{35}S, which was applied to the fibre after equilibration with the same but unlabelled dye. Both types of molecule are redistributed to give a steady-state situation in which diffusion of the labelled species is independent of free energy changes associated with concentration-dependent effects.

By applying the same techniques to drawn and undrawn nylon 6.6 filaments, the initial concave curvature of the C_t/C_∞ *vs* $t^{\frac{1}{2}}$ plots can be interpreted in terms of the cross-sectional anisotropy of individual filaments with an increase in barrier layer thickness on drawing [97].

5.4 ISO-REACTIVE DYEING

The determination of diffusion coefficients and theoretical aspects of dye diffusion processes as described in this chapter are based mainly on dye absorption rates obtained at constant temperature. The need for greater efficiency and the advent of computer program-controlled dyeing machinery have meant that dyeing is generally conducted under non-isothermal conditions. By controlling the rate of temperature rise it is theoretically possible to achieve a constant rate of dye sorption according to

$$\frac{dC_t}{dt} = \frac{C_\infty}{t_\infty} = \frac{C_t}{t_i} \tag{5.76}$$

where C_t is the amount of dye absorbed at some instantaneous time t_i and the suffix ∞ refers to the equilibrium exhaustion value for C at time t. Such a system is claimed to give the highest probability of dye levelling, thus maximising efficiency. The process has been described as *iso-reactive dyeing* [98] although, to avoid confusion with the dyeing of reactive dyes, a better term should be sought. The dyeing times in an iso-reactive dyeing system are not identical with those obtained under isothermal conditions. This is best illustrated in Figure 5.21, which is taken from the explanation given by Cegarra [99].

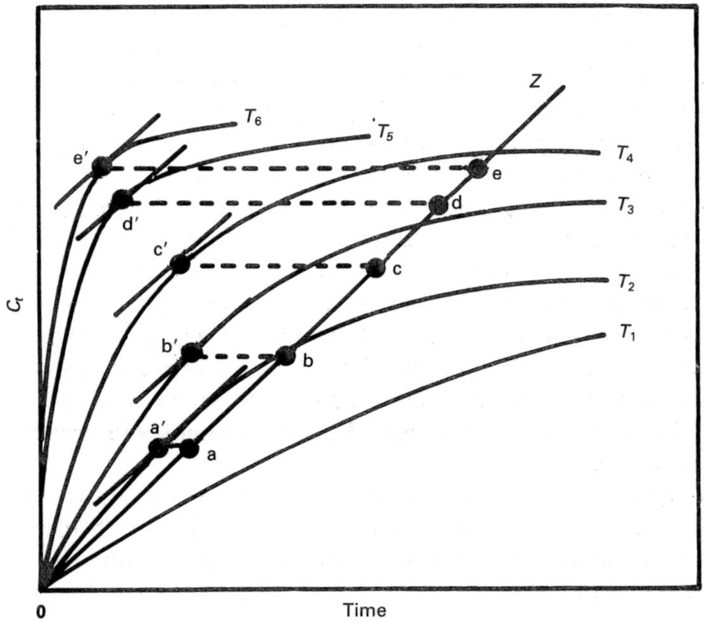

Figure 5.21 – Isothermal rate-of-dyeing curves and OZ iso-reactivity line

In this figure the isothermal dyeing-rate curves at temperatures T_1 to T_6 are shown as continuous lines. The slope of the line OZ represents a constant dyeing rate. By determining the points a′ to e′ on the isothermal rate curves whose tangents correspond to the slope OZ it can be seen that all have the same rates equivalent to the slope OZ. The corresponding points a to e on

OZ have the same constant dyeing rate but the times required to reach C_t, i.e. the iso-reactive dyeing times t_i, are greater than the times for isothermal dyeing.

By combining equations representing both the kinetics and the thermodynamics of the process Cegarra *et al.* [98] have derived an expression giving the temperatures and times required to achieve a constant dyeing rate. From Eqn 5.77, relating the amount of dye absorbed to the dyeing rate constant K, postulated earlier by Cegarra [100],

$$\ln\left[1 - \frac{C_t^2}{C_\infty^2}\right] = -(Kt)^a \tag{5.77}$$

where a is a constant, and the Arrhenius equation

$$K = K_0 \exp(-E/RT) \tag{5.78}$$

where E is the activation energy of dyeing, the expression

$$T = \frac{E}{R}\left\{\ln\frac{a}{2} + \ln K_0 + \ln\frac{t_\infty{}^2 - t_i^2}{t_i} + \frac{a-1}{a}\ln\left[-\ln\left(\frac{1-t_i}{t_\infty{}^2}\right)^2\right]\right\}^{-1} \tag{5.79}$$

can be derived. This equation allows the calculation of the dyeing temperatures T required at different dyeing times t_i, up to the time t_∞ when the degree of exhaustion reaches its maximum value. E, a and K_0 are deduced from dyeing-rate curves and the Arrhenius equation (Eqn 5.78) and t_∞ from the exhaustion rate of the system. In practice, however, the results show a slight positive deviation from constant dyeing rates. The validity of the method has been questioned on the grounds that Eqn 5.77 applies to isothermal conditions only, and that the rate constant is time-dependent. An alternative solution based on an empirical kinetic equation and the Williams–Landel–Ferry equation (see section 3.6), replacing Eqns 5.77 and 5.78 respectively, was therefore proposed [101]. Both alternative solutions were subsequently criticised by Hoffmann [102], on the grounds that iso-reactive dyeing systems are not necessarily those which produce the most effective level dyeing and also that the temperature changes necessary to maintain a constant dyeing rate could alter boundary conditions and hence diffusion rates. The topic remains open-ended, and in practical systems the method would necessitate a degree of precision in temperature control not yet attainable.

5.5 ACTIVATION ENERGIES OF DIFFUSION

The effect of temperature on rates of diffusion may be expressed quantitatively by determining the activation energy of diffusion. Diffusion, taking place by random motion from regions of high concentration to regions of lower concentration, increases on raising the temperature, as a consequence of a reduction in the time spent by a molecule in the vicinity of a reaction site or on the surface of capillaries. Consider our swarm of 'super bees' moving along a greatly magnified tube (section 5.1). As the thermal motion (temperature) of the wall and its surroundings increases, more and more resting 'bees' will acquire sufficient minimum energy for them to fly to a new position on the tube wall. The number in flight at any instant in time will depend on the total number present and on the term exp $(-E/RT)$, where R is the gas constant and T the thermodynamic temperature. This factor represents the fraction of the number of 'bees' possessing sufficient minimum energy E for take-off. In dyeing processes, therefore, E is the amount of energy of an activated dye molecule in excess of the average energy of the molecules. In addition to influencing the rate of diffusion, it is apparent therefore that the number of adsorbed dye molecules at equilibrium will be reduced as the temperature is raised. The effect is a consequence of the exothermic nature of the dyeing process. It is illustrated in the idealised dyeing situation shown in Figure 5.22. Provided that dyeing is sufficiently prolonged, e.g. to point

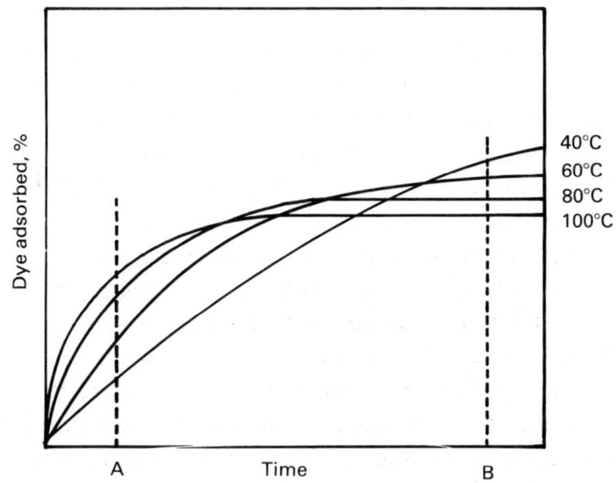

Figure 5.22 – Effect of temperature on rate of dyeing and equilibrium saturation values

B in time, the final amount of dye adsorbed is less as the temperature increases. This effect is not usually observed during commercial dyeing, since economic considerations demand dyeing times that are shorter and may extend only as far as point A.

Since in dyeing processes diffusion is always accompanied by absorption

of dye molecules to a greater or less extent, the above exponential term can be included in the temperature-dependent diffusion coefficient D_T, and consequently

$$D_T = D_0 \exp(-E/RT) \qquad (5.80)$$

where D_T is the diffusion coefficient at temperature T and D_0 is a constant. Taking natural logarithms of this equation,

$$\ln D_T = \ln D_0 - \frac{E}{RT} \qquad (5.81)$$

Eqn 5.81 is linear in form and, by converting to common logarithms, the activation energy of diffusion E may be obtained directly from the slope of the plot of log D_T against $1/T$, i.e.

$$E = -\,2.303\,R\frac{\mathrm{d}(\log D_T)}{\mathrm{d}(1/T)} \qquad (5.82)$$

The constant D_0 is determined from the intercept of this curve on the log D_T axis. It must be assumed that between the experimental temperature limits no change in the structure of the substrate occurs. Such changes can take place if, for example, a second-order phase transition intervenes. Figure 5.23 illustrates the results obtained [103] for diffusion of azobenzene vapour into cellulose acetate film over the temperature range 130–150 °C with a constant vapour concentration of 5.83×10^{-3} g l^{-1}. Since the values of the diffusion coefficients are apparent or mean values determined when $C_t/C_\infty = 0.5$, the energy of diffusion is the *apparent* activation energy of diffusion and in this example was found to be 96.1 kJ (23.0 kcal) mol^{-1}. It must be emphasised that this apparent activation energy of diffusion is equivalent to the true activation energy only when the diffusion coefficients are independent of dye concentration within the substrate. The general method so far has not been modified to take into account concentration-dependent diffusion coefficients.

Linear relationships between $1/T$ and ln D_T or ln K, where K is the velocity constant for rate of dyeing, have been found for several dyeing systems. A comparison of activation energies for different classes of dyes on different fibres is shown in Table 5.3, which is an extended version of that given by Bird [104]. It is seen that the approximate magnitudes of the activation energies of diffusion are related to the different mechanisms of diffusion for the various substrates already described. Thus in cellulose sheet, Heliotrope 2B gives [105] an apparent activation energy of 58.5 kJ (14 kcal) mol^{-1}, and

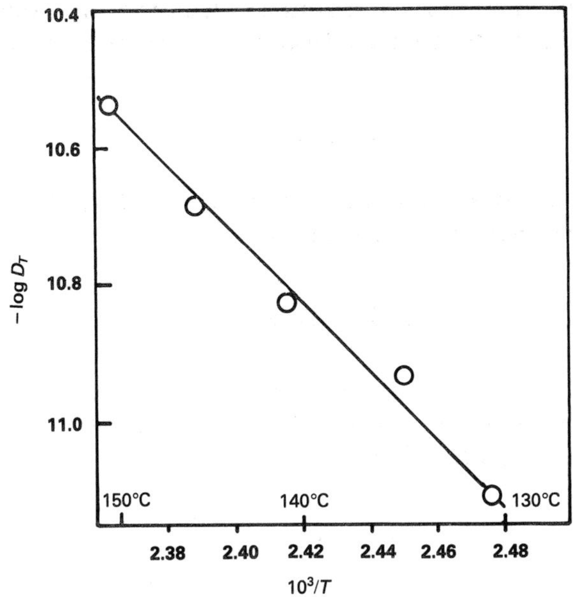

Figure 5.23 – Log D_T versus $10^3/T$ plot for azobenzene vapour in cellulose acetate film

as many as 17 out of 22 disperse dyes examined on nylon [106] have an activation energy lying between 83 and 100 kJ (20 and 24 kcal) mol^{-1}. With some modern fibres which possess a more compact structure, it is necessary to increase dyeing rates by using temperatures up to 140 °C or by using carriers to increase the space within the structure available for diffusion. In the former case raising the temperature to this high value markedly increases the diffusion coefficient, with the result that the activation energy of diffusion is also increased. Thus disperse dyes on polyester fibres [53] give values of the order of 250 kJ (60 kcal) mol^{-1}, the actual value depending both on the dye used and on the physical modifications and pre-history of the fibre structure.

From Table 5.3 it can also be seen that in those dyeing systems having a comparatively low activation energy the generally accepted mechanism of diffusion is one of dye transfer from aqueous solution by adsorption on, and migration over, capillary surfaces. For values of E in excess of approximately 125 kJ (30 kcal) mol^{-1} the activation energy of diffusion may in part be considered as a measure of the necessity to overcome the cohesive energy of the polymer chains and form a 'hole' through which the dye may diffuse. The activation energy is thus related to the energy necessary to disrupt the intermolecular forces between lengths of polymer chains.

The foregoing assumes a constant activation energy over the temperature

TABLE 5.3

Activation energies of diffusion

Type of dye	Fibre	Activation energy /kJ mol^{-1}	/kcal mol^{-1}
Azoic coupling components	Viscose rayon	42	10
Vat dyes	Viscose rayon	52	12.5
Direct dyes	Viscose rayon	59	14
Equalising acid dyes	Wool	92	22
Milling acid dyes	Wool	121	29
Disperse dyes	Nylon	92	22
Disperse dyes	Polypropylene	105	25 [107]
Disperse dyes	Polyester	167	40 [53]
Basic dyes	Acrylic (Orlon 42)	251	60 [108]

range of measurement. Gupta *et al.* [109] have found that the activation energy of diffusion can decrease in the dyeing of polyester which had been heat-treated with and without tension, due to relaxation processes occurring during dyeing, and Alberghina *et al.* [110] found a similar energy decrease in applying disperse dyes to cellulose acetate in the presence of swelling agents.

Thermodynamical considerations (see Chapter 4) indicate tht the activation energy of diffusion E is related to the enthalpy ΔH_{diff} and entropy ΔS_{diff} of diffusion according to

$$E = \Delta H_{\text{diff}} - T\,\Delta S_{\text{diff}} \tag{5.83}$$

It follows that Eqn 5.80 can be expressed as

$$D_T = D_0\,[\exp(-\Delta H_{\text{diff}}/RT)][\exp(\Delta S_{\text{diff}}/R)] \tag{5.84}$$

By assuming that diffusion is accompanied by a negligible change in volume, Eqn 5.84 becomes

$$D_T = D_0\,[\exp(\Delta S_{\text{diff}}/R)][\exp(-E/RT)] \tag{5.85}$$

From the theory of rate processes developed by Eyring [111] it can be shown that

$$D_T = \mathrm{e}\lambda^2\frac{kT}{h}[\exp(\Delta S_{\text{diff}}/R)][\exp(-E/RT)] \tag{5.86}$$

where λ is the distance between successive equilibrium positions of the diffusing molecule, and k and h are respectively the Boltzmann constant and the Planck constant. Thus, from Eqns 5.85 and 5.86,

$$D_0 = e\lambda^2 \frac{kT}{h} \exp\left(\Delta S_{diff}/R\right) \tag{5.87}$$

and since D_0 and E can be determined experimentally, the value of ΔS_{diff} can be calculated if an arbitrary value is chosen for λ. Patterson and Sheldon [53] used a value of 1.5 nm for λ, this being the approximate length of the diffusing dye molecules, and obtained very large and positive entropy changes ranging from 355 to 544 J (84.5 to 129.5 kcal) K^{-1} mol^{-1}, and indicating hole formation over considerable lengths of polymer chains.

REFERENCES

1. J H de Boer, *The dynamical character of adsorption*, 2nd edn (London: Oxford University Press, 1968), 38.
2. A Fick, *Ann. Phys. Leipzig*, **170** (1855) 59.
3. J B Fourier, *Théorie analytique de la chaleur* (Oeuvres de Fourier, 1822).
4. H Daynes, *Proc. Roy. Soc.*, **97A** (1920) 286.
5. J Crank, *The mathematics of diffusion*, 1st edn (Oxford: Clarendon Press, 1956), 47.
6. C L Bird, F Manchester and D Scott, *J.S.D.C.*, **72** (1956) 49; D Scott, MSc thesis (Leeds University, 1956), 105.
7. H H Sumner and B Taylor, *J.S.D.C.*, **83** (1967) 445.
8. E Sada, H Kumazawa and T Ando, *J.S.D.C.*, **100** (1984) 97.
9. J Crank, in *Diffusion in polymers* (ed. by J Crank and G S Park) (London: Academic Press, 1968), 16.
10. T Vickerstaff, *The physical chemistry of dyeing*, 2nd edn (London and Edinburgh: Oliver and Boyd, 1954), 136.
11. Ref. 5, p. 45.
12. H K Frensdorff, *J. Polymer Sci.*, **2** (1964) 341.
13. J W McBain, *Z. Physikal. Chemie*, **68** (1909) 471.
14. Ref. 10, p. 131.
15. L Peters, private communication.
16. A B Newman, *Assocn. Inst. Chem. Eng. J.* (1931) 203.
17. Ref. 5, p. 49.
18. Ref. 10, p. 132.
19. R H Peters, in *Diffusion in polymers* (ed. by J Crank and G S Park) (London: Academic Press, 1968), 318.
20. R McGregor and R H Peters, *J.S.D.C.*, **81** (1965) 393.
21. P C Carman and R A W Haul, *Proc. Roy. Soc.*, **222A** (1954) 109.
22. A H Wilson, *Phil. Mag.*, **39** (1948) 48.
23. Ref. 5, pp. 54 and 71.
24. N A Barrett and C C Cook, *J.S.D.C.*, **93** (1977) 335.
25. J N Etters and A Urbanik, *Text. Research J.*, **53** (1983) 598.
26. A Urbanik, *J.S.D.C.*, **92** (1976) 377.
27. *Idem*, *Textilveredlung*, **13** (1978) 278.
28. T Shibusawa, *J.S.D.C.*, **95** (1979) 175.
29. J N Etters, *Text. Chem. Col.*, **12** (1980) 140.
30. J Cegarra and P Puente, *Textilveredlung*, **12** (1977) 50.
31. J Boulton and T Morton, *J.S.D.C.*, **55** (1939) 481.

32. Ref. 10, p. 151.
33. J Boulton, *J.S.D.C.*, **60** (1944) 5.
34. S M Neale, *ibid.*, **52** (1936) 252.
35. J Militky and J Rais, *ibid.*, **93** (1977) 346.
36. R McGregor and R H Peters, *ibid.*, **84** (1968) 267.
37. J Crank and M E Henry, *Trans. Faraday Soc.*, **45** (1949) 639, 1119.
38. J Crank and G S Park, *ibid.*, **45** (1949) 240.
39. W M Garvie and S M Neale, *ibid.*, **34** (1938) 335.
40. J Boulton and T H Morton, *J.S.D.C.*, **55** (1939) 481.
41. J O Warwicker, *J. Polymer Sci. A*, **1** (1963) 3105.
42. B Kramer, *Melliand Textilber.*, **35** (1954) 518.
43. W Luck, *ibid.*, **36** (1954) 924, 1028.
44. F L Goodwin and S Rosenbaum, *Text. Research J.*, **35** (1965) 439.
45. M Sekido, T Iijima and A Takahashi, *Kogyo Kagaku Zasshi*, **68** (1965) 524.
46. R H Peters and J H Petropoulos, *Bull. Inst. Text. France*, **84** (1959) 49.
47. R McGregor, *Trans. Farday Soc.*, **58** (1962) 771, 1045, 1054.
48. C Matano, *Jap. J. Phys.*, **8** (1932) 109.
49. Ref. 5, pp. 148 and 232.
50. R H Peters, J H Petropoulos and R McGregor, *J.S.D.C.*, **77** (1961) 704.
51. J Crank, *Phil. Mag.*, **39** (1948) 362.
52. Ref. 19, p. 341.
53. D Patterson and R P Sheldon, *Trans. Faraday Soc.*, **55** (1959) 1254.
54. R McGregor, R H Peters and C R Ramachandran, *J.S.D.C.*, **84** (1968) 9.
55. T Vickerstaff, *Hexagon Digest*, **20** (1954) 7.
56. M J Schuler, *Text. Research J.*, **27** (1957) 352; W R Remington and H E Schroeder, *ibid.*, **29** (1959) 177.
57. O Glenz, W Beckmann and W Wunder, *J.S.D.C.*, **75** (1959) 141.
58. E Elöd, *Melliand Textilber.*, **41** (1960) 195.
59. G A F Roberts and R K Solanki, *J.S.D.C.*, **95** (1979) 427.
60. R J Pace and A Datyner, *Text. Research J.*, **50** (1980) 34.
61. D F Hemmings, J D Collins and A Datyner, *J. App. Poly. Sci.*, **20** (1976) 597.
62. C Oschatz, U Ruf and F Somm, *J.S.D.C.*, **92** (1976) 73.
63. C Cometto, MSc thesis (Leeds University, 1971).
64. J G Blacker and D Patterson, *J.S.D.C.*, **85** (1969) 598.
65. B Milićević and R McGregor, *Helv. Chim. Acta*, 49 (1966) 1302, 1319.
66. T Hori, Y Sato and T Shimizu, *J.S.D.C.*, **97** (1981) 6.
67. S M Neale and W A Stringfellow, *Trans. Faraday Soc.*, **29** (1933) 1167; *J.S.D.C.*, **59** (1943) 241.
68. E Valko, *Trans. Faraday Soc.*, **31** (1935) 278; T H Morton and S M Neale, *ibid.*, **31** (1935) 262.
69. P B Weisz, *ibid.*, **63** (1967) 1801; P B Weisz and J S Hicks, *ibid.*, **63** (1967) 1807; P B Weisz and H Zollinger, *ibid.*, **63** (1967) 1815; **64** (1968) 1693.
70. Ref. 10, p. 351.
71. A N Derbyshire and R H Peters, *J.S.D.C.*, **71** (1955) 530.
72. R H Peters, *ibid.*, **61** (1945) 95; H J Palmer, *J Textile Inst.*, **49** (1958) T33.
73. D Balmforth, C A Bowers and T H Guion, *J.S.D.C.*, **80** (1964) 577.
74. W Beckmann, *ibid.*, **77** (1961) 616.
75. M E Hopper, R McGregor and R H Peters, *ibid.*, **86** (1970) 117.
76. J Komiyama, J H Petropoulos and T Iijima, *ibid.*, **93** (1977) 217.
77. E Sada, H Kumazawa and T Ando, *ibid.*, **98** (1982) 121.
78. *Idem*, *ibid.*, **100** (1984) 97.
79. R J Harwood, R McGregor and R H Peters, *ibid.*, **88** (1972) 288.
80. E Sada, H Kumazawa and T Ando, *J. Appl. Poly. Sci.*, **28** (1983) 3817.
81. R S Asquith, H S Blair and N Spence, *J.S.D.C.*, **94** (1978) 49.
82. Z Gur-Arieh, W Ingamells and R H Peters, *ibid.*, **92** (1976) 332.
83. P V Danckwerts, *Trans. Faraday Soc.*, **46** (1950) 300; **47** (1951) 1014.

84. I D Rattee, *J.S.D.C.*, **85** (1969) 23.
85. H H Sumner and B Taylor, *ibid.*, **83** (1967) 445.
86. I Rusznak, J Zobor and G Levai, *Dyes and Pigments*, **2** (1981) 285.
87. J Lindberg, *Text. Research J.*, **23** (1953) 573.
88. F Townend, *J.S.D.C.*, **61** (1945) 144.
89. C L Bird, *The theory and practice of wool dyeing*, 4th edn (Bradford: Society of Dyers and Colourists, 1972) 42.
90. P Alexander and R F Hudson, *Text. Research J.*, **20** (1950) 481.
91. E Valko, *Colloid Chem.*, **6** (1946) 594.
92. M W Andrews and J A Medley, *Text. Research J.*, **29** (1959) 398.
93. Ref. 5, p. 34.
94. M L Wright, *Disc. Faraday Soc.*, **16** (1954) 58.
95. G M Hampton and I D Rattee, *J.S.D.C.*, **95** (1979) 396.
96. D K Al-Hariri, J A Coates and I D Rattee, *ibid.*, **95** (1979) 432.
97. J A Coates, V Ellard and I D Rattee, *ibid.*, **96** (1980) 14.
98. J Cegarra, P Puente and J Valdeperas, *ibid.*, **92** (1976) 327.
99. *Idem*, *ibid.*, **95** (1979) 336.
100. J Cegarra and P Puente, *Text. Research J.*, **37** (1967) 343.
101. J Militky and J Rais, *J.S.D.C.*, **94** (1978) 348.
102. F Hoffmann, *ibid.*, **95** (1979) 153.
103. F Jones and R Seddon, *Text. Research J.*, **35** (1965) 334.
104. Ref. 89, p. 23.
105. W M Garvie, L H Griffiths and S M Neale, *Trans. Faraday Soc.*, **30** (1934) 271.
106. Ref. 10, p. 161.
107. C L Bird and A M Patel, *J.S.D.C.*, **84** (1968) 560.
108. A Tokaska, *J. Soc. Text. Cellulose Ind. Japan*, **20** (1964) 100.
109. V B Gupta, A K Gupta, V V P Rajan and N Kasturia, *Text. Research J.*, **54** (1984) 54.
110. G Alberghina, M L Longo and M Torre, *Dyes and Pigments*, **4** (1983) 49.
111. S Glasstone, K J Laidler and H Eyring, *The theory of rate processes* (London: McGraw Hill, 1941), 524.

CHAPTER 6

Reactive dye–fibre systems

PAUL RYS AND HEINRICH ZOLLINGER

6.1 INTRODUCTION

6.1.1 Definition of dye–fibre reactions

There are three ways in which dyes can be attached to a substrate: by (a) physical adsorption, (b) mechanical retention, and (c) chemical reaction. Dyeing of cotton with direct dyes and of man-made fibres with disperse dyes are examples of (a), mass coloration (dope dyeing) with pigments exemplifies (b) and dyeing with vat, sulphur and azoic dyes is a sequence of (a) and (b).

There is no clear borderline between physical and chemical interactions between molecules. Van der Waals forces and hydrophobic bonding are usually considered to be physical processes, whereas hydrogen bonding and Coulombic attraction of ions can be considered as physical or chemical phenomena. On the other hand, the formation of covalent bonds is definitely a chemical process [1].

Reactive dyes are coloured compounds which have suitable groups capable of forming covalent bonds between a carbon or phosphorus atom of the dye ion or molecule and an oxygen, nitrogen or sulphur atom of a hydroxyl, an amino or a mercapto group, respectively, of the substrate.

This definition excludes mordant dyes and 1:1 chromium–dye complexes of the Neolan type which, in dyeing protein fibres, may form covalent bonds between the metal ion and the nucleophilic groups of the fibres. Although these dyes have been in use for fifty years, the covalent nature of their interaction with fibres was not discussed before the mid-1950s, when reactive dyeing proper became known. It must be mentioned at this point that differentiation between covalent and ionic metal–ligand bonds is difficult [2].

6.1.2 History of reactive dyes

Chemical reactions for attaching dyes to substrates were recognised clearly as such and became technologically important only in 1956, i.e. 100 years after the discovery of Mauvein, the first synthetic dye, in 1856. Cross and Bevan (in 1895) [3] were probably the first scientists who thought of fixing a dye to a textile substrate via a covalent ('chemical') bond. They made coloured esters of cellulose by a six-step synthesis which was, of course, too complicated for practical application. In 1906, Schroeter [4] tried to synthesise azo dyes containing sulphochloride groups which he wanted to react with cellulose, but he was unsuccessful in the synthesis of the sulphochlorides, a problem that was solved only in the 1950s in the course of the synthesis of azo compounds of relatively large r.m.m. used as pigment dyes [5].

In 1925, Günther found that azo derivatives of isatoic acid anhydride (e.g. I) react with cellulose under mild conditions in the presence of sodium carbonate [6].

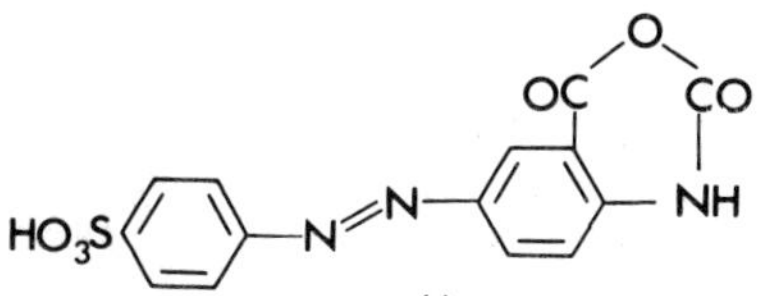

(I)

In 1929, Haller and Heckendorn [7] substituted cellulose with cyanuric chloride (2,4,6-trichloro-1,3,5-triazine) in organic aprotic solvents. The modified cellulose reacted with dyes containing amino groups, which is an example of the reactive group being supplied by the substrate and not by the dye. Guthrie [8] reported in 1952 that cellulose fibres can be dyed with dyes of type II from an aqueous dyebath but, owing to the high concentrations of sodium hydroxide necessary for fixation, this group of dyes was not investigated further.

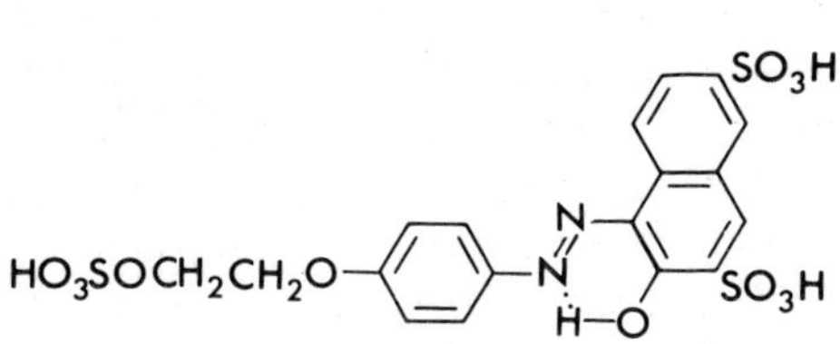

(II)

Industrially successful reactive dye technology began in the field of wool dyeing around 1930 [9] with the introduction of chloroacetylamino ($ClCH_2CONH$—) groups in dyes. Dyes with sulpho-β-chloroethyleneimide (—$SO_2NHCH_2CH_2Cl$) groups were also used [10]; related compounds became more important some 20 years later – a fact which demonstrates that old discoveries may become very interesting at a later date in the context of new experience! The same can be said for the first wool-reactive Remalan dyes of Farbwerke Hoechst, as well as for the wool-reactive Cibalan Brilliant dyes of CIBA. The former are sulphuric acid esters of the β-hydroxyethylsulphone ($—SO_2CH_2CH_2OSO_3H$) group, which is a precursor of a vinylsulphone group. The latter contain monochlorotriazine groups. These two reactive dye classes were introduced in the early 1950s but were not very successful, for reasons which are discussed later. They were very important as 'precursors' for reactive dyes for cotton, however, as they helped these two firms to enter the field of reactive dyeing of cellulose fibres shortly after the introduction of reactive dyes for cotton by ICI in 1956, whereas for other dyestuff manufacturers the situation was much more difficult owing to patent problems.

The introduction of reactive dyes for cellulose by ICI was the most important event in the history of reactive dyeing. These dyes are based on the discovery made in 1954 by Rattee and Stephen [11], who claimed that, by applying water-soluble dyes containing a dichlorotriazine group to cellulose from a neutral dyebath and then increasing the pH value, covalent bonds were formed between a triazine carbon atom and an oxygen atom of a cellulose hydroxyl group. An example of this first group of reactive dyes is Procion Orange MX-G (C.I. Reactive Orange 1) (III):

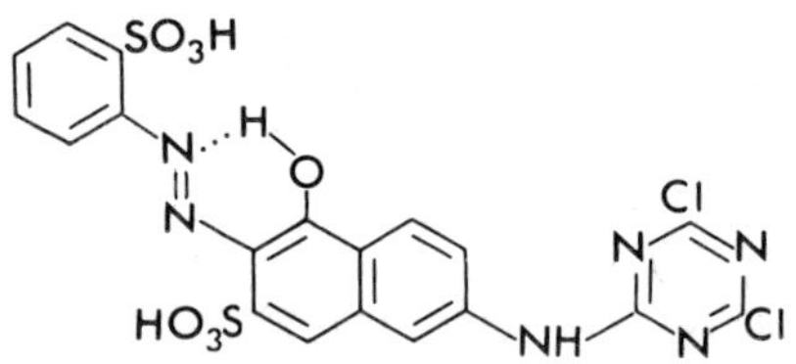

(III)

All major dyestuff manufacturers began investigations of new reactive groups and of application methods for cellulosic fibres immediately after the introduction of the Procion dyes. CIBA found that monochlorotriazine dyes also could be used as reactive dyes for cotton when the dyeing temperature and the pH value of the dyebath for the second part of the (batch) dyeing operation were increased. CIBA and ICI introduced the monochlorotriazine dyes jointly in 1957 as Cibacron and Procion H (H for 'Hot') dyes, respectively.

Although it was mentioned in the original Remalan patents of Farbwerke Hoechst [12] in 1949 that, besides wool, other substrates, e.g. cellulosic fibres, could be dyed with vinylsulphone dyes, the possibility of reactive dyeing of cellulose fibres was obviously not (or at least not carefully enough) studied by Hoechst chemists. Their claims in the original patents, however, enabled Hoechst to introduce in 1958 cotton-reactive dyes containing sulphuric acid ester groups of β-hydroxyethylsulphone under the trade name Remazol dyes. The relatively late introduction of this reactive group is astonishing when one considers that divinylsulphone and its derivatives were mentioned as crosslinking reagents for cellulosic fibres as long ago as 1950 [13]. The probable explanation for this delay is that, in the early fifties, very few dye chemists realised that the chemical (and physical) properties of fibres are factors of major importance and have to be taken into account when synthesising a dye. Similarly, textile chemists were slow in accepting the idea that the major interaction in easy-care treatments of cellulosic fibres with formaldehyde, *NN′*-dimethylolurea and related compounds is the chemical reaction with the fibre (crosslinking).

In subsequent years various dyestuff manufacturers introduced further reactive groups, the most important and commercially useful of which are shown in Table 6.1. A similar list is given by Abeta and Imada [14]. Following the introduction of the Procion dyes in 1956, at least one new reactive group was introduced commercially every year (except 1969) until 1971. Between 1972 and 1988, however, only five series with reactive groups not used before were introduced. Obviously the development of new reactive systems lost some of its impetus. It became difficult to find new reactive groups that could be introduced in dyes economically *and* that showed advantages in the methods of application and fastness properties over existing reacting systems. Moreover, the pioneering patents for the first reactive dyes invented in the 1950s began to expire. As a result dichlorotriazine reactive dyes, for example, are found today not only in ICI's series of Procion dyes, but also in the commercial dyes of at least 23 other dyestuff manufacturers around the world, while monochlorotriazines are now produced commercially by more than 30 manufacturers. On the other hand, some of the reactive dye classes introduced in the late 1950s and 1960s are no longer manufactured. These dye classes were introduced because at that time their manufacturers were not allowed to make dyes protected by patents of their competitors. After such patents expired manufacturers discontinued the production of dyes with reactive groups developed by themselves and used from then on the older (but obviously better) reactive groups of their competitors.

The monofluorotriazine dyes, introduced by Ciba-Geigy in 1978, are historically an interesting case. Dyes with this reactive group were synthesised and patented first by Bayer in 1957 [15], but Bayer never produced them commercially. The reason was the difficulty of substituting two fluorine atoms in cyanuric trifluoride by amines without partial hydrolysis. If a technical monofluorotriazine dye contains significant percentages of hydrolysed dye, the latter cannot, of course, form a covalent bond to cellulose; it is lost and causes waste water problems. Fluorine in mono-, di- or trifluorotriazine is

TABLE 6.1
Reactive groups in reactive dyes of the major dye manufacturers of the world (1988)

Reactive group	First sold	Trade name and original manufacturer[a]	Used for
(D)—$SO_2CH{=}CH_2$ + (D)—$SO_2CH_2CH_2OSO_3H$	1952	Remalan (HOE)	Wool
(D)—$NHCOCH_2Cl$ and 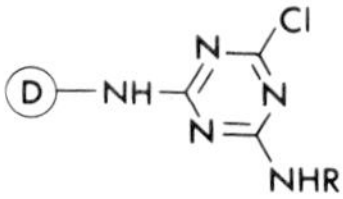	1954	Cibalan Brilliant (CIBA) (now Irganol, CGY)	Wool
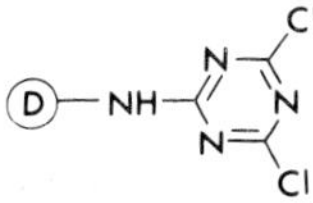	1956	Procion MX (ICI)	Cellulose
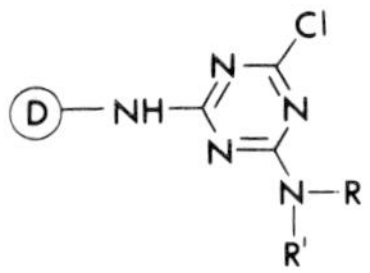	1957	Cibacron (CIBA)	Cellulose
ditto	1957	Procion H (ICI)	Cellulose
(D)—$SO_2CH_2CH_2OSO_3H$	1958	Remazol (HOE) †Remalan Fast (HOE)	Cellulose Wool
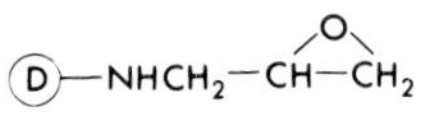 and others (disperse dyes)	1958	Procinyl (ICI)	Polyamide
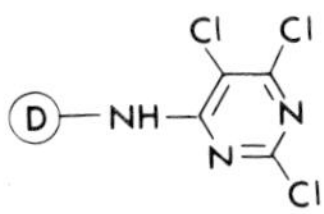	1960	Drimarene Z and X (S)	Cellulose
ditto	1960	Reactone (Gy) (now Cibacron T, CGY)	Cellulose
(D)—$SO_2NHCH_2CH_2OSO_3H$	1960	Levafix J (BAY)	Cellulose
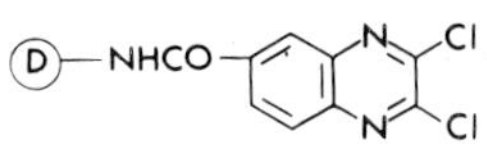	1961	Levafix E (BAY)	Cellulose
(D)—$NHCOCH_2CH_2Cl$ + (D)—$NHCOCH_2CH_2OSO_3H$ + (D)—$NHCOCH{=}CH_2$	1961	Primazine (BASF)	Cellulose
(D)—$SO_2NHCH_2CH_2SSO_3H$	1962	†Wool Fast Turquoise Blue SW (HOE)	Wool
(D)—$CH_2NHCOC(Cl){=}CH_2$	1962	Lanasyn Pure Blue FBL (S)	Wool
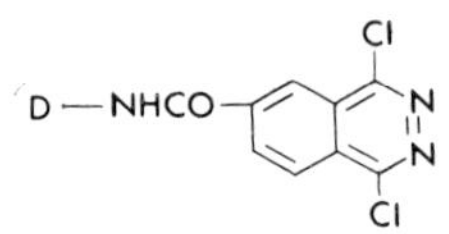 + 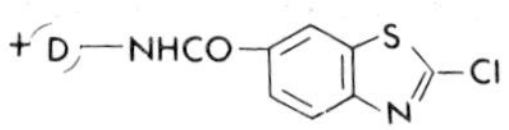 (this group given up later)	1963	Reatex (Fran) (now Elisiane)	Cellulose

Reactive group	First sold	Trade name and original manufacturer[a]	Used for
(D)—$NHCOCH=CH_2$ (metal complexes)	1964	Procilan (ICI)	Wool
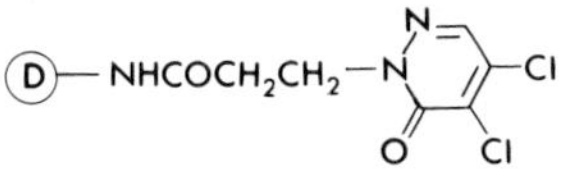	1964	†Primazine P (BASF)	Cellulose
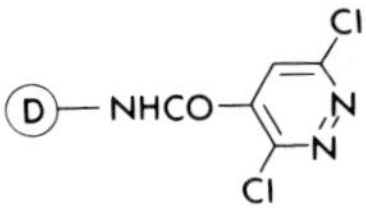 + (D)—$NHCOCH_2CH_2SO_2R$	1964	Solidazol (Fran)	Cellulose
(D)—$SO_2CH_2CH_2NR_2$	1965	Remazol H (HOE)	Cellulose
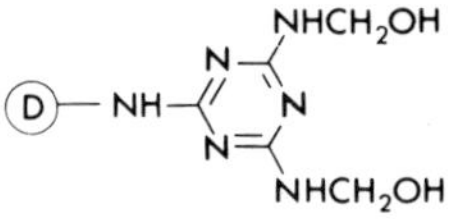	1965	†Calcobond (ACY)	Cellulose
(D)—$NHCO-C(Br)=CH_2$	1966	Lanasol (CIBA)	Wool
(D)—$CH_2NHCOCH_2CH_2SO_2CH_2CH_2OSO_3H$	1967	†Lanafix (NSK)	Wool
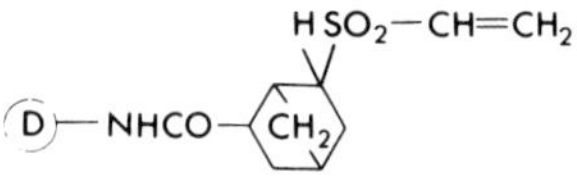	1967	Solidazol N (Fran)	Cellulose
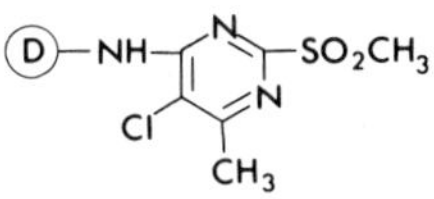	1967	Levafix P (BAY)	Cellulose
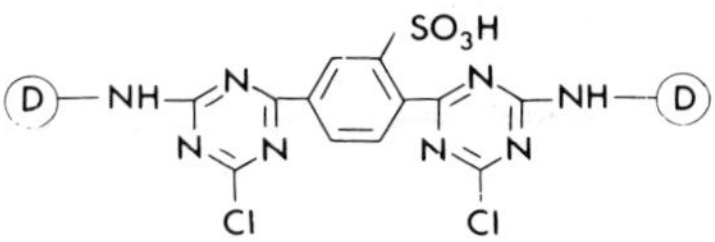	1967	Procion Supra (ICI)	Cellulose
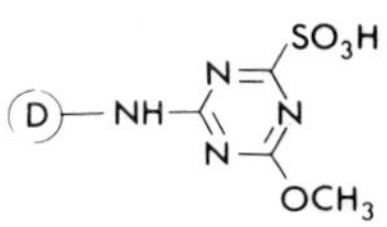 + 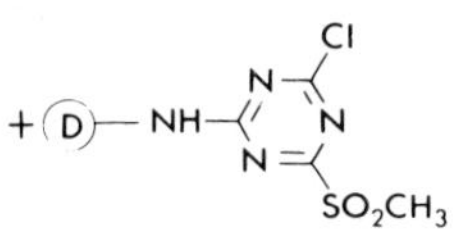	1967	Procion SP (ICI)	Cellulose
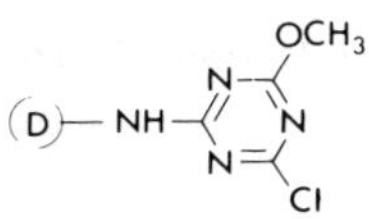	1968	Cibacron Pront (CIBA)	Cellulose

Reactive group	First sold	Trade name and original manufacturer[a]	Used for
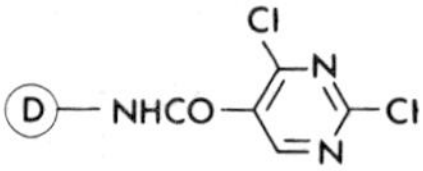	1968	†Reactofil (Gy)	Cellulose
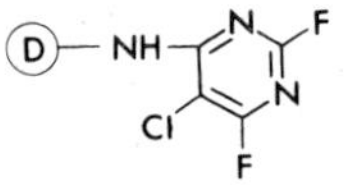	1970	Verofix (BAY)	Wool
ditto	1970	Levafix P–A (BAY)	Cellulose
ditto	1970	Drimalan F (S)	Wool
(D)—$SO_2CH_2CH_2N(CH_3)CH_2CH_2OSO_3H$	1970	Hostalan (HOE)[b]	Wool
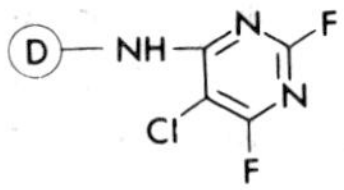	1971	Drimarene R and K (S)	Cellulose
(D)—PO_3H_2	1978	Procion T (ICI)[c]	Cellulose
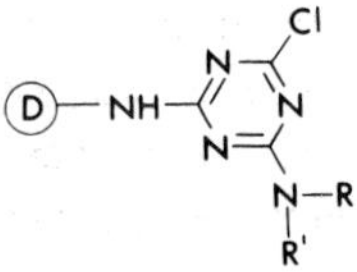	1978	Cibacron F (CGY)	Cellulose
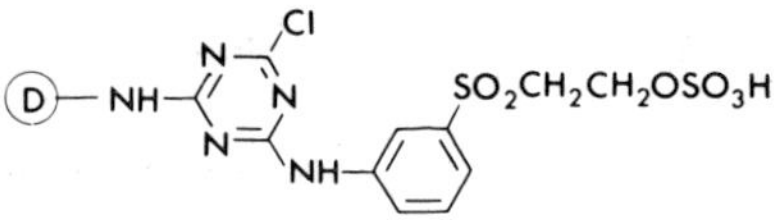	1980	Sumifix Supra (NSK)	Cellulose
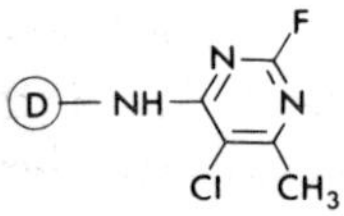	1981	Levafix PN (BAY)	Cellulose
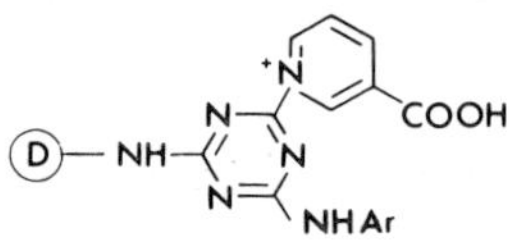	1984	Kayacelon React (KYK)	Cellulose
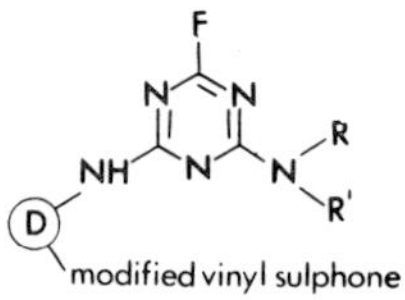	1988	Cibacron C (CGY)	Cellulose

[a] C.I. abbreviations of presently existing manufacturers, old abbreviations only for companies no longer existing as such. Dyes marked † are no longer produced.
[b] Most Hostalan dyes contain, however, one of the reactive groups —$SO_2CH_2OSO_3H$ or —$SO_2CH_2CH_2Cl$.
[c] Now marketed in mixtures with disperse dyes as Procilene dyes.

substituted faster by water or by alcohols (ROH) than by primary or secondary amines (even though amines are stronger bases), whereas the reverse situation is observed for the respective chlorinated triazines; underlying this observation is the concept of 'hard' and 'soft' acids and bases put forward by Pearson [16]. In a nutshell, this regards so-called 'hard' nucleofugic leaving groups as those that are substituted rapidly by 'hard' nucleophiles, but slowly by 'soft' nucleophiles (and vice versa for the substitution of 'soft' nucleofugic leaving groups by 'soft' and 'hard' nucleophiles). Fluoride ions, as well as water, alcohols and their respective anions, are 'hard' nucleofugic leaving groups and 'hard' nucleophilic reagents; chloride ions are less 'hard' than fluoride ions, and verge on 'softness'. Amines are 'soft'. For the same reason it is possible to synthesize cyanuric trifluoride from cyanuric trichloride simply by treatment with sodium fluoride. The chlorine–fluorine exchange in chlorinated pyrimidines used as intermediates for various reactive classes is based on the same principle. The chemistry of fluorinated heterocyclic intermediates was summarised recently in the context of reactive dye synthesis by Chambers [17] and by Stead [18].

In the mid-1970s research chemists of Ciba-Geigy found ways to minimise hydrolysis during the synthesis of monofluorotriazine dyes. It is now known that, in order to obtain a pure dye, as little salt as possible should be present in the solution where the last synthetic step is made. Therefore the dye cannot be precipitated by addition of salt, but the solution has to be evaporated at low temperature *in vacuo*. Recognition of this enabled the firm to start producing such dyes on a technical scale.

Research and development of dyes containing *two* reactive groups had started already in the late 1950s. For obvious reasons higher fixation ratios are obtainable with such dyes (the *fixation ratio* or *degree of fixation* is the ratio of covalently bonded to total dye present on the fibre). The first commercial dye of that type was probably Remazol Black B (C.I. Reactive Black 5) (IV). Later ICI developed the Procion Supra dye range, with two monochlorotriazine groups per molecule.

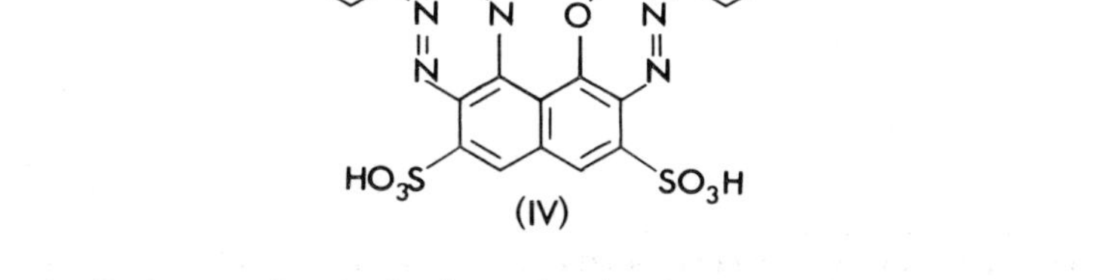

(IV)

Historically interesting is the introduction in 1980 of Sumitomo's Sumifix Supra dyes, which contain two *different* reactive groups. They are basically monochlorotriazine dyes characterised by a substitution of the third chlorine atom of the original cyanuric chloride by the sulphuric acid ester of 3-aminophenyl-β-hydroxyethylsulphone, i.e. by a combination of two reactive groups. (This type of structure was the subject of a Hoechst patent as long ago as 1961 [19]!). The advantages of the mixed bifunctional reactive system of Sumifix Supra dyes with respect to dyeing and fastness properties was discussed by Fujioka and Abeta [20]. In the 1980s there has been increased activity in synthesis of commercial dyes containing two reactive groups, which may be identical or different groups.

Nearly all the reactive dye classes for cellulose fibres use basically the original dyeing process of Rattee and Stephen, i.e. alkaline conditions depending on the reactivity of the reactive groups and the temperature. The two exceptions are the Calcobond dyes of American Cyanamid and the Procion T dyes of ICI. The Calcobond dyes form dye–fibre bonds under acidic conditions. Their reactive groups are comparable to the *N*-methylol derivatives of urea and cyclic ureas used for crosslinking of cellulose fibres; those reagents are also applied under acidic conditions.

The Procion T dyes are principally of interest because they are the only class of reactive dyes that were discovered by chemists who were employed not by dye manufacturers but by a textile company, Burlington Industries [21]. These dyes contain phosphonic acid groups, i.e. groups in which a carbon–phosphorus bond is present. The dyes are designed to be used on cellulose–polyester blends in the Thermosol process, together with disperse dyes for the polyester part of the fabric. The padding liquor contains cyanamide or dicyandiamide besides the disperse dye and the Procion T dye and, if appropriate, textile auxiliary products. The fabrics are padded and cured at temperatures around 200 °C. Cyanamide and dicyandiamide are dehydrating reagents at this temperature, forming urea under these conditions; the —PO_3H_2 residues of the reactive dye form esters with the hydroxyl groups of cellulose, e.g.:

$$\textcircled{D}-PO_3H_2 + \text{CellOH} \rightarrow \textcircled{D}-\underset{\displaystyle \mid \atop OH}{\overset{O \atop \uparrow}{P}}-\text{OCell} + H_2O$$

Today Procion T dyes are marketed in mixtures with disperse dyes under the name Procilene dyes.

The various dye ranges are characterised by widely differing reactivities. The spread in reactivity of the major ranges for dyeing cellulosic fibres is shown in Figure 6.1. In order to obtain maximum fixation yield appropriate

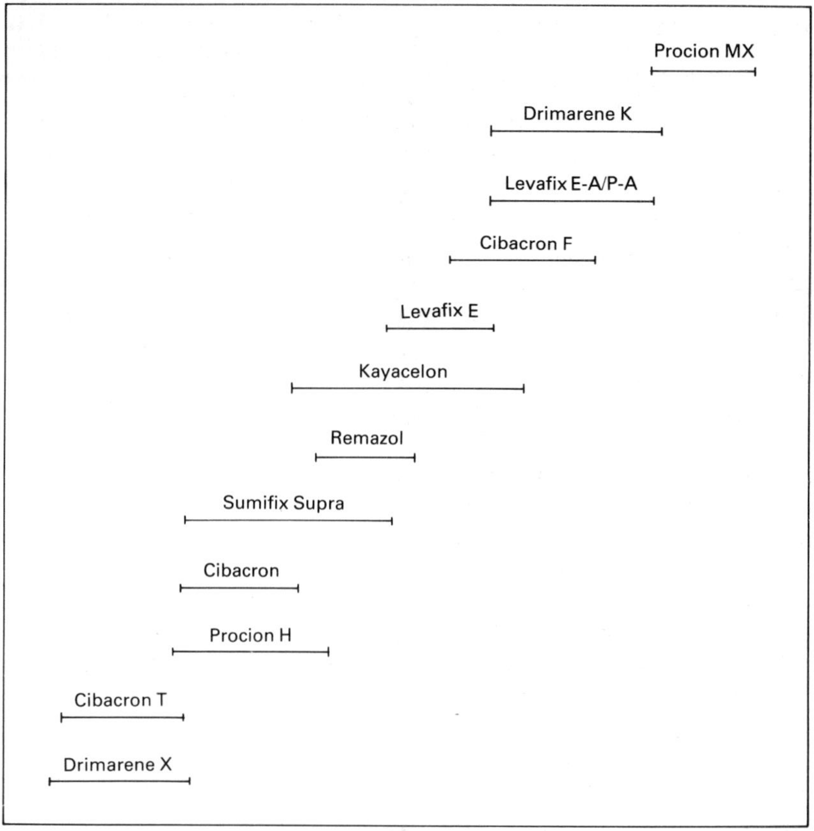

Figure 6.1 – Relative reactivities of reactive dyes for cellulose fibres

fixation temperatures and pH values are chosen, i.e. high-reactivity dyes are dyed at low temperatures (Procion MX at 30 °C) and relatively low pH (pH 8–9), while dyes at the other extreme are dyed at boiling temperature and pH 10–11.

Although reactive dyeing began with wool, reactive dyes did not become important for wool until 1967, when CIBA introduced the Lanasol dyes. These dyes contain an α-bromoacrylamide (—NHCOC(Br)═CH_2) group, The success of the Lanasol dyes is due to their high fixation ratio. Dyes containing a difluorochloropyrimidyl group [e.g. Drimalan (S) and Verofix (BAY) dyes] introduced in 1969 show a similar high fixation ratio.

Reactive dyeing of man-made fibres is possible only when the fibres contain functional groups that can react with a reactive group of a dye by one of the mechanisms discussed in section 6.3. It is obvious that regenerated-cellulose fibres, particularly viscose rayon, can be dyed in essentially the same way as cotton. With hydrophobic man-made fibres the situation is less favourable. For example, polyamides contain amino end-groups which are potentially suitable for reactive dyeing, but the low concentration of these groups (*ca* 40 mmol kg^{-1} fibre) prevents the attainment of full depths with reactive dyes; this is obviously why the Procinyl dyes of ICI, which are disperse dyes containing either monochlorotriazine or epoxy reactive groups, never became technologically important.

Three monographs on reactive dyes are currently available: those by Beech in English [22], by Lukos and Ornaf in Polish [23] and by Krichevski in Russian [24]. Venkataraman devoted a whole volume in his series *The chemistry of synthetic dyes* to reactive dyes [25]. It contains comprehensive chapters on reactive groups, on chromophoric systems, and on application and properties written by three chemists from Bayer, namely Siegel, Schündehütte and Hildebrand. These authors give the largest number of literature references on reactive dye structure, application and properties ever published – about 900 patents and 700 articles from scientific and technical journals. Another volume of the Venkataraman series [26] contains a chapter by Rattee on the physico-chemical aspects of dye fixation and dye–fibre bond hydrolysis.

Hildebrand [27], in a series of papers published in 1970–71, has surveyed 578 papers and patents related to reactive dyes and their application. Rosenthal [28] continued this compilation by reviewing some 360 papers and patents published in 1971–75. Recently Rattee [29] and Ramsay [30] reviewed the general development of reactive dyes for cellulose and Lewis [31] that of wool reactive dyes. The history of reactive dyes was reviewed by Brunnschweiler and Siegrist [32].

In this chapter we discuss in general only reactive groups which are or were used in commercial dyes: their number (20 to 30) is relatively small compared to the total number of reactive groups described in the patent literature (>200, see [25]).

6.2 STRUCTURE OF REACTIVE DYES

6.2.1 General classification

The characteristic structural features of a reactive dye (V) are shown below. In some cases the reactive group is attached directly, i.e. without a bridge, to the chromogenic system.

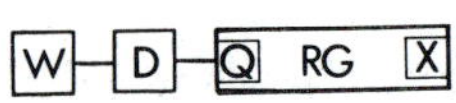

(V)

RG = reactive group
Q = bridge link
X = leaving group
D = chromogen
W = water-solubilising group

In the following sections we shall discuss in turn reactive groups, the leaving group, the bridge link, the chromogen and the water-solubilising group.

6.2.2 Reactive groups

All reactive dyeing is based on the fact that cellulose, protein and nylon fibres contain groups (HY) that are able to react as nucleophiles, i.e. hydroxyl, amino or mercapto groups, and reactive groups, chosen accordingly, may be classified according to their reaction mechanisms.

Class A: Groups that react by the nucleophilic addition–elimination (heteroaromatic) substitution mechanism [33]

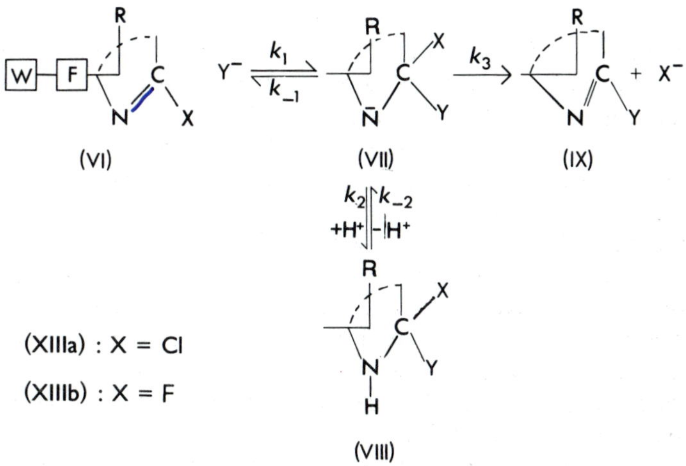

(VI) (VII) (IX) (VIII)

(XIIIa) : X = Cl

(XIIIb) : X = F

The characteristic features of the mechanism illustrated above are as follows:

(a) Attack on the heterocyclic carbon of the reactive group by the most basic form of the nucleophilic group (Y) of the substrate (e.g. the $CellO^-$ groups of cellulose), i.e. specific base catalysis of the first reaction step (k_1). This catalysis is the essential part of Rattee and Stephen's invention of the first reactive dyes for cellulose. In the dyeing process the pH value, and therefore the equilibrium concentration of cellulosate anions [$CellO^-$], has to be increased for the formation of the dye–fibre bond.

(b) If VIII accumulates rapidly, general base catalysis is observed (k_2) [34–38].

(c) The last step (k_3) does not take place in the reverse direction with the ions X^- (Cl^-, F^-, $^-SO_2CH_3$) under dyeing conditions.

The following are important representatives of these reactive groups (commercial names are given in parentheses; R = H or, in a few cases, CH_3):

- di- or monohalogenotriazines (X, Procion MX; XIa, Cibacron and Procion H; XIb, Cibacron F; XIc, Cibacron Pront; XId, Procion SP; XIe, Sumifix Supra; XIf, Cibacron C; XII, Procion Supra);
- 2,4,5-trihalogenopyrimidines (XIIIa, Drimarene X and Z, Cibacron T; XIIIb, Verofix, Levafix P-A, Drimalan F, Drimarene R and K);
- 2-methylsulphonyl- and 2-fluoro-4-methyl-5-chloropyrimidine (XIVa and XIVb, Levafix P and PN, respectively);
- 3,6-dichloropyridazone (XV, Solidazol);
- 2,3-dichloroquinoxaline (XVI, Levafix E);
- 1,4-dichlorophthalazine (XVII, Reatex);
- 4,5-dichloropyridazone (XVIII, Primazin P).

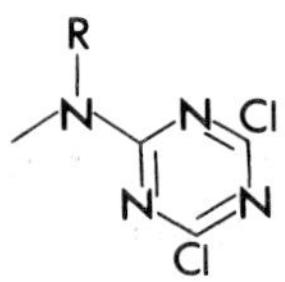

(X)

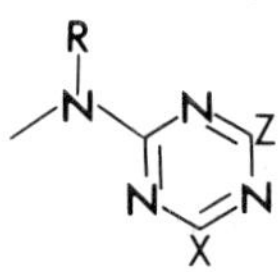

(XI)

(XIa) : X = Cl; Z = NHR, NHAr, NR_1R_2, NRAr

(XIb) : X = F; Z = NR_1R_2, NRAr

(XIc) : X = Cl; Z = OR

(XId) : X = Cl; Z = SO_3H

(XIe) : X = Cl, Z = NH-C_6H_4-3-$SO_2CH_2CH_2OSO_3H$

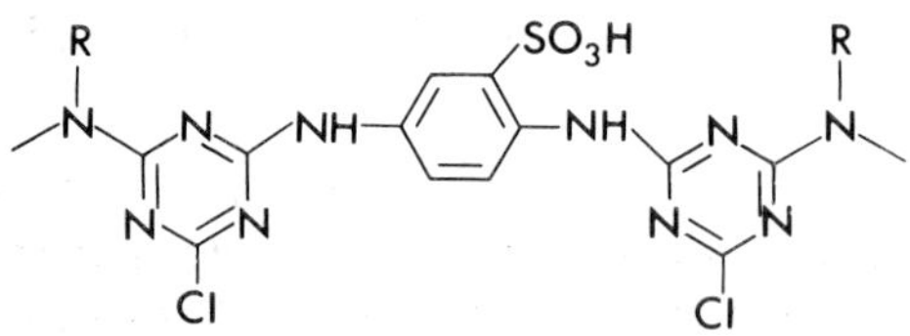

(XII)

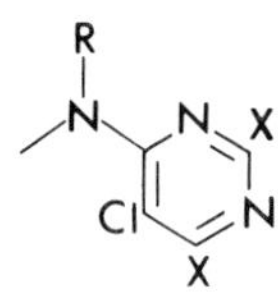

(XIIIa: X = Cl)

(XIIIb: X = F)

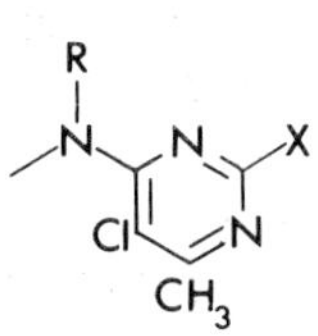

(XIVa) : X = SO_2CH_3

(XIVb) : X = F

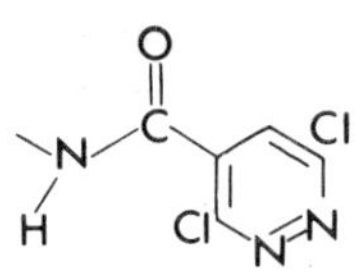

(XV)

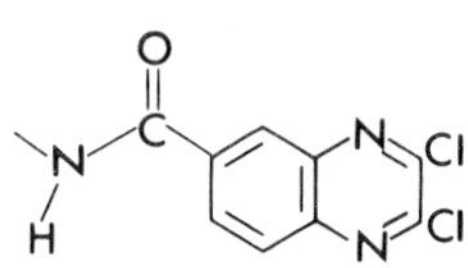

(XVI)

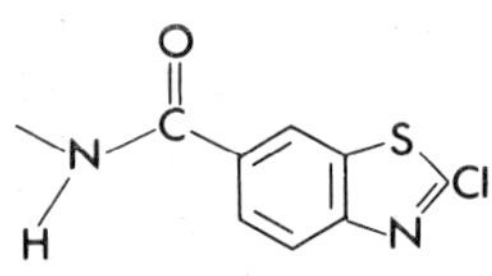

(XVII)

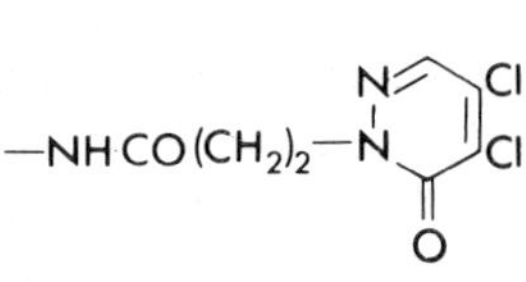

(XVIII)

It can be seen that in most of the reactive systems X–XVIII, X in VI is a chlorine atom, but there can be other nucleofugal leaving groups such as fluorine, alkyl sulphones, sulphonic acid groups or, as in XIX (Kayacelon), quaternary ammonium groups.

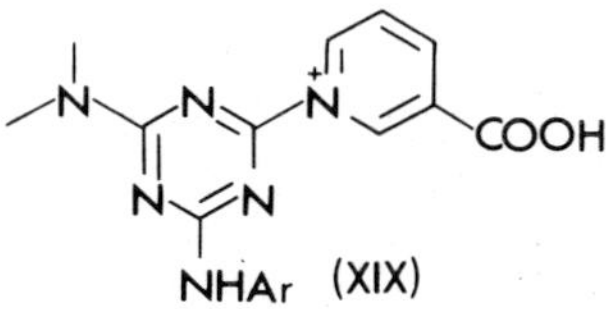

(XIX)

The reactivity of these *N*-heterocyclic reactive groups is greater – in cases where the addition reaction (k_1 in the reaction scheme above) has a limiting effect on the rate – when the electrophilic carbon atom is made more positive. This can be achieved by:

(a) additional electron-attracting cyclic nitrogen atoms (cf. the higher reactivity of chlorotriazine derivatives relative to the comparable pyrimidines),
(b) introduction of substituents R or Z which are less electron-releasing in VI (cf. the increase in reactivity XIa < XIc < X),
(c) the use of strongly electron-attracting leaving groups such as quaternary ammonium groups (XIX) or fluoride instead of chloride (XIa < XIb, XIIIa < XIIIb).

The addition of catalytic amounts of tertiary aliphatic amines such as 1,4-diazabicyclo[2,2,2]octane (DABCO) or 1,1-dimethylhydrazine to dye liquors with monochlorotriazine reactive dyes was recommended some 20 years ago [39]. Rapid substitution of the chlorine takes place, producing a triazine with a quaternary ammonium group, i.e. a significantly better leaving group is formed *in situ*. Unfortunately, this catalysis decreases the ratio of reaction with the fibre relative to that of hydrolysis, and the method has been abandoned.

In the cases of (a) and (b), hydrolysis of the covalent dye–fibre bond of a reactive dyeing (e.g. IX, Y = OCell) is also accelerated, so in general the strength of the covalent dye–fibre linkage diminishes with increasing reactivity of the reactive group [40]. Monofluorotriazine dyes (XIb) and corresponding monochlorotriazine dyes (XIa) with the same groups Z in XIa and XIb have different reactivities (Xb > Xa), but the reaction products with cellulose are identical; the dye–fibre bond strengths are therefore the same for both types.

Cibacron Brilliant Red 3B (XX; C.I. Reactive Red 4) is a typical dye with an *N*-heterocyclic reactive group of type XIa. Procion Blue MX-R (XXI; C.I. Reactive Blue 4) is an example of type X.

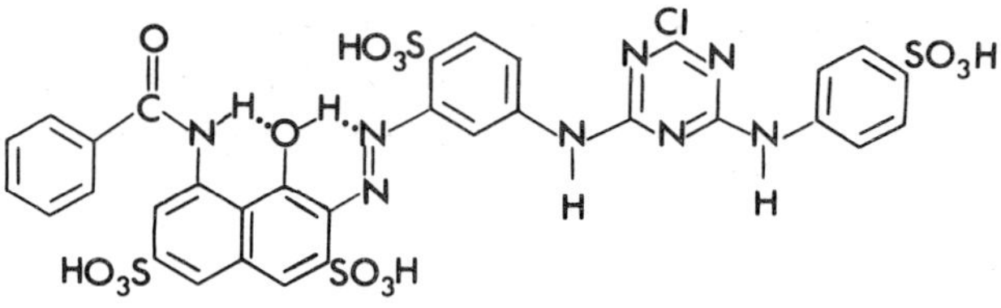

(XX)

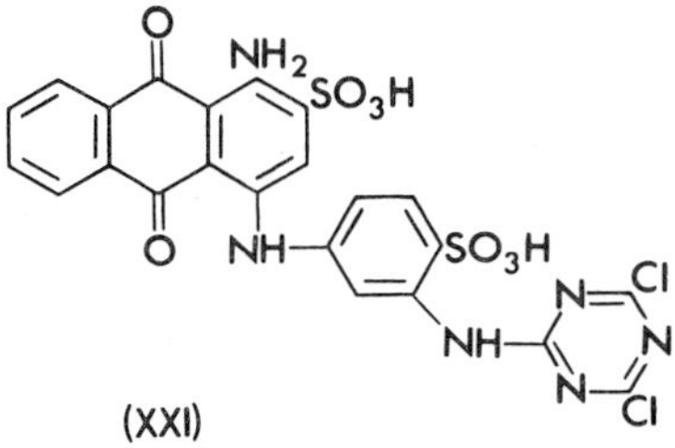

(XXI)

The principle underlying the synthesis of dyes such as XXI is first to prepare the chromogenic part containing at least one free amino group, which can then be reacted with, for example, cyanuric chloride to give the dichlorotriazine dye of type X. Type XIa dyes can be prepared from X by reaction with ammonia or a primary or secondary aliphatic or aromatic amine. It is more common, however, to react cyanuric chloride with a colourless amine as the first step, and then to react the resulting dichlorotriazine derivative with the amino group of the dye. Finally, in cases such as XX, where the chromogenic part containing a free amino group is not easily obtained, cyanuric chloride is reacted in two steps with two amines, the second one containing an additional diazotisable amino group (for XX the first reaction is with sulphanilic acid, the second with 2,4-diaminobenzene-sulphonic acid). The compound formed in this way is diazotised and coupled with an appropriate coupling component (for XX, *N*-benzoyl-H-acid, i.e. 1-benzoylamino-8-naphthol-3,6-disulphonic acid). Dyes with other *N*-heterocyclic reactive groups are prepared similarly. If one chlorine atom of cyanuric chloride has to be substituted by an alkoxy group instead of a colourless amine (dyes of type XIc), the reaction with the alcohol takes place always as the first substitution.

Class B: Groups that react by addition of the nucleophilic group of the substrate to a C═C double bond on the reactive group

The double bond of the reactive group is in most cases not present in the reactive dye, but is formed later in the process of dyeing in the first, general base-catalysed step of the reaction mechanism shown below. The addition of the nucleophilic fibre group HY is, as with reactive groups of type VI, specific base-catalysed, i.e. it is the basic equilibrium form Y^- which is added to the β-carbon, followed by a fast (not rate-limiting) protonation of the α-carbon:

$$\underset{\text{(XXII)}}{-\,ZCH_2CH_2X} \underset{k_{-1}}{\overset{k_1}{\rightleftharpoons}} \underset{\text{(XXIII)}}{\underset{(+HX)}{-\,ZCH{=}CH_2}} \underset{k_{-2}}{\overset{k_2(+Y^+)}{\rightleftharpoons}} -\,ZCHCH_2Y \underset{k_{-3}}{\overset{k_3(+H^+)}{\rightleftharpoons}} -\,ZCH_2CH_2Y$$

Probably some dyes with the general reactive group XXII may react also by a one-step (S_N2) mechanism of substitution which is also specific base-catalysed:

$$—ZCH_2CH_2X + Y^- \rightarrow \left[—ZCH_2\overset{X}{\underset{Y}{CH_2}}\right] \rightarrow —ZCH_2CH_2Y + X^-$$

Representatives of the general type XXII include:

(a) sulphuric acid esters of β-hydroxyethylsulphones ($—SO_2CH_2CH_2OSO_3H$) which form vinylsulphones in the dyebath by abstraction of sulphuric acid (Remalan, Remazol, Sumifix Supra and Cibacron C dyes), and the related Sumitomo dyes containing sulphatoethylsulphonacrylamide groups ($—NHCOCH_2CH_2SO_2CH_2CH_2OSO_3H$), which today are no longer produced;
(b) *N*-substituted β-aminoethylsulphones which also form vinylsulphones ($—SO_2CH_2CH_2N(CH_3)CH_2CH_2SO_3H$, Hostalan dyes);
(c) sulphuric acid esters of β-hydroxypropionamides ($—NHCOCH_2CH_2$-OSO_3H) and β-chloropropionamides ($—NHCOCH_2CH_2Cl$) which – at least formally – may form acrylamide groups (Primazin dyes);
(d) epoxy groups and their precursor 2-chloro-1-hydroxyethyl groups (Procinyl dyes);
(e) aziridine and 2-chloroethyl-1-amino groups, mentioned in patents, were probably not used for commercial products. Related to the 2-chloroethylamino reactive groups are their sulphonamides ($—SO_2NHCH_2CH_2Cl$) as well as the corresponding sulphuric acid esters of the respective 2-hydroxyethylsulphonamides ($—SO_2NHCH_2CH_2OSO_3H$), which are used in Levafix dyes.

An important difference between the fixation mechanisms of Class A and Class B dyes is that with Class A dyes the functional group of the textile fibre participates in the addition as well as in the elimination step, whilst with Class B dyes the first elimination step is independent of the substrate. Therefore, with the reactive systems of Class B dyes there is in principle the possibility, at constant dyebath pH, of optimising, relative to each other, the rate of formation of the true reactive compound XXIII and the rate of diffusion of the dye by varying the concentration and type of buffer; this may be of particular significance in obtaining level reactive dyeings. Furthermore, the degree of fixation of Class B dyes, in contrast to those of Class A, must always depend on the type and concentration of buffer used in the dyeing process (general base-catalysed reaction rate k_{-3} in the reaction mechanism shown for Class B dyes on p. 443).

For the synthesis of reactive dyes containing groups (a) to (e) above, a chromogenic system with a free amino group can be reacted with a compound supplying the reactive group, e.g. a β-substituted chloropropane for (c), epichlorohydrin (3-chloro-1,2-epoxypropane) for (d).

Procinyl Blue RS (XXIV, C.I. Reactive Blue 6) is an example of type (d).

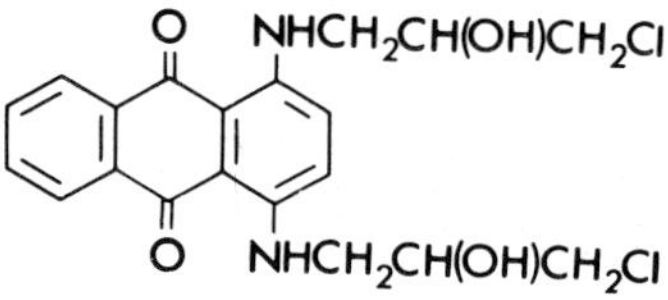

(XXIV)

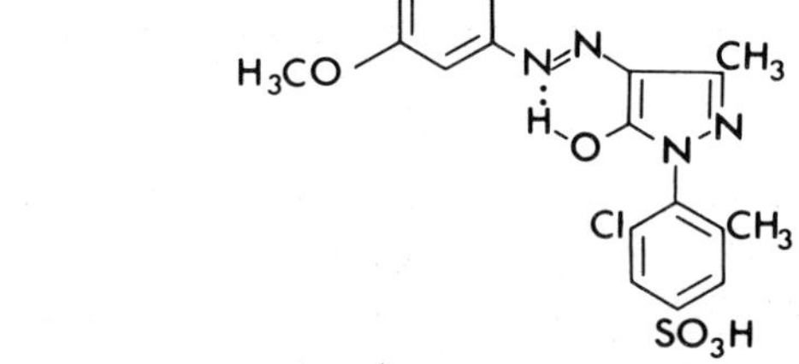

(XXV)

In dyes of types (a) and (b), an intermediate containing a β-hydroxyethylsulphone group is used for the synthesis of the chromogenic system, which, at the end of the synthesis only, is esterified with sulphuric acid. Remazol Golden Yellow G (XXV; C.I. Reactive Yellow 17), a typical representative of this class of reactive dyes, is made by coupling diazotised 4-amino-2,5-dimethoxyphenyl-β-hydroxyethylsulphone with 1-(2′-chloro-4′-sulpho-6′-methyl)phenyl-3-methyl-5-pyrazolone. The resulting dye is then esterified by dissolving it in concentrated sulphuric acid.

The β-hydroxyethylsulphone derivatives are obtained by transforming an intermediate product into the sulphinic acid and reacting it with ethylene oxide (epoxyethane), or by oxidising β-hydroxyethylsulphide, prepared by reacting a thiol with ethylene oxide.

Class C: Groups that react via several addition and elimination steps with the functional group HY
The mechanism of reaction is shown below:

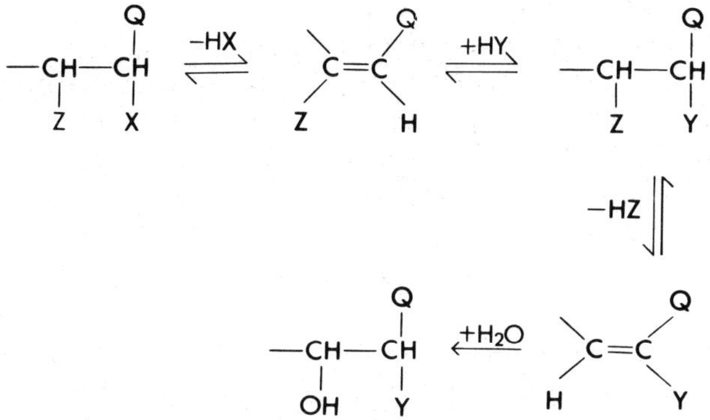

The α,β-dibromopropionamides (—$NHCOCHBrCH_2Br$) are typical of this class and are used in Lanasol dyes, as are the α-bromoacrylamides (—$NHCOCBr{=}CH_2$); the corresponding α-chloroacrylamide group is present in Lanasyn Brilliant Blue FBL (C.I. Acid Blue 227). Reactive dyes with these groups can be synthesised in the same way as those of type (c) of Class B above.

Class D: Groups that react with the dye under acidic conditions
These contain *N*-methylol (*N*-hydroxymethyl) groups, e.g. obtainable by reacting dyes containing melamine (1,3,5-triaminotriazine) groups with formaldehyde to give 2,4-di(hydroxymethylimino)triazines (Calcobond dyes). They are no longer produced, as American Cyanamid has discontinued the production of dyes of all classes, not only reactive dyes.

Class E: Reactive dyes with phosphonic acid groups and phosphoric acid ester groups
Commercial dyes of this type (Procion T and Procilene dyes) contain only phosphonic acid groups ($-PO_3H$), but not phosphoric acid ester groups (—OPO_3H_2), although both types are mentioned in patents. Dyes with the former group give better results with respect to fixation ratio and fastness properties [41].

Aromatic phosphonic acids can be synthesised from diazonium salts with phosphorus trichloride in the presence of copper(I) ions (Sandmeyer reaction), or the chlorine atom in benzyl chloride can be substituted by a —$P(O)(OR)_2$ group using phosphorus trialkyls in an Arbusov reaction, followed by hydrolysis of the alkoxy groups.

Class F: 'Reactive dyes' with thiosulphato (—SSO_3H) groups

Two ranges of reactive dyes (Dykolite dyes and Hydrosol dyes) contain thiosulphato groups directly substituted in aromatic nuclei of the dye, and some dyes of Hoechst (Inthion dyes; Wool Fast Turquoise Blue SW; C.I. Acid Blue 228) contain alkylthiosulphato groups ('Bunte salt dyes'). Osterloh [42] concluded that these dyes attack cysteine groups when applied to wool, and the asymmetrical disulphide formed largely disproportionates into symmetrical dye disulphides which are relatively insoluble. These dyes, therefore, are borderline cases with respect to the definition of reactive dyes. They are discussed more fully in Chapter 7.

Class G: Dyes with two reactive groups

The first commercial dye with two reactive groups, Remazol Black B (IV), and the more recently introduced Sumifix Supra and Cibacron C dye ranges with two different reactive groups (XIe and XIf) have already been mentioned. Reactive dyes with two monochlorotriazine groups have been quite common since ICI's introduction of Procion Supra dyes (XII) in the late 1960s. Some Cibacron E dyes have also two monochlorotriazine groups. Cibacron Brilliant Red 4G-E (C.I. Reactive Red 120) (XXVI) is an example of the same structural type as the Procion Supra dyes.

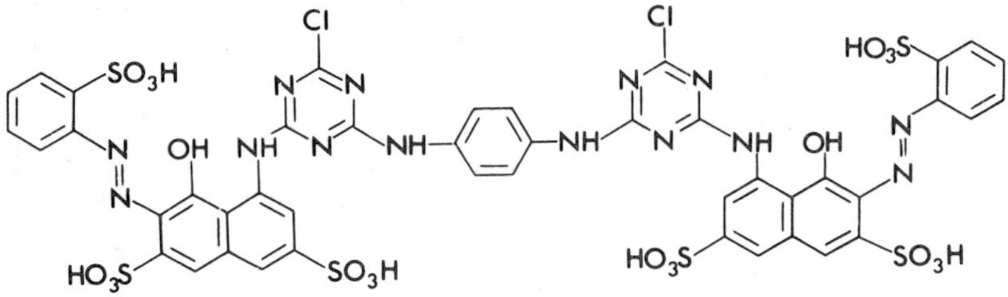

(XXVI)

Some commercial reactive dyes for wool also have two reactive groups, e.g. Lanasol Red G (C.I. Reactive Red 83) contains two α-bromoacrylamide groups [47].

Chekalin's research group [48] reported that fixation of dyes of type XXVI was 80 per cent, significantly higher than that of the corresponding dyes with only one monochlorotriazine group.

In recent years all major dyestuff manufacturers investigated syntheses and dyeing properties of dyes with two reactive groups, but none of them made available scientific information on that work with the exception of the papers of Abeta and Imada [14] and of Chekalin *et al.* [48]. Zollinger *et al.* have, however, investigated the fixation properties of dyes with two different reactive groups (Sumifix Supra dyes) on cellulose [49] and of various dyes with two identical reactive groups on wool and silk [50].

6.2.3 The leaving group

With a given reactive group (RG in the general formula V, p. 438) the reactivity can be influenced significantly by a change of the nucleofugic leaving group. We will discuss briefly the characteristics for that large majority of reactive dyes which react with cellulosic hydroxyl groups under neutral to alkaline conditions or with the various nucleophilic groups of protein fibres (amino, aliphatic and aromatic hydroxyl and mercapto groups), i.e. the conditions for all reactive dyes of Classes A and C (and, in part, of Class B) described in section 6.2.2.

In any nucleophilic substitution it is desirable that the nucleofugic leaving group should form a stable ion or molecule. This criterion is fulfilled for all leaving groups in practical use, i.e. for the halogens, which form F^-, Cl^- and Br^- ions, and for the sulphuric acid esters of β-hydroxyethylsulphones, which form sulphate dianions (the proton of the —OSO_3H group is already dissociated under dyeing conditions). The methylsulphone group of the Levafix P dyes dissociates as the methanesulphonate anion, and the sulphonic group attached to the triazine ring of some Procion Supra dyes forms sulphite ions.

As mentioned in section 6.1.2, the concept of 'hard' and 'soft' bases (and acids) [16] is also relevant, although because of the heterogeneity of the chemical structures of reactive groups it is difficult to rationalise the practical experience. As an example, the success of the Lanasol dyes for wool dyeing is understandable since both the bromine to be substituted and the amino groups of wool keratin are relatively 'soft' nucleophiles. This is likely to be the reason why α-bromoacrylamido and not the corresponding α-chloroacrylamido groups are used. On the other hand, 2,4-difluoro-5-chloropyrimidine groups (XIIIb) are more suitable for reactive dyes for wool than are 2,4,5-trichloropyrimidine groups. On the basis of the 'hard'/'soft' concept one would expect the reverse result! It has to be emphasised, however, that the very complex competitive system of reactive dye fixation versus hydrolysis is still to a large degree incompletely understood and therefore such structural concepts as 'hardness' cannot yet be applied for reliable predictions.

6.2.4 The bridge link

The bridge link is, in most cases, necessary for synthetic reasons. It influences three principal factors.

(a) *The reactivity of the reactive system* Dissociation of the imino bridge (—NH— → —N⁻—) of types X–XIV may reduce the reactivity of the reactive groups by several powers of ten [43–45]. This, in many cases, also entails a diminution of substantivity, and can lead to lower fixation [34, 35, 37].

(b) *The degree of fixation (selectivity)* In reactive groups of types X–XIV the proton of the imino bridge is in rapid tautomeric equilibrium with one or several isomers containing the proton at a heterocyclic nitrogen atom, e.g. XXVII.

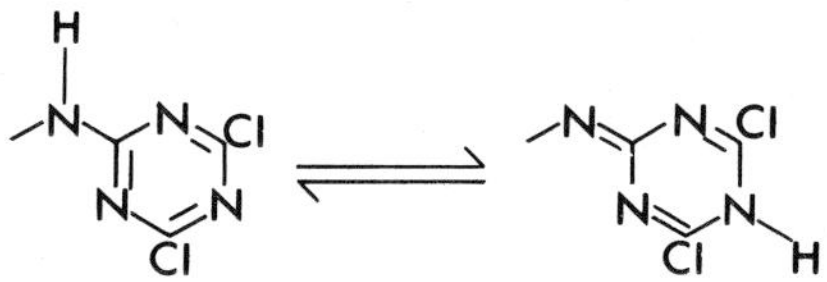

(X) (XXVII)

For these reasons a few reactive dyes contain a methylated imino bridge (—N(CH_3)—); Procion Orange MX-2R (C.I. Reactive Orange 4) and Procion Scarlet H-RN (C.I. Reactive Red 33) are examples.

Tautomeric forms such as XXVII have a lower selectivity [36, 37], i.e. a lower ratio of reactivities with cellulose and with water. Therefore a given reactive group has the highest selectivity when the imino bridge in X is alkylated. On the other hand, alkylation generally, but not always, reduces the substantivity for the substrate and hence the degree of fixation. The work of Omura and Kenmochi [51] on Sumifix Supra dyes (XIe) showed that methylation of the amino group between the triazine ring and the benzene ring containing the sulphone group can increase the degree of fixation. Changes in the length and flexibility of the bridge also affect fixation, e.g. when an imino bridge is replaced by a α,ω-di-iminoalkylene chain [—NH(CN_2)$_n$NH—].

(c) *The stability of the reactive dyeing* [52] The link between chromogen and reactive group can be split, as well as the dye–fibre bond.

6.2.5 The chromogen

The coloured part of a reactive dye (i.e. the chromogen) may belong, at least in principle, to any of the chemical classes of commercial dyes, provided it has the necessary substantivity for the fibre. In general, the following are used: for yellow, orange and red, simple, metal-free monoazo dyes; for ruby, violet and navy, cupriferous mono- and dis-azo dyes; for bright blue, derivatives of 1-amino-4-arylaminoanthraquinone-2-sulphonic acid and phthalocyanine derivatives; for blacks, disazo dyes. The influence of the chromogen on reactivity has been studied [34, 35, 37]. Dyes with a high association tendency, because of association pre-equilibria, reduce the rate of hydrolysis and frequently improve fixation owing to their higher substantivity. In some cases, however, undesirable dimers may form [53]. Reactive dyes which are applied at a higher temperature because of a lower reactivity of the reactive group normally require larger molecules in order to provide sufficient substantivity for the substrate at the dyeing temperature.

6.2.6 Water-solubilising groups

Reactive dyes for cellulose and protein fibres require from one to three (in a few cases four) sulphonic acid groups. Reactive disperse dyes for nylon fibres do not generally require special solubilising groups of the types used for other disperse dyes, however, because the reactive groups themselves confer sufficient aqueous solubility for the disperse-dyeing process.

6.2.7 Fixation of dyes with crosslinking and other reagents

In the late 1950s, i.e. at the beginning of the reactive dye era, it was found that dyes with suitable nucleophilic functional groups (e.g. non-hydrogen-bonded amino groups) can be indirectly bonded to cellulose with crosslinking reagents. The Basazol dye range of BASF resulted from those investigations [54]. In 1982 Sandoz introduced the range of Indosol dyes for dyeing cellulose fibres. These are selected direct dyes without specific nucleophilic groups. They are applied together with a crosslinking reagent (e.g. dimethyloldihydroxyethyleneurea, DMDHEU) and a fixation reagent made by condensation of polyamines (e.g. diethylenetriamine) with dicyanodiamide or related compounds [55]. Details of the chemistry and, in particular, of the mechanism of fixation were disclosed in 1985, but the technical application papers published [56] do not contain reliable and scientifically convincing information on the chemical background. It may be that ionic interactions between the anionic direct dyes and the cationic fixation reagent are important. It seems unlikely, however, that the Indosol system forms a new class of reactive dyeing as defined at the beginning of this chapter. The fastness properties of Indosol dyeings are slightly lower than those of most reactive dyeings, but definitely higher than those with direct dyeing even with subsequent treatment with conventional cationic reagents.

6.3 MECHANISMS OF REACTIONS OF REACTIVE GROUPS

6.3.1 Hydrolysis by water molecules and by hydroxide and hydroxonium ions

For technological and scientific reasons, hydrolysis of reactive dyes is a very important process:

(a) Aqueous solutions of reactive dyes become unreactive towards textile substrates through hydrolysis.
(b) Hydrolysis is the same type of reaction as fixation to the functional groups of fibres, in particular to cellulose hydroxyl groups (R—OH: cellulose, R = Cell; water, R = H).

From (a) it would seem advisable to make reactive dyes very stable against hydrolysis, but (b) indicates that this would be undesirable. What is required is that the ratio of the reactivity towards functional groups of the fibre to that of hydrolysis should be as high as possible.

Knowledge of the kinetics and mechanism of hydrolysis is useful for an understanding of the mechanism of covalent-bond formation with the textile substrate. As this is a heterogeneous reaction, its mechanism is intrinsically difficult to investigate. Investigations of the mechanism of reactions with hydroxyl groups of water and with the hydroxide ion in homogeneous solutions are therefore useful model reactions for elucidating the fixation of reactive dyes to hydroxyl groups of cellulose and other substrates. A better model reaction than hydrolysis is alcoholysis of reactive dyes, which is discussed later.

Although no detailed investigations have been published for acidic solutions, it can be concluded indirectly from studies of dye–fibre bond stabilities [40] that, for dyes with heteroaromatic reactive groups (Class A), the rate of hydrolysis increases also on the acidic side of the pH scale (below *ca* pH 5). Stability in acidic solutions is not a problem for the practical dyer; it is, however, important for the storage of highly reactive dyes, in particular for the dichlorotriazine (Procion MX) and the monofluorotriazine (Cibacron F) dyes. Moisture may hydrolyse some dye and, therefore, form some hydrochloric acid. This acid protonates one of the triazine nitrogen atoms, giving the triazinium ion (e.g. XXVIII), which is more reactive than a triazine group. Acid hydrolysis is therefore autocatalysed. For this reason, as recommended by Stephen [57], Procion MX and Cibacron F dyes contain disodium phosphate to neutralise the hydrochloric acid formed. It can be shown [58] that buffered solutions of these dyes are much more stable than those without additives.

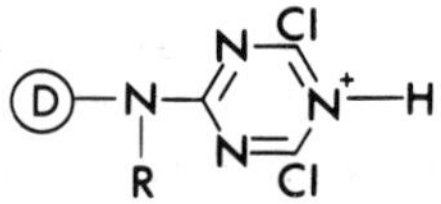

(XXVIII)

Investigations of hydrolysis in neutral to alkaline solutions are more important, because these conditions are closely related to those of fixation of reactive dyes. In general, hydrolysis of all classes of reactive dyes is faster the higher the alkalinity. Detailed and quantitative investigations have been described only for dyes of Class A, in particular for mono- and dichlorotriazines and pyrimidine derivatives.

Before discussing specific experimental results, some kinetic principles will be briefly explained. The substitution of a nucleofugal leaving group (e.g. Cl) by a base (e.g. H_2O, ^-OH or ^-OR) consists in the attack of an ion or molecule with a lone pair of electrons on the carbon atom to which the leaving group is attached [33]. This carbon atom is electrophilic (electron-attracting) because of the presence of electronegative nitrogen atoms in the heteroaromatic ring, e.g. the triazine ring. In an alkaline aqueous system, water molecules and hydroxide ions may be the nucleophilic reagent; hydroxide ions are, however, the more reactive by many powers of ten (measured by the respective rate constants). The reaction with water molecules therefore has to be considered only in cases of very low concentrations of hydroxide ions.

If only the attack of hydroxide ions on the reactive group is rate-determining, a second-order rate law is to be expected, i.e. the rate of concentration decrease of the reactive group will be proportional to the concentration of the reactive group as well as to that of the hydroxide ions:

$$\text{rate} = k_{\text{hy}}[\text{D}][\text{OH}] \qquad (6.1)$$

where k_{hy} is the second-order rate constant of hydrolysis, [D] is the concentration of reactive dye, and [OH] is the hydroxide ion concentration (for simplicity's sake, the ionic charges are omitted from the symbols). Since hydroxide ions will be used up during the reaction, their concentration will decrease (decrease of pH). This effect can be virtually eliminated by the use of a buffer solution. In this case [OH] in Eqn 6.1 is a constant, which can be included in the rate constant k_{hy}, which then becomes the pseudo-first-order rate constant $k_{\text{hy}}^{\text{ps}}$ where

$$\text{rate} = k_{\text{hy}}^{\text{ps}}\ [\text{D}] \qquad (6.2)$$

From Eqns 6.1 and 6.2 it is clear that the relationship between log $k_{\text{hy}}^{\text{ps}}$ and pH should be linear with unit slope. Several authors [34–38, 43–46, 59–61] have investigated this problem and their results are summarised schematically in Figure 6.2. Curve A illustrates the simple linear relationship, but Figure 6.2 shows the existence of a variety of other relationships between the logarithm of the pseudo-first-order rate constant and the pH value of buffer solutions.

The relationship A was found only for model compounds [43, 44], for

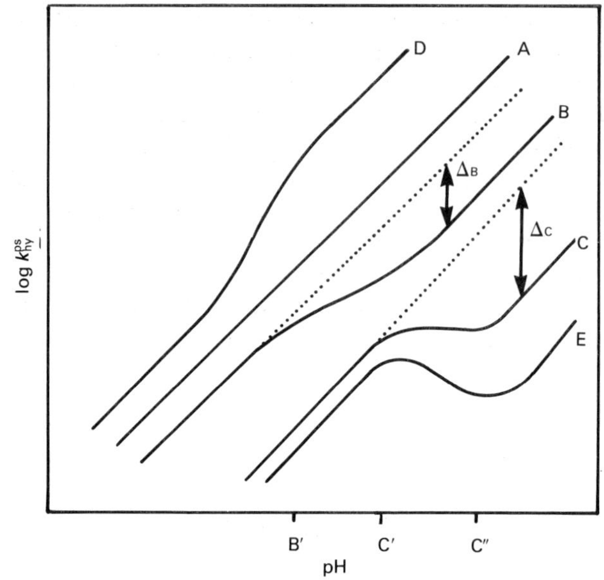

Figure 6.2 – Relation between log k_{hy}^{ps} and pH

monohalogenotriazine dyes [46] and for dyes that are not in commercial use [34–37], all of which have an *N*-alkyl bridge, instead of an imino (—NH—) group, between the heteroaromatic ring (IX–XIII) and the chromogenic part of the dye. It is probable that dyes with other reactive groups, belonging to Class A and other classes (pp. 438), also show a linear dependence of the hydrolysis rate on hydroxide ion concentration but, as mentioned previously, quantitative measurements do not seem to have been published.

When the alkylimino bridge is replaced by an imino group, rate-of-hydrolysis relationships illustrated by curves B, C, D and E are obtained. Curves B and C are explainable, as shown by Ackermann and Dussy [43], Horrobin [44], Rys *et al.* [34–37] and Aspland and Johnson [45, 60], by an acid–base equilibrium of the reactive dyes, which have pK_a values corresponding to the pH values B′ and C′ for dyes B and C (Figure 6.2) respectively. The basic equilibrium forms of the dyes (B^- and C^-) have rate constants (k_{hy}^{B} and k_{hy}^{C}), the logarithms of which are smaller by the amounts Δ_B and Δ_C than the rate constants k_{hy}^{BH} and k_{hy}^{BH} of the acid equilibrium forms:

$$\Delta_B = \log k_{hy}^{BH} - \log k_{hy}^{B} \qquad (6.3)$$

$$\Delta_C = \log k_{hy}^{CH} - \log k_{hy}^{C} \qquad (6.4)$$

For clarity in typesetting, minus signs have been omitted from superscripts and from some concentration terms in square brackets.

In the pH range immediately above pH = C′, the rate of hydrolysis of dye C is independent of pH, because a given increase of pH decreases the equilibrium concentration of the more reactive form CH by the same percentage as the hydroxide ion concentration increases. At pH = C″ the concentration of the less reactive basic form becomes so significant that its reactivity contributes significantly to the total rate. For dye B the same mechanism is effective but, since Δ_B is smaller, the horizontal part of the curve is not present.

The rate equation for curve B (as well as for C) can be derived from the mechanistic equations for the acid–base equilibrium of the reactive dye BH and the hydrolysis of the two equilibrium forms:

$$BH + OH^- \overset{K}{\rightleftharpoons} B^- + H_2O$$

$$BH + B^- = B_{total}$$

$$BH + OH^- \xrightarrow{k_{hy}^{BH}} \text{products of hydrolysis}$$

$$B^- + OH^- \xrightarrow{k_{hy}^{B}} \text{products of hydrolysis}$$

Taking the (constant) concentration of water in the first of these equations into the equilibrium constant, one obtains the basicity constant $K' = K[H_2O]$ and, since the total concentration of dye [D] is the sum of the equilibrium concentrations ([B] + [BH]), then for [BH] and [B] respectively:

$$[BH] = \frac{K'[D]}{K' + [OH]} \tag{6.5}$$

$$[\bar{B}] = \frac{[D]\,[OH]}{K' + [OH]} \tag{6.6}$$

Applying Eqn 6.1 for the hydrolysis reactions, one obtains the overall rate equation for the simultaneous hydrolysis of BH and B^-:

$$-\frac{d[D]}{dt} = -\frac{d[BH]}{dt} - \frac{d[B]}{dt}$$

$$= \frac{k_{hy}^{BH}K' + k_{hy}^{B}[OH]}{K' + [OH]}[D^-].[OH] \tag{6.7}$$

The sum of the logarithm of the ratio on the right-hand side of Eqn 6.7 (which corresponds to the observed second-order rate constant) and the hydroxide ion concentration plotted against pH corresponds to curves B and C in Figure 6.2 If $k_{hy}^{BH} \gg k_{hy}^{B}$, curves such as C are obtained. Curve B corresponds to a case where k_{hy}^{BH} is only slightly higher than k_{hy}^{B}.

Since hydrolysis of types B and C occurs with compounds with an imino bridge, most authors have explained the kinetic results by deprotonation of the imino group. The attack by hydroxide ions of the anionic basic equilibrium form (XXIX) of, for example, a monotriazine reactive dye is electronically – owing to mesomeric structures like XXIXb – and electrostatically more difficult:

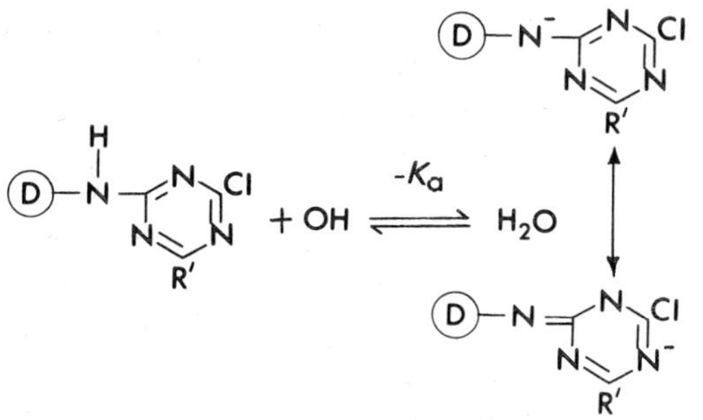

The pK_a values for equilibria of this type were determined for model compounds by Horrobin [44], by Dussy and Ackermann [43] and by Luttringer and Dussy [46]. They lie between 10 to 14, i.e. the equilibria are shifted to the right-hand side at pH 10–14. Hildebrand and Beckmann [61] demonstrated that not only these, but also deprotonations at other positions of the dye, influenced the ratio in the same way. On the other hand, Rattee and Murthy [62] questioned whether a deprotonation of the imino group is indeed involved.

As shown by Datyner, Rys and Zollinger [35, 37], rate dependences of type D occur when in the lower pH range dye–dye aggregates are present, since these have a lower reactivity. When the pH is increased deprotonation takes place. In the case investigated by these authors, it was not the imino bridge but a relatively remote naphtholic hydroxyl group of the chromogenic part of the dye that was deprotonated. The increase in charge causes a disaggregation through which the more reactive monomeric dye is formed. Here, obviously, the deactivating effect of deprotonation discussed by Hildebrand and Beckmann is overcompensated by the activating influence of disaggregation of dye–dye aggregates.

Ingamells, Sumner and Williams [59] found curve E for the hydrolysis of Procion Rubine MX–B (XXX; C.I. Reactive Red 6). Their results were later confirmed by work of Fisichella *et al.* [63]. A mechanistic interpretation for these results was given by Rys [37].

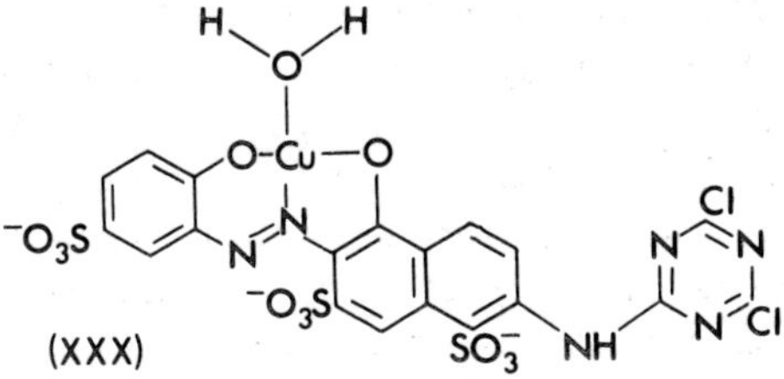

(XXX)

Basically, curve E corresponds to a mechanism of type C, i.e. a mechanism with an acid–base pre-equilibrium. It was observed that the rate maximum in curve E is more pronounced at higher dye concentration. The maximum is subject to general base catalysis (e.g. by carbonate–hydrogencarbonate buffer systems) at low dye concentrations; at high dye concentrations, however, an increase in buffer concentration decreases the rate in the pH range of that maximum.

In an evaluation of kinetic data obtained from the alcoholysis of reactive dyes, Rys [37] demonstrated that triazine reactive dyes with imino bridge links exist in two or more tautomeric forms, e.g. XXXI and XXXII, and that these tautomers have different reactivities with nucleophiles. In addition, their aggregation equilibria will be different. Neglecting the latter for a moment, the hydrolysis in dilute dye solutions can be explained as follows:

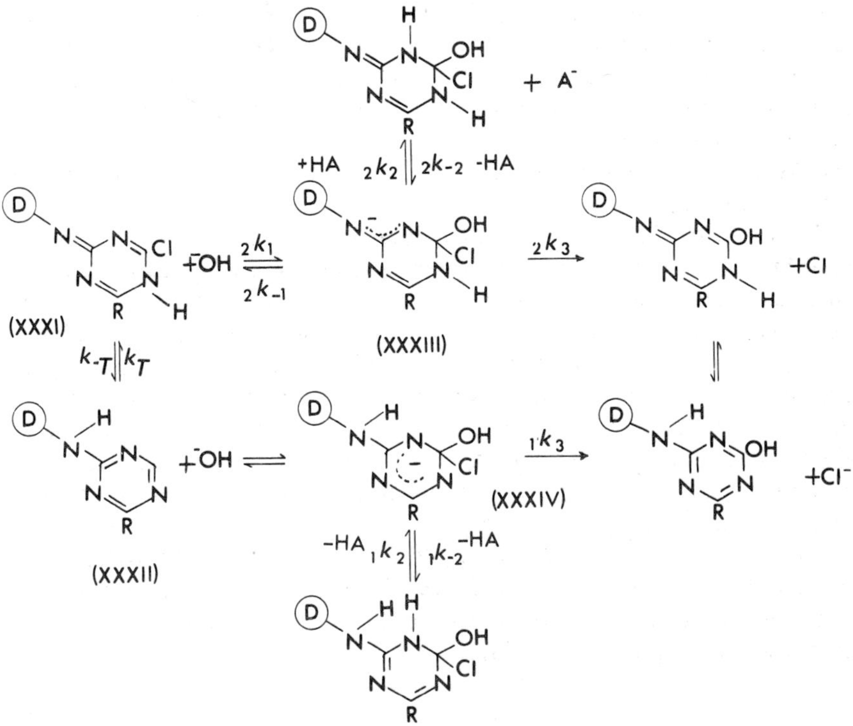

The conventional addition intermediates XXXIII and XXXIV of a nucleophilic aromatic substitution are involved in side-equilibria with the general acid HA. During the reaction, part of the dye is trapped in the side-equilibrium; the formation of the end-product of hydrolysis from the trapped side-product is general base-catalysed (steps with rate constants ${}_1k_{-2}$ and ${}_2k_{-2}$).

Rys, Schmitz and Zollinger [38] carried out a complete kinetic analysis with an analogue computer for two model compounds. At high dye concentrations aggregation equilibria are superimposed on the scheme above. If, for example, the tautomeric equilibrium between XXXI and XXXII is on the side of XXXII, but XXXI has a higher aggregation constant, XXXI is formed in the disaggregation process in a higher concentration than that which corresponds to the tautomeric equilibrium XXXI $\rightleftharpoons$ XXXII. If ${}_1k_1 < {}_2k_1$, the base-catalysed tautomerisation XXXI $\rightarrow$ XXXII will compete with the hydrolysis of XXXI. This results in a decrease of the rate of hydrolysis with increasing buffer concentration [34]. This example demonstrates the complex character of the reactions of at least some reactive dyes. It is therefore impossible to predict the preferred reaction mechanisms of a specific dye. For this reason an empirical or a physico-organic investigation of its behaviour is required.

6.3.2 Alcoholysis of reactive dyes

Alcohols are better model compounds than water for studying the reaction of reactive dyes with cellulose. D-Glucose (XXXV), the parent monosaccharide of cellulose (XXXVI), is not a suitable model, however. As shown by Stamm [64], glucose reacts with reactive dyes by forming glucosides, i.e. at the hydroxyl group in position 1, whereas cellulose can react only in positions 2, 3 and 6 (except for the anhydroglucose units at the ends of the polymeric chains):

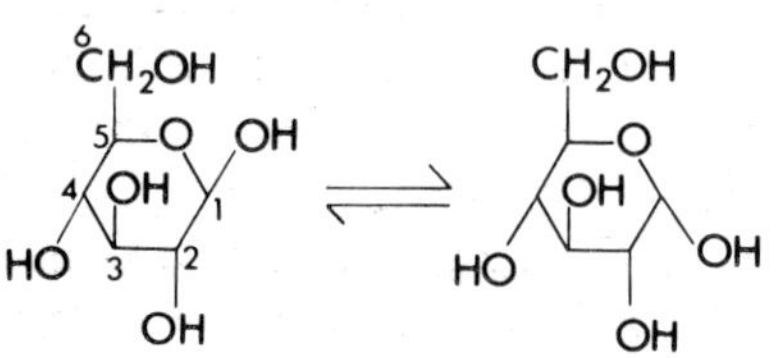

(XXXV)

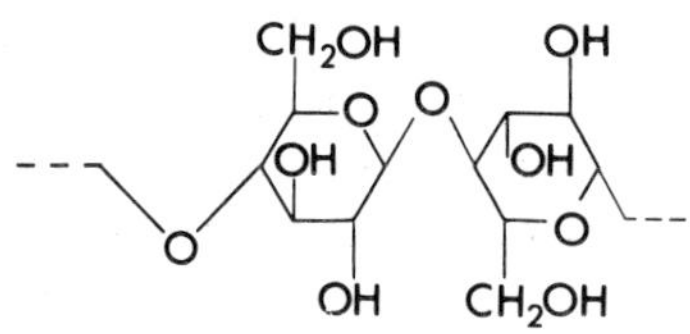

(XXXVI)

Glucosides formed with vinylsulphone reactive dye can be hydrolysed easily, in contrast to reaction products of these dyes with truly alcoholic hydroxyl groups, which are particularly stable to acid hydrolysis. Polyhydric alcohols such as mannitol (XXXVII) and sorbitol (XXXVIII) are therefore preferable as model compounds.

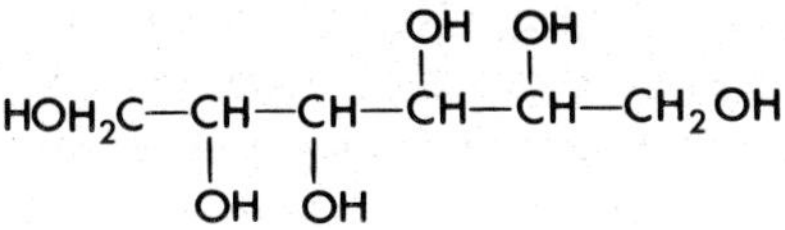

(XXXVII)

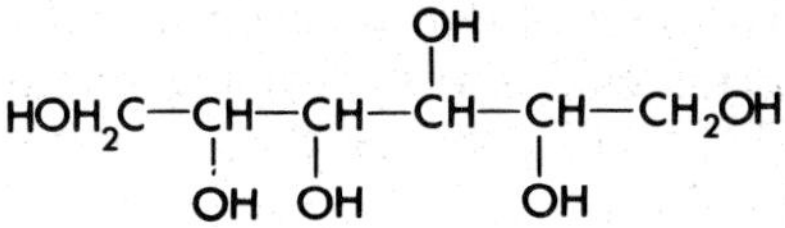

(XXXVIII)

Preston and Fern [65] demonstrated that the ratio of the measured overall pseudo-first-order rate constants for mannitol and for water reacting with monochlorotriazine dyes is not constant over a range of alkalinity. The ratio is, however, constant when the constants are calculated on the basis of hydroxide ion and mannitol anion concentrations, thus indicating that mannitol reacts, like water, essentially only in its basic (anionic) form. Depending on the dye structure, the mannitol anion is more reactive than the hydroxide ion by one to two powers of ten. Sorbitol and α-methyl-D-glucoside have also been used for this type of investigation [36, 37, 46, 59, 65].

If in hydrolysis and in alcoholysis the attack of the nucleophilic anion is the rate-determining step, the product ratio is linearly dependent on alcohol concentration. There are mono- and dichlorotriazine dye reactions, however, where this is not so [34–37]. These cases can be explained if two (or more) tautomeric forms of the reactive dye are present and if they have different rate-limiting steps, e.g. ${}_1k_1$ and ${}_2k_{-2}$ in the reaction scheme on p. 455.

The reactions of reactive dyes with thiols are not referred to here. They are discussed later in connection with the reactive dyeing of wool.

6.3.3 Aminolysis of reactive dyes

Reactions of reactive dyes with ammonia, alkylamines, amino acids and (water-soluble) peptides have been studied as model reactions for the reactive dyeing of wool. Most significant in this field of research is the work of Shore [66, 67], who surveyed comprehensively the literature on the mechanism of reactions of proteins with reactive dyes [68]. Although amino groups are, besides the amide groups, the most significant functional groups in wool, mercapto groups (in cysteine), as well as alcoholic and phenolic hydroxyl groups (in serine and tyrosine respectively), have also to be considered as reactive sites. Shore [66] therefore investigated the rate of reaction of a monochlorotriazine dye, Procion Red H-3B (C.I. Reactive Red 3), with water, ethanol, propan-2-ol, phenols, ethanethiol, ammonia, various alkylamines, α- and β-alanine, piperidic acid, heterocyclic amines (pyrrolidine, imidazole and indole), n-alkylamides and n-alkylguanidines.

In spite of the complex relation between reaction rate and hydrogen ion concentration of the aqueous solution, Shore was able to confirm that, in hydrolysis and alcoholysis, the respective anions (HO^-, RO^-) react with the dye. This is also the case for thiols. Amines in water react as free bases, even when the equilibrium is far on the side of the corresponding ammonium ion.

Of great interest and importance for the understanding of wool dyeing is the relative reactivity of the major functional groups (Table 6.2). The

TABLE 6.2

Relative reactivities of functional groups with Procion Red H-3B at 20 °C in water [66]

pH	C_2H_5SH	$C_2H_5NH_2$	C_2H_5OH	C_6H_5OH
4	485	2.8	1	0.86
5	510	5.5	1	0.87
6	540	10.5	1	0.89
7	570	20	1	0.90
8	600	39	1	0.92
9	635	76	1	0.93
10	665	150	1	0.94
11	700	285	1	0.96
12	735	550	1	0.97

mercapto group is the most reactive. The apparent increase of the reactivity of the amino group at higher pH values is probably due primarily to the shift of the amine $\rightleftharpoons$ ammonium ion equilibrium. It should be emphasised, however, that the change of relative amine reactivities of Table 6.2 does not correspond *quantitatively* solely to the influence of that equilibrium. Other pH-dependent processes are obviously superimposed on it.

Shore [67] also investigated reactions of reactive dyes with water-soluble proteins. The results confirmed the conclusions reached with the simple model compounds mentioned above.

6.3.4 Reactions with textile fibres

The fundamental problem of reactive dyeing is that the reaction of the reactive dye with water (hydrolysis) competes with the formation of the desired covalent bond between the dye and the textile substrate (fixation reaction). As the hydrolysed dye cannot react with the fibres under normal dyeing conditions, the extent to which hydrolysis occurs during dyeing directly governs the practical usefulness of a reactive system or the suitability of a particular dyeing method. Therefore, for the most efficient application of reactive dyes, dyeing conditions must be chosen so that good exhaustion of the dye on to the substrate is assured before hydrolysis in the dyebath takes place. Under such conditions the ratio of fixation to hydrolysis is determined only by the extent of hydrolysis in the liquid-filled pores of the textile fibres. However, it must be borne in mind that efficiency of fixation is not the only requirement for a good overall result; other criteria such as levelness are also important factors.

It is difficult if not impossible to give a unified quantitative relationship between the efficiency of reactive dyeings and all the parameters governing them under a wide variety of practical conditions. Nevertheless, one can gain useful practical hints concerning the behaviour of reactive dyes in textile fibres by making simplifying assumptions and using a simple model to describe the dyeing mechanism. For chemical processes to occur inside a textile substrate, a flow of reagents to the interfaces as well as into the interior of the substrate is necessary. In reactive dyeing one of the reaction partners is present in the substrate in the form of sites at which adsorption occurs first and reaction later, while the other partner, the reactive dye, must first be transported to the interior of the substrate. It is obvious that the kinetics of fixation as well as hydrolysis in the liquid-filled pores of the textile fibre are partly determined by the diffusion behaviour of the dye to and within the substrate.

The behaviour of a reactive dye moving from the dyebath into the interior of the textile substrate depends primarily upon four processes: the convectional diffusion in the dyebath to the fibre surface, the molecular diffusion through the hydrodynamic boundary layer and within the fibre, the adsorption and the chemical reaction. The latter two processes cause immobilisation of the diffusing species on the polymer chains of the textile fibres. The overall effect of these four steps determines the levelness of the dyeing, the spatial distribution of the fixed dye throughout the fibre cross-section and also the ratio of fixation to hydrolysis. Methods have been developed to study the diffusion and the chemical reactions within textile substrates simultaneously [69].

In a more detailed analysis of reactive dyeing we have derived, neglecting hydrodynamic effects in the dyebath, an efficiency expression E [70]. E is defined as the rate ratio of the fixation and the hydrolysis and varies between the limits of E_R, which arises when reactivities are the only factors, and E_D, which arises when fixation is diffusion-controlled. Which factors determine the efficiency in reactive dyeings depends upon the magnitude of the ratio of relaxation times (the time required for a process to reach $(1 - 1/e)$ of its final value) of the diffusion and chemical processes (fixation and hydrolysis). When this ratio is much less than unity, the efficiency is dominated by the chemical reactions (*reaction-controlled efficiency*) and

$$E = E_{\mathrm{R}} = \frac{Pk_{\mathrm{r}}^{\mathrm{ps}}}{Vk_{\mathrm{hy}}^{\mathrm{ps}}}Kn_{\mathrm{r}}\left(\frac{1}{1+\dfrac{P}{V}\dfrac{n_{\mathrm{r}}n_{\mathrm{hy}}}{m}}\right)$$

$$= \frac{P}{V}\frac{k_{\mathrm{r}}^{\phi}[\mathrm{Y}^{\phi}]}{k_{\mathrm{hy}}^{\sigma}[\mathrm{OH}^{\sigma}]}Kn_{\mathrm{r}}\left(\frac{1}{1+\dfrac{P}{V}\dfrac{n_{\mathrm{r}}n_{\mathrm{hy}}}{m}}\right) \tag{6.8}$$

When the ratio of relaxation times greatly exceeds unity, for sufficiently large values of V/P the role of diffusion becomes dominant (*diffusion-disguised efficiency*) and

$$E = E_{\mathrm{D}} = 2\frac{P}{V}+\frac{k_{\mathrm{r}}^{\mathrm{ps}}}{k_{\mathrm{hy}}^{\mathrm{ps}}}Kn_{\mathrm{r}}\left(\frac{D}{L^{2}(k_{\mathrm{r}}^{\mathrm{ps}}K+k_{\mathrm{hy}}^{\mathrm{ps}}\,n_{\mathrm{hy}}/m)}\right)^{\frac{1}{2}}$$

$$= 2\frac{P}{V}\frac{k_{\mathrm{r}}^{\phi}[\mathrm{Y}^{\phi}]}{k_{\mathrm{hy}}^{\sigma}[\mathrm{OH}^{\sigma}]}Kn_{\mathrm{r}}\left(\frac{D}{\mathrm{L}^{2}(k_{\mathrm{r}}^{\phi}[\mathrm{Y}^{\phi}]K+k_{\mathrm{hy}}^{\phi}[\mathrm{OH}^{\phi}])}\right)^{\frac{1}{2}} \tag{6.9}$$

where $k_{\mathrm{r}}^{\mathrm{ps}} = k_{\mathrm{r}}^{\phi}[\mathrm{Y}^{\phi}]$

$k_{\mathrm{hy}}^{\mathrm{ps}} = k_{\mathrm{hy}}^{\sigma}[\mathrm{OH}^{\sigma}]$

$n_{\mathrm{r}} = [\mathrm{D}_{\mathrm{M}}^{\phi}]/[\mathrm{D}_{\mathrm{M}}^{\sigma}] = \exp(-Ze\psi/kT)$

$n_{\mathrm{hy}} = [\mathrm{OH}^{\phi}]/[\mathrm{OH}^{\sigma}] = \exp(-e\psi/kT)$

Z = charge number of the dye ion

e = charge of the electron

ψ = surface potential

kT = Boltzmann energy

$m = k_{\mathrm{hy}}^{\sigma}/k_{\mathrm{hy}}^{\phi}$

k_{r}^{σ} = second-order rate constant of the fixation reaction

k_{hy}^{ϕ} = second-order rate constant of the hydrolysis in the pore liquid

k_{hy}^{σ} = second-order rate constant of the hydrolysis in the dyebath

$[\mathrm{Y}^{\phi}]$ = concentration of a nucleophilic functional group of the textile fibre

$[\mathrm{OH}^{\phi}]$ = concentration of the hydroxide ions in the pore liquid

$[\mathrm{OH}^{\sigma}]$ = concentration of the hydroxide ions in the dyebath

L = radius of the cylindrical fibre (or yarn)

P = porosity

K = inner adsorption equilibrium constant

D = concentration-independent diffusion coefficient (for the physical meaning of D see refs. [71] and [72], and also Chapter 5)

V = liquor-to-goods ratio and

V/P = ratio of the volume of the dyebath and the volume of pore liquid.

Although the simplifying approximation of neglecting hydrodynamic effects does not represent a situation encountered in practice, the assumptions made for the derivation of the limiting efficiencies E_R and E_D do not limit the validity of the semi-quantitative conclusions about the influence of various parameters on the efficiency of reactive dyeing. In the following, the most important factors that can influence these parameters and thus the efficiency E are discussed. Where experimental data are available, a comparison of the predicted with the observed influence on the efficiency is made.

pH

pH influences the ratio $[Y^{\phi}]/[OH^{\sigma}]$ and thus the ratio k_r^{ps}/k_{hy}^{ps}; the higher this ratio, the greater is the efficiency. In dyeing wool the optimum for this ratio occurs between pH 5 and pH 7, but when cellulosic fibres are dyed the pH has to be increased in order to produce cellulosate ions (Y^{ϕ} = $CellO^-$). The increase of $[CellO^-]$ in the fibre produces an increased negative charge on the fibre, resulting in an increase of the surface potential ψ and thus in a decrease in the distribution coefficients n_r and n_{hy}. Sumner [73] has shown that the pH inside the fibre is always lower than the external pH of the dyebath. He determined the ratio $[CellO^-]/[OH^{\sigma}]$ as a function of pH and of salt concentration in the dyebath. It is evident that this ratio (and hence k_r^{ps}/k_{hy}^{ps}) decreases as the pH increases and passes the pK_a value of the cellulose. Owing to repulsion of the cellulosate ions and the dye anions, an increase in pH will also lead to a decrease in K. Finally an increase of $[CellO^-]$ leads to the decrease in the ratio of the bulk-like and the Donnan phase within the fibre [74, 75] and thus to a decrease in the porosity P effective for the diffusion process.

In summarising the expected influence of the pH on the efficiency one can say that an increase in pH should decrease E_R as well as E_D. This prediction was recently verified experimentally by Morita *et al.* [76]. The authors investigated the pH-dependence of k_{Cell} (the symbols k_{Cell} and k_w used by Morita *et al.* [76] correspond in our nomenclature to $k_r^{ps}K(= k_{Cell}^{ps}K)$ and $k_{hy}^{ps}n_{hy}/m(= k_{hy}^{\phi}[OH^{\phi}])$.) and of the ratio k_w/k_{Cell} for several dichlorotriazine dyes. In Figure 6.3 their results for the determination of the logarithmic pseudo-first-order rate constants of six of these dyes are plotted as a function of pH.

The relationships between log k_{Cell} and pH fall into two classes; they are, however, closely related to those for hydrolysis observed by Ingamells *et al.* [59], Rys *et al.* [34–37] and others, as summarised in Figure 6.2 (p. 452) of this chapter. The curves for log k_{Cell} *vs* pH of the two dyes C.I. Reactive Red 1 and C.I. Reactive Red 2 are similar to curve B, while all the others have a resemblance to curve C in Figure 6.2. The inflection points of these relationships were found by Morita *et al.* near pH 10 except for C.I. Reactive

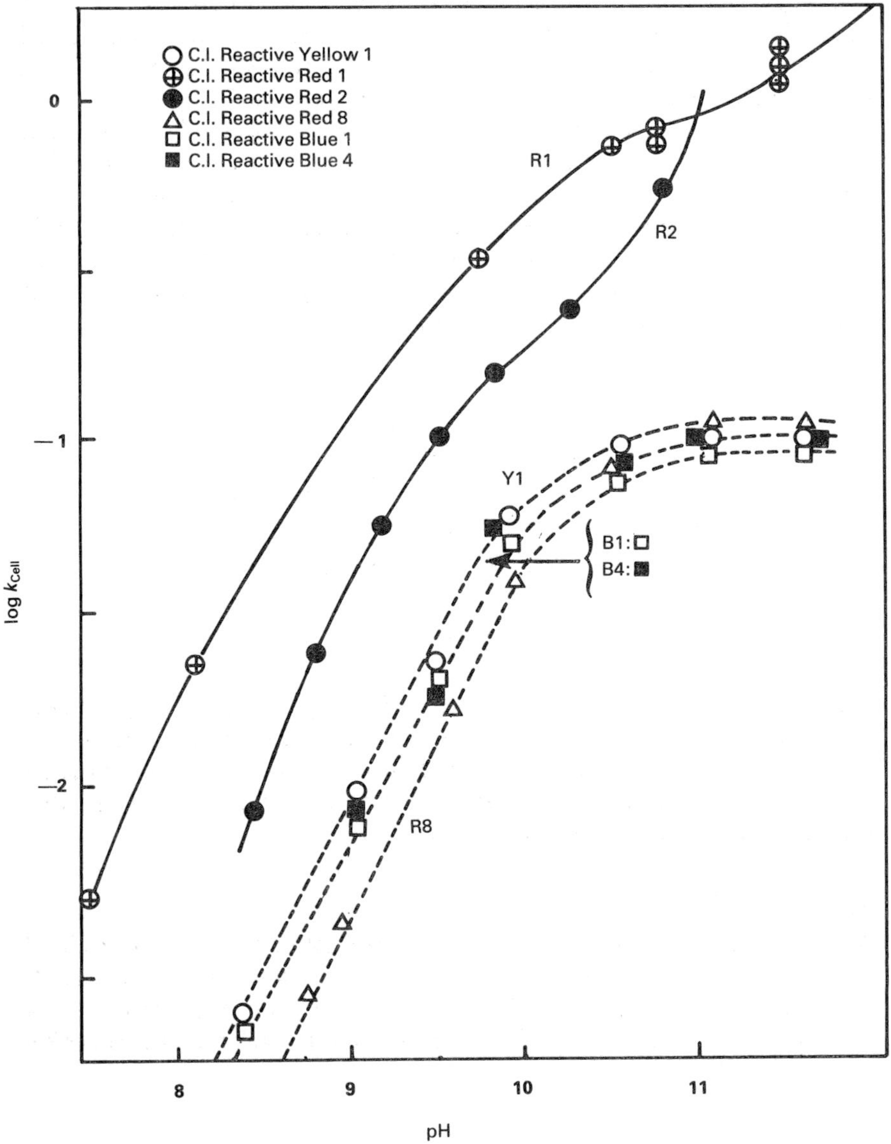

Figure 6.3 – Relationships between log k_{Cell} and pH for dichlorotriazine dyes at 30 °C and ionic strength $I = 0.15$ (with permission of Morita et al.*) [76]*

Red 1 (pH 9.8), Red 2 (9.5) and Red 8 (10.4). They coincide with the pK_a values for these dyes. These pK_a values were obtained by Aspland and Johnson [45] who analysed the results reported by Ingamells *et al.* for the hydrolyses of these dyes [59] (except C.I. Reactive Blue 4).

Morita's observation that the reactivities of C.I. Reactive Red 1 and Red 2 are characteristic of Class B dyes, as defined in Figure 6.2, is probably a consequence of the structure of these two dyes: C.I. Reactive Reds 1 and 2 are the only dyes which are based on the H acid-monoazo structure (XXXIX). It is characteristic that the imino bridge between the naphthalene and the triazine nuclei is in the 1,8-position to a hydroxyl group to which it may form a hydrogen bond (as shown in XXXIX).

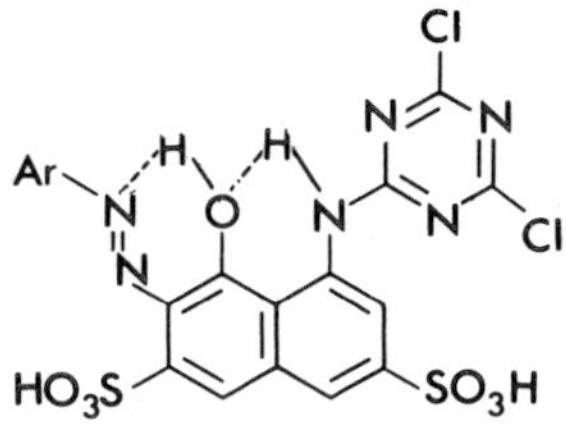

(XXXIX)

The ratio k_{hy}/k_{Cell} ($= k^{ps}_{hy} n_{hy}/k^{ps}_{Cell} Km$) determined by Morita *et al.* [76] under the experimental condition $1 \ll Pn_r n_{hy}/Vm$ (see Eqn 6.8) is given for five dyes in Figure 6.4 as a function of pH. This is consistent with the prediction and the experience of the practical dyer, namely that dichlorotriazine dyes can be dyed efficiently only at pH 7 to 8.

Temperature

A decrease in temperature is predicted to lead not only to an increase in K, but also to a decrease in the reactivities and thus to an increase in the selectivity ratio $k^{\phi}_{r}/k^{\phi}_{hy}$. From this it follows that E_R and E_D should increase with decreasing temperature. The experimental verification can be recognised from Figure 6.4.

Electrolyte concentration

When the neutral salt concentration in the dyebath is increased, the electrostatic repulsion forces decrease. As a consequence an increase in K, n_r, n_{hy} and thus in E_R as well as in E_D can be foreseen. This prediction is consistent with the experience in practice and with the experimental results of Morita *et al.* [76]: increasing the ionic strength from $I = 0.15$ to $I = 1.00$

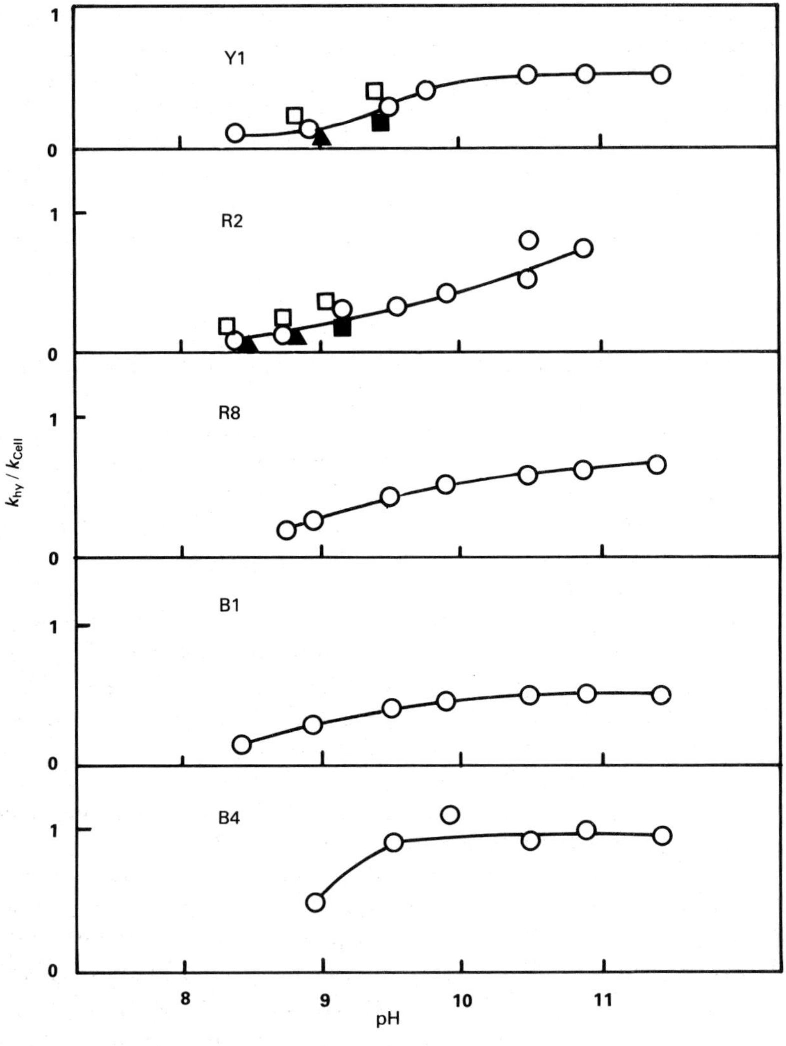

Figure 6.4 – The ratio k_{hy}/k_{Cell} obtained by Morita et al. *[76] from dyeing experiments with reactive dyes (see Figure 6.3) at $I = 0.15$ and temperatures of 35 °C (○), 30 °C (□) and 25 °C (▲) (with permission of Morita* et al.*)*

decreases the values of k_w/k_{Cell} for all eight dyes investigated by factors of between 1.08 (C.I. Reactive Yellow 4) and 2.00 (C.I. Reactive Red 2).

Dye structure

The overall structure of the dye molecule regulates the extent of aggregation, the affinity for the fibre as well as the selectivity ratio k_r^ϕ/k_{hy}^ϕ. As aggregation can either decrease [35, 37] or increase [34, 35] the reactivity towards water, depending on the structure, its influence on the efficiency cannot be generalised. If washing-off of the hydrolysed dye and good levelness are assured, the efficiency can be improved by increasing the adsorption constant K, e.g. by varying the dye structure. The selectivity ratio k_r^ϕ/k_{hy}^ϕ depends on the structure of the reactive groups (see section 6.2). Finally, particularly high efficiencies can be obtained by introducing more than one reactive group into the dye molecule.

Fibre structure

The influence of the porosity of wool and of cellulosic fibres on reactive dyeing kinetics is large, as it was shown by Weisz and co-workers [71] that diffusion in these fibres takes place through pores and not through the polymer matrix. As the pore-size distribution of these textiles substrates determines the effective porosity P and thus the 'dyebath-to-pore liquid' ratio V/P, it influences also the efficiencies E_R (Eqn 6.8) and E_D (Eqn 6.9).

Convection in the dyebath

The importance of hydrodynamic factors in dyeing processes has been argued by Peters and McGregor [77]. Sada and Kumazawa [78] have presented a mathematical model which can simulate the influence of agitation in the dyebath on the efficiency and dyeing rate in reactive dyeing. It correlates the unlevelness of dyeings with the degree of mixing.

The influence of all these variables on the efficiency of practical dyeing systems has been discussed in more detail by Rattee [26].

In summary, these investigations explain why the large excess of water present (5–20 ml for each gram of cellulose) does not favour hydrolysis as the overwhelming reaction in the competition between fixation to cellulose and hydrolysis of the reactive dye. There are essentially two effects which swing the reaction in favour of fixation in this heterogeneous process. First, the concentration of dye in the fibre phase is up to 500 times greater than the concentration of dye in solution, because of dye adsorption on the fibre; secondly, the ready ionisation of hydroxyl groups in cellulose leads to about a 25-fold excess of cellulosate anions over hydroxide ions inside the fibre. These two effects completely outweigh the vast excess of water in the system.

From the point of view of high efficiency, all application methods are designed in such a way as to let diffusion precede fixation and to avoid wasteful hydrolysis outside the fibre by approaching the optimum percentage of fixation as closely as possible.

In *batch dyeing* a reactive dye is applied first from a neutral aqueous solution. In due course the dye diffuses into the fibre and is partly adsorbed at the fibre–water interface. To obtain well-penetrated and level dyeing at this stage no fixation should take place. After diffusion and adsorption equilibria have been established the chemical reaction with the fibre can take place. In dyeing cellulosic fibres this is achieved by increasing the concentration of the cellulosate ions, which causes an increase in k^{ps}_{Cell}. Owing to the sensitivity of wool to alkali, reactive dyeing of wool has to be performed at a neutral or slightly acid pH. This is possible because many nucleophilic functional groups, e.g. amino groups, are present and sufficiently reactive at a neutral pH.

For a given textile substrate and reactive system the appropriate ratio of rate of diffusion to chemical reaction can also be optimised by varying the temperature. The reason for performing the dyeing process in two stages now becomes obvious. If most of the reactive dye can be transported into the fibre before the chemical reaction, little of it will be hydrolysed in the dyebath and hence E_R will be approached more closely. At first sight it might seem desirable to endow reactive dyes with high substantivity. However, for reasons associated with levelness as well as difficulties of subsequent removal of hydrolysed dye, a compromise is necessary and reactive dyes usually possess comparatively low substantivities.

There are papers in the literature in which correlations between various parameters of reactive dyes (e.g. affinity, diffusion rate, fixation rate, aggregation) and between such parameters and dyeing conditions (e.g. dye concentration, pH, salt concentration, temperature) are discussed. They demonstrate the enormous complexity of the reactive dyeing processes. For some of these papers scientific evaluation is unfortunately impossible as the descriptions given are superficial (e.g. [79]).

In a *semi-continuous* process such as pad–batch dyeing, the fabric is padded with a solution of dye and, in the case of cellulosic fibres, also with alkali. It is then batched either hot or cold before soaping off.

In *continuous* dyeing the fabric is impregnated with dye (and with cellulosic fibres also with alkali) and then subjected to heating and/or steaming to fix the dye.

All these application methods are based on the chemical and physico-chemical principles discussed in previous chapters. The precise conditions are modified to suit the peculiarities of particular dyes or reactive systems, or to provide for a particular type of production. In this sense textile printing, a very important field of application of reactive dyes, is not fundamentally different from dyeing. Various other aspects of these dyeing processes have been discussed in more detail by several authors [80, 81].

Wet-fastness properties of reactive dyeings on cellulosic fibres are primarily a function of the dye–fibre bond strength, as demonstrated by an investigation by Zollinger *et al.* [40] of dyeings obtained with a series of dyes covering six different reactive groups. All dyeings show maximum stability near the middle of the pH range; with triazine- and pyrimidine-type reactive dyes it lies between pH 6 and 7, while for vinylsulphone dyes it is at pH 4.5. On both sides of this maximum the rate of hydrolysis increases, in general, tenfold per pH unit. This pH-dependence proves, therefore, that hydrolysis of reactive dyeings is acid- and base-catalysed. In base catalysis the rate-determining step is the nucleophilic substitution of the reactive dye group on cellulose by a hydroxide ion. Acid catalysis involves a pre-equilibrium protonation of the *N*-heterocycle of the reactive dye group with dyes of Class A (p. 439). With vinylsulphone dyeings and, partly, in pyrimidyl dyeings, acid catalysis entails hydrolysis of the glucosidic bonds of cellulose, yielding not the hydrolysed dye but a dye particle bonded to a glucose unit hydrolysed from the fibre polymer chain. In addition, side-reactions within the dye molecule itself occur during the splitting of dye–fibre bonds. With dyes containing a monochlorotriazine *and* a vinylsulphone group (Sumifix Supra dyes, XIe) the maximum stability is expected to extend about two pH units between pH 5 and pH 7, corresponding to the combination of stabilities of both reactive groups.

More recently [41], the dye–fibre bond stability of phosphonic acid reactive dyes was determined. These reactive dyes have stabilities which are similar to those of the older reactive dyes on cellulose.

In wool, in contrast to cellulose, dye–fibre bond stabilities do not govern wet-fastness properties. It has been shown [82] that the rate of splitting of these bonds is much slower than the degradation of wool. Decisive for the suitability of a specific reactive dye or a specific dyeing method for wool is the requirement that the fixation ratio of the dyeing process (p. 459) should be extremely high. At least 90 per cent of the dye on the fibre must be fixed by covalent bonds; the remaining dye is adsorbed only as an acid dye and so has only low wet-fastness properties. Whereas with cellulose it is easy to wash out the hydrolysed, non-covalently bonded dye, the removal of an acid dye of this type from wool is difficult, especially on the large scale in a textile mill. Methods for the determination of the fixation ratio of reactive dyes on wool have been described by Datyner *et al.* [83], Rattee *et al.* [84] and Zollinger *et al.* [82].

A recent investigation of reactive dye–fibre bond stabilities for silk [47] demonstrated that these bonds are stable enough for excellent wet-fastness. In this respect reactive dyeings on silk are comparable to those on wool.

Recently the dye–fibre bond stabilities on cellulose, wool and silk were compared as a function of the pH value [85].

The exact determination of the fixation ratio was important for a critical evaluation of various methods described in the technical literature for aftertreatments of reactive wool dyeings. It was claimed in some technical papers that in these treatments unreacted dye would react with wool and thus increase the fixation ratio, i.e. improve wet-fastness properties. In two investigations [86, 87] it was shown that no 'secondary reaction of dyes with wool' takes place, but that these aftertreatments are more or less efficient in washing out hydrolysed reactive dye. The percentage of dye bonded to wool relative to the total amount of dye on the fibre is therefore higher after these treatments. Their efficiency is a simple function of the pH value: the higher the pH value, the more hydrolysed dye is removed during a given time and at a given temperature. Increasing pH, time and/or temperature of the treatments will increasingly damage the wool, however, as pH values of 6.0 and higher are applied. Therefore in practice a careful optimisation of the aftertreatment is necessary.

A study of the products of hydrolysis of difluorochloropyrimidyl dyes (Drimalan and Verofix dyes, XIIIb) has shown that the high fixation ratio of these dyes on wool is probably due to the fact that the second of the two fluoro groups hydrolyses only slowly and is therefore suitable for reaction with wool even after long dyeing times [88].

Wool dyed with reactive dyes has a higher stability towards hydrolytic degradation and removal of wool gelatine than undyed wool [89]. It is likely that reactive dyes react with suitable proteins in the non-keratinous part of the wool fibre and that they do not enter the cortex [90].

Reviews of reactive wool dyeing were published in 1976 [91], 1982 and 1986 [31].

6.4 REACTIVE SITES IN TEXTILE FIBRES

6.4.1 Methods for the identification of dye-reactive sites

Determination of the nucleophilic groups of textile fibres with which reactive dyes form covalent bonds is primarily a scientific problem, but it also has practical importance, particularly for wool dyeing. Four basic methods of identification are discussed below.

1. If the particular covalent bond formed is characterised by a specific reaction, this reaction can be used for the determination. This method was used when the covalent character of reactive dye–cellulose bonds was proved for the first time in a direct chemical way. For example, if dyes with sulphonylfluoride reactive groups [92] (which are not used commercially) react with hydroxyl groups of cellulose, they form sulphonic acid ester groups. It is known from sugar chemistry that such ester groups can be substituted by iodine from sodium iodide in propanone only if primary hydroxyl groups are involved:

$$\begin{array}{c} CH_2OSO_2R \\ | \\ \ldots Glu \ldots \\ | \\ OSO_2R \end{array} \xrightarrow{Na^+I^-} \begin{array}{c} CH_2I \\ | \\ \ldots Glu \ldots \\ | \\ OSO_2R \end{array} + RSO_3^-Na^+$$

R = *p*-tolyl or reactive dye residue

By using this method, it was shown that 75 per cent of a sulphonylfluoride dye had reacted with hydroxyl groups of cellulose in the 6-position [93].

2. In cases where the reactive dye–fibre bond is stable against acid hydrolysis, the dyed fibre can be degraded by aqueous acid to its monomeric units. Some of these units will contain reactive dye as a substituent. After appropriate separation of the monomers (by thin- or thick-layer chromatography, etc.), the coloured products are identified by instrumental analysis, preferably by comparison with identical compounds synthesised by independent methods.

For studies of reactive dyeings on *cellulose*, the synthesis of the respective glucose derivatives for identification purposes is very time-consuming in many cases, so an alternative method is preferable, viz. methylation of all free hydroxyl groups in the coloured degradation products. The dye–fibre bonds and the methylglucoside bond (i.e. at carbon atom 1) are then split, by alkaline hydrolysis and acid hydrolysis respectively, and the various isomeric trimethylglucoses are identified. The presence in the mixture of 2,3,4-trimethylglucose, for example, indicates – because of the free hydroxyl group in the 6-position – that the reactive dye has formed a covalent bond in the 6-position. This method was used on cellulose dyeings by Chekalin [94] and Stamm [64]. The main difficulty with all methods of determination of the relative reactivity of the hydroxyl groups in the 2-, 3- and 6-positions of the anhydroglucose unit is the low degree of substitution in a reactive dyeing of medium strength, e.g. a dyeing containing 2–4 per cent of a commercial dye. A reliable activity ratio can be obtained only after a quantitative determination of all the products. This was not carried out in some published investigations.

For reactive dyeings on *wool*, the following procedures are applicable for dyes which form acid-stable dye–fibre bonds: acid degradation of dyed wool is followed, first by chromatographic separation of the various free amino acids and those which are attached to a dye molecule, and then by identification of the latter by comparison with synthesised reaction products of amino acids with the respective reactive dye. For amino acids containing a side-chain nucleophilic group (lysine, serine, cysteine, etc.) in addition to the α-amino group two reference compounds are necessary, one containing the reactive dye at the α-amino group and one containing it at the side-chain. In some cases, e.g. with reactive dyes containing acryloylamino groups, it is only the reactive group of the dye which stays with the amino acid after hydrolysis [95, 96]. The amino acid analysis of the hydrolysate of dyed wool can also be compared with that of the undyed wool. A decrease in the peaks for, say, lysine and histidine indicates that the dye has reacted with the side-chain amino groups of these amino acids [95, 97]. Baumann [98] has also isolated dyed wool proteins by mercaptolysis according to Goodard and Michaelis's method [99].

3. When the dye–fibre bond is not stable under strong acid conditions analysis is much more difficult, especially with dyeings on cellulose. With cellulose, microbiological or enzymic degradation of the dyed material is possible, provided that the degree of polymerisation is not too high. Such methods of degradation were used by Stamm *et al.* [100].

In his study of reactive-dyed wool, Hille [101] blocked all the free nucleophilic groups with 2,4-dinitrofluorobenzene, which yields acid-stable bonds in proteins. After acid hydrolysis, the resultant dinitrophenyl amino acid derivatives could be estimated quantitatively. From the observed decrease (relative to undyed wool) in the amounts of cysteine, lysine, histidine, tyrosine and serine, Hille concluded that these amino acids had reacted with the chlorotriazine dye used. Hille applied extremely high dye concentrations which are not representative for practical dyeings.

In addition, Asquith *et al.* [102] showed later that 2,4-dinitrofluorobenzene does not react with all amino groups of wool if the wool was dyed before with (anionic) reactive or acid dyes. This observation was explained by the hypothesis that, if a dye molecule has been fixed covalently to a hydroxyl group of wool (serine, threonine or tyrosine), the solubilising anionic sulpho groups of this dye may then *ionically* interact with ammonium groups of lysine and histidine within the fibre; therefore these groups become inaccessible for 2,4-dinitrofluorobenzene. Even non-covalently bonded acid dyes block a considerable number of wool amino groups towards attack by 2,4-dinitrofluorobenzene.

For cellulose, an analogous method to the 2,4-dinitrofluorobenzene treatment of wool is permethylation of all hydroxyl groups not reacted with a reagent covalently. It has been applied successfully for the identification of the reactivity ratio of the 2-, 3- and 6-positions of cellulose with crosslinking reagents. Its application to reactive-dyed cellulose is possible in principle, but owing to the low degrees of substitution it would be extremely difficult to obtain reliable data.

4. Reactive dyeings on nylon were investigated by Scott and Vickerstaff [103] by titrating the amino end-groups of the dyed fibre dissolved in a mixture of phenol and methanol.

6.4.2 Results with cellulose

Results with cellulose dyeings are in conflict with one another, probably in part for the reasons mentioned in the previous section. It should also be borne in mind that different application conditions, e.g. differences in pH, can change the ratio of primary and secondary cellulosate anions, which in turn leads to a different positional selectivity.

With Remazol Brilliant Blue R (C.I. Reactive Blue 19), Stamm [100] found that, in the acid degradation of cellulose with a medium DP, the derivative was substituted by the dye mainly in the 2-position. In the microbiological degradation product of viscose powder of DP 32, however, he found that 57 per cent of the dye had reacted in the 2-position, and 19 per cent and 24 per cent were fixed at the end of the cellulose chain in the 1- and 4-positions, respectively. Obviously, the absolute reactivity of the chain ends is higher.

On the other hand, Chekalin [94], using colourless β-sulphatoethylsulphonyl model compounds, obtained a reactivity ratio of about 7:3 for the 6- and 2-positions, respectively. This result is consistent with work by Rowland *et al.* [104] on the reactivity ratio for methyl vinylsulphone. Venkataraman [105] found n.m.r. spectroscopic evidence for the reaction of vinylsulphone dyes at the 6-position.

Dawson, Fern and Preston [106] produced semi-quantitative evidence for the reactivity ratio of dichlorotriazine dyes, this being between 3:1 and 7:1 for the 6- to the sum of the 2- and 3-positions.

6.4.3 Results with wool and silk

Work by Derbyshire and Tristram [95], Zahn and Reinert [96], Baumann [97, 98] and Corbett [107] with vinylsulphone dyes indicates that thiolate groups (anionic cysteine groups), free amine and hydroxide ion groups have an approximate reactivity ratio of 10^4:10^2:1. This is consistent with the finding that reactive dyes react primarily with thiolate groups. Because the thiolate groups are normally present only in very small amounts, however, ε-amino groups of lysine and the heterocyclic amino group of histidine, as well as terminal amino groups of polypeptide chains, are the most important reactive sites in wool. On the basis of reactions with model compounds of amino acids, Wirnik and Chekalin [108] came essentially to the same conclusion. Reaction with serine hydroxyl groups obviously takes place only with very heavy dyeings (see p. 470). These results are in agreement with those of Shore [66, 67], who summarised the above investigations [68] and who also examined the relative reactivities of Procion Red MX-3B (C.I. Reactive Red 3) with model compounds and water-soluble proteins (see p. 458). In the critical examination of the 2,4-dinitrofluorobenzene method [102] mentioned above, Asquith *et al.* showed that with the same wool, five reactive dyes with five different reactive groups had significantly different reactivities with lysino and histidino groups. Unfortunately Asquith *et al.* did not include dyes with reactive groups which give high fixation ratios.

Baumann *et al.* [109] investigated the relative reactivity of the morphological components of wool with a vinylsulphone reactive dye. For this purpose the dyed wool was treated with ultrasonic waves in 98 per cent formic acid which separated the wool into cuticular cells, dissolved proteins, cortex cells, cell fractions and residues of various origin. The content of dye in the isolated morphological units was determined; the soluble proteins were found to have the highest dye content, followed by that of the cuticular and the ortho cortex cells. The para cortex cells have the lowest dye content.

Knowledge of the reactivity of the functional groups of wool in relation to the type of reactive dye and to the method of dyeing is important for the fundamental understanding of reactive wool dyeing. By reducing the wool first, i.e. by increasing the number of mercapto (cysteine) groups, the fixation ratio can be increased, as shown by Zahn *et al.* [110]. The wool is damaged by such a treatment, however. Consequently, although reactive dyeing of wool was first carried out in the early 1950s (see p. 430), it was only in the late 1960s, with the development of the α-bromoacrylamido and the difluoromonochloropyrimidine reactive dyes (Lanasol, Verofix and Drimalan types), which give a high fixation ratio (> 90 per cent), that reactive dyes for wool became technologically significant.

In contrast to wool, silk cannot be dyed efficiently with reactive dyes in a slightly acidic or in a neutral dyebath; sufficient fixation takes place only at pH 8–9. This indicates that, instead of the relatively few amino groups in silk protein (and only traces of cysteine groups), hydroxyl groups are the major reaction sites. Indeed, work of Xia [111] with a difluorochloropyrimidine dye demonstrates that when that dye is applied at 40 °C to silk, it reacts with *two* tyrosine residues. The reaction with two tyrosines is indicated by the significantly decreased solubility of silk in 49 per cent lithium thiocyanate after dyeing. As fibroin contains tyrosine at the end of the crystalline segment, many tyrosine side-chains are present in close proximity to one another at the end of these segments. This structure facilitates crosslinking with dyes which contain bifunctional reactive groups as they are present in the dye used by Xia.

Other bifunctional reactive dyes and dyes with two reactive groups crosslink silk as well as wool, as shown by Zollinger and co-workers [112] by solubility tests. It is particularly interesting to see that even dyes with one α-bromoacrylamide group can crosslink wool by reaction with two amino groups of keratin (XLII) (silk to a very small degree only). This result corroborates a claim of Lewis [31], previously postulated by Mäusezahl [113], that proteins react with α-bromoacrylamide dyes either at the α- or at the β-

carbon atom (XXXIX and XL respectively) or at the aziridine derivative (XLI).

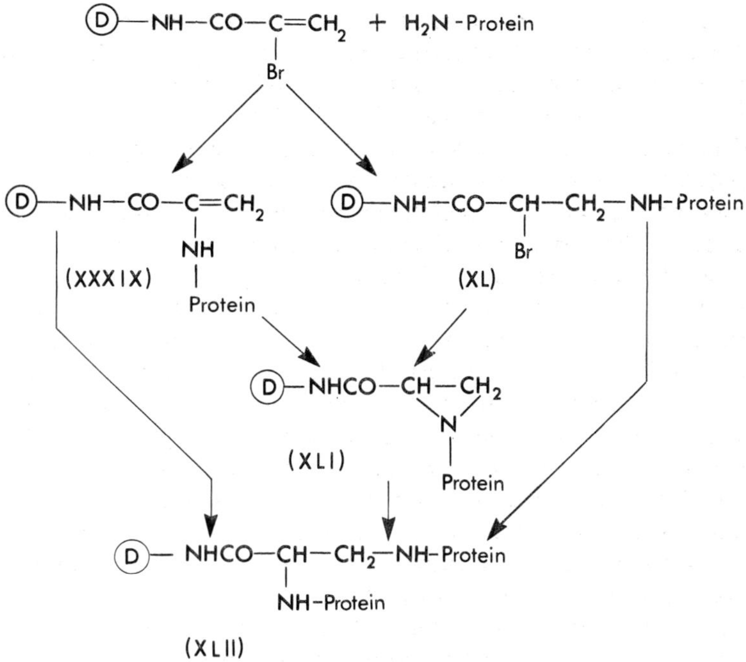

In the context of discussing reactive sites in wool dyeing, the use of reactive dyes for the identification of active sites in enzymes should be mentioned. Enzymes are inactivated by reaction with reactive dyes. By enzymatic degradation and subsequent analysis of the coloured fraction the active sites can be determined. In this way Small *et al.* [114] showed that Procion Blue MX-R (C.I. Reactive Blue 4, a dichlorotriazine dye) is bonded to the mercapto group of cysteine 174 of horse liver alcohol dehydrogenase.

Another well-known use of reactive dyes in biochemical analysis is affinity chromatography (see monograph of Scouten [115]). This method is, however, outside the subject of this chapter.

REFERENCES

1. For further details of intermolecular forces see Chapter 2.
2. See, e.g. H Zollinger, *Chemie der Azofarbstoffe* (Basel: Birkhäuser Verlag, 1958), Chapter 13; L F R Hartley, *Australian J. Chem.*, **23** (1970) 287.
3. C F Cross and E J Bevan, *Researches on cellulose (1895–1900)*, 34.
4. G Schroeter, *Ber.*, **39** (1906) 1559.
5. M Schmid, R Mory and CIBA, German P 1,026,750 (1954).
6. J Günther and I.G. Farbenindustrie, USP 1,567,731 (1925).

7. R Haller, A Heckendorn and CIBA, German P 554,781 (1929), 560,035 (1930), 583,398 (1932); BP 342,167 (1931), 393,914 (1933); USP 1,896,892 (1933), 2,025, 660 (1935).
8. J D Guthrie, *Amer. Dyestuff Rep.*, **41** (1952) 13.
9. I.G. Farbenindustrie, BP 341,461 (1930), 499,868 (1940).
10. H Schweizer and O Bayer, cited by O Bayer, *Angew. Chem.* **73** (1961) 343.
11. I D Rattee, W E Stephen and ICI, BP 772,030 774,925, 871,930 (1954); see also ref. 58.
12. J Heyna, W Schumacher and HOE, German P 965,902 (1949).
13. D L Schoene, V S Chambers and Goodrich Co., USP 2,524,399 (1950).
14. S Abeta and K Imada, *Senshoku Kogyo*, **32** (1984) 93.
15. Farbenfabriken Bayer, German P 1,044,091, 1,076,696 and 1,125,875 (1957).
16. R G Pearson, *J. Am. Chem. Soc.*, **85** (1963) 3533, **89** (1967) 1827; *J. Chem. Educ.* **45** (1968) 581, 643.
17. R D Chambers, *Dyes and Pigments*, **3** (1982) 183.
18. C V Stead, *Dyes and Pigments*, **3** (1982) 161.
19. Hoechst, USP 3,223,470 (1961).
20. S Fujioka and S Abeta, *Dyes and Pigments*, **3** (1982) 281.
21. B L McConnell, L A Graham and R A Swidler, *Text. Research J.*, **49** (1979) 458.
22. B L Beech, *Fibre-reactive dyes* (London: Logos Press, 1970).
23. A Lukos and W Ornaf, *Barwniki reaktywne* (Warsaw: Wydawnictwo Przemyslu Lehkiego, 1966).
24. G E Krichevski, *Aktionüje krasitjeli* (Moscow: Jzdatjelstvo Industrija, 1968).
25. E Siegel, K-H Schündehütte and D Hildebrand, in *The chemistry of synthetic dyes* (ed. K Venkataraman), vol. 6 (New York: Academic Press, 1972).
26. I D Rattee, in *The chemistry of synthetic dyes* (ed. K Venkataraman), vol. 8 (New York: Academic Press, 1978) 1–36.
27. D Hildebrand, *Textil Praxis* (1970) 351, 428, 531, 621; (1971) 45, 433, 564.
28. P Rosenthal, *Rev. Progr. Coloration*, **7** (1976) 23.
29. I D Rattee, *Rev. Progr. Coloration*, **14** (1984) 50.
30. D W Ramsay, *J.S.D.C.*, **97** (1981) 102.
31. D M Lewis, *J.S.D.C.*, **98** (1982) 165; D M Lewis, *Melliand Textilber.*, **67** (1986) 717.
32. E Brunnschweiler and G Siegrist, *Textilveredlung*, **19** (1984) 305.
33. J Miller, *Aromatic nucleophilic substitution* (Amsterdam: Elsevier, 1969).
34. P Rys and H Zollinger, *Helv. Chem. Acta*, **49** (1966) 749.
35. A Datyner, P Rys and H Zollinger, *ibid.*, **49** (1966) 755.
36. P Rys and H Zollinger, *ibid.*, **49** (1966) 761.
37. P Rys, *Textilveredlung*, **2** (1967) 95.
38. P Rys, A Schmitz and H Zollinger, *Helv. Chim. Acta*, **54** (1971) 163.
39. T L Dawson, *J.S.D.C.*, **80** (1964) 134.
40. R Ch Senn and H Zollinger, *Helv. Chim. Acta*, **46** (1963) 781; R Ch Senn, O A Stamm and H Zollinger, *Melliand Textilber.*, **44** (1963) 261.
41. M Gisler and H Zollinger, *Text. Research J.*, **50** (1980) 519; T N Vorozheva, M G Romanova, K V Solodova and V D Stepanenko, *Zhur. Priklad. Khim.*, **58** (1985) 1204; G Alberghina, M E Amato, S Fisichella and I D Rattee, *Dyes and Pigments*, **7** (1986) 33.
42. F Osterloh, *Melliand Textilber.*, **44** (1963) 57.
43. H Ackermann and P Dussy, *ibid.*, **42** (1961) 1167; *Helv. Chim. Acta*, **45** (1962) 1683.
44. S Horrobin, *J.C.S.* (1963) 4130.
45. J R Aspland and A Johnson, *J.S.D.C.*, **81** (1965) 425, 477.
46. J P Luttringer and P Dussy, *Melliand Textilber.*, **62** (1981) 84.
47. U Meyer, J-Z Wand, Y Xia, J-Z Yang and H Zollinger, *J.S.D.C.* **102** (1986) 6.
48. K O Stepanenko, N A Zhukova, M G Romanova and M A Chekalin, *Zh. Prikl. Khim. (Leningrad)*, **56** (1983) 624; *Chem. Abs.* (1984) 6814j.
49. M Matsui, U Meyer and H Zollinger, *J.S.D.C.*, **102** (1986) 6.
50. U Meyer, J-Z Yang and H Zollinger, *Textilveredlung*, **20** (1985) 354; P Ball, U Meyer and H Zollinger, *Text Research J.*, **56** (1986) 447.
51. T Omura and H Kenmochi, paper presented at 13th Congress IFATCC (London, 1984).
52. J Benz, *J.S.D.C.*, **77** (1961) 734; O Thumm and J Benz, *Angew. Chem.*, **74** (1962) 212.

53. P Rys and O A Stamm, *Helv. Chim. Acta*, **49** (1966) 2287.
54. U Baumgarte, *Melliand Textilber.*, **49** (1968) 1432.
55. T Robinson, W Knobel and B Kissling, and Sandoz, DOS 3105897 (1981) and DOS 3330120 (1983).
56. T Robinson and W B Egger, *Text. Chem. Col.*, **15** (1983) 189.
57. W E Stephen and ICI, BP 838,337 and 842,933.
58. W E Stephen, *Chimia*, **19** (1965) 261.
59. W Ingamells, H H Sumner and G Williams, *J.S.D.C.*, **78** (1962) 274.
60. J R Aspland, A Johnson and R H Peters, *ibid.*, **78** (1962) 453.
61. D Hildebrand and W Beckmann, *Melliand Textilber.*, **45** (1964) 1138.
62. I D Rattee and K S Murphy, *J.S.D.C.*, **85** (1969) 368.
63. S Fisichella, S Occhipinti, G Alberghina and M E Amato, *Text. Research J.*, **51** (1981) 683.
64. O A Stamm, *Helv. Chim. Acta*, **46** (1963) 3008, 3019.
65. C Preston and A S Fern, *Chimia*, **15** (1961) 177.
66. J Shore, *J.S.D.C.*, **84** (1968) 413.
67. *Idem*, *ibid.*, **84** (1968) 545.
68. *Idem*, *ibid.*, **84** (1968) 408.
69. Z Morita and H Motomura, *Sen-i Gakkaishi*, **42** (1986) T-626; **43** (1987) 198, 204; Z Morita, T Saito and H Motomura, *J. Appl. Polymer Sci.*, **32** (1986) 3929.
70. P Rys and P Sperb, *Text. Research J.*, submitted.
71. P B Weisz, *Trans. Faraday Soc.*, **63** (1967) 1801; P B Weisz and J S Hicks, *ibid.*, **63** (1967) 1807; P B Weisz and H Zollinger, *ibid.*, **63** (1967) 1815; **64** (1968) 1693; *Melliand Textilber.*, **48** (1967) 70.
72. P Rys, *Text. Research J.*, **43** (1973) 24.
73. H H Sumner, *J.S.D.C.*, **76** (1960) 672.
74. D G Tsimboukis and J H Petropoulos, *J. Chem. Soc. Faraday I*, **75** (1979), 705, 717.
75. T Hori, R J Ott and P Rys, *Polym. Eng. Sci.*, **20** (1980), 264; H H Sumner and C D Weston, *Amer. Dyestuff Rep.*, **52** (1963) 442.
76. H Motomura and Z Morita, *J. Appl. Polym. Sci.*, **21** (1977) 487; *Bull. Chem. Soc. Japan*, **51** (1978) 1332; Z Morita, I Nishikawa and H Motomura, *Sen-i Gakkaishi*, **39** (1983) T-485.
77. R H Peters and R McGregor, *J.S.D.C.*, **81** (1965) 393.
78. E Sada and H Kumazawa, *J Appl. Polymer Sci.*, **27** (1982), 2987, 4745; **30** (1985), 4113.
79. J F Landré and A Schaub, *Textilveredlung*, **21** (1986) 425.
80. I D Rattee, *Melliand Textilber.*, **45** (1964) 782.
81. I D Rattee, *J.S.D.C.*, **85** (1969) 23; J A Fowler and W J Marshall, *ibid.*, **80** (1964) 358; J A Fowler, W J Marshall and I Seltzer, *Dyer*, **128** (1962) 153; W J Marshall, *J.S.D.C.*, **82** (1966) 169; M Capponi and R Ch Senn, *ibid.*, **82** (1966) 8; D M Lewis and I Seltzer, *ibid.*, **84** (1968) 501.
82. H K Rouette, J F K Wilshire, I Yamase and H Zollinger, *Text. Research J.*, **41** (1971) 518.
83. J R Christoe and A Datyner, *J.S.D.C.*, **86** (1970) 210.
84. K R Cockett, I D Rattee and C B Stevens, *ibid.*, **85** (1969) 113; **86** (1970) 211.
85. P Ball, U Meyer, J-Z Yang and H Zollinger, *Textilveredlung*, **20** (1985) 88.
86. E Finnimore, U Meyer and H Zollinger, *J.S.D.C.*, **94** (1978) 17.
87. F Carlini, U Meyer and H Zollinger, *Melliand Textilber.*, **60** (1979) 587.
88. A Datyner, E Finnimore and U Meyer, *J.S.D.C.*, **93** (1977) 278.
89. H Baumann, in *Fibrous proteins, scientific, industrial and medical aspects* (eds. D A D Perry and L K Cramer), vol. 1 (London: Academic Press, 1979), 323.
90. H Baumann and L D Setigwan, *Proc. 7th Int. Wool Text. Res. Conf. Tokyo*, **5** (1985), 1; T Hori and H Zollinger, *Melliand Textilber*, **68** (1987) 675.
91. H Zollinger, *Proc. 5th Int. Wool Text. Res. Conf. Aachen*, **1** (1976) 167.
92. ICI, German P 1,052,946 (1958); Cassella, BP 827,570 (1958); CIBA, French P 1,192,485 (1958); HOE, Belgian P 578,517 (1958); ICI, BP 824,690 (1958).
93. B Krazer and H Zollinger, *Helv. Chim. Acta*, **43** (1960) 1513.
94. M A Chekalin, *Tekstil. Prom.*, **21**(1) (1961) 40; **23**(8) (1963) 67.

95. A N Derbyshire and G R Tristram, *J.S.D.C.*, **81** (1965) 584.
96. G Reinert, K Mella, P F Rouette and H Zahn, *Melliand Textilber.*, **49** (1968) 1313.
97. H Baumann, *Textilveredlung*, **5** (1970) 506.
98. *Idem*, *Appl. Polymer Symp. no. 18* (1971) 307.
99. D Goodard and J Michaelis, *J. Biol. Chem.*, **106** (1934) 605; **112** (1935) 367.
100. O A Stamm, *Helv. Chim. Acta*, **46** (1963) 3008, 3019; P Hagen, E T Reese and O A Stamm, *ibid.*, **49** (1966) 2278.
101. E Hille, *Textil Praxis*, **17** (1962) 171.
102. R S Asquith and D K Chan, *J.S.D.C.*, **87** (1971) 181.
103. D F Scott and T Vickerstaff, *J.S.D.C.*, **76** (1960) 104.
104. V O Cirino, A L Bullock and S P Rowland, *J. Polymer Sci.*, A-1 **7** (1969) 1225.
105. K Venkataraman, *Colourage*, **15**(1) (1968) 59.
106. T L Dawson, A S Fern and C Preston, *J.S.D.C.*, **76** (1960) 210.
107. J F Corbett, *Proc. Internat. Wool Textile Res. Conf. Paris*, **3** (1965) 321.
108. A D Wirnik and M A Chekalin, *Tekhnol. Tekstil. Prom.*, (1961) (1) 95; *Textil Praxis*, **17** (1962) 577.
109. H Baumann, V G Kulkarni and A Bobrowski, *Proc. 6th Int. Wool Text. Res. Conf. Pretoria* (1980), 519.
110. H Zahn and P F Rouette, *Textilveredlung*, **3** (1968) 211.
111. Y Xia, *J.S.D.C.*, **99** (1983) 56.
112. P Ball, U Meyer and H Zollinger, *Text. Research J.*, **56** (1986) 447.
113. D Mäusezahl, *Textilveredlung*, **5** (1970) 839.
114. D A P Small, C R Lowe, A Atkinson and C J Bruton, *Eur. J. Biochem.*, **128** (1982) 119.
115. W H Scouten, *Affinity chromatography* (New York: John Wiley, 1981).

CHAPTER 7

Chemistry of formation of coloured molecules in the fibre

ALAN JOHNSON

The first five chapters of this book are concerned with the various aspects of physical chemistry which are involved in the transfer of dye molecules from an external phase into a fibre phase and their adsorption within the fibre phase, whilst in Chapter 6 the role of reactional chemistry leading to the formation of covalent bonds between dye and fibre molecules was discussed. In this final chapter we are again concerned with chemical reactions, but this time covalent bonds to the fibre molecule are not involved. Rather, we are concerned with those classes of dyes in which chemical changes brought about during application lead to the formation of water-insoluble coloured products, which are therefore more in the nature of pigments than dyes.

Nevertheless, the precursors of the final pigment are applied from solution and are subject to the same laws of physical chemistry as are other classes of dyes. They are worthy of equal attention but, since little relevant work has yet been published, only brief mention of these matters can be made, and the chapter will mainly be concerned with the organic and physico-organic chemistry involved in the production of pigments within fibres.

7.1 INSOLUBLE AZO COLOURS OR AZOICS

The ideal of coloration by synthesising water-insoluble azo compounds directly on cotton textiles was conceived by Thomas and Robert Holliday in 1880 [1], and has been considerably extended and exploited since then. Their first attempts led to the precipitation of benzene-azo-2-naphthol and, although this was soon superseded as a useful colouring matter, it can serve our purpose here to illustrate the following four essential features of the process.

1. *Naphtholating* The cloth is padded through an alkaline solution of 2-naphthol and carefully dried.

2. *Diazotisation* Aniline is diazotised by treatment with sodium nitrite and an excess of hydrochloric or sulphuric acid to produce the corresponding diazonium salt:

3. *Coupling* The naphtholated cloth is passed through the buffered diazo solution. The diazonium ions diffuse more rapidly into the fibre than naphtholate ions diffuse out, and diazonium and naphtholate ions react to produce the insoluble benzene-azo-2-naphthol. This reaction,

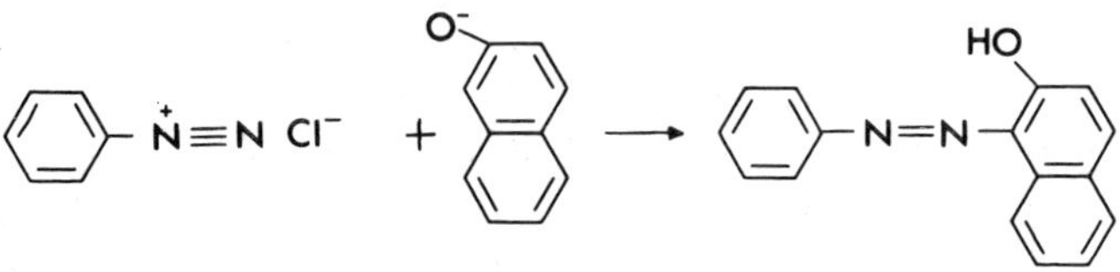

which is universally used to introduce the azo group in all classes of dye, is known as coupling.

4. *Soaping* The pigmented cloth is treated with boiling soap solution, which removes surface pigment and causes aggregation of the remainder.

Of these four stages the two chemical reactions (diazotisation and coupling) have received by far the greatest amount of attention and indeed are of the greatest importance. They will therefore be discussed first and at the greater length. References to actual application to textiles are to be understood to mean application to cellulosic materials, for which these colouring matters were first introduced. Although their usage is now declining, cellulosic materials still furnish their major outlet. Application to non-cellulosic textiles will be dealt with in section 7.1.9.

7.1.1 Diazotisation

This is a reaction, leading to the production of diazo compounds, undergone by primary aromatic amines. The simplest conditions of the reaction are exemplified in paragraph 2 above, the amine being dissolved in mineral acid (usually diluted), cooled and treated with sodium nitrite, when diazotisation occurs rapidly. Ease of diazotisation is affected by substituents in the aromatic ring, electron-donating (base-strengthening) substituents, e.g. *p*-Me and *p*-MeO, leading to easier diazotisation, whereas electron-accepting (acid-strengthening) substituents, e.g. *p*-NO_2, *p*-Br, make diazotisation more difficult – indeed, very weakly basic amines cannot be diazotised by the simple procedure outlined above but have to be treated in concentrated sulphuric acid or mixtures of this with glacial acetic acid.

In the early days of azoic dyeing diazotisation had to be carried out by the dyer, often under far from satisfactory conditions. The resulting difficulties led to the introduction of stabilised forms of the diazo compounds, which can be prepared under properly controlled conditions by the dyestuff manufacturer and need only be weighed out and dissolved for use, so that nowadays diazotisation in the dyehouse is less usual. Nevertheless the reaction remains of importance to the dyer, not only for applying azoics but also for aftertreating other classes of dyes by diazotisation and coupling, and must be considered here, even if not in great detail. (For a comprehensive account, Zollinger's book [2] is outstanding.)

The diazo reaction was discovered in 1858 by Griess, who treated picramic acid with nitrous fumes. The versatility of the reaction and the great reactivity of the resulting diazo compounds were soon recognised and exploited, particularly in the emerging dyestuffs industry. In 1894 Bamberger [3] put forward the suggestion, later to be revised, that diazotisation was a particular example of nitrosation, which in the case of primary aromatic amines continued beyond the nitrosamine stage:

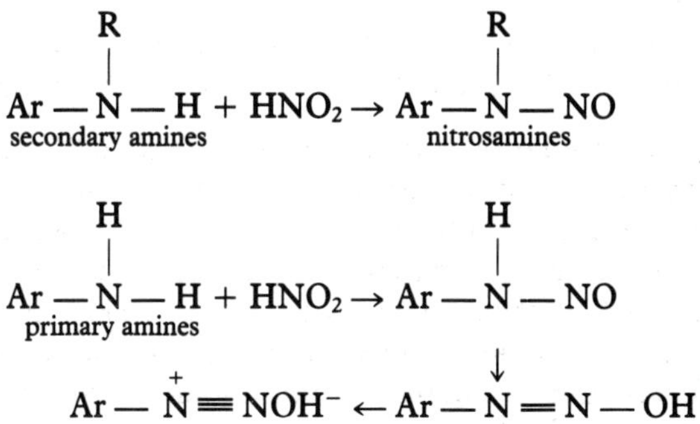

Bamberger offered no experimental evidence, however, and it was left to Hantzsch and Schumann [4] in 1899 to attempt the first investigation of the kinetics of diazotisation. Hantzsch and Schumann worked with dilute solutions in very dilute acid (0.002M HCl) and found that the reaction was, overall, of the second order. They formulated the rate equation as

$$\text{rate} = k[\text{Ar}\overset{+}{\text{N}}\text{H}_3][\text{HNO}_2] \qquad (7.1)$$

i.e. the reaction was regarded as bimolecular between the anilinium ion and undissociated nitrous acid. They also found that, under their conditions, all the amines they investigated diazotised at the same rate. It should be noted that their rate equation was not proved experimentally.

From 1899 until 1935 the reaction was reinvestigated by several workers [2], all of whom found the reaction to be overall of the second order, but their results varied as to the effect on the reaction rate of the basicity of the amine and of the concentration of acid. In 1936 Schmid [5], using higher acidities than before (0.2M), found that diazotisation was of the third order and established the rate expression

$$\text{rate} = k[\text{ArNH}_2][\text{HNO}_2]^2 \tag{7.2}$$

This rate expression had already been found in 1928 by Taylor [6], for deamination and for the nitrosation of aliphatic secondary amines, but no attempt was made at the time to associate the two or to explain the discrepancy from the earlier results. In 1940, however, Hammett [7] suggested that nitrosation was a reaction of dinitrogen trioxide (N_2O_3), the formation of which would require two molecules of nitrous acid.

Elucidation of the mechanism of diazotisation was finally provided by Hughes, Ingold and Ridd in 1958, and an authoritative review has been given by Ridd [8] (see also ref. 2). It should be stressed that they arrived at their conclusions not only as a result of careful experimentation, but also because by that time the principles of electrophilic substitution were understood. Nitrosation was recognised as an electrophilic substitution which could be expected to be similar to the much more extensively studied reaction of nitration. Thus it was likely that the substitution proper was brought about by a cation such as the nitrosonium ion (NO^+) or nitrous acidium ion ($H_2NO_2^+$) or, in the more general sense, by a nitrosation carrier NOX, where X is a group attached to the nitroso group (NO), but which is eliminated during the substitution. The carrier NOX must arise from the reaction solution, so that X could vary according to the composition of the solution, and differing results might be expected in the presence of, say, chloride than in the presence of acetate. A very important conclusion from the assumption that a carrier NOX is responsible for nitrosation is that the observed kinetic form could be due to either (a) the formation of NOX or (b) the reaction of NOX with the species being nitrosated, according to which was the slower, i.e. rate-determining process.

With these thoughts in mind, we can now look at the work carried out on the diazotisation of aniline in aqueous perchloric acid, brought about by the addition of sodium nitrite. The original papers should be consulted for precise experimental details, it being sufficient here to note that allowance was made for the formation of anilinium perchlorate, the concentration of free aniline only (i.e. $C_6H_5NH_2$ and not $C_6H_5NH_3^+$) being used in the kinetic calculations. A second point of importance is that it was known that the salt of nitrous acid and perchloric acid (nitrosyl perchlorate) is fully dissociated, i.e. it exists completely as $NO^+\ ClO_4^-$, so that $NOClO_4$ could be discounted as the nitrosation carrier NOX.

Rates of diazotisation were measured for varying concentrations of amine and nitrite at constant acidity, enabling the overall kinetic order to be established, and then repeated at higher acidities ranging from 0.002M to 0.05M acid. It was found that, at the lower acidities, an order of 2 was obtained and then the order began to increase until an order of 3 was obtained at higher acidities. This result reconciled the observations of second-order kinetics obtained by the earlier workers with the third-order kinetics of Schmid, whilst the finding of fractional orders increasing from 2 to 3 in the intermediate range clearly pointed to the presence of two kinetically significant processes, one being rate-determining at low and the other at higher acidity.

It was now possible to examine the results on the assumption that the reaction was essentially a nitrosation brought about by a nitrosation carrier NOX. Firstly, the possible carriers must be established and, secondly, the kinetic expression that would result from each if (a) its formation or (b) its reaction with the amine were rate-determining. Under the conditions used there are only four possible nitrosation carriers, and these will now be considered in turn.

1. Molecular nitrous acid, HO.NO ($X = OH^-$)

(a) Rate of formation is not relevant, since it is produced instantaneously from sodium nitrite and perchloric acid.

(b) A reaction between HO.NO and $ArNH_2$ would lead to the expression

$$\text{rate} = k[ArNH_2][HNO_2] \qquad (\text{order} = 2) \qquad (7.3)$$

2. Nitrous acidium ion, $H_2NO_2^+$ ($X = H_2O$)

(a) Again, the rate of formation is not relevant.

(b) Reaction between $H_2NO_2^+$ and $ArNH_2$ would give

$$\text{rate} = k[H^+][ArNH_2][HNO_2] \qquad (\text{order} = 3) \qquad (7.4)$$

3. Dinitrogen trioxide, N_2O_3 ($X = NO_2^-$)

(a) Dinitrogen trioxide is formed from two molecules of nitrous acid by dehydration:

$$2HNO_2 \rightleftharpoons N_2O_3 + H_2O$$

leading to a rate equation

$$\text{rate} = k[HNO_2]^2 \qquad (\text{order} = 2) \qquad (7.5)$$

(b) Reaction of N_2O_3 with $ArNH_2$ would give

$$\text{rate} = k[HNO_2]^2[ArNH_2] \qquad (\text{order} = 3) \qquad (7.2)$$

4. Nitrosyl ion, NO^+

(a) There are two possible routes to the production of NO^+:

$$HNO_2 + H^+ \rightarrow H_2\overset{+}{N}O_2 \rightarrow NO^+ + H_2O$$

$$\text{rate} = k[HNO_2][H^+] \qquad (\text{order} = 2) \tag{7.6}$$

and

$$2HNO_2 \rightleftharpoons N_2O_3 \rightleftharpoons NO^+ + NO_2^-$$

$$\text{rate} = k[HNO_2]^2 \qquad (\text{order} = 2) \tag{7.7}$$

(b) The corresponding expressions for a reaction between NO^+ and $ArNH_2$ are

$$\text{rate} = k[HNO_2][H^+][ArNH_2] \qquad (\text{order} = 3) \tag{7.4}$$

and

$$\text{rate} = k[HNO_2]^2[ArNH_2] \qquad (\text{order} = 3) \tag{7.2}$$

Experimentally, it was found that the order increased from 2 to 3 with increasing acidity. Moreover, the third-order reaction at high acidities was already known, from the work of Schmid, to obey the equation

$$\text{rate} = k[HNO_2]^2[ArNH_2] \tag{7.2}$$

From the above considerations the only nitrosating agent that can give Eqn 7.2 is either N_2O_3 or NO^+ if this is in turn derived from N_2O_3. Both of these give an order of 3 if their attack on the amine is rate-determining, and an order of 2 if their formation is rate-determining. The kinetic expressions for the second-order process are not the same for the two cases, however.

An overall order of 2 had already been obtained by earlier workers in low acidities and Hantzsch had proposed the equation

$$\text{rate} = k[Ar\overset{+}{N}H_3][HNO_2] \tag{7.1}$$

but this had not been proved. It was tested by Ridd *et al.* by the simple expedient of carrying out the reaction at low acidity with a constant concentration of nitrous acid, but with the amine concentration varying tenfold. According to Hantzsch's expression this should have led to a tenfold increase in rate, but in fact the rate scarcely increased at all. It must be concluded that, at low acidities, the rate is independent of the amine, i.e. is essentially the rate of formation of the nitrosation carrier and, since the overall order is 2, is described by the equation

$$\text{rate} = k[HNO_2]^2 \tag{7.7}$$

The only nitrosation carrier which can lead to the rate equations 7.2 and 7.7 is N_2O_3, leading to the final conclusion that in dilute perchloric acid the overall reaction is represented by

$$2HNO_2 \rightarrow N_2O_3 + H_2O$$

$$Ar\text{—}NH_2 + N_2O_3 \rightarrow \underset{\displaystyle NO}{Ar\text{—}\overset{}{N}H} + HNO_2$$

The first of these reactions is applicable at very low acidities, because under these conditions very little amine salt is present (i.e. the value of $[ArNH_2]$ is relatively high), so N_2O_3 is removed as rapidly as it is formed and it is the rate of formation of N_2O_3 that is rate-determining. At high acidities most of the amine is present as $ArNH_3^+$, so $[ArNH_2]$ is now very small; N_2O_3 therefore accumulates and its rate of reaction with the amine becomes rate-determining.

These reactions, of course, lead to the formation of a primary aromatic nitrosamine. This then decomposes rapidly to the diazonium salt, as originally suggested by Bamberger. Also, these reactions explain the observations that different amines diazotise at the same rate at very low acidities, since the amine is not kinetically involved under these conditions.

In applying this mechanism to actual diazotisations at acidities high enough for the second reaction to play some part, it must be noted that the influence of the acid is to affect the protonation of the amine, i.e. to determine the ratio $[ArNH_2]/[ArNH_3^+]$. For a weakly basic amine this ratio will be larger than for one which is more strongly basic, and this can lead to a faster conversion of the less basic amine, even though this is less susceptible to electrophilic attack than the more basic amine. For very weakly basic amines, which are diazotised in concentrated sulphuric acid by the addition of solid sodium nitrite, the nitrosation carrier will be NO^+ (from $NO^+\ SO_4H^-$). This consideration, however, is not relevant to the application of insoluble azoic dyes since such amines are not used.

7.1.2 Coupling

Coupling is the reaction of an aromatic diazo compound with a second compound (the coupling component) resulting in substitution at a carbon atom of the coupling component with the formation of an azo compound. The production of Para Red is a typical example:

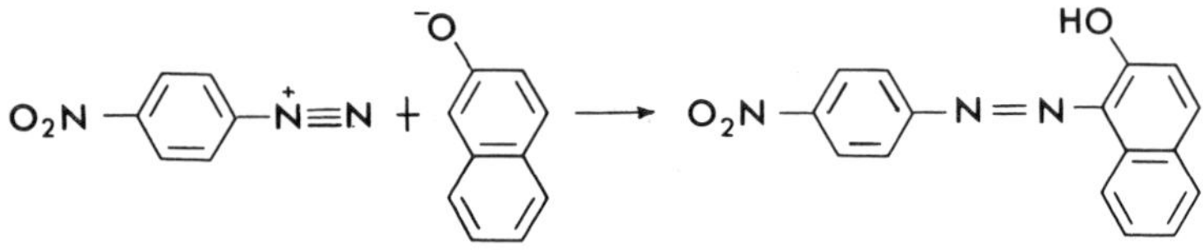

Para Red

This formulates the reaction correctly as occurring between the diazonium cation and the naphtholate anion, but the coupling reaction was known and exploited for almost a century before the correct mechanism was successfully elucidated.

In considering the chemistry of azoic dyes, coupling is more important than diazotisation, since coupling must be carried out at the point of application, whereas diazotisation is now more often done by the dye manufacturer, who then supplies the diazo compound in a stabilised form. It is, therefore, pertinent that the mechanism of coupling should be discussed here in some detail. Before doing this, however, it is necessary to give a brief summary of the acid–base equilibria in which diazo compounds and coupling components can take part. Fuller accounts of these, and of coupling itself, are given elsewhere [2, 9].

Diazo equilibria

These may be summarised as follows:

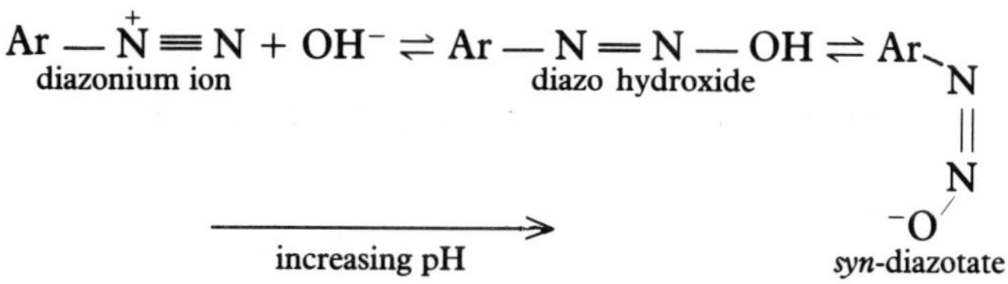

It is usual to say that all these three forms can undergo coupling, but this idea dates from the time when the nature of the equilibria was not fully understood. Since all three forms will always be present to some extent over the whole pH range covered by the equilibria, it is only necessary for one form actually to take part in the coupling reaction. The pH range required to go from virtually all diazonium ion to virtually all *syn*-diazotate naturally varies with substituents in the aromatic nucleus. There is a further step beyond the *syn*-diazotate, which takes place at high pH and leads to the formation of the *anti*-diazotate:

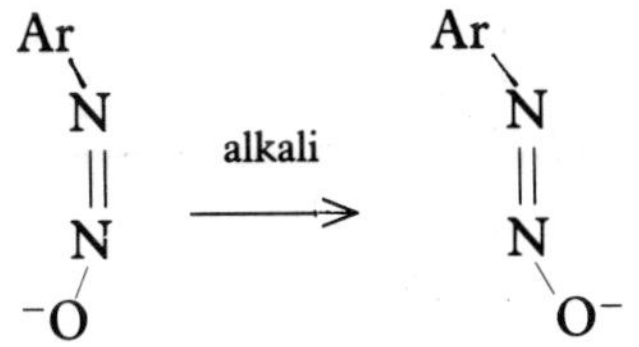

This change was formulated by Hantzsch as an example of geometric isomerism [10] and this formulation has subsequently been shown to be correct [9].

anti-Diazotates do not couple and are only slowly converted back to diazonium ions on acidification, so the pH range over which coupling can occur extends up to the pH of formation of the *anti*-diazotate. This pH is lower for diazo compounds with electron-accepting substituents than for those with electron-donating substituents.

Equilibria of coupling components

The three types of coupling component important for making azo dyes are (a) aromatic hydroxy compounds, (b) compounds with activated methylene groups and (c) aromatic amines, although only the first two of these are used in producing azoic dyes. All three groups of compounds show acid–base equilibria, as follows:

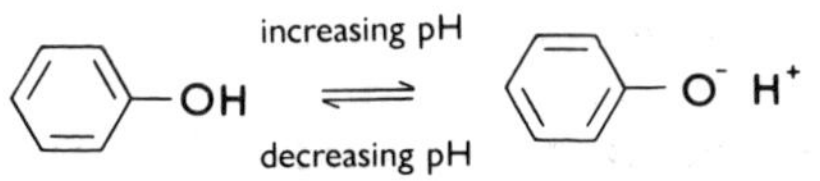

aromatic hydroxy compounds, i.e. phenols and naphthols

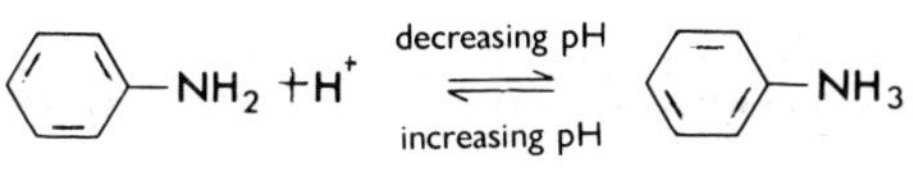

aromatic amines

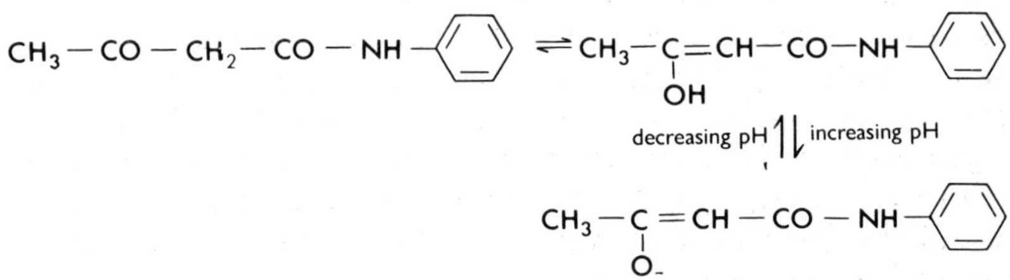

Thus aromatic hydroxy compounds and compounds with an activated methylene group exist either as neutral molecules or, at higher pH, as anions, whilst aromatic amines exist as neutral molecules or, at lower pH, as cations.

With the knowledge of these acid–base equilibria involving both partners of the coupling reaction, and with the reasonable supposition that only one equilibrium form in each case will undergo the reaction, it is not surprising that coupling is very sensitive to pH and that for each coupling reaction there is a relatively narrow pH range for coupling activity. This has been everyday knowledge ever since the coupling reaction was first used. Much of the earlier work on coupling is unhelpful, however, because the equilibria were not known, particularly that of the diazo component, and hence the need to work in buffers was not appreciated. Indeed, much early work was carried out before buffers themselves were known. Again, as with diazotisation, a full understanding of the reaction was not possible until the nature of electrophilic substitution, and a fuller appreciation of the power of kinetic studies, had been realised. For these reasons much of the earlier work will not be discussed here. A fuller account is given in ref. 2.

It is clear, however, that the central problem is that of deciding which of the equilibrium forms of the two components actually takes part in the coupling reaction. An investigation of importance is that carried out by Conant and Peterson in 1930 [11], who measured the rates of coupling of a range of diazotised amines with various phenols and naphthols in buffered solutions. They found that, for diazo compounds, the rate of coupling increased as substituents became increasingly electron-accepting (i.e. as the amines themselves became weaker bases), whilst the reverse situation applied to the phenols and naphthols. They concluded that coupling was a reaction between the undissociated diazo hydroxide and phenol or naphthol. In 1941, however, Hauser and Breslow [12] carried out coupling reactions using solid diazonium salts and 2-naphthol dissolved in dry pyridine. Coupling was rapid and, since the diazo hydroxide could not have been formed in the absence of water, they concluded that it was the diazonium ion that coupled. In the same year a very important investigation was carried out by Wistar and Bartlett [13], who studied the coupling of various diazo compounds with naphthylamine sulphonates. They recognised the acid–base equilibria in both coupling components and therefore considered the four possibilities:

(a) diazo hydroxide + ammonium ion ($H_3\overset{+}{N}-Ar$),
(b) diazonium ion + free amine,
(c) diazonium ion + ammonium ion,
(d) diazo hydroxide + free amine.

Since reasonable values of the dissociation constants for both the diazo components and the coupling components were known, i.e. for the equilibria

$$Ar-\overset{+}{N}\equiv N + OH^- \rightleftharpoons Ar-N=N-OH$$

and

$$Ar-\overset{+}{N}H_3 \rightleftharpoons ArNH_2 + H^+$$

it was possible to calculate the fractional concentrations of the four possible species at different pH. Thus, if the free amine is A, then its fractional concentration γ_A is given by

$$\gamma_A = \frac{[ArNH_2]}{[ArNH_2] + [Ar\overset{+}{N}H_3]} \tag{7.8}$$

Similarly, for the ammonium ion B,

$$\gamma_B = \frac{[Ar\overset{+}{N}H_3]}{[ArNH_2] + [Ar\overset{+}{N}H_3]} \tag{7.9}$$

for the diazo hydroxide C,

$$\gamma_C = \frac{[ArN{=}NOH]}{[ArN{=}NOH] + [Ar\overset{+}{N}{\equiv}N]} \tag{7.10}$$

and for the diazonium ion D,

$$\gamma_D = \frac{[Ar\overset{+}{N}{\equiv}N]}{[ArN{=}NOH] + [Ar\overset{+}{N}{\equiv}N]} \tag{7.11}$$

The reaction is

diazo component + coupling component → azo compound

and the rate is given by

rate = k[diazo component][coupling component]

In the actual experiments, however, the concentrations measured were [total diazo] and [total amine] and substitution in the rate equation would give not the true second-order rate constant k, but an apparent constant k_{app}.

For the four possible reactions the apparent rate constant would be related to the true rate constant as follows:

1. $k_{app1} = k\gamma_C\gamma_B$; $\log k_{app1} = \log k + \log \gamma_C + \log \gamma_B$ (7.12)

2. $k_{app2} = k\gamma_D\gamma_A$; $\log k_{app2} = \log k + \log \gamma_D + \log \gamma_A$ (7.13)

3. $k_{app3} = k\gamma_D\gamma_B$; $\log k_{app3} = \log k + \log \gamma_D + \log \gamma_B$ (7.14)

4. $k_{app4} = k\gamma_C\gamma_A$; $\log k_{app4} = \log k + \log \gamma_C + \log \gamma_A$ (7.15)

From the calculated values of γ, curves were plotted showing the variation of ($\log \gamma_C + \log \gamma_B$), ($\log \gamma_D + \log \gamma_A$), ($\log \gamma_D + \log \gamma_B$) and ($\log \gamma_C + \log \gamma_A$) against pH; these curves have the same shape as the corresponding curves of $\log k_{app}$ against pH, being displaced by the constant $\log k$.

The four calculated curves are shown in Figure 7.1. The curve of $\log k_{app}$

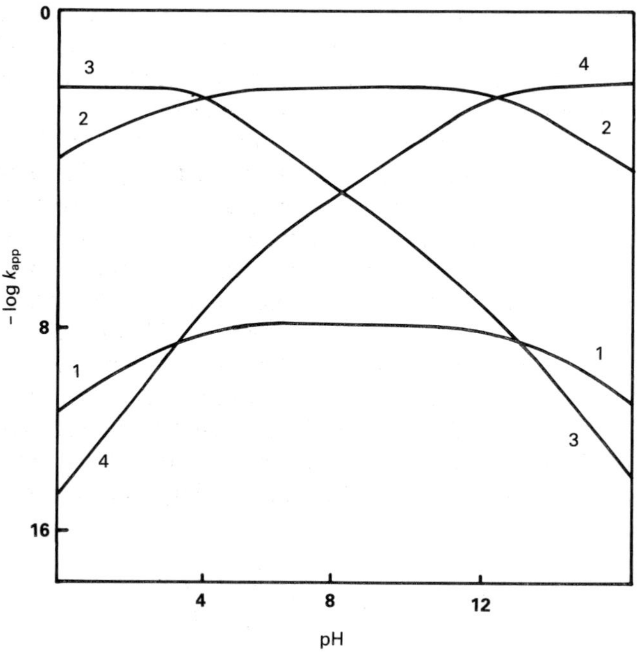

Figure 7.1 – Theoretical plots of log k_{app} versus pH

versus pH, from the actual experimental results, had the same shape as curves 1 and 2, so the actual reaction must be one of these two. Strong support for reaction 2 rather than 1 comes from a consideration of the variation in rate on either side of the pK value of the amino group in the coupling component for the equilibrium

$$Ar\overset{+}{N}H_3 \rightleftharpoons ArNH_2 + H^+$$

The pK is the pH at which the amine salt is half-dissociated. Lowering the pH by one unit from pK would result in a tenfold decrease in concentration of $ArNH_2$, whilst increasing the pH by one unit above pK would give a tenfold decrease in the concentration of $ArNH_3^+$. There would be corresponding variations in the rates of reactions 2 and 1 respecitvely. In fact, for 1-naphthylamine-4-sulphonate as the coupling component, for which the pK of the amino group is *ca* 5, decreasing the pH of coupling from 5 to 4 brought about the expected tenfold decrease in rate, whilst raising the pH to 6 caused only a slight increase. The weight of all the evidence is therefore that coupling with amines is a reaction between the diazonium ion and the free amine, i.e. reaction 2.

By analogy, coupling with phenols and naphthols would be expected to be between the diazonium cation and the phenolate or naphtholate anion, rather than between the diazo hydroxide and undissociated phenol as suggested by Conant and Peterson. In fact, treatment of these results by the method of Wistar and Bartlett leads to this conclusion.

If coupling is a reaction between diazonium ion and either free amine or phenolate ion, then it would fit clearly into the category of an electrophilic substitution, i.e. an attack by a positive ion at a site of high electron density. Of the possible diazo species the diazonium ion is the strongest electrophile, and the free amine and the phenolate ion respectively would yield the highest electron density in the coupling component. In fact, in the years following the work of Wistar and Bartlett, a good deal of evidence came to light to support the idea of coupling being an electrophilic substitution.

The final conclusive evidence, however, was provided in a most elegant way by Zollinger [14], making use of the Brønsted equation, which relates the rate constant of a reaction with the charges on the reactants and the ionic strength I of the reaction solution. For the coupling reaction, this equation can be written as

$$\log k_1 = \log k_0 + \frac{2Z_D Z_C \alpha I^{\frac{1}{2}}}{1+\beta\alpha I^{\frac{1}{2}}} \qquad (7.16)$$

where k_1 and k_0 are the rate constants at ionic strength 1 and 0 respectively, Z_D and Z_C are the net charges on the diazo and coupling components and α and β are constants.

Zollinger studied the coupling of diazotised *p*-toluidine with 2-amino-naphthalene-6-sulphonate. The two components and the corresponding values of Z are as shown below, since at the pH of the experiments the sulphonic acid group was fully ionised.

Diazo component:

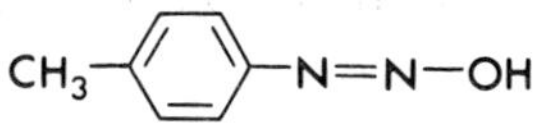 or 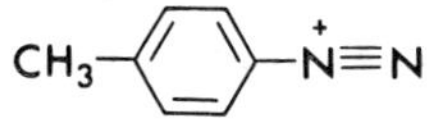

$Z_D = 0$ $Z_D = +1$

Coupling component:

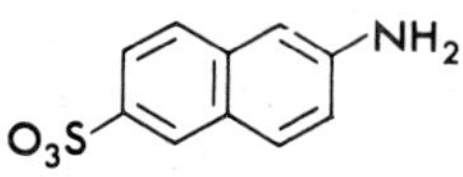 or 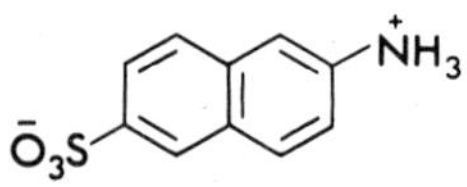

$Z_C = -1$ $Z_C = 0$

For a reaction of the diazo hydrate ($Z_D = 0$) with either coupling component or for a reaction of the anilinium ion ($Z_C = 0$) with either diazo component, the second term on the right-hand side of the Brønsted equation would be zero, i.e. the rate would be independent of ionic strength. However, for reaction between diazonium ion ($Z_D = +1$) and the free amine ($Z_C = -1$), this term would have a negative value, and hence the rate would decrease with increasing ionic strength (negative salt effect). This is what was actually found to happen. In confirmation, the coupling of diazotised sulphanilic acid and of diazotised 2,4-disulphoaniline was studied. As the diazonium ions these would have the forms:

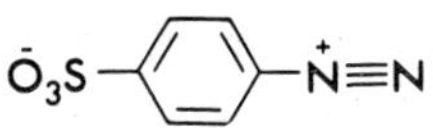 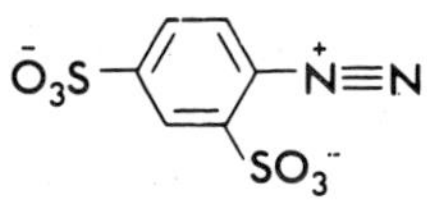

$Z_D = 0$ $Z_D = -1$

leading to zero salt effect and a positive salt effect respectively. Both these effects were observed, with very close agreement between the actual rates and those predicted from the Brønsted equation.

Thus there is now no doubt that coupling is an electrophilic substitution by the diazonium ion into either a free amine or a phenolate or naphtholate ion, and is therefore facilitated by electron-withdrawing substituents in the diazonium ion and electron-donating substituents in the coupling component. In many cases a series of substituted diazonium ions reacting with the same coupling component show a linear relationship between log (rate constant) and the Hammett σ value for the substitutents [15].

For the production of insoluble azo colours, where only hydroxy coupling components are used, it is thus clear that the final coupling conditions on the cloth must be alkaline so that the ionised active form of the coupling component will be present, but at the same time too high a pH must be avoided, since it is desirable that the diazo equilibrium is as far on the side of the diazonium ion as possible. A point to bear in mind is that coupling liberates acid into the solution, e.g.

$$Ar{-}\overset{+}{N}{\equiv}N\ Cl^- + {}^-O{-}Ar' \rightarrow Ar{-}N{=}N{-}Ar'{-}O^- + HCl$$

so there is a natural tendency for the pH to fall as coupling proceeds, but under practical applications this is far outweighed by the excess alkali present in the solution in order to maintain the coupling component fully ionised.

The coupling components commonly used have p*K* values in the region of 10–11, i.e. in this pH range they are ionised to the extent of 50 per cent. For complete ionisation the pH must be raised by two units, i.e. to 12–13, and in practice it is common for coupling components to be applied in solutions approximately 0.1 *M* in sodium hydroxide. This may be a tenfold excess over the concentration of the coupling component itself.

These difficulties are overcome in practice by first neutralising the diazo solution by addition of sodium acetate and then adding a quantity of acetic acid appropriate to the particular coupling component being used. This acid is usually referred to as an alkali binder; it may, on occasion, be replaced by an acid salt such as aluminium sulphate. The overall result is that the pH on the cloth as coupling actually takes place will be approximately 10. A more recent innovation to avoid both difficulties in pH control and high electrolyte concentrations is the introduction of liquid forms of coupling components and of bases for diazotisation. These are either solutions in non-volatile alcohols or dispersions [16]. Manufacturers of azoic combinations have gone to considerable trouble to determine the best conditions of application and their recommendations should be followed. Nevertheless, an understanding of the fundamental processes involved is invaluable if these recommendations are to be used to the best advantage.

In the benzene series, coupling takes place *para* to the activating group and no *ortho*-substitution occurs. For 1-naphthol and its derivatives, however, both *ortho*- and *para*-coupling is observed. Coupling in the *para*-position is subject to general base catalysis whilst *ortho*-coupling is not, so that the proportions of the two isomers may be controlled by varying the concentration and nature of base present [15]. For derivatives of 2-naphthol, which are all important for insoluble azo colours, coupling takes place only in the 1-position, so that *ortho*-hydroxyazo compounds are always produced.

7.1.3 Naphtholating

The process of applying the coupling component to the cloth is of considerable practical importance. If this can be properly controlled and if an excess of diazo component is used at the coupling stage, then it is easy to see that naphtholating determines the depth of colour finally attained. The coupling components are, of necessity, applied from alkaline solution, often with an addition of sodium chloride. They may be applied by exhaust methods or, more commonly, by padding.

The first insoluble azo colours, as mentioned earlier, had 2-naphthol as the coupling component. This, however, suffers from several defects, the most important from the point of view of dyeing being that it has no substantivity for cellulose, so cloths had to be carefully dried before coupling. This situation was improved in 1912 by the introduction of 2-hydroxy-3-naphthoic acid anilide (I) as coupling component [17]:

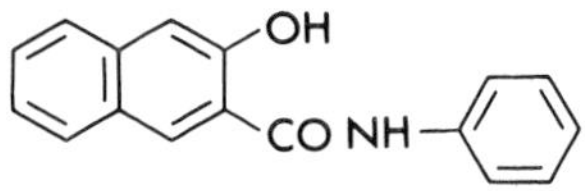

(I)

This compound has substantivity for cellulose, most probably associated with the anilide grouping, which also occurs in some direct dyes and vat dyes. Following its introduction, many other coupling components containing essentially the same structural feature have been introduced. A selection of these, and of diazo components, is given in Table 7.1 below.

In application, these coupling components behave in the same way as do direct dyes, showing the same dependence on electrolyte and temperature; it is generally accepted that they are adsorbed by cellulose by the same mechanism, but very little work has been crried out on fundamental aspects of their adsorption.

Reference may be made, however, to the work of Dunn [18] and of Daruwalla and Mehta [19]. These workers found that the standard affinities of various coupling components for cellulose, measured essentially by the method of Marshall and Peters [20], were significantly lower than for direct dyes, typical values of $-\Delta\mu^{\ominus}$ being of the order of 12kJ (3 kcal) per mole. Daruwalla and Mehta obtained values slightly lower than those of Dunn, but these differences are not significant. In fact, it would be expected that only approximate values would be obtained, since the temperature of application (25 °C) is much lower than that used for direct dyes, so that it is extremely likely that considerable aggregation would occur. This probably accounts for the observation that the results did not fit the model equations of Marshall and Peters nearly so well as did results with direct dyes (p. 301).

Apart from these results, it is a matter of everyday experience that the affinity of coupling components for cellulose is appreciably lower than that of direct dyes. Since the affinity of direct dyes is associated with the relatively large size and linear shape of their molecules, it is not surprising that the smaller and generally more compact molecules of coupling components have lower affinities.

Small molecules, however, are usually more mobile than large ones and this, together with their lower affinity, should lead to the coupling components diffusing more rapidly into cellulose than do direct dyes. Again, it is a matter of everyday experience that this is so, coupling components normally coming to equilibrium with cotton in about 30 min at 25 °C. Dunn [21] carried out some measurements of diffusion coefficients, and found values about 100 times greater than those of direct dyes.

It is unfortunate that more fundamental work has not been carried out. Nevertheless, a considerable amount of practical information has been compiled by dyestuff manufacturers, and this well repays study.

7.1.4 Stabilised diazo compounds

In the foregoing sections it was assumed that the process of diazotisation was an essential part of the practical application of azoic dyes, i.e. that diazotisation had to be carried out by the dyer himself. This is still done in many cases, but there are obvious difficulties in achieving a satisfactory result. Not all aromatic primary amines are easy to diazotise on the large scale; efficient cooling is necessary because of the high instability of many diazo solutions, and because of this instability the full yield of diazo compound may not be attained; finally, the resulting diazo solution is acidic and must be neutralised before use. The great instability of simple diazonium salts in the solid state precludes their isolation by the manufacturer.

However, it was found that some diazonium salts were sufficiently stable to be isolated; 4-nitrobenzenediazonium sulphate, for example, is so although the chloride and nitrate are not. This diazonium sulphate was in fact marketed towards the end of the last century and was a great convenience for dyers in applying azoic combinations involving 4-nitroaniline. It was prepared by diazotising in sulphuric acid, carefully concentrating the solution under vacuum at a low temperature, and then adding anhydrous sodium and aluminium sulphates to absorb the remaining water and sulphuric acid. A dry powder was thus obtained containing an inert salt which probably serves to absorb heat generated by any diazo decomposition which may occur.

Another, and earlier, observation of diazo stabilisation was that of Griess [22], viz. that benzenediazonium chloride formed stable complex salts with gold(III) or platinum (II) chloride; other workers also reported such stable complexes with chlorides of the noble metals. These, of course, are too expensive for technical use, but in 1923 BASF made available the double salts of diazonium chlorides with zinc chloride. The zinc chloride double salts of diazonium chlorides are stable and are now widely used by dyers. These salts are bisdiazonium tetrachlorozincates of general formula

$$[Ar-\overset{+}{N}\equiv N]_2[ZnCl_4]^{2-}$$

Some borofluoride complexes are also used. The structure of these complexes has not been determined but, bearing in mind that the β-nitrogen atom of a diazonium ion possesses an unshared pair of electrons ($Ar-N\equiv N$:), it is possible that this can function as a ligand and so be bound to a zinc or other suitable metal ion (see section 7.6 for a discussion of chelate formation).

A further method of stabilising some diazonium salts stems from the observation of Becker [23] that the addition of 1- or 2-naphthalenesulphonic acids to some diazonium solutions gave stable precipitates of the corresponding diazonium naphthalenesulphonates which could subsequently be redissolved to yield active diazonium solutions. Technically, the monodiazonium salts of naphthalenedisulphonic acids

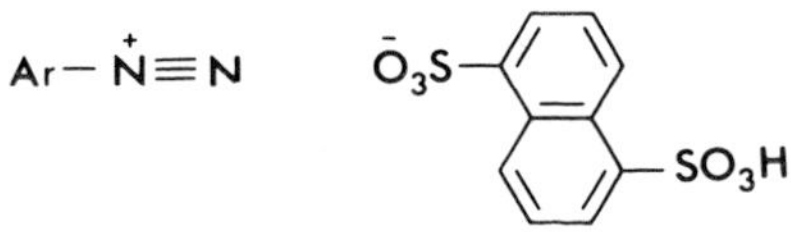

are more useful because of their higher solubility. Use can also be made of the dispersing action of naphthalenesulphonic acids, an excess of the acid serving to hold in solution a sparingly soluble diazonium naphthalenesulphonate. Unfortunately, little is known of the reasons for the stability of these salts.

In spite of the lack of fundamental knowledge, however, diazonium salts stabilised by one or other of these methods are of considerable technical importance. The dyer is freed from the often difficult operation of large-scale diazotisation, needing only to dissolve the appropriate amount of the stable salt. Furthermore, it is possible, by the addition of neutral diluents, to prepare the stable salts in a standard strength, so that uncertainties about the coupling capacity of the diazo solution are removed. It is also usual to include suitable buffers in the dry powder so that diazo solutions of the correct pH are obtained, and the dyer needs to add only the alkali binder (e.g. acetic acid) appropriate to the coupling component being used.

These stabilised diazonium salts are marketed as Fast Colour Salts, whilst the aromatic amines from which they are derived are known as Fast Colour Bases. Fast Colour Salts are usually equivalent to 20 per cent of the corresponding Bases. Examples are listed in Table 7.1, which also includes the most important coupling components.

7.1.5 Non-coupling mixtures with stable diazo compounds

The simple procedure of applying the coupling component and the diazo

TABLE 7.1

Azoic components

Diazo components: Fast Colour Base	Structure	Fast Colour Salt
Scarlet GG	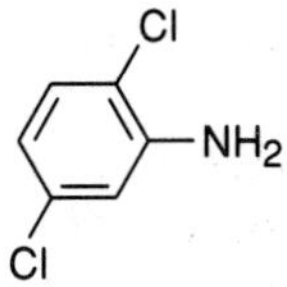	$ZnCl_2$ double salt
Red KB	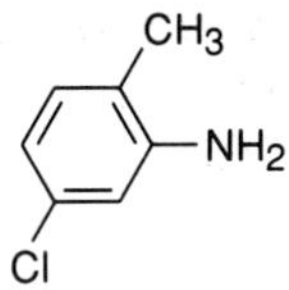	aryl sulphonate
Red 3GL	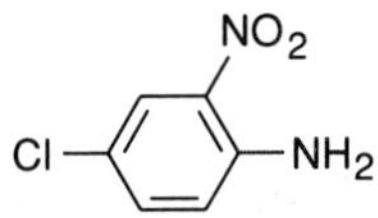	$ZnCl_2$ double salt
Red B	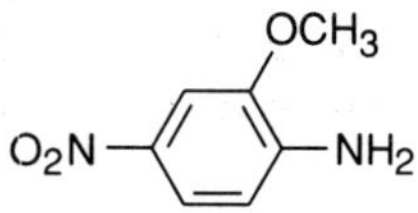	aryl sulphonate
Orange GC	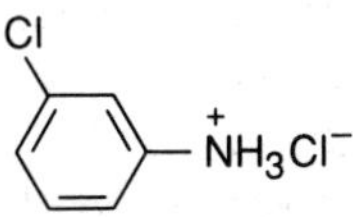	borofluoride
Variamine Blue B	–	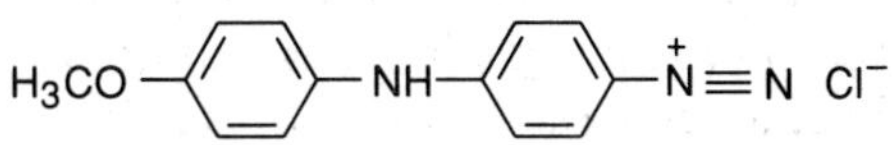
Yellow GC	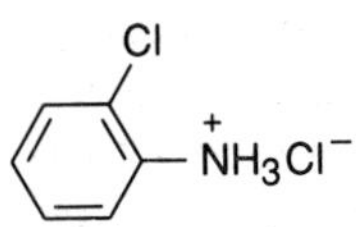	$ZnCl_2$ double salt

Coupling components:	Structure
Naphtol AS	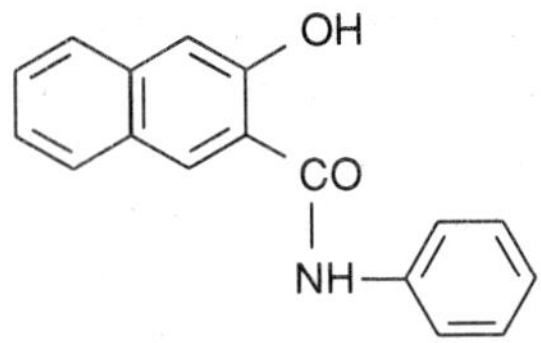
Naphtol AS-PH	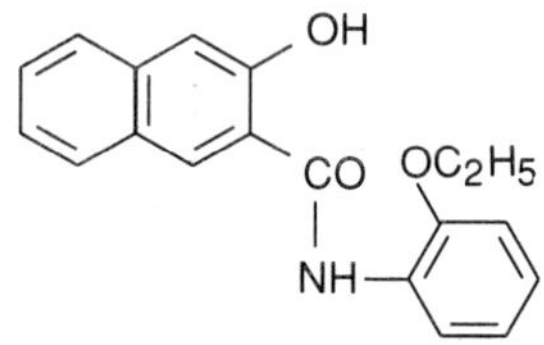
Naphtol AS-GR	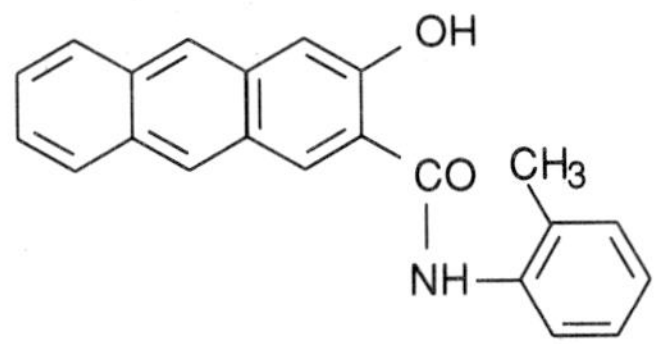
Naphtol AS-L4G	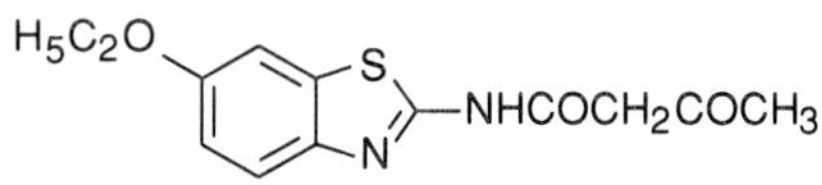
Naphtol AS-G	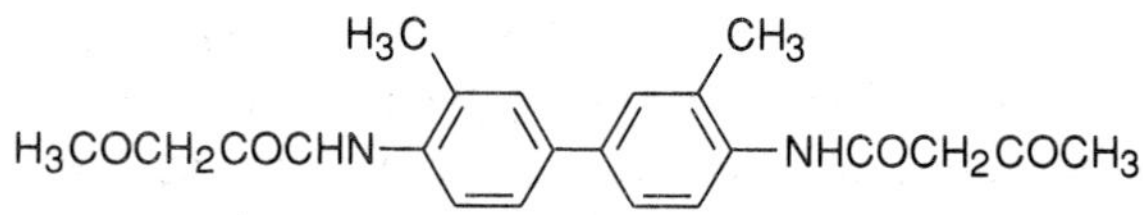
Naphtol AS-LB	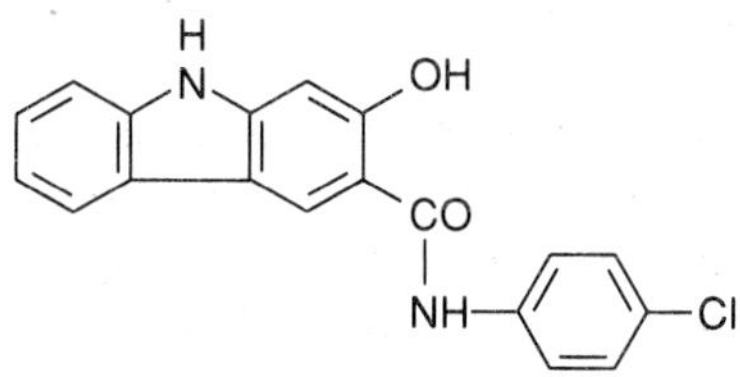

component separately has disadvantages, particularly in textile printing. If a cloth has been prepared with one coupling component, only those colours obtainable from that component can be produced in one passage through the printing machine. If further colours from a second coupling component are then required, the cloth must be washed off and then treated with the new component before a second run through the printing machine. Not only is this time-consuming and wasteful of coupling component, but difficulties in register of the pattern are encountered during the second passage.

To avoid this sort of difficulty, methods have been sought for converting diazo compounds into stable non-coupling derivatives which can be mixed with coupling components, the mixtures then being applied as required. After application a simple aftertreatment (development) regenerates the active diazonium salt under conditions which still permit coupling to occur.

7.1.6 Mixtures with *anti*-diazotates

As mentioned earlier, if diazo compounds are subjected to high pH the equilibria are forced beyond the *syn*-diazotate stage, resulting in the formation of *anti*-diazotates:

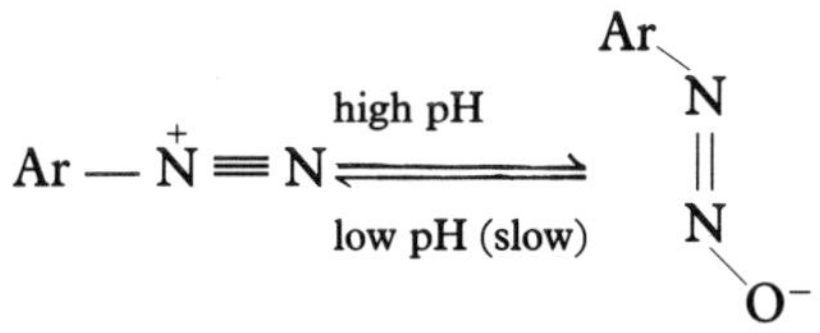

anti-Diazotates are the most stable of the diazo bodies and do not couple, i.e., under alkaline conditions the reverse reaction back to the diazonium ion is so slow as to be negligible. Under acidic conditions, however, the diazonium ion is re-formed and coupling power is thereby restored. The mechanism by which this reverse reaction takes place has been the subject of several investigations [2], but no completely satisfactory explanation has yet been provided.

From the point of view of the suitability of *anti*-diazotates for mixing with coupling components, however, the reaction mechanism as such is not important. It is sufficient that non-coupling *anti*-diazotates can be mixed with coupling components and these mixtures applied to textiles, usually by printing. A subsequent treatment with acid will liberate the diazonium ion so that coupling can then occur. These mixtures are marketed as Rapid Fast Colours; some examples are listed in Table 7.2.

Nevertheless, even though the detailed mechanism of formation of the diazonium ion is not important in this context, the relative ease with which this conversion occurs is of considerable significance, as also is the ease of

TABLE 7.2

Rapid Fast Colours

Rapid Fast	Diazo component (as *anti*-diazotate)	Naphtol
Yellow GH	Fast Scarlet GC	AS-G
Scarlet IRH	Fast Red KB	AS-PH
Blue B	Dianisidine	AS
Olive Brown IBH	Fast Red RC	AS-LB

preparation of the *anti*-diazotate by the dyestuff manufacturer. As with other reactions of aromatic compounds, these changes are determined by the influence of other substituents in the aromatic nucleus. Substituents that readily accept electrons facilitate the formation of *anti*-diazotates, which can often be produced by simple addition of sodium hydroxide solution, whereas diazonium salts containing electron-donating substituents often require very concentrated alkali solutions and high temperatures to produce *anti*-diazotates. The relative stabilities of the *anti*-diazotates reflect these effects, those in the first group reverting back to the diazonium ion only with difficulty, whilst those with electron-donating substituents are very readily converted.

In selecting *anti*-diazotates for mixtures with coupling components, a compromise must therefore be adopted. The least stable *anti*-diazotates are unsuitable because they are prone to premature development, whilst the most stable would only slowly regenerate the diazonium ion. It must be remembered that acidic conditions are required for the diazonium ion to be produced at all rapidly, but such conditions mitigate against coupling. Thus, only a few diazo compounds and coupling components are suitable for these mixtures.

The production of acidity for development is most readily accomplished by injecting acetic acid into the steamer; alternatively, ammonium salts, which lose ammonia in steaming, can be included in the print paste.

7.1.7 Mixtures with diazoamino compounds

The coupling reaction with phenols and with amines has been discussed earlier in terms of the formation of azo compounds in which the azo group is joined to two carbon atoms, as a result of electrophilic attack by the diazonium ion at a carbon atom of high electron density in the coupling component. It is also possible, however, under suitable conditions, for the diazonium ion to attack the oxygen or nitrogen atom to produce diazoethers or diazoamino compounds respectively:

$$\mathrm{Ar{-}\overset{+}{N}{\equiv}N + HOR \rightleftharpoons \underset{\text{diazoether}}{Ar{-}N{=}N{-}OR} + H^+}$$

$$\mathrm{Ar{-}\overset{+}{N}{\equiv}N + H_2NR \rightleftharpoons \underset{\text{diazoamino compound}}{Ar{-}N{=}N{-}NHR} + H^+}$$

Diazoethers are very unstable and are of no value to dyeing technology. Diazoamino compounds are stable, however, and can be used in non-coupling mixtures; they are produced at slightly higher pH values than those at which normal coupling occurs. The reaction is reversible under the influence of acid, and it is this scission which forms the basis of their use, the diazonium ion coupling with the coupling component with which the diazoamino compound had been previously mixed, i.e.

$$\mathrm{Ar{-}N{=}N{-}NH{-}R + HO{-}Ar' \xrightarrow{acid} Ar{-}\overset{+}{N}{\equiv}N + HO{-}Ar' + H_2N{-}R}$$
$$\downarrow$$
$$\mathrm{Ar{-}N{=}N{-}Ar'{-}OH + H^+}$$

However, there are some drawbacks to this simple scheme. The most obvious is that the amine used in the preparation of the diazoamino compound (RNH_2, known as the stabilising amine) is itself liberated alongside the diazonium ion, so coupling may take place with the stabilising amine rather than with the coupling component with which the diazoamino compound had been mixed. The stabilising amine must therefore be one which does not readily couple, and for this reason an aliphatic amine is often chosen. A further difficulty arises if the stabilising amine is a primary amine, leading to a primary diazoamino compound. Such compounds readily tautomerise, so that two modes of scission are possible, leading to two diazo compounds:

$$\begin{array}{ccc} \mathrm{Ar{-}N{=}N{-}NH{-}R} & \rightleftharpoons & \mathrm{Ar{-}NH{-}N{=}N{-}R} \\ \downarrow \mathrm{H^+} & & \downarrow \mathrm{H^+} \\ \mathrm{Ar{-}\overset{+}{N}{\equiv}N + H_2N{-}R} & & \mathrm{ArNH_2 + N{\equiv}\overset{+}{N}{-}R} \end{array}$$

The simplest means of avoiding this is to use a secondary amine as stabiliser; alternatively, the tautomeric equilibrium can be forced to the left if the stabilising amine is less basic than the diazotised amine. Finally, it must be remembered that the diazoamino compound must be soluble, whilst the final azo compound must be insoluble, so water-solubilising groups are always present in the stabilising amine.

The ease with which diazoamino compounds are split is determined by the structures of both the diazo residue and the stabilising amine. In practice this affords a greater degree of control than is possible with *anti*-diazotates, since diazo compounds which tend to give more labile diazoamino compounds can be stabilised with less reactive amines, and conversely. This dual control permits the production not only of a range of diazoamino compounds of approximately similar reactivities in acidic media, but also of diazoamino compounds that are more reactive and will split under more or less neutral conditions. Such compounds can be printed alongside vat dyes, which cannot tolerate acidic steaming conditions.

It is a matter of general experience that the introduction of more strongly electron-attracting substituents into the diazo residue facilitates splitting by acid, the reverse effect being shown by variation in basicity of the stabilising amine, the least basic leading to the lowest reactivity [1]. The scission of diazoamino compounds has been studied by many workers [2] and it is clear that protonation of the diazoamino group is the first step in the overall reaction. It is most likely that the resulting cation is then attacked by a base (e.g. solvent water or hydroxide ion) [24]. In aqueous solution this would lead to the formation of the diazo hydroxide and hence the diazonium ion:

$$\mathrm{Ar{-}N{=}N{-}NH{-}Ar' + H^+ \rightleftharpoons Ar{-}N{=}N{-}\overset{+}{N}H_2{-}Ar'}$$

$$\mathrm{Ar{-}N{=}N{-}\overset{+}{N}H_2{-}Ar' + OH^- \rightleftharpoons Ar{-}N{=}N{-}OH + H_2N{-}Ar'}$$

$$\mathrm{Ar{-}N{=}N{-}OH \rightleftharpoons Ar{-}\overset{+}{N}{\equiv}N + OH^-}$$

Support for this mechanism is given by a study of the comparative rates of scission of a series of diazoamino compounds obtained from diazotised 4-aminodiphenyl and a range of stabilising amines of increasing basicity [25a]. The basicity of the stabilising amine would influence both the rate of protonation and the rate of attack by the base, a strongly basic amine facilitating the first reaction but inhibiting the second, whilst a weakly basic amine would have the reverse effect. Either reaction can therefore be rate-determining. It was found that the lowest rates of scission were obtained with 4-nitroaniline (protonation slow) and with dimethylamine (slow attack by base) as the stabilising amines. Siegrist [25b] has suggested that two mechanisms are possible, one involving protonation and the other direct attack of the diazoamino compound by water:

$$\mathrm{Ar{-}N{=}N{-}NH{-}R + H^+ \rightleftharpoons Ar{-}N{=}N{-}\overset{+}{N}H_2{-}R}$$

$$\mathrm{Ar{-}N{=}N{-}\overset{+}{N}H_2{-}R \rightarrow Ar{-}\overset{+}{N}{\equiv}N + H_2N{-}R}$$

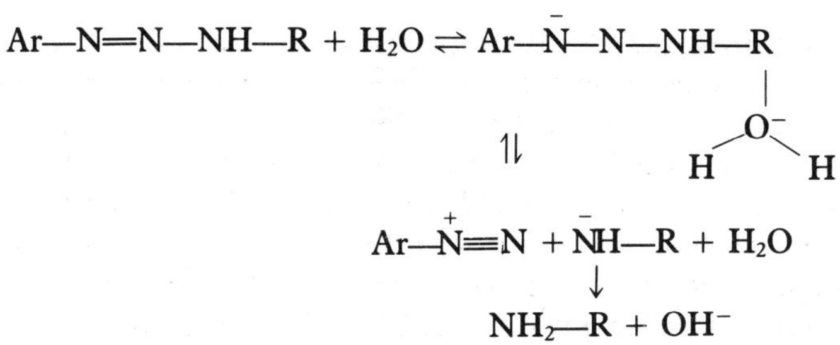

Whatever the exact mechanism of scission may be, from the practical point of view it is sufficient to know that the reactivity can vary considerably, and that this variation can be controlled by suitable choice of the stabilising amine. Thus, two groups of non-coupling mixtures have become available.

The first, exemplified by the Rapidogens (BAY), are mixtures of coupling components with diazoamino compounds which require acidic conditions for development, and in which the stabilising amines are chiefly sarcosine, *N*-methyltaurine, 4-sulphoanthranilic acid, 2-methylamino-5-sulphobenzoic acid and 2-ethylamino-5-sulphobenzoic acid.

Rapidogens are similar in application to the Rapid Fast Colours (FH), development being induced by injecting acid into the steamer or by the inclusion of hydrolysable ammonium salts in the print paste. Examples are listed in Table 7.3.

TABLE 7.3

Rapidogens

Rapidogen	Diazo component	Stabiliser	Naphtol
Yellow G	Fast Red KB	*N*-methyltaurine	AS-G
Scarlet IR	Fast Red KB	sarcosine	AS-PH
Navy Blue IB	Variamine Blue B	2-methylamino-5-	AS
Brown IRRN	Fast Red RL	sulphobenzoic acid	AS-LB

The second group introduced later (1953) are sufficiently reactive to develop in neutral steam and thus avoid the drawbacks associated with acid steaming. Examples are the Neutrogene (Fran) and Rapidogen N (BAY) colours. These use as stabilising amines such compounds as 2-carboxyphenylglycine and some of its derivatives and various other substituted anthranilic acids.

In fact, there is no very obvious difference between the stabilisers of the two groups, and it may be that the difference in reactivity also is not so great as may be supposed. Rapidogens are often mixed with sodium hydroxide in the printing paste, so much of the acid in the steamer is required to neutralise this. So far as is known, no measurements of the pH of the printed mark during steaming have been made. It is interesting to note the use of Developer N, which was introduced to avoid acid steaming conditions before colours such as the Neutrogenes were available. Developer N is a volatile strong base (2-diethylaminoethanol) which replaces sodium hydroxide in the printing paste. During steaming the base evaporates and development occurs, but the pH cannot be lower than neutral.

7.1.8 Mixtures with diazosulphonates

These mixtures, marketed as Rapidazols (FH), constitute only a short range. Diazosulphonates are prepared by treating diazonium salts with concentrated sulphite solutions. It is now established that they exist in two stereoisomeric forms:

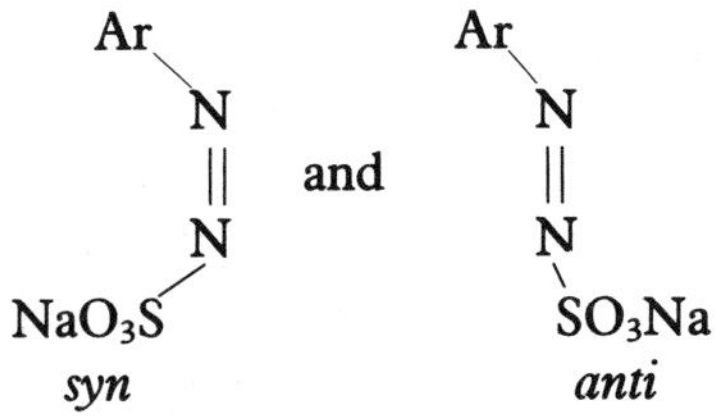

as do the diazotates [26], the *anti*-form being the more stable.

Rapidazol mixtures differ from the other two groups in that they require oxidation for development, e.g. with sodium chromate, leading to the formation of the diazonium sulphate ($ArN^+{\equiv}N\ Na^+\ SO_4^{2-}$). They are also sensitive to light, which causes ionisation and the production of diazonium sulphite ($ArN^+{\equiv}N\ Na^+\ SO_3^{2-}$) [27], and for this reason they have been used for photographic printing processes. Cleavage of the $N—SO_3^-$ bond by oxidation and by photolysis may be analogous to the similar cleavage of the $O—SO_3^-$ bond in the sulphuric esters of leuco vat dyes.

7.1.9 Application to non-cellulosic fibres

Although the main application of azoic dyes is to cellulose, they find a significant outlet on the cellulose acetate, nylon and polyester fibres, particularly for full depths which are difficult to obtain with disperse dyes.

In their application to cellulose, both components of the azoic combination are applied as ions from aqueous solution, as indeed are all dyes for cellulosic fibres. The similarity between coupling components and direct dyes has already been pointed out. The Fast Colour Bases themselves, however, and the un-ionised coupling components under neutral conditions, are only very slightly soluble in water and in fact resemble disperse dyes. Not surprisingly, therefore, the un-ionised components are adsorbed by the more hydrophobic fibres as though they were disperse dyes, and can be applied in much the same way as disperse dyes, whilst the ionised forms are in general not adsorbed.

The usual procedure is to apply the free base, either alone or in some cases with the un-ionised coupling component, from an aqueous dispersion, under conditions similar to those used for disperse dyes. Polyester fibre therefore needs the highest temperature, and temperatures above 100 °C, using pressurised vessels, as well as carriers, may be used. Diazotisation is carried out on the fibre, after adsorption, by treatment with nitrous acid solutions. Naturally, diazotisation is slower than under the normal conditions in solution, but higher temperatures are possible, e.g. temperatures as high as 85 °C can be used for polyester fibre. If both components are applied together, coupling takes place rapidly once diazotisation has been effected, since the diazo compound and coupling component are at relatively high concentration within the polymer network and are also effectively immobilised. If the free base is applied alone, the textile material must be treated with a solution of coupling component after diazotisation.

The limitations on molecular size which apply to the adsorption of disperse dyes apply also to coupling components, so that only selected members can be successfully used, particularly on cellulose acetates. The most commonly used coupling component is 2-hydroxy-3-naphthoic acid, which is also used for developing some disperse dyes that contain a primary amino group, by diazotisation and coupling after dyeing.

7.2 VAT DYES

Whereas the application of azoic combinations leads to the synthesis of the final insoluble molecule by chemical combination within the fibre, in the case of vat dyes the insoluble pigment is prepared by the manufacturer and is presented to the dyer in this form. For application it is necessary for the dyer to convert the insoluble pigment into a soluble derivative having substantivity for the textile and then, after application, to regenerate the pigment within the fibre by the reverse reaction.

Vat dyes may be regarded as falling into two main chemical classes: (a) indigoid dyes, of which indigo itself (II) may be taken as an example, and (b) anthraquinonoid dyes, for example 1,5-bis(benzoylamino)anthraquinone (III) (Caledon Golden Yellow 3G).

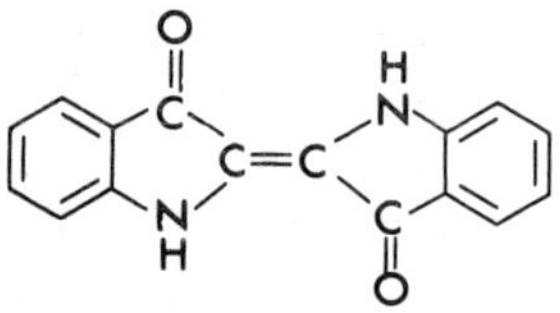

(II)

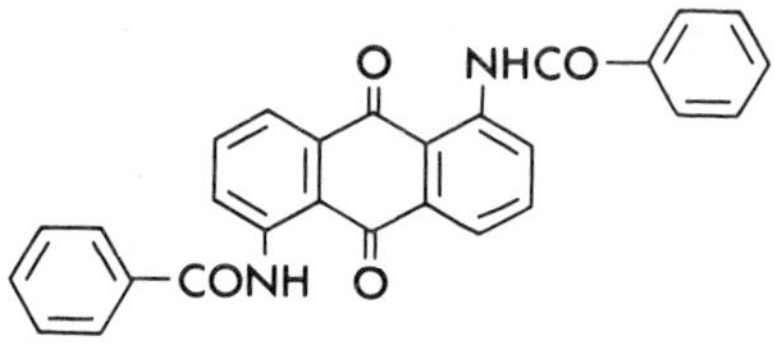

(III)

The anthraquinonoid class contains many important members which are derivatives of polycyclic quinones that are more complex than anthraquinone, but since a detailed chemical classification is not called for here, it is unnecessary to give their structures. In fact, only two chemical features are essential within these two classes for a coloured compound to be regarded as a vat dye: firstly, that the compound should not contain a water-solubilising group and, secondly, that it should contain at least two carbonyl groups connected by a conjugated chain. This last feature is of the greatest importance, since it is the basis of the temporary solubilisation necessary for the application of vat dyes.

The conjugated dicarbonyl system is capable of easy reduction, accompanied by a change in conjugation, so that the reduction is a two-electron change only, and the resulting conjugated dihydric alcohol is also readily reoxidised:

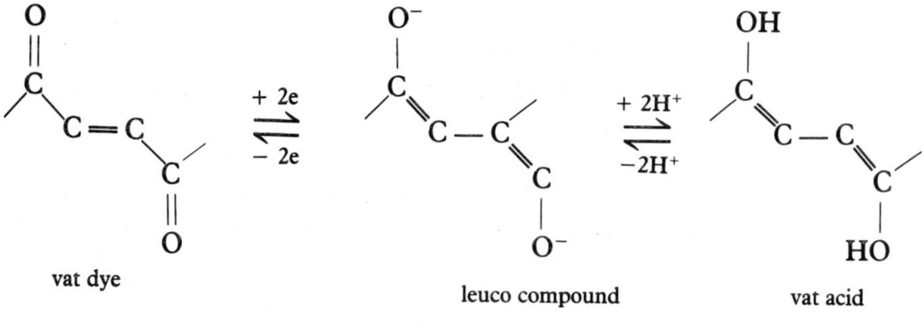

This formulation shows the vat dye as one partner of a simple redox system, the other partner – the reduced form – being known as the *leuco compound*. Reduction, of course, is the normal 'vatting' process by which the vat dye is brought into solution before application, and oxidation represents the regeneration, or development, of the insoluble vat dye after application. The above equation also shows the leuco compound as a di-anion derived from a dibasic acid – often referred to as the 'vat acid' or 'acid-leuco'. The vat acids are weak acids with pK_a values usually in the range 9–11; they are also virtually insoluble in water, so that alkaline conditions are necessary in practice in order to retain the leuco compound in the soluble ionised form.

Two matters of considerable significance in the application of vat dyes, and in particular in attempts to classify vat dyes, are therefore the relative ease of reduction and reoxidation of the redox partners and the relative pK_a values of the vat acids. Also of importance, of course, is the affinity of the leuco compound di-anion, together with its diffusion properties, tendency to aggregate, etc.

7.2.1 Redox behaviour

The vat dye–leuco compound system has all the appearances of a typical reversible redox system (cf. section 1.4), which would be expected to show the normal electrode behaviour of such a system.

The partner at the higher oxidaton level (the vat dye) would abstract electrons from a suitable indicating electrode, whilst the other partner (leuco compound) would donate electrons. The potential developed at the electrode would therefore be a function of the relative concentrations of the two forms, given by

$$E = E^{\ominus} + \frac{RT}{zF} \ln \frac{[\mathrm{Ox}]}{[\mathrm{Red}]} + \frac{RT}{F} \ln\,[\mathrm{H}^+] \qquad (7.17)$$

where E is the observed potential (measured against a hydrogen electrode as standard), [Ox] and [Red] are the concentrations of the oxidised and reduced forms, and z is the number of electrons involved in the redox change. $E^{\ominus}$ is therefore the potential corresponding to the situation when [Ox] = [Red] at unit activity of hydrogen ions (both logarithm terms zero) and is known as the *standard redox potential* (defined as being measured at 25 °C). Electrode systems including the redox system have been discussed in detail in Chapter 1 and need not be discussed further here, it being sufficient to reiterate that measurement of $E^{\ominus}$ for a redox system is a sensitive and fundamental measurement of the tendency of the system to exist in one or other of the two forms. The lower the value of $E^{\ominus}$, the more readily is the reduced form oxidised and, conversely, the more difficult is reduction of the oxidised form. It is the value of $E^{\ominus}$ for a redox system which governs the choice of reducing agent since, to be capable of reducing the oxidised partner, the reducing agent must give a potential lower than that of the redox system it is intended to reduce.

In the light of considerations such as these, it is not surprising that the redox behaviour of vat dye–leuco vat dye systems has been investigated, and a review of some of the results of these investigations is necessary. To begin with, it must be stressed that the essential molecular feature under investigation is the conjugated dicarbonyl–conjugated diol system, whether or not the compound containing it is actually a vat dye.

For compounds in which all forms are readily soluble in aqueous solutions no difficulties are encountered, for example with sulphonated anthraquinones. Such compounds readily yield curves such as those in Figure 7.2 when a solution of the oxidised form is titrated with a suitable reducing agent or, alternatively, a reduced solution is titrated with an oxidising agent. Sigmoidal curves are obtained with an inflection at the mid-point of the titration which corresponds to $E^{\ominus}$. (In some cases two steps are obtained, one for each of the two one-electron changes, as shown by the dotted curve, but this need not concern us here.) Vat dyes, however, are virtually insoluble in water at all pH values and leuco compounds are soluble only at high pH, so simple redox titrations in aqueous conditions are impossible. For this reason redox titrations of some vat dyes, and other polycyclic quinones, have been carried out in non-aqueous systems [28, 29]. There are difficulties in such work, particularly in devising suitable electrodes but, with care, normal redox curves of the shape shown in Figure 7.2 can be obtained. The results of such titrations show that the vat dyes cover a very wide range of reducibility, Durindone Blue 4BC (C.I. Vat Blue 5) having a value of $E^{\ominus}$ of −70 mV, whilst at the other extreme $E^{\ominus}$ for Caledon Brilliant Violet R (C.I. Vat Violet 17) has a value of −480 mV. The significance of this difference is brought

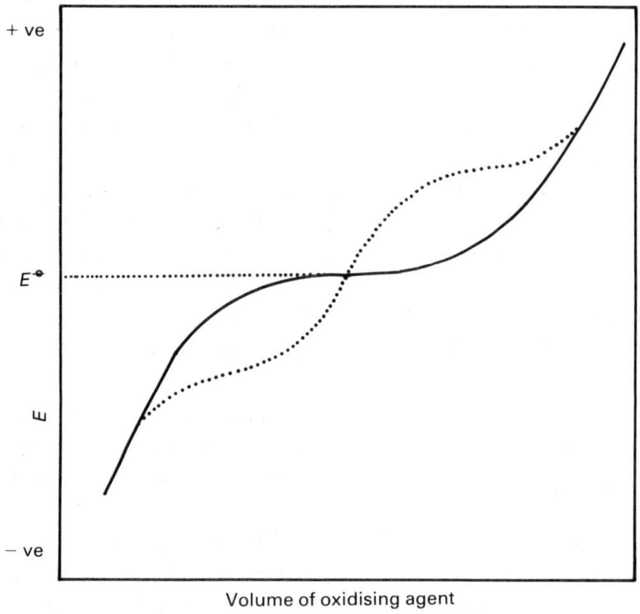

Figure 7.2 – General shapes of redox titration curves

out when it is realised that for a vat dye (or any other two-electron redox system) a tenfold change in the ratio [Ox]/[Red] corresponds, at 30 °C, to a change in the observed potential E of *ca* 30 mV, so that, whereas at −70 mV C.I. Vat Blue 5 is 50 per cent reduced ([Ox]/[Red] = 1), for C.I. Vat Violet 17 the ratio would be *ca* $1{:}10^{-14}$ and reduction would barely be measurable.

Although work in non-aqueous conditions is of value in understanding the fundamental redox behaviour of compounds such as vat dyes, it cannot be accepted as a guide to their behaviour in typical aqueous systems, at least until sufficient measurements in aqueous systems have been made to show that a reliable degree of comparability obtains. Unfortunately this is not the case, titrations in aqueous solutions yielding curves which differ markedly in shape from the typical curves in Figure 7.2 and which do not permit a value for $E^{\ominus}$ to be obtained.

Redox curves for vat dyes in aqueous solutions are best obtained by first fully reducing the dye and then titrating the leuco compound with a suitable oxidising agent, e.g. potassium hexacyanoferrate (III) (potassium ferricyanide). This is run in from a burette, the potential being read after each addition in the usual way. The first part of the titration is simply the oxidation of excess reducing agent, and then follows the titration of the leuco compound itself. Finally, when this is fully oxidised, the potential of the excess oxidising agent is recorded. Titration curves obtained in this way have the shape shown in Figure 7.3, A marking the beginning of the titration of leuco compound and B showing the potential of excess oxidising agent. The shape of the curve between A and B is markedly different from that of the normal curve in Figure 7.2, the difference arising from the fact that the vat dye, which is produced at point A, has an extremely low solubility. Thus, almost from the beginning, the solution is saturated with vat dye, the remainder precipitating out, so that in Eqn 7.17 the term [Ox] has a constant, very small value over virtually the whole of the titration. The shape of the resulting curve is therefore determined entirely by the decreasing value of [Red], rather than by this value and the correspondingly increasing value of [Ox], as in Figure 7.2. The shape of the curve in Figure 7.3 may also, to some extent, be influenced by occlusion of leuco compound on the precipitated vat dye [30].

The point A, therefore, effectively marks the potential at which the vat dye first begins to form and to precipitate. It has been designated the *leuco potential* [31] to differentiate it from the unobtainable standard redox potential.

In earlier work, Appleton and Geake [32] carried out redox titrations of reduced vat dyes in aqueous solutions and recorded the mid-point at 50 per cent oxidation, which corresponds to the position of $E^{\ominus}$ in a system in which both oxidised and reduced forms are soluble.

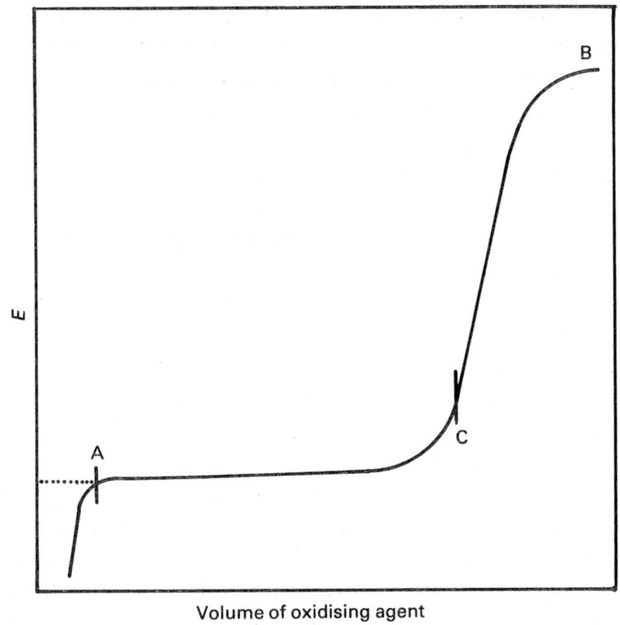

Figure 7.3 – Redox titration curve of a typical vat dye

Thus, many redox titrations of different vat dyes have been carried out under a variety of conditions and different points on the curves have been recorded. A comparison of some of the results has been given by Vickerstaff [31] and is reproduced in Table 7.4.

TABLE 7.4

Comparison of various redox measurements

Vat dye	Leuco potential (in water) /mV	Mid-point potential (in water) /mV	Mid-point potential (in 50% pyridine) /mV	$E^{\ominus}$ (in pyridine) /mV
Khaki 2G	−1000			
Indanthrone	−955			−223
Pyranthrone	−950	−660	−449	−300
Dibenzanthrone	−900	−640	−328	−222
Jade Green X	−890	−575	−370	−264
Dibenzpyrenequinone		−487	−340	−242
Flavanthrone	−770	−233	−391	−230

There are obviously wide discrepancies between these different sets of values. Nevertheless, the two sets of results under purely aqueous conditions (leuco potential and mid-point potential in water) are in the same order, but there is no correlation between these and the values in 50 per cent pyridine and pure pyridine. It is perhaps significant that there is only poor correlation between the values in the last two columns, both of which refer to 50 per cent reduction, but for the last column the vat dye remained completely dissolved, whereas for the preceding column this was not the case. The effect of vat dye precipitation is clear.

More recent work indicates that it may after all be possible to obtain genuine $E^{\ominus}$ values for vat dyes in water [33]. If the sulphuric esters of leuco vat dyes (see section 7.3) are irradiated, or oxidised with stoichiometric amounts of oxidising agents, in very dilute solutions so that production of vat dye is very slow then the vat dye does not precipitate but remains dissolved in water; presumably, as a result of the slow rate of formation single vat dye molecules are hydrated before they can aggregate. These solutions may be titrated against alkaline dithionite solutions, yielding in some cases redox titration curves of the standard form (Figure 7.2) with apparently no significant formation of semiquinone. Unfortunately it is not possible to compare values obtained in this way with leuco potentials, mid-point potentials or characteristic potentials for particular dyes, but values between −700 and −800 mV have been obtained for two thioindigo derivatives.

In another investigation Nabar and Shenai [34] attempted to define a more fundamental potential in aqueous systems than the leuco potential. The basis of this attempt is that in aqueous systems at constant pH, where precipitation of the vat dye occurs, the terms [Ox] and ln H^+ in Eqn 7.17 are constant and can therefore be included with $E^{\ominus}$. The equation then simplifies to

$$E = E_c - \frac{RT}{zF} \ln [\text{Red}] \qquad (7.18)$$

where E_c is defined as the *characteristic potential*, and is the potential at which [Red] is unity (i.e. the logarithm term is zero). Since the titrations are carried out at much lower concentrations than this (*ca* 10^{-5}M), E_c cannot be located on the titration curve, which is of course that of Figure 7.3, but has to be calculated from points on the curve. It is possible to locate the point C at which the leuco compound is completely oxidised, so the value of [Red] at points lower down the curve can be calculated from the stoichiometry of the titration. Substitution of these values and the corresponding values of E in Eqn 7.18 permits E_c to be calculated. Reasonably constant values were obtained, but the method is obviously sensitive to the accuracy of the titration and of the plotting of the curve. Although values of E_c are possibly of greater theoretical value than are leuco potentials their practical value may be questioned, since they refer to molar solutions of leuco compounds which, in practice, are never attainable. Nevertheless, the order of E_c for a series of vat dyes is very close to that of the leuco potentials.

7.2.2 Practical considerations

It might, *a priori*, be expected that there would be some correlation between the redox potential (or equivalent) of a vat dye and its ease of reduction in practical applications. In a qualitative way, such a correlation has been obtained under printing conditions. Printing pastes of a series of vat dyes were prepared containing varying amounts of reducing agent and the prints obtained were steamed for varying lengths of time. The degrees of fixation obtained were assessed visually, and on this basis the dyes were placed in a numerical order of ease of printing. A rough correlation was found between this order and the order of $E^{\ominus}$ found for the same dyes in pyridine solution [35]. Work of this kind has not been extended, however; bearing in mind that generally the agreement found between $E^{\ominus}$ measured in a non-aqueous system and leuco potential is not very good, it is doubtful if meaningful correlations could be expected.

A most important reason why leuco potentials, measured in aqueous systems, bear little relation to practical vatting behaviour is that leuco potentials measure the ease of oxidation of the soluble leuco anion (since the titrations are carried out from a fully reduced solution) rather than the ease of reduction of the insoluble vat dye. It is scarcely to be expected that an insoluble particle, of which only the surface molecules are in contact with the reducing agent, would behave in exactly the same way as a solution of the same compound, where all the molecules are uniformly accessible to the reducing agent. In the practical vatting of vat dyes, however, we are concerned entirely with the reduction of insoluble particles.

In such a system the rate of reduction is determined not only by the intrinsic chemical reactivity of the dye, but also by the physical features of the particles. These are firstly the particle size, since small particles have a greater specific surface (i.e. area/unit mass) than large particles and would be expected to react more rapidly. In any sample of vat dye there will be a variety of particle sizes, however, and size distribution is also important, since a sample with a larger proportion of small particles would be expected to reduce more rapidly than one with predominantly large particles. Superimposed on this factor, however, are the effect of crystal habit, which will determine the orientation of the reducible carbonyl groups at the surface, and also the perfection of crystallinity – a badly formed crystal containing imperfections and surface fissures would be expected to reduce more rapidly than a perfect crystal of the same habit and size. Several experiments have shown the influence of these factors, and some results are shown in Tables 7.5 to 7.7.

The rate decreases with increasing particle size, but not to the extent expected from the change in surface area (calculated values). This is thought to be due to an increasing number of imperfections in the larger particles [36] (Table 7.5).

The increase in rate, particularly to almost complete (95 per cent) reduction, clearly increases as the proportion of smallest particles increases (Table 7.6).

The particle size decreases continually with increase in the time of milling,

TABLE 7.5

Rates of reduction and particle size of pyranthrone

Mean particle size/μm	< 0.7	0.7–1.0	1.0–3.0	> 3.0
Time of half-reduction (obs)/s	50	75	68	120
Time of half-reduction (calc)/s	53	75	176	350

TABLE 7.6

Effect of particle size distribution [37]

Vat dye	wt % of Stokes diameter < 0.25 μm	wt % of Stokes diameter < 3 μm	Time of reduction/s[a] t_{50}	Time of reduction/s[a] t_{95}
Caledon Olive R	49.5	72.0	12	145
	71.5	99.0	10	35
	90.5	98.0	7	20
Caledon Khaki 2G	48.0	89.5	17	125
	62.0	99.0	12	32
	80.0	99.0	10	28
Caledon Dark Brown 3R	43.5	65.0	24	250
	69.5	91.5	14	42
	90.0	99.5	9	20

[a] t_{50} and t_{95} are times for 50% and 95% reduction.

TABLE 7.7

Effect of milling on rate of reduction of Caledon Jade Green XBN [37]

Milling time/h	0.5	4	10	20	30	40	50
Time of t_{50}/s	50	47	47	44	54	64	63
Reduction t_{95}/s	120	120	130	140	175	245	250

but the rate of 50 per cent reduction passes through a maximum with a marked fall in the rate after about 20 h of milling, whilst the rate for 90 per cent reduction falls over the whole range of milling time. These effects are thought to be due to induced crystallisation after long periods of milling, less perfect crystals being replaced by more perfect ones (Table 7.7).

It is apparent from these results (and others of a similar nature) that rate of reduction is inextricably related to the physical form of the vat dye as well as to its chemical structure. From a practical point of view both are of importance, but in rather different ways. The chemical structure determines the value of the redox (or equivalent) potential and therefore absolutely determines whether or not a particular vat dye can be reduced by a given reducing agent, irrespective of the time required. Fortunately, this consideration does not in practice loom large, since the reducing agents most commonly used (sodium dithionite – 'hydrosulphite' – or formaldehyde–sulphoxylate) give potentials which, although not clearly defined, are sufficiently low to bring about reduction of all commercially available vat dyes. This may simply be an alternative way of saying that quinones incapable of being reduced by these agents are not marketed as vat dyes. For practical vat dyes, i.e. those reducible by the commonly available reducing agents, rate of reduction is the most important consideration, and it is evident from the results given above that reduction will occur most rapidly if the particles are small and are uniformly small. Rapid reduction is, of course, at a premium in printing applications and in the more modern continuous methods of dyeing, e.g. padding and flash ageing, this is possible solely because dyestuff manufacturers are now able to supply vat dyes in a suitable state of subdivision and in so-called liquid form. The conditions recommended for vatting – chiefly alkali concentration and temperature – also influence rate of reduction, the more slowly reduced dyes requiring the more vigorous conditions. To meet the need for rapid reduction necessary for modern continuous and automatic application other reducing agents, such as thiourea dioxide (formamidine sulphinic acid) and sodium borohydride have been used, sometimes with the use of accelerators such as sodium nickel cyanide for borohydride reduction. Accelerators for dithionite (for example, cobalt dioximes) have also been suggested. There are clear environmental difficulties associated with some of these.

Dyeing conditions also vary among the vat dyes, particularly the amount of alkali (usually sodium hydroxide) required. Although this is possibly related to ease of reduction, it must also be connected with the pK_a of the acid leuco compound, i.e. the strength of this as an acid and hence its ability to exist as the soluble di-anion. Very little work in this field has been reported, but it is probable that leuco vat dyes having high pK_a values will require more alkaline dyebaths than those with lower pK_a values. In practice it is just as important to ensure that the dyebath is maintained at the correct alkalinity as it is to ensure the presence of excess reducing agent, since precipitation of the insoluble vat acid can be as troublesome as precipitation of the vat dye itself. The reducing agent most widely used, sodium dithionite, is oxidised to hydrogensulphate, according to the equation

$$2Na_2S_2O_4 + 3O_2 + 2H_2O \rightarrow 4NaHSO_4$$

i.e. an acid salt is produced. Consequently, the entry of oxygen from the atmosphere during dyeing will not only remove excess reducing agent but will also lead to a fall in pH.

7.2.3 Affinity

The thermodynamic treatment of dye adsorption is dealt with in detail elsewhere in this book (Chapter 4). It is sufficient here to mention only briefly some of the main features found with vat dyes. An extensive account is given by Vickerstaff [31].

In the application of a vat dye to cellulose the operation is carried out with a solution of the ionised leuco compound containing excess sodium hydroxide and reducing agent, as well as sodium chloride, the leuco compound being adsorbed. The system is reminiscent of that for direct dyes in that the species being adsorbed is an anion and there are no specific sites for adsorption in the cellulose. Not surprisingly, therefore, the diffuse adsorption model with a Donnan distribution of ions, which has served well for direct dyes, has also been applied to vat dyes. Unfortunately, experimental difficulties have prevented the same wealth of data being accumulated, largely because leuco vat dyes exhibit a rapid 'strike', i.e. rapid adsorption of a large part of the dye in the early stages. This is followed by much slower diffusion through the fibre before a true equilibrium is reached, a process that might take longer than a week. In spite of these difficulties some reliable results have been obtained which accord as well with the Donnan model (p. 305) as do those of direct dyes. These results give values of affinity ranging from *ca* 8 to 25 kJ (2 to 6 kcal) mol^{-1}. These values are somewhat lower than those for direct dyes and are, at first sight, difficult to reconcile with the much higher exhaustion normally encountered with vat dyes. The reason for this apparent anomaly is that the electrolyte concentration in vat dyebaths is much higher than in direct dyebaths. One important difference between direct dyebaths and vat dyebaths is the presence of sodium hydroxide in the latter. This functions as an electrolyte, and its presence has been taken into account in the application of the Donnan model referred to above. Following more recent work with reactive dyes, however (Chapter 6), it is now well established that sodium hydroxide solutions lead to a significant ionisation of cellulose, with a consequent effect on the ionic concentrations in the fibre phase. Indeed, in an extensive study with 14 vat dyes Liddell, Weedall and McKay [38] have shown that allowance for both ionised carboxyl and hydroxyl groups can be made as well as for dithionite and sulphite ions to give consistent values for affinity over a wide range of dyebath concentrations. The values obtained ranged from around 18 to 25 kJ (4.5 to 6 kcal) mol^{-1}. Moreover, the use of these values in predicting the outcome of practical dyeings gave very good agreement with the actual results.

7.2.4 Soaping

It is the usual practice, after the application of a vat dye, to give the textile material a final treatment in a hot soap or detergent solution. This not only removes surface dye, but often brings about a change in hue of the fixed dye. Again, this topic is not of special importance in this chapter and will be discussed only briefly.

Vat dyes are adsorbed by the textile material in the form of their leuco anions. These are believed to be monodisperse, and will certainly be adsorbed in an orientation determined by their interactions with the cellulose molecules. On oxidation, therefore, the insoluble vat dye will be produced essentially as discrete molecules orientated in much the same way as were the leuco anions. It is not to be expected that this orientation will be the same as that taken up by vat dye molecules when they are allowed to crystallise. The changes which take place during soaping are essentially those of crystallisation of the vat dye into a more stable crystal habit. Evidence for some change in orientation comes from the observation of changing dichroism on soaping, as well as changes in absorption spectrum; in some instances, X-ray examination of dyed and soaped cellulose films has shown the X-ray pattern of the vat dye superimposed on that of the cellulose [31].

7.3 SULPHURIC ACID ESTERS OF LEUCO VAT DYES

These compounds were introduced in the mid-1920s as a means of overcoming difficulties associated with the application of vat dyes by the conventional vatting method. This was particularly required at that time for the application of vat dyes to wool, because of the severe degradation brought about by the alkaline reducing conditions of the vat. There were also advantages to be gained in their application to cotton since, with many vat dyes, it is difficult to obtain levelness in pale depths owing to their rapid 'strike', whilst in printing, alkaline printing pastes can cause damage to copper printing rollers. In fact, the major use of the sulphuric esters of leuco vat dyes has been on cellulosic textiles, including polyester–cellulose blends.

The sulphuric esters of leuco vat dyes provide a good example of the successful application of fundamental chemistry to solve a technical problem but unfortunately, once some technical success had been achieved, the fundamental work was neglected. The main work of invention was done by Marcel Bader [39] in collaboration with Sunder. They recognised that one way of avoiding both the need to vat in the dyehouse, as well as the use of alkaline reducing conditions, would be for the dyestuff manufacturer to produce a derivative of the leuco compound which was sufficiently stable for normal commercial handling, which was water-soluble and could be applied in a simple way to textiles (principally wool), and which could then be easily converted back to the leuco compound. Oxidation of this to the parent vat dye would cause no difficulty, since it could be done under either acidic or alkaline conditions. Recognising that the hydroxyl groups in leuco vat dyes are essentially alcoholic, they turned naturally to consider the simple derivatives of alcohols, namely ethers and esters, finally deciding that esters would be more useful because they are fairly readily hydrolysable, whereas ethers are more stable. Furthermore, esters produced from dibasic acids, such as sulphuric acid, should be particularly useful, since these would also produce monosodium salts which should be water-soluble, i.e.

$$ROH + H_2SO_4 \rightarrow H_2O + ROSO_3H \rightarrow ROSO_3^-Na^+$$

In fact, such esters derived from phenol and also from quinol (which can be regarded as the simplest leuco vat dye) had already been obtained by Baumann and Thiemann [40], who had shown that they were fairly readily hydrolysed by aqueous mineral acids. Furthermore, and more directly relevant, von Baeyer [41] had attempted to esterify leuco indigo with potassium pyrosulphate and had obtained a solution from which indigo was precipitated by the action of acidic iron(III) chloride. If von Baeyer had in fact esterified leuco indigo, then the reaction with acidic iron(III) chloride could be a demonstration of the kind of dyeing system which Bader and Sunder were searching for, viz.

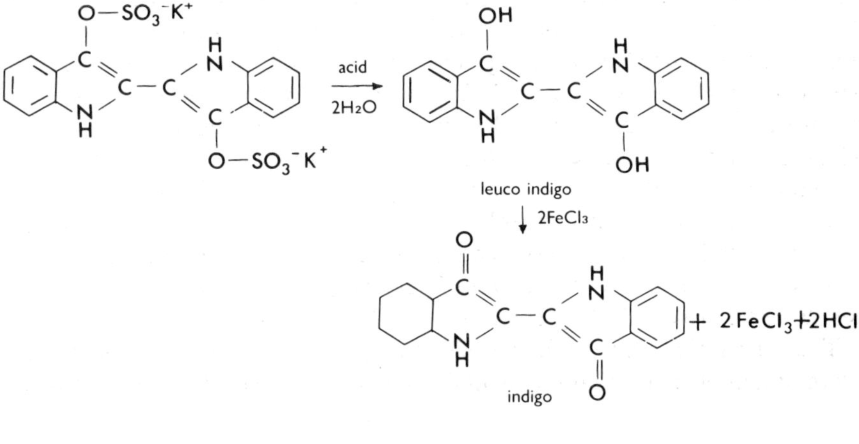

Following such deliberations, Bader and Sunder prepared the disulphuric esters of the leuco compounds of some indigoid vat dyes and found them to be reasonably stable solids (particularly if mixed with some alkali) which dissolved readily in water, giving solutions from which the parent vat dyes were precipitated on treatment with an acid and an oxidising agent. Their solutions in dilute acid containing some reducing agent were stable and could be used for dyeing wool; after dyeing, the vat dye was produced on the wool by treatment with an acidified dichromate solution.

These, and similar, observations led to the marketing of the disulphuric esters of leuco indigoid vat dyes under the name Indigosol by the firm of Durand & Huguenin AG. They were rapidly followed by corresponding derivatives of anthraquinonoid vat dyes (Scottish Dyes Ltd), produced by a better method, and generally accepted. Also accepted was the development mechanism above, postulated by Bader. Work was concentrated on devising improved methods of application, including better oxidising agents, and little attention was paid to the development reaction itself, even though Bader pointed out that his proposed scheme was probably incorrect [42, 43].

7.3.1 Chemical development

The essential features of the development mechanism have now been worked out, however; in particular, the technically important method of nitrite development has been explained. The first work in this field was a study of the development of the sulphuric acid esters of leuco-tetrasulphonated indigo and of leuco-2-sulphoanthraquinone, by means of acidified solutions of hydrogen peroxide [44]. The sulphonated compounds were used because they led to water-soluble products which made absorptiometric analysis of the reaction solutions easy, and hydrogen peroxide was chosen because its mode of action as an oxidising agent was much better understood than those of the more commonly used technical agents. The most important findings of this work were as follows.

(a) The leuco esters reacted very much more slowly in dilute acid solutions containing no hydrogen peroxide or dissolved oxygen than they did in solutions of the same acidity containing hydrogen peroxide. Under these non-oxidising conditions hydrolysis only could occur, so this reaction is much slower than the overall process in the presence of hydrogen peroxide. It follows, therefore, that development could not occur by Bader's two-stage mechanism above (with hydrogen peroxide in place of iron(III) chloride) because, by this mechanism, the overall process cannot be faster than the first step.

(b) In a series of experiments carried out at the same concentration of acid, the rate of production of the oxidised form passed through a maximum with increasing concentration of hydrogen peroxide. In these experiments, however, most of the hydrogen peroxide was consumed, even though it was originally very much in excess of the leuco ester. This observation implies that some other process leading to the consumption of hydrogen peroxide can occur.

It was already known [45, 46] that, in acidic oxidations with hydrogen peroxide, one or both of the radicals $HO\cdot$ or $HO_2\cdot$ are produced and also that the hydroxyl radical ($HO\cdot$) can lead to the spontaneous decomposition of hydrogen peroxide [47] by the following reactions:

$$H_2O_2 + HO\cdot \rightarrow H_2O + HO_2\cdot$$

$$H_2O_2 + HO_2\cdot \rightarrow O_2 + H_2O + HO\cdot$$

which form a chain, the $HO\cdot$ radical produced in the first reaction taking part in the second. Thus, a reaction mechanism for the leuco esters which involved the formation of radicals would explain the observed results.

(c) It seemed very likely that quinols (i.e. leuco vat dyes) would produce $HO\cdot$ radicals from hydrogen peroxide, and this was substantiated in experiments in which small amounts of quinol were added to the solutions. A marked increase in both the rate of oxidation of the leuco ester and the consumption of hydrogen peroxide was observed.

(d) The rate-of-reaction curves in the absence of added quinol were sigmoidal, strongly indicative of an autocatalytic reaction, i.e. some product produced slowly from the beginning of the reaction initiated a more rapid process. Since hydrolysis must occur, and is slow, some quinol (or mono ester) must be produced and from paragraph (c) this could well be a catalyst initiating a more rapid reaction with hydrogen peroxide via $HO\cdot$ or $HO_2\cdot$.

(e) Some measurements of the rate of liberation of sulphate were made. However, it was shown later [48] that these measurements were subject to error. They will not, therefore, be considered here.

The most important conclusions to be drawn from this work with hydrogen peroxide can be summed up as follows:

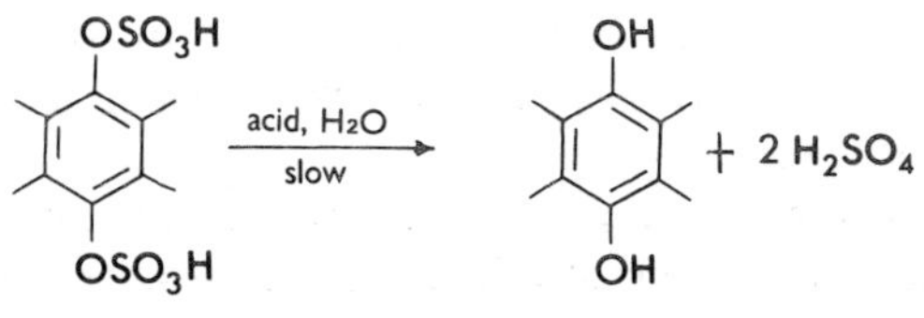

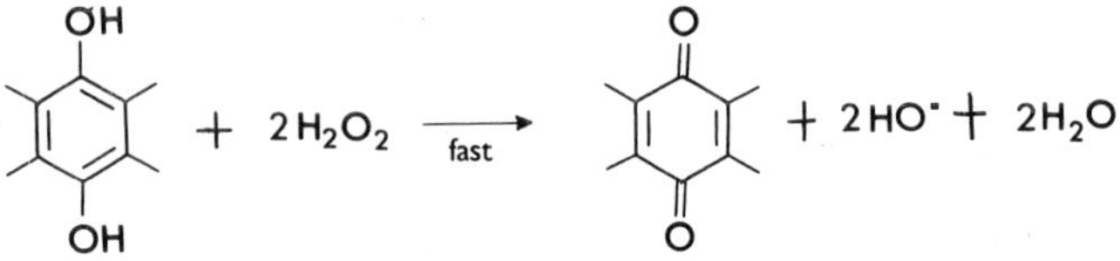

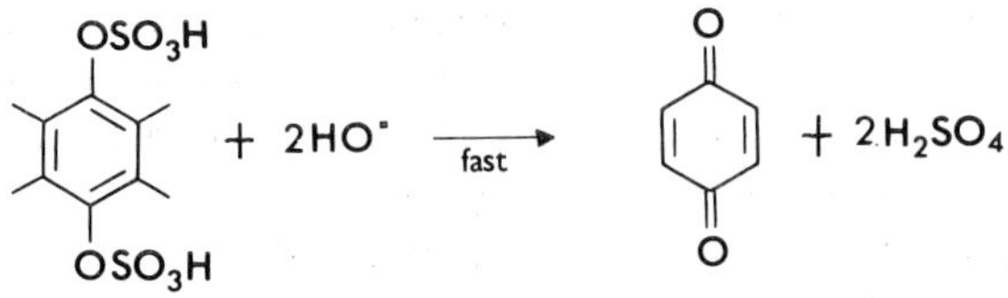

$$H_2O_2 + HO\cdot \rightarrow H_2O + HO_2\cdot$$

$$H_2O_2 + HO_2\cdot \rightarrow O_2 + H_2O + HO\cdot$$

This is not a complete scheme and does not account for all the observed effects; a full scheme is given in the original paper [44]. However, there can be no doubt that sulphate groups can be removed directly by oxidation by hydroxyl radicals (see the third equation in the sequence above), and this is the only conclusion needed in order to be able to consider the more important mechanism of oxidation by nitrous acid, which is extensively used in practice.

This reaction has been studied by Johnson and Lockett [49], who were able to arrive at a satisfactory explanation and, as a result, were able to suggest a simple means of improving the technical process. The rate of reaction of the sulphuric ester of leuco anthraquinone-2-sulphonic acid was followed in acidic solutions containing varying concentrations of the ester, nitrous acid and oxygen, it being known from earlier work [44] that oxygen could play a part in the development reaction with nitrous acid, leading to a greater rate, but that oxygen was rapidly consumed. It was shown first that, in the presence of oxygen, the rate increased with increasing concentration of nitrous acid until the oxygen had been consumed; after that, the rate was independent of nitrous acid and was equal to the rate of hydrolysis. It was therefore possible to conclude that, in the absence of oxygen, the original two-stage mechanism envisaged by Bader actually occurred, i.e.

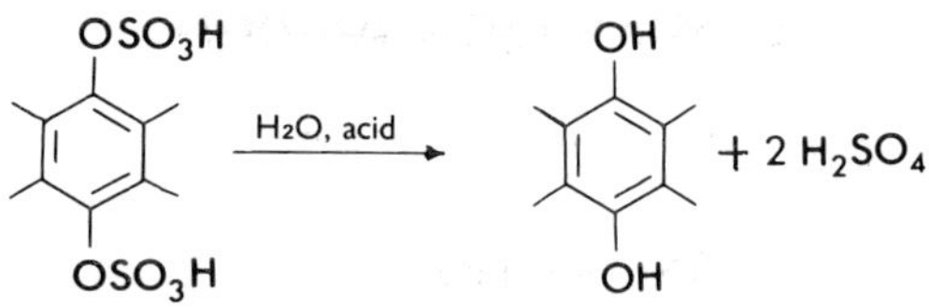

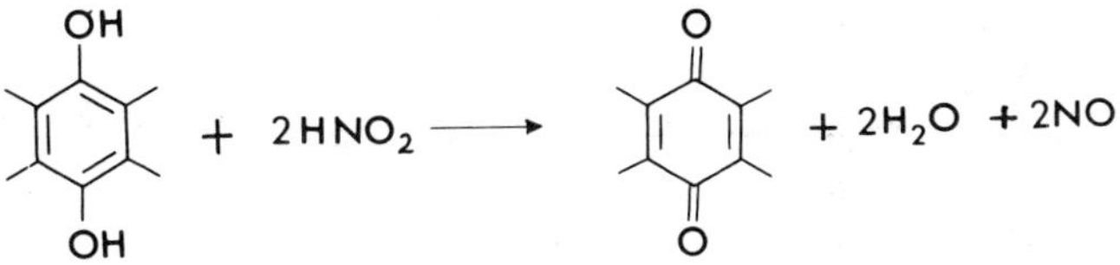

An observation of importance was that the second of these reactions was also slow and, indeed, at low concentrations of nitrous acid could be slower than the first, so that it was possible for the leuco compound to accumulate in the reaction solution. (Of course, for technical development, these reactions are not significant, because oxygen is always present.)

The important problem, therefore, was to establish the role of oxygen, i.e. the mechanism by which this leads to considerably increased rates of development. The first problem was to determine with which of the various species present (either initially or produced during reaction) oxygen actually reacted, and it was possible to show that this was the leuco compound which, of course, is continually but very slowly produced by hydrolysis. This reaction between a leuco compound and oxygen leads to the formation of hydrogen peroxide, i.e.

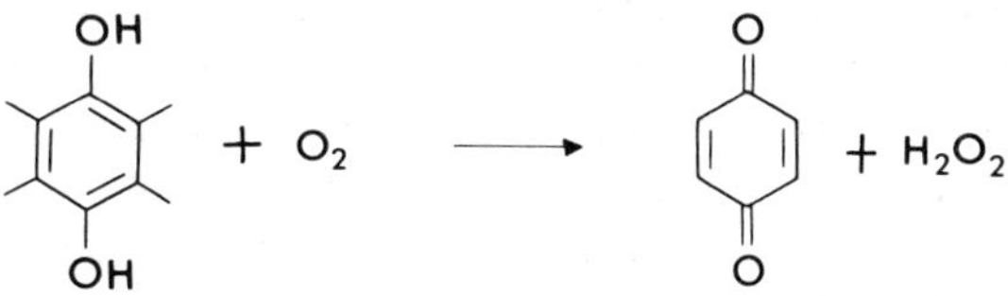

Thus, in a development with nitrous acid in the presence of oxygen, hydrogen peroxide is produced at a rate that cannot exceed the rate of hydrolysis. In the presence of oxygen, however, the total rate of development is many times higher than the rate of hydrolysis. The main problem, therefore, is to explain why a few molecules of hydrogen peroxide can lead to the oxidation of many molecules of the leuco ester by means of nitrous acid, i.e. to identify the mechanism by which hydrogen peroxide acts as a catalyst (or initiator) for the direct oxidation with nitrous acid.

The reaction between hydrogen peroxide and nitrous acid had already been investigated [50], for equimolar concentrations of the two reagents. This reaction can be represented as follows:

$$H_2O_2 + HO.NO \rightarrow H_2O + HO.ONO$$

$$HO.ONO \rightarrow HO\cdot + \cdot ONO$$

$$\cdot ONO \rightarrow \cdot NO_2$$

$$HO\cdot + \cdot NO_2 \rightarrow HONO_2\ (HNO_3)$$

i.e. nitrous acid is oxidised to nitric acid via the formation of pernitrous acid (HO.ONO). Nitric acid cannot be produced by radical recombination until nitrogen dioxide has undergone rearrangement to the radical $\cdot NO_2$.

However, in the development of sulphuric acid esters of leuco vat dyes, the concentration of nitrous acid is very much in excess of that of hydrogen peroxide (which is produced only slowly), and under these conditions the nitrogen dioxide produced by the scission of pernitrous acid can react with the excess nitrous acid before the rearrangement can take place, as follows:

$$HO.NO + \cdot ONO \rightarrow HO.ONO + NO$$

Thus, a further molecule of pernitrous acid is produced, which once again dissociates. Dissociation and re-formation of pernitrous acid thus constitute a reaction chain leading to the continuous production of hydroxyl radicals ($\cdot OH$). These, in turn, are then available to oxidise the leuco ester directly to the vat dye.

These observations enable the development with nitrous acid under normal working conditions to be explained. The cloth is padded with a solution of leuco ester and sodium nitrite, passed through a bath of dilute acid and then 'skied' so as to enable development to occur. The acid causes slow hydrolysis, thus producing a few molecules of leuco compound which are oxidised on exposure to the air during 'skying', resulting in the production of an equal number of molecules of hydrogen peroxide. These in turn react with nitrous acid, thus initiating the chain reaction, so that a copious supply of hydroxyl radicals ($\cdot OH$) is available to oxidise the bulk of the leuco ester.

It can now be seen that the considerable excess of nitrous acid normally recommended for the nitrite process is necessary to ensure that the $\cdot ONO/HNO_2$ reaction takes place in preference to $\cdot ONO$ rearrangement (which would stop the chain). It can also be seen that the technical process could be improved by the deliberate addition of some hydrogen peroxide, so that the slow hydrolysis reaction would not be the only source of hydrogen peroxide. The rate of formation of $\cdot ONO$ would thereby be increased and it would no longer be necessary to use excessive amounts of nitrite. These deductions were confirmed experimentally.

7.3.2 Development by light

Sulphuric esters of leuco vat dyes are sensitive to light irradiation under normal conditions, leading to the formation of the parent vat dyes. In normal dyeing and printing applications this is a disadvantage, since it tends to lead to irregular development because of uneven illumination. It is possible to exploit the effect to produce prints by a photographic method, in which a cloth impregnated with a solution of leuco ester containing a weak acid and a weak oxidising agent is exposed to an intense light source through a negative, but no commercially viable process has been introduced.

Work on the mechanism of photo-development indicated in the first place that irradiation causes hydrolysis of the ester to yield the leuco compound [51], i.e. light simply replaces the action of acid in normal chemical development. The critical evidence for this conclusion was the observation that irradiation in solutions which had been completely freed from all oxidising capacity by scouring with hydrogen in the presence of palladium led to the formation of the leuco compound in stoichiometric yield. The conclusion may, however, be criticised, because hydrogen remained in the solutions during irradiation so that leuco compound could rise by reduction of vat dye produced directly by photolysis. The irradiations were therefore repeated using solutions from which hydrogen had been displaced by nitrogen, i.e. there was neither oxidising nor reducing capacity. When this was done, using a solution of the thioindigo derivative C.I. Solubilised Vat Red 1, the vat dye itself was produced, eventually in 100 per cent yield. At the same time sulphate equivalent to about 60 per cent of the original combined sulphur (measured as $BaSO_4$) was liberated [52]. Interestingly, the vat dye produced in this way remained dissolved in water and the solutions were stable for several days before precipitation occurred.

The only possible route from the leuco ester to the vat dye by direct photolysis is by the following reaction, which shows only the essential conjugated system:

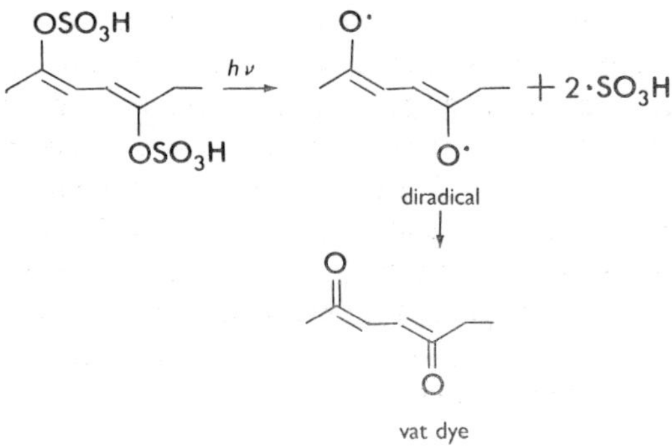

The leuco ester is cleaved at the oxygen–sulphur bonds to give the diradical (which then rearranges to the vat dye) and two sulphite radicals. These radicals must then react further and the following possibilities may be considered:

$$2\cdot SO_3H \rightarrow H_2S_2O_6 \quad (1)$$

$$\cdot SO_3H + H_2O \rightarrow HOSO_3H + \cdot H \quad (2)$$

$$\upharpoonleft\downharpoonright$$

$$\cdot OSO_2H + H_2O \rightarrow HSO_3H + \cdot OH \quad (3)$$

The radical nature of the photolysis proper was clearly demonstrated by using the radical-capture reagent diphenylpicryl hydrazyl, which readily combines with free radicals (·R) as follows, its colour changing from purple to yellow:

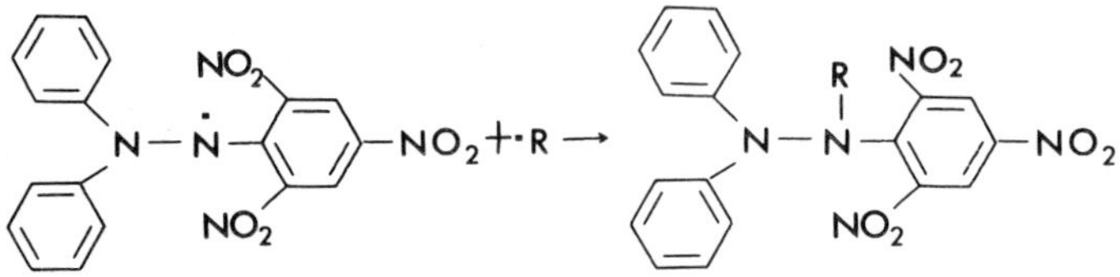

In irradiations of leuco ester in the presence of the hydrazyl, yellow products containing combined sulphur from the capture of sulphite radicals were obtained.

It was also possible to show that dithionate (reaction 1), sulphate (reaction 2) and sulphite (reaction 3) were all produced in the absence of any oxidising capacity, whilst only sulphate is produced if oxygen or other oxidising agents are present. The reactions (2) and (3) are alternative reactions with water because the sulphite radical can exist with the free electron on either sulphur or oxygen. Hydroxyl radicals produced by reaction (3) may either oxidise further molecules of leuco ester or oxidise sulphite ions or radicals to sulphate. In the presence of oxygen sulphite radicals are oxidised before they can dimerise to give dithionate.

The radical nature of the photolysis of leuco esters means that they may be used as photo-initiators for other radical reactions. One obvious consequence of this is that if a mixture of two or more leuco esters is irradiated, photolysis of only one of them will lead to the oxidation of them all. For this reason it will not be possible to produce multicolour patterns photographically by selective irradiation of individual components of a mixture of leuco esters.

7.4 SULPHUR AND CONDENSE DYES

These dyes date from about 1873, when Croissant and Bretonnière prepared brown dyes for cellulose by fusing various vegetable and animal waste products with solutions of sodium sulphide or polysulphides. An improvement was Vidal Black, prepared in 1893 by Vidal by fusing 4-aminophenol or *p*-phenylenediamine with sulphur, and since that time sulphur dyes have been produced from intermediates of known structure. Nevertheless, the process remains one of heating the intermediates either in molten sulphur or sodium polysulphide or in a solution of sodium sulphide or polysulphide. For this reason the structures of sulphur dyes are largely unknown, although in some cases the presence of thiazole (IV), thiazone (V) or thianthrene (VI) residues has been demonstrated, and it is reasonably certain that all sulphur dyes possess sulphur-containing heterocyclic rings. Of greater importance from a practical point of view is the presence of chains of sulphur atoms at various places in the molecule.

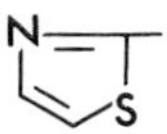

(IV)

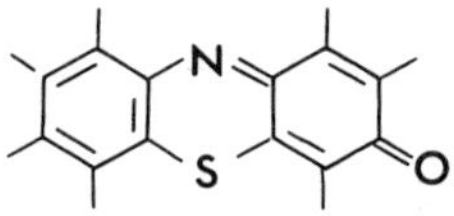

(V)

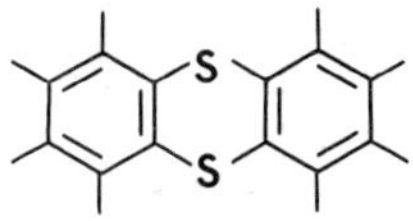

(VI)

Sulphur dyes, like vat dyes, are applied from alkaline reducing solutions, the reducing agent normally being sodium sulphide rather than dithionite. The reduced (or leuco) form is alkali-soluble, whereas the oxidised form which is finally produced on the (invariably cellulosic) textile material is insoluble. Reduction affects disulphide (or possibly disulphoxide) bonds which join the main aromatic and heterocyclic units of the dye molecule, the resultant thiols being alkali-soluble:

$$\mathrm{Ar{-}S{-}S{-}Ar'} \underset{\text{oxidation}}{\overset{\text{alkaline reduction}}{\rightleftharpoons}} \mathrm{Ar{-}S^{-} + {}^{-}S{-}Ar'}$$

Although this is essentially a reversible redox system, as with vat dyes, it differs in that reduction is a cleavage, whilst oxidation is, in effect, an oxidative condensation reaction. With sulphur dyes, therefore, the leuco compounds have r.m.m.s that are only a fraction of those of the oxidised forms. It is evident that, if the main aromatic units of the leuco compounds contain two (or more) mercapto groups, the oxidation process amounts to a polymerisation. The r.m.m.s of sulphur dyes are, indeed, very high. Because of the uncertainty of the molecular structure of sulphur dyes and the virtual impossibility of obtaining pure compounds, no fundamental results comparable with the measurement of redox or leuco potentials for vat dyes have been obtained.

Some of the combined sulphur in many sulphur dyes is designated as 'loosely combined'. It probably exists in the form of polysulphide chains and is fairly readily removed by oxidation. With many of the earlier sulphur dyes, particularly sulphur blacks, this sulphur gave rise to tendering of the dyed textile material, because oxidation in the presence of atmospheric water led to the production of sulphuric acid and consequent hydrolysis of the cellulose. Treatment of sulphur dyeings in the laboratory with alkaline chlorate(I) (hypochlorite) solutions leads to the quantitative production of sulphuric acid, which can be estimated as barium sulphate. This test can be used to assess the likelihood of tendering of a particular dyeing, although improved methods of dyestuff manufacture have virtually eliminated this fault. The suitability of many sulphur dyeings for 'topping' with basic dyes may also be due to oxidation of some of the sulphur to sulphonic or other acidic groups.

Because sulphur dyes are manufactured under alkaline reducing conditions the dyes are first produced as the soluble leuco compounds. The usual procedure is then to oxidise them, e.g. by blowing air, so as to precipitate the insoluble oxidised form. It is however possible to isolate the reduced form, still containing excess alkali and sulphide, giving so-called 'soluble' sulphur dyes, which can in fact be readily dissolved in the dyebath, only a small addition of sodium sulphide being necessary. An alternative, and far more elegant, method of achieving the same end is the introduction of thiosulphuric acid groups ($—S.SO_3Na$) into the unreduced form of the dye. Such compounds have adequate aqueous solubility, but reduction is still required in order to impart affinity for cellulose. Subsequently, they are oxidised to produce dyeings equivalent to those of conventional sulphur dyes.

7.4.1 Condense or polycondensation dyes

The device of producing dyes which are water-soluble because of the presence of thiosulphate groups has been exploited since about 1960 in the Inthion dyes (FH) and more recently in the Dykolite range (Southern Dyestuff Co.). They may be regarded as a development in the sulphur dye field, since the final form of the dye is insoluble and consists of dye units linked by disulphide bonds, as with conventional sulphur dyes. With the Dykolite range the final disulphide bonds link two aromatic residues, as in conventional sulphur dyes, so the dyeings can, if necessary, be reduced with sodium sulphide solutions (forming thiols, as described above) in order to bring about levelling [48]. The Inthion dyes, on the other hand, produce dyeings in which the disulphide bond is linked to two methylene groups, i.e. $—CH_2SSCH_2—$, and these dyeings are fast to sodium sulphide [53].

The mechanism of conversion of the original thiosulphate groups to the final disulphide bond is most probably the same in all cases and depends on the susceptibility of the thiosulphate groups to nucleophilic attack. Normally, the dyes are applied from an alkaline solution containing sodium sulphide, in which the sulphide ion acts as a nucleophile. Based on this assumption and supported by evidence from experiments in which sodium sulphide labelled with ^{35}S was used, the following reaction scheme has been proposed [54, 55]:

$$\text{D-S.SO}_3^- + {}^*\text{S}^{2-} \rightleftharpoons \text{D-S.}{}^*\text{S}^- + \text{SO}_3^{2-}$$

$$\text{D-S}^*\text{.S}^- + \text{D-S.SO}_3^- \rightleftharpoons \text{D-S.}{}^*\text{S.S-D} + \text{SO}_3^{2-}$$

$$\text{D-S.}{}^*\text{S.S-D} + \text{SO}_3^{2-} \xrightarrow{\text{Na}_2\text{S}} \text{D-S.S-D} + {}^*\text{S.SO}_3^{2-}$$

where *S is the labelled sulphur. The last of these reactions has been formulated by Minoura [56] in two stages as:

$$\text{D-S.}{}^*\text{S.S.-D} + \text{SO}_3^{2-} \xrightarrow{\text{Na}_2\text{S}} \text{D-S.}{}^*\text{S.SO}_3^- + {}^-\text{SD}$$

$$\text{D-S.}{}^*\text{S.SO}_3^- + {}^-\text{SD} \rightarrow \text{D-S.S-D} + {}^*\text{S.SO}_3^{2-}$$

The role of sodium sulphide in the first stage of this reaction is not clear. An alternative formulation that does not require formation of a trisulphide is

$$\text{D-S.SO}_3^- + {}^*\text{S}^{2-} \rightarrow \text{D-S}^- + {}^*\text{S.SO}_3^{2-}$$

$$\text{D-S.SO}_3^- + \text{D-S}^- \rightarrow \text{D-S.S-D} + \text{SO}_3^{2-}$$

For the time being some uncertainty must persist as to the actual mechanism, but it is certain that disulphide bonds are finally introduced, thereby linking the sites to which the thiosulphate groups were originally attached. If the original dye contains only one thiosulphate group, the result will be the formation of a dimer:

$$2\text{D-S.SO}_3^- \xrightarrow{\text{Na}_2\text{S}} \text{D-S.S-D}$$

but if two groups are originally present, a long-chain polymeric dye, or a cyclic dye, will be produced:

$$\text{O}_3\text{S.S-D-S.SO}_3^- \xrightarrow{\text{Na}_2\text{S}} \text{-D-S-[S-D-S-]}_n\text{-S-D-}$$

or

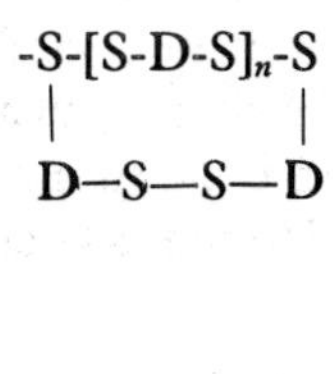

In either case, the water-solubilising thiosulphate group is lost and an insoluble dye of high fastness is produced.

The reactions shown above refer to the application of condense dyes to cellulose, and Luttringhaus has stated that no reaction occurs between thiosulphate dyes and cellulose. It is also possible, however, to apply these dyes to wool, and in this case there is a distinct possibility of reaction with the fibre, leading to dye–fibre bonds. This is because of the presence in wool of a large number of disulphide bonds in the cystine residues. It is well known that these bonds are susceptible to attack by nucleophiles, leading to the formation of cysteine (i.e. thiol) residues, which are themselves strong nucleophiles. Such fission of cystine bonds in wool can occur readily in aqueous solutions,

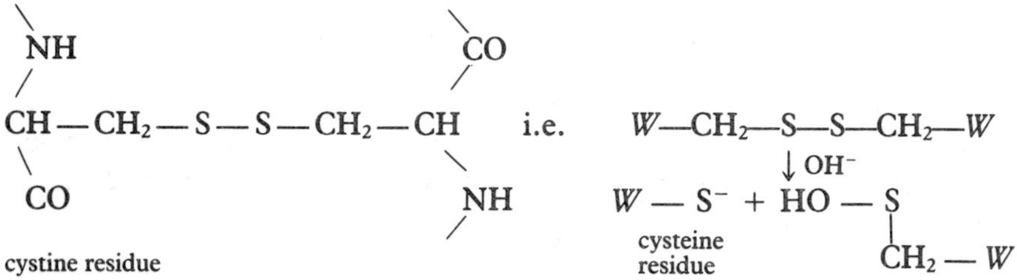

where *W* represents the keratin chain, so that some cysteine residues are always present in aqueous treatments of wool, particularly at pH values above 7.

The possibility that cysteine residues could interact with disulphide dyes was first suggested by Milligan and Swan [57], who used dyes of the form D-NH.CO.CH_2.S-SO_3^-. They found that, although the main product was the dye disulphide, i.e. D-NH.CO.CH_2.S.S.CH_2.CO.NH-D, some mixed disulphide with wool cysteine was also produced, by the reaction

$$W\text{-}S^- + {}^-O_3S.S.CH_2.CO.NH\text{-}D \rightarrow W\text{-}S.S.CH_2.CO.NH\text{-}D + SO_3^{2-}$$

The possibility of reaction between wool and thiosulphate dyes has been more extensively investigated by Osterloh [58], who studied the interactions between wool and three dyes containing respectively one, two and three thiosulphate groups. He also studied the reactions of the dyes with the isolated amino acids, cystine and cysteine. Reaction of the thiosulphate dyes with cystine was very slow, leading to the formation of dye disulphide after about 5 h. Chromatographic evidence during the course of the reaction showed, however, that the mixed dye–cysteine disulphide was produced, reaching a maximum concentration after about 2 h. This observation suggests that cysteine may be produced from cystine and plays a part in the reaction, i.e.

$$\underset{\text{cystine}}{\text{Cy-S-S-Cy}} \rightarrow \underset{\text{cysteine}}{\text{2Cy-S}^-}$$

$$\text{Cy-S}^- + {}^-\text{O}_3\text{S.S-D} \rightarrow \underset{\substack{\text{mixed}\\\text{disulphide}}}{\text{Cy-S-S-D}} + \text{SO}_3^{2-}$$

This participation of cysteine was confirmed in experiments carried out in the presence of a thiol blocker ($CH_2{=}CH.SO_2R$), which would react instantaneously with any cysteine to produce a thioether:

$$\text{Cy-SH} + \text{CH}_2{=}\text{CH.SO}_2\text{R} \rightarrow \text{Cy-S.CH}_2\text{.CH}_2\text{.SO}_2\text{R}$$

As expected, the thioether was produced, confirming the formation of cysteine during the reaction. In the absence of thiol blocker, *S*-sulphocysteine ($Cy\text{-}S.SO_3^-$) was produced. All these observations can be explained by the following reactions:

$$\text{Cy-S}^- + \text{D.S.SO}_3^- \rightarrow \text{Cy-S.S.-D} + \text{SO}_3^{2-}$$

$$\text{Cy-S-S-Cy} + \text{SO}_3^{2-} \rightarrow \text{Cy-S.SO}_3^- + \text{Cy-S}^-$$

$$\text{Cy-S-S-D} + \text{Cy-S}^- \rightarrow \text{Cy-S-S-Cy} + \text{D-S}^-$$

$$\text{Cy-S-S-D} + \text{D-S}^- \rightarrow \text{D-S-S-D} + \text{Cy-S}^-$$

$$\text{Cy-S-S-Cy} + 2\ \text{D.S.SO}_3^- \rightarrow 2\text{Cy-S-SO}_3^- + \text{D-S-S-D}$$

The first two of these equations together represent a reaction chain by which the adventitious presence of a few cysteine residues (CyS^-) leads to the production of the mixed cysteine–dye disulphide (Cy-S-S-D) and *S*-sulphocysteine ($Cy\text{-}S.SO_3$). The next two equations similarly represent a chain by which the mixed disulphide is converted to dye disulphide. The overall conversion is represented by the fifth equation.

With wool, the evidence suggests that similar reactions occur, with $W\text{-}CH_2\text{-}S\text{-}S\text{-}CH_2\text{-}W$ and $W\text{-}CH_2\text{-}S^-$ replacing Cy-S-S-Cy and $Cy\text{-}S^-$ respectively. Again, the presence of a thiol blocker demonstrated the important influence of cysteine residues ($W\text{-}CH_2\text{-}S^-$), dyeings being produced which had only the wet-fastness associated with an equalising acid dye; i.e., reaction of the thiosulphate groups of the dye was inhibited, no insoluble dye–

disulphide was produced and the dye was adsorbed by virtue of normal ionic attraction. In the absence of the thiol blocker, dyeings of considerably greater fastness were obtained; moreover, maximum fixation occurred in the pH range 8–9, a result which would be impossible to account for solely by ionic adsorption. Finally, in dyeings on silk, which contains virtually no cystine, the thiosulphate dyes behaved in exactly the same way as normal acid dyes containing sulphonic acid groups in place of the thiosulphate groups.

The overall conclusion is that thiosulphate dyes can be applied to wool, leading to fast dyeings of insoluble dye disulphides (or polymeric dyes), under relatively mild conditions which do not lead to serious degradation of the wool. The application is possible because of the presence in wool of a large number of cystine residues which can take part in its reaction sequence. Some cystine bonds are thereby broken, but the number is small (equal to only half the number of dye molecules fixed), so no serious degradation is produced.

7.5 PHTHALOCYANINES

The most important member of this class is copper phthalocyanine (VII) which was introduced as a pigment in 1936 and was soon followed by chlorinated, and other, derivatives:

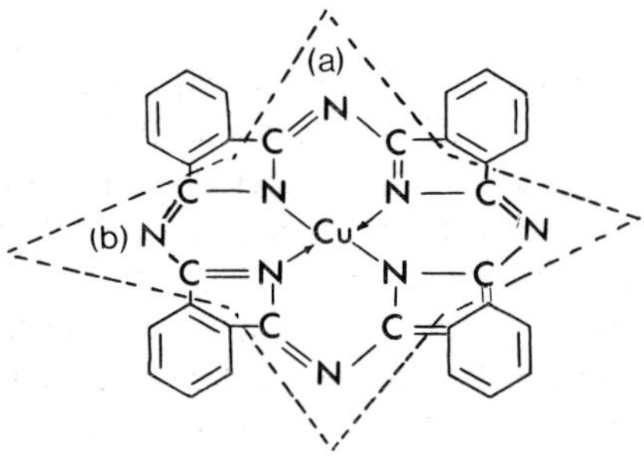

(VII)

The 16-membered heterocyclic ring within the dotted line is aromatic; the single and double bonds are not localised.

The manufacture of phthalocyanines is beyond the scope of this chapter and is fully described elsewhere [e.g. 59], but the following reaction sequence is given to outline the route from phthalic anhydride by the urea method of Wyler [60]:

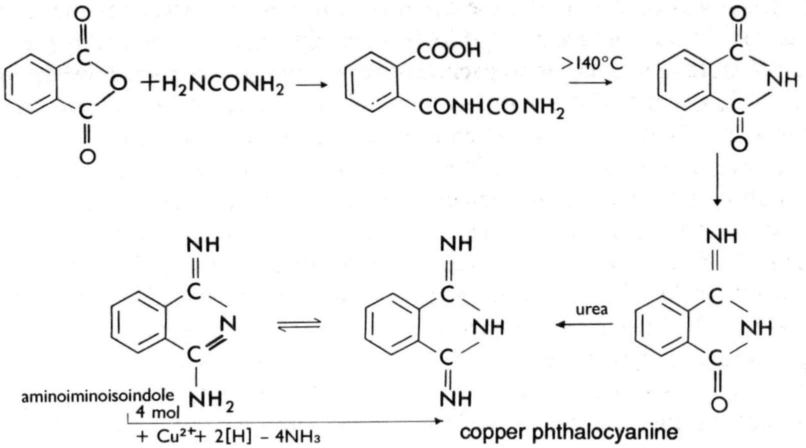

The phthalocyanine pigments are bright blue or greenish blue and are characterised by very high stability; they have inevitably found many outlets in the pigment field. These include a limited amount of textile application in the mass coloration of manufactured fibres and for resin-bonded pigments, particularly in printing. Not surprisingly, methods have been developed to permit their wider application to textiles by more conventional means from aqueous solutions.

The most obvious method for imparting aqueous solubility is the introduction of sulphonate groups, and sulphonated phthalocyanines are available as direct dyes. Although useful, these dyes do not possess high fastness to washing, because the sulphonate group is still present in the dyed material and the dyes therefore remain water-soluble. Much more satisfactory are methods that lead to the formation of an insoluble phthalocyanine pigment within the fibre, and this can be achieved (a) by the introduction of temporary solubilising groups which are lost during application, and (b) by the synthesis of the phthalocyanine on the fibre from water-soluble intermediates.

7.5.1 Temporary solubilising groups

The most important method for imparting temporary aqueous solubility depends on the ability of tertiary bases to form quaternary salts with halogenoalkanes (or alkyl sulphates), and the similar ability of compounds contining doubly bonded sulphur to form sulphonium salts, i.e.

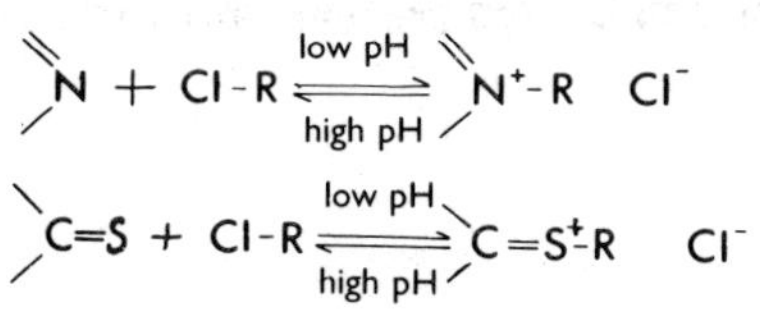

In both cases the salts (onium salts) are stable under mildly acidic conditions, and are soluble, but revert back at higher pH values to produce the original tertiary base and halogenoalkane, neither of which need be water-soluble. This simple scheme can therefore form the basis of a method for producing an insoluble phthalocyanine in a textile fibre.

Two partners are required for the formation of an onium salt, and the phthalocyanine can be either one of them. The first successful exploitation seems to have used tetrapyridyl phthalocyanine, i.e. pyridine (a tertiary base) is substituted into each of the benzene rings of phthalocyanine, the tetrapyridyl derivative giving a quaternary salt with dimethyl sulphate, i.e.

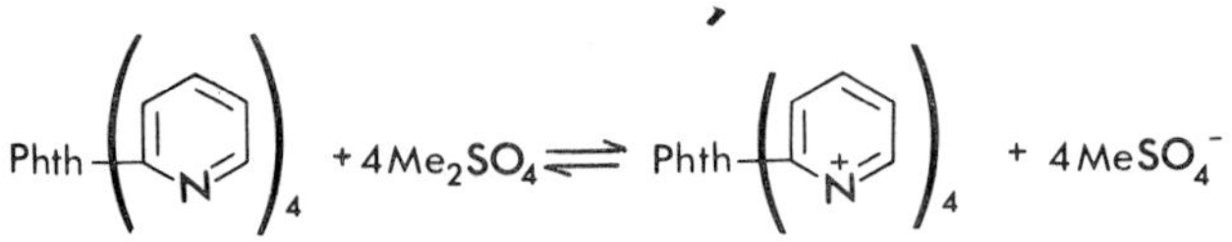

where Phth represents the phthalocyanine molecule. More success, however, is achieved if the phthalocyanine is converted into a halogenoalkane by the introduction of a chloromethyl group [61] by the action of di(chloromethyl) ether:

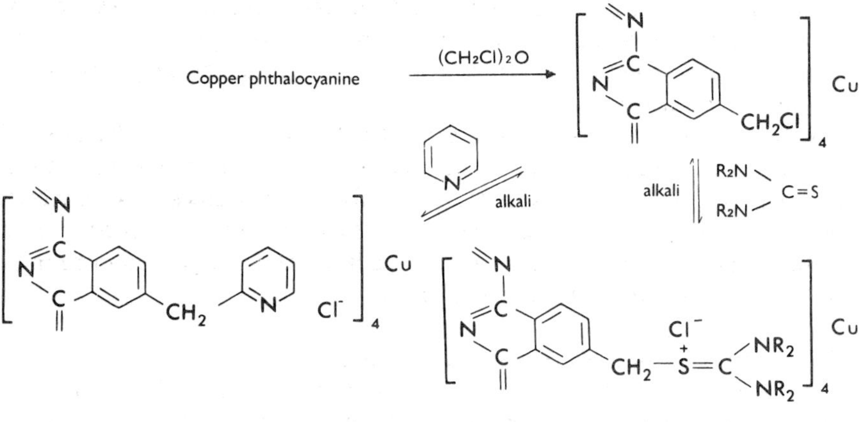

The resulting chloromethylated phthalocyanine can then give an onium salt with a tertiary base such as pyridine or with a tetra-alkylthiourea.

Monastral Fast Blue (copper phthalocyanine) reacts in this way with thiourea to give the important Alcian Blue 8G, which is usually applied by printing a paste containing sodium acetate and a volatile acid such as acetic or lactic acid. Sufficient acid is lost during steaming to allow the pH to rise to a value high enough for the chloromethyl phthalocyanine to be precipitated in the fibre.

The ability of the chlorine atoms in chloromethylated phthalocyanines to be readily displaced by nucleophilic substitution has also been utilised to produce onium dyes, as shown below:

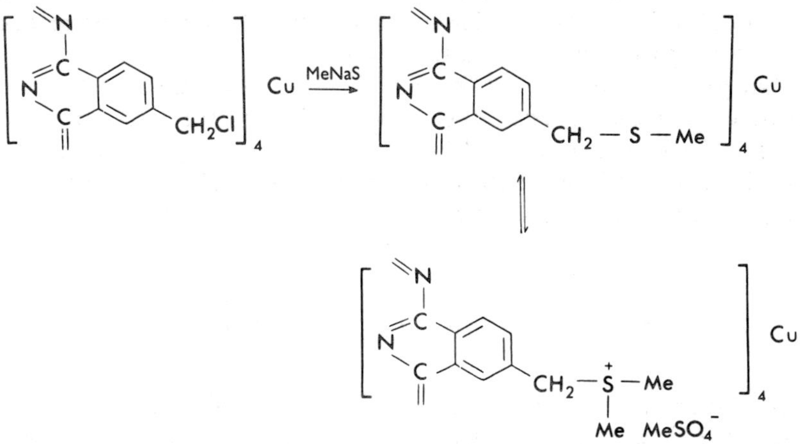

The Alcian range of phthalocyanine dyes employs one or other of these devices for producing onium salts. The chemical principle is always the same: the onium salt is water-soluble and can therefore be applied from aqueous solution; subsequently the pH is raised and the salt decomposes into its insoluble constituents.

Although these methods have proved extremely valuable, the insoluble phthalocyanine finally produced on the fibre still contains the substituents necessary for onium salt formation, and this can sometimes have an adverse effect on fastness properties [58]. As a result, methods for producing an unsubstituted phthalocyanine on the fibre have been sought, culminating in the introduction of Phthalogen Brilliant Blue.

7.5.2 Synthesis of phthalocyanine on the fibre

Phthalogen dyes employ the procedure of applying water-soluble intermediates or precursors from which the final phthalocyanine is synthesised on the fibre. The method is derived from Wyler's urea synthesis of phthalocyanines (p. 529) and the observation of Baumann *et al.* [59] that the tautomeric pair, di-iminoisoindoline and aminoiminoisoindole, can be isolated in good yield. Aminoiminoisoindole reacts readily with copper salts under reducing conditions to yield copper phthalocyanine (Monastral Fast Blue B) at a temperature of about 100 °C, i.e. under conditions appropriate to textile application.

The exploitation of this reaction for textile application is not as simple as carrying out the reaction on the bulk scale. The components must be presented in forms that are readily accepted by the textile fibre, so the fibre itself should be as receptive as possible, and the reactions must take place in the correct sequence within the fibre, and at a reasonable rate.

Aminoiminoisoindole has only a low solubility in water at room temperature, but raising the temperature to increase its solubility is unsatisfactory because this leads to hydrolysis, which prevents subsequent formation of phthalocyanine:

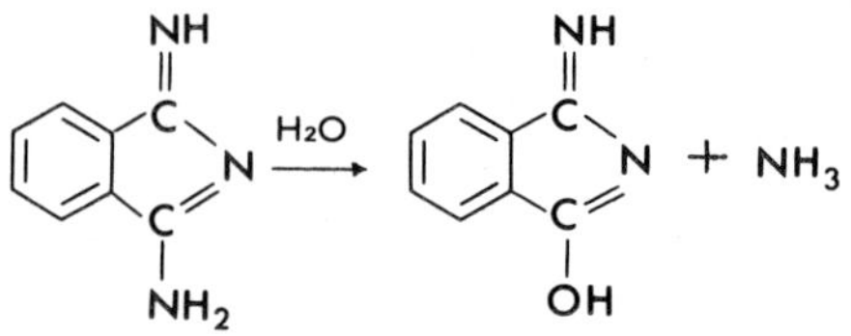

However, solubility is much increased in mixtures of water with polar organic solvents such as ethanediol, formamide and dimethylformamide. The last two solvents are particularly useful, since they are also mild reducing agents and hence ensure the reducing conditions necessary for the formation of phthalocyanine. They also probably act as swelling agents for cellulose.

A further difficulty is that aminoiminoisoindole readily forms an insoluble complex with copper (II) ions; hence, if the isoindole and a simple copper salt were included together in a printing paste or dyebath, the complex would precipitate and penetration of the textile material would not take place. The solution to this difficulty is to immobilise the copper in an alternative, but soluble, complex from which copper (II) ions are only slowly liberated during application. Chelating agents such as mono- or di-methylaminoacetic acid are used for this purpose.

The overall application procedure may be summarised as follows:

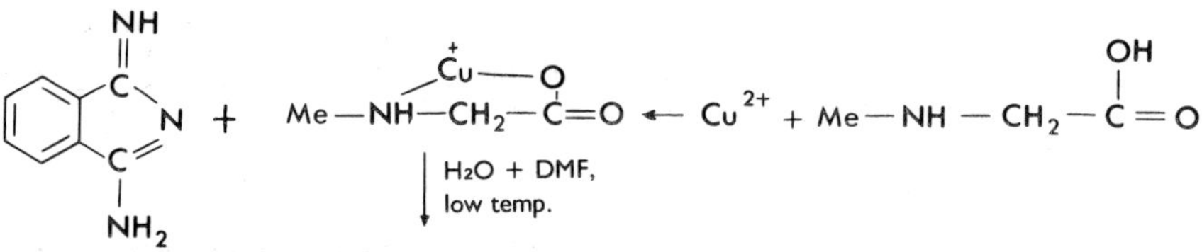

1. Apply to textile material and dry below 80 °C, i.e. remove water but not DMF.
2. Steam or bake at about 100 °C, giving

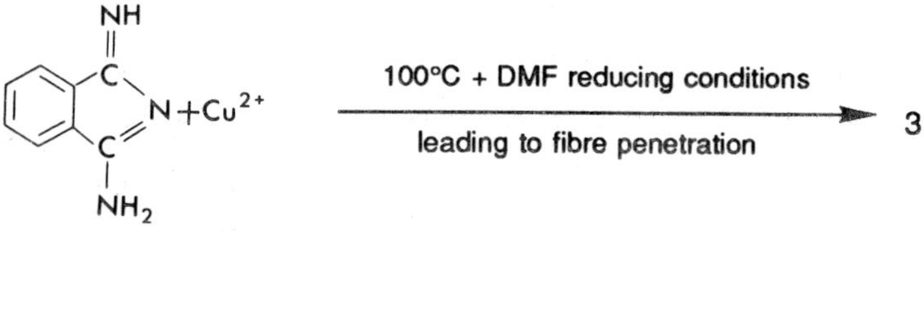

3. Formation of copper phthalocyanine in the fibre.

In a similar way, starting from the 6-phenyl derivative of aminoiminoisoindole, the greener tetraphenylcopper phthalocyanine is produced.

7.5.3 Miscellaneous applications

It is possible to introduce mercapto groups into phthalocyanine, yielding compounds that can be linked by oxidation to give di- or poly-phthalocyanine disulphides:

$$2\ \text{Phth-SH} \xrightarrow{[O]} \text{Phth-S-S-Phth}$$

$$n\ \text{HS-Phth-SH} \xrightarrow{[O]} \text{—[S-Phth-S-S-Phth-S-]}_n\text{—}$$

which are essentially sulphur dyes. The disulphides may be prepared by the manufacturer and reduced at the application stage by means of alkaline sulphide solutions, as for conventional sulphur dyes:

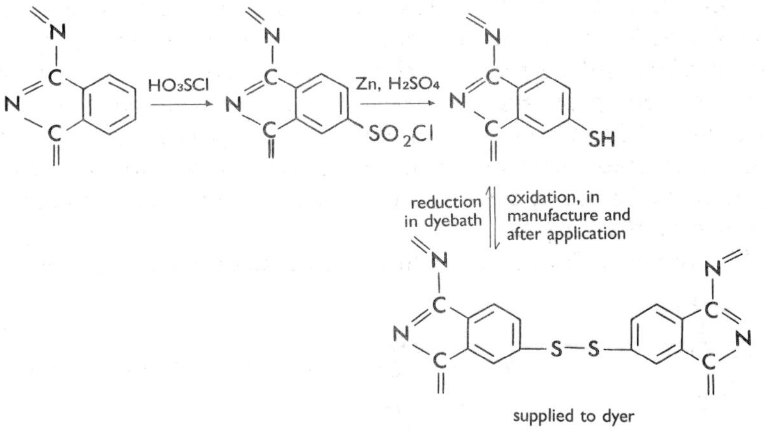

Alternatively, the dyer may be supplied with a phthalocyanine thiocyanate which is converted to thiol in the dyebath, again by means of sulphide, but acting in this case as a nucleophile:

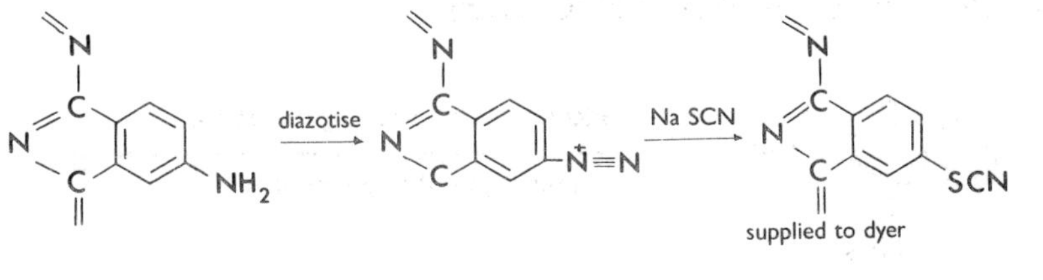

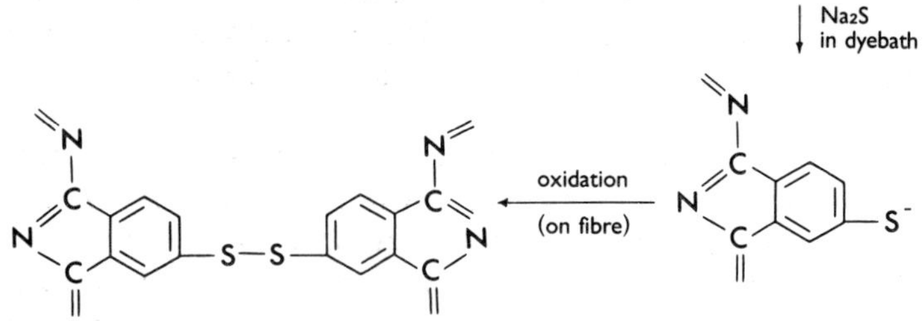

An isolated application of some interest is provided by cobalt phthalocyanines. These are capable of ready reduction to dihydrocobalt phthalocyanine, which gives a water-soluble sodium salt, i.e. reduction of cobalt phthalocyanine in alkaline solution yields the equivalent of a leuco vat dye. Since this is easily reoxidised and also has substantivity for cellulose, cobalt phthalocyanine can be applied by the vat dye method. Reduction occurs at the two nitrogen atoms (a) and (b) in structure VII.

Finally, it should be mentioned that phthalocyanines can be substituted by groups capable of reaction with textile fibres, especially cellulosic fibres, and hence appear in the important group of reactive dyes. The chemistry involved in the application of reactive dyes has been dealt with in Chapter 6 and will not be discussed here.

7.6 CHELATED OR METAL-COMPLEX DYES

The enhancement of one or more properties of a colouring matter by the device of chelate formation is virtually as old as the art of dyeing itself, since many of the natural colouring matters made use of chelate formation, both to give better fixation and – most importantly - very often to give colour itself. A well-known example is the famous Turkey Red, in which the colouring matter alizarin (extracted from the root of the madder plant) was applied to cloth previously prepared with aluminium salts. In this way the yellow non-substantive alizarin (1,2-dihydroxyanthraquinone) was converted into a red alizarin–aluminium complex of high fastness. Nowadays, very many synthetic dyes which can bind a metal ion in the same way are available; some, like alizarin, are hydroxyanthraquinones, but the majority contain the *ortho*-hydroxyazo system. The most common metal is chromium, followed by copper and a few examples of other metals such as cobalt. In application, the metal and dye may be applied separately with dye–metal bonding taking place within the textile (mordant dyeing), or the dye–metal bonding may be carried out by the dye manufacturer, the resulting metal-complex dyes then being regarded as acid dyes. Chemically, however, there is no distinction.

The formation of covalent bonds between a metal ion and a dye molecule (or ion) is an example of the widespread phenomenon of complex formation.

More detailed discussions of the subject in relation to the chemistry and application of dyes are given elsewhere [62–64]; a simple outline of the mode of formation of these bonds and the factors affecting their stability is considered to be sufficient here.

A covalent bond between two atoms within a molecule consists of two electrons which are shared between the two participating atoms. If ionisation of the bond can occur the two atoms move apart, one of them (the anion) taking both of the originally shared electrons with it and hence bearing a negative charge, the other (the cation) being denuded and hence having a vacant orbital. The reverse of ionisation is the formation of a covalent bond between the separated atoms by the donation of the pair of electrons by the anion into the vacant orbital of the cation. This donation of a pair of electrons into a vacant orbital to produce a bond is what is meant by complex formation. The bond thus produced is sometimes described as a *coordinate bond* or a *dative covalency bond*, but these special names describe the formation of the bond rather than the bond itself. The species donating the electron pair is usually referred to as the *ligand*.

The formation of a covalent bond by the reverse of ionisation is the simplest example of complex formation. However, the essential feature of the process, i.e. the donation of a pair of electrons into a vacant orbital, does not require that the participants be ions. In fact, the majority of examples entail donation by neutral molecules (or groups within a molecule), although the recipient is almost invariably a metal ion. The reverse of ionisation of acetic acid and the formation of the cuprammonium ion are given below for comparison, only the donated electrons being shown:

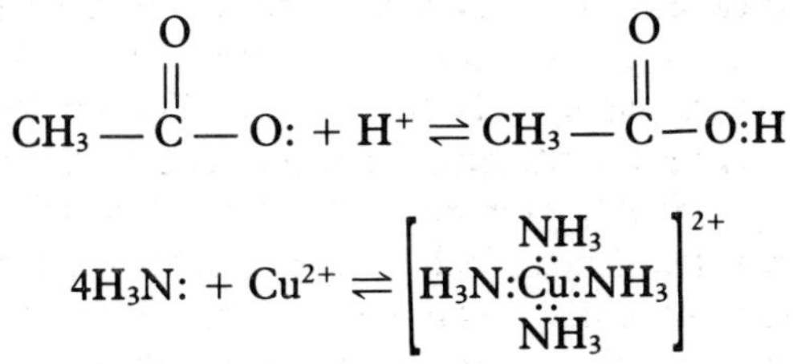

Each process is an equilibrium and can therefore be characterised by an equilibrium constant which may be either a dissociation constant, as is usual for the ionisation of an acid:

$$K_a = \frac{[CH_3COO^-][H^+]}{[CH_3COOH]} \quad (7.19)$$

or the reciprocal, viz, a formation constant, as is more usual for a complex:

$$K = \frac{[Cu^{2+}(NH_3)_4]}{[Cu^{2+}][NH_3]^4} \qquad (7.20)$$

The formation constants of complexes are usually much greater than those of even weak acids, i.e. complexes dissociate to a much smaller extent and are normally regarded as being stable. Even greater stability would be expected if more than one pair of electrons could be donated from the ligand to the receiving metal ion, resulting in the metal ion being bound by two or even three covalent bonds. Complexes in which this occurs are known as *chelates*, and it is with chelates that this section is concerned.

7.6.1 Coordination number

This is the number of covalent bonds that can be formed with a metal ion, i.e. the number of pairs of electrons which can be received or, alternatively, the number of unfilled orbitals within the metal ion which are accessible for donation. The factors affecting the value of the coordination number are various, so that over the whole field a wide range of values (between 2 and 9) [65] has been recorded. Most commonly the value is 4 or 6, however, and these are the only values with which we need be concerned in this chapter. An important factor is certainly the availability within the electronic structure of the metal ion of unfilled orbitals of almost equal energy which are capable of hybridisation, so that a symmetrical system of equal bonds can be established around the metal atom. The examples of copper (coordination number 4) and chromium (coordination number 6) are given below (Table 7.8).

In the first copper(II) complex the donated electrons (shown as o) enter the 4*s* and the three 4*p* orbitals, so that the resulting bonds are sp^3 hybrids (as in the case of the covalent bonds to carbon) and are directed from the central copper(II) ion towards the corners of a regular tetrahedron, whereas in the second case the hybridisation is dsp^2 and the bonds are directed from the copper(II) ion towards the corners of a square. In the vast majority of cases copper(II) chelates and complexes have the square planar configuration, as in copper phthalocyanine.

For the chromium(III) complex of coordination number 6, the donated electrons are as shown and the bonding is therefore d^2sp^3, which has the octahedral configuration. The same is true of cobalt(III) complexes and chelates. A consequence of spatially directed bonds is that stereoisomers may be obtained. This is certainly true of metal complexes and chelates and has aroused a good deal of attention, but is not relevant to our purposes in this chapter. (For a fuller review and bibliography see refs. 64 and 66.)

TABLE 7.8

Electronic configuration for copper and chromium and their ions and complexes

Metal, ion or complex	Electronic orbitals: 3*d* (filled with 5 pairs)	4*s* (one pair)	4*p* (3 pairs)
Cu	xxxxx	x	
	xxxxx		
Cu^{2+}	xxxxx		
	xxxx		
$Cu^{2+}L_4$	xxxxx	o	ooo
	xxxx	o	ooo
$Cu^{2+}L_4$	xxxxo	o	oox
	xxxxo	o	oo
Cr	xxxxx	x	
Cr^{3+}	xxx		
$Cr^{3+}L_6$	xxxoo	o	ooo
	oo	o	ooo

7.6.2 Stability of chelates

A chelate, as defined earlier, is simply a ligand–metal complex in which more than one pair of electrons is donated from each ligand molecule to the central metal. A metal ion of coordination number 4 can therefore bind two molecules of a ligand capable of donating two pairs – a so-called *bidentate* ligand giving a 1:2 chelate which can be represented as ML_2. Similarly a metal of coordination number 6 could bind three molecules of such a ligand or two molecules of a *tridentate* ligand; the resulting metal–ligand chelates are represented as ML_3 and ML_2 respectively and are known as 1:3 and 1:2 chelates*. It should be pointed out here that in ordinary solutions containing metal ions the coordination number of the ion is satisfied by water molecules, each donating a pair of electrons, so that complex and chelate formation are in reality the replacement of water by the entering ligand. An important consequence is that chelates can be formed containing fewer ligand molecules than are demanded by the coordination number. Thus, from a metal of coordination number 6 and a tridentate ligand, the two chelates ML and ML_2 (1:1 and 1:2 respectively) are possible, but it must be borne in mind

*The numbering of chelates containing more than one ligand molecule varies from source to source, the chelate ML_2 being referred to as either a 1:2 or 2:1 chelate. The chelates used as dyes are described in the *Colour Index* as metal-complex dyes and are numbered with the metal first, i.e. ML_2 is described as 1:2. This is a logical consequence of the term 'metal complex' and this system will be used in this chapter, although 'metal complex' does not necessarily indicate the presence of a chelate structure.

that the correct representation of ML should be $ML(H_2O)_3$. In practice, the existence of these complexed water molecules can be largely ignored.

In the field of dyes the most important chelates are those formed between the Cr^{3+} and Cu^{2+} ions and tridentate ligands, which are usually *o,o′*-dihydroxyazo compounds, i.e. they contain the structure

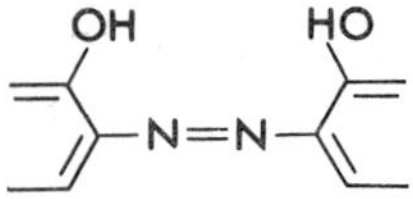

(VIII)

Thus for Cr^{3+} (CN = 6) either 1:1 or 1:2 chelates may be formed, but for Cu^{2+} and Ni^{2+} (CN = 4) only 1:1 chelates are possible. With bidentate chelates, e.g. *o*-hydroxyazo compounds, 1:2 chelates with metals of coordination number 4 are also possible, however, and for reasons of simplicity, these have been the most extensively studied [67–69]. All of these ligands are characterised by having one or two weakly acidic (usually hydroxyl) groups and a neutral group that can act as an electron donor. Chelate formation requires first of all ionisation of the hydroxyl group before donation to the metal ion can occur.

Thus with Cu^{2+} and an *o*-hydroxyazo compound, the following reversible reactions take place:

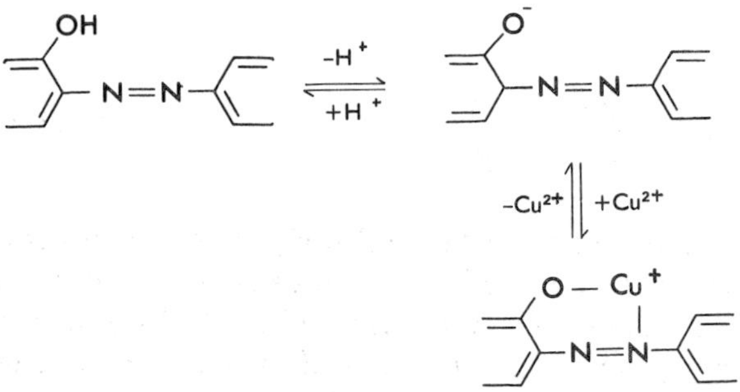

This formulation shows that donation from the ionised oxygen to Cu^{2+} is simply an alternative to donation to H^+ (i.e. reverse of ionisation). Since one ligand group is an anion, chelation results in the neutralisation of one of the positive charges of Cu^{2+}. Naturally, if a second ligand molecule were chelated, the resulting chelate would be uncharged.

If the *o*-hydroxyazo compound is represented as HL, formation of the chelate may be written as:

$$HL \rightleftharpoons H^+ + L^-$$

$$M^{n+} + L^- \rightleftharpoons M^{(n-1)+}L$$

$$M^{(n-1)+}L + L^- \rightleftharpoons M^{(n-2)+}L_2$$

where M^{n+} represents any suitable metal ion.

For these three equilibria, the three equilibrium constants may be written as

$$K_a = \frac{[H^+][L^{-1}]}{[HL]} \tag{7.21}$$

$$K_1 = \frac{[M^{(n-1)+}L]}{[M^{n+}][L^-]} \tag{7.22}$$

$$K_2 = \frac{[M^{(n-2)+}L_2]}{[M^{(n-1)+}L][L^-]} \tag{7.23}$$

The overall chelate-formation process may be written

$$M^{n+} + 2L^- \rightleftharpoons M^{(n-2)+}L_2$$

giving

$$K = \frac{[M^{(n-2)+}L_2]}{[M^{n+}][L^-]^2} = K_1K_2 \tag{7.24}$$

K_a is, of course, a dissociation constant, whereas K_1, K_2 and K are formation or stability constants. For a particular chelate to be satisfactory as a dye, it is necessary for K to be large, and in practice values of at least 10^{15} are obtained [e.g. 68]. The corresponding values of K_a would be of the order 10^{-7}, giving 10^7 for the formation constant of the un-ionised hydroxyl group, i.e. the formation or stability constant for the binding of the metal ion is some 10^8 times greater than that of the binding of a proton.

Measurement of the stability constant K may be achieved in various ways, each enabling the values of the concentration terms on the right-hand side of Eqn 7.2.4 to be measured. This, in turn, means that the value of K_a must also be known, as well as the pH (i.e. $[H^+]$) in order to calculate $[L^-]$ from [HL].

It is not appropriate to give details of the various methods here, but they are given elsewhere [70]. It is sufficient to say that they include the measurement of chelate concentration ($[M^{(n-2)+}L_2]$), either by measuring the extent to which the ionisation of HL is disturbed (potentiometric method) or directly by absorptiometry. Alternatively, the equilibrium concentration of unchelated ligand may be measured by absorptiometry – a method that depends, of course, on there being a sufficient difference in absorbance between the chelate and the unchelated ligand. Direct measurement of the concentration of unchelated metal is also possible, for which purpose polarography is a most convenient technique.

Whichever method is used, it must be borne in mind that chelation is a reaction of the ligand ion L^- rather than HL, so that the value of K_a must also be known. Furthermore, it is necessary to know the composition of the chelate, i.e. the number of ligand molecules bound per metal ion, for which Job's method of continuous variations is most suitable [70].

In relation to the large number of dyes used commercially as chelates, there has been comparatively little fundamental work on their stability and the factors affecting it. This may be partly because dyes for chelate formation are selected, like all other dyes, on the basis of their commercial performance. Those chelates which are satisfactory as metal-complex or chrome mordant dyes have such high stability that accurate measurement becomes difficult and numerical values are probably not very accurate. Nevertheless, one would expect that stability of a chelate would be related to the ability of the ligand to donate lone pairs of electrons, as well as to the ability of the metal ion to receive them [68, 69, 71, 72].

The effect of the ligand is most easily demonstrated because it is possible to study a series of ligands which vary systematically, usually by varying one substituent. If the binding of the metal ion to the ligand is essentially the same as the binding of a proton to produce the un-ionised hydroxyl group, a clear correlation should be found between the two equilibrium constants K_a and K (see earlier). That this is so is demonstrated, for example, by the results of Calvin and Wilson [73] with substituted salicylaldehydes and by those of Johnson *et al.* [68] for the substituted *o*-hydroxyazo compounds (IX).

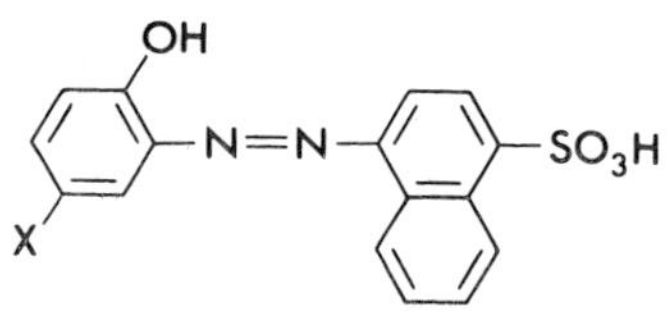

(IX)

These latter results are illustrated in Figure 7.4, in which pK_a) (i.e. $-\log K_a$) is plotted against $\log K$. This direct linear relationship indicates strongly

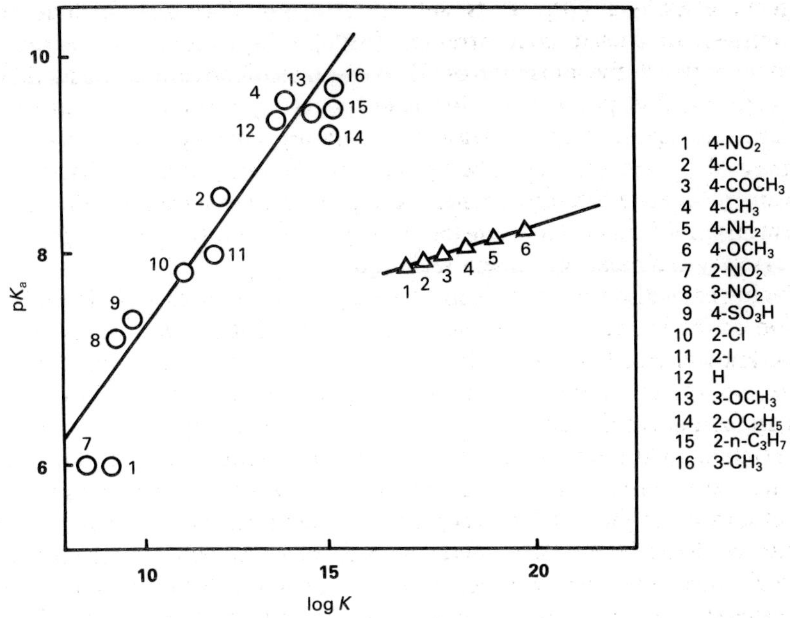

Figure 7.4 – Relation between pK_a and log K for substituted salicylaldehydes (○) and the o-hydroxyazo compounds IX (△)

that the two processes are essentially the same and that the bonding of the proton is also a chelation, i.e. electrons are donated from two sources. Thus, for *o*-hydroxyazo compounds the process would be represented as follows:

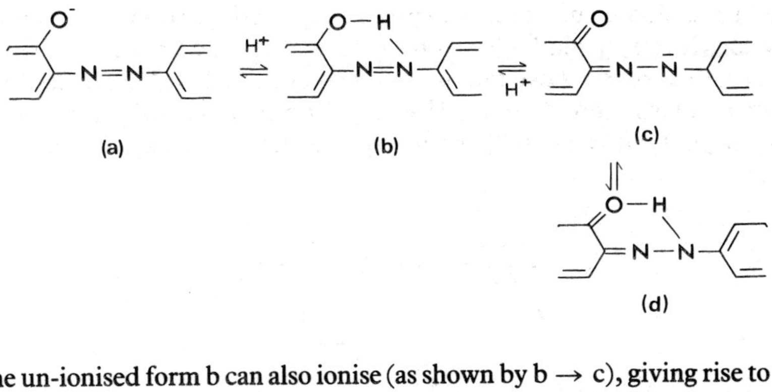

The un-ionised form b can also ionise (as shown by b → c), giving rise to the quinone-phenylhydrazone tautomer c↔d. The existence of such tautomeric equilibria was firmly established by u.v. spectral analysis [74] and has since been confirmed by Raman, n.m.r. and mass spectroscopy [75]. There is no reason why tautomeric forms of metal chelates should not also exist; indeed, the results of Johnson *et al.* [68] could be taken to support this contention.

Although these and other fundamental studies are of considerable interest and importance in elucidating the fundamental processes underlying chelation, it must be admitted that they play little part in our understanding of the role of chelation in the application of chrome mordant dyes, where chelation is an integral part of the dyeing operation. Work in this important area has been reviewed recently by Hartley [76], so the salient points can be covered by a brief synopsis of this review.

In the application of chrome dyes by the chrome mordant and afterchrome methods the wool is treated with dichromate solutions ($Cr_2O_7^{2-}$), and in the metachrome process with chromate (CrO_4^{2-}). In both cases chromium is in the anion and is hexavalent, although for chelate formation a chromium(III) cation (Cr^{3+}) is needed. The use of anionic chromium rather than solutions of chromium (III) salts is simply because the adsorption of anionic chromium occurs easily and uniformly, essentially by ionic attraction to positively charged basic groups in the wool, as with other inorganic anions. On the other hand, chromium(III) ions interact with the carboxyl groups in the wool where they are in fact chelated, the carbonyl oxygen acting as electron donor:

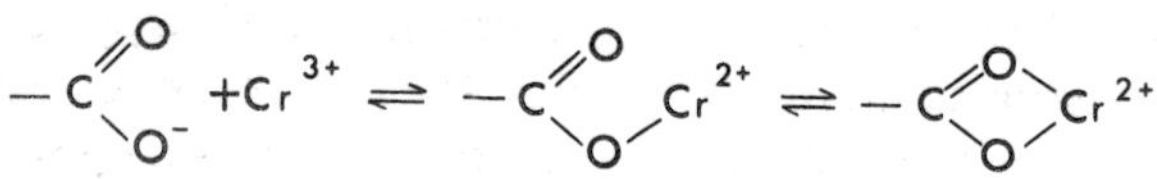

They are thus more tightly bound than the anions. Furthermore, although the reaction is reversible, the activation energy for the loss of Cr^{3+} is very high, so this process is extremely slow. The overall consequence is that level application of chromium is more difficult to obtain with Cr^{3+} than it is with and anions. However, if chromium(VI) is applied, it is necessary for it to be reduced to Cr^{3+} before chelation with a dye can occur. Such reduction in practice involves the cystine bonds of wool in one of two ways: either by direct reaction with dichromate (or chromate) in which cystine is oxidised to cysteic acid, or by hydrolysis of cystine bonds to give cysteine groups and sulphenic acid followed by oxidation of the cysteine by the dichromate. Either way, some cystine bonds are broken, leading to some decrease in the wet strength of the wool. At the same time the pH of the bath rises owing to the release of alkali, arising from the reduction of CrO_7^{2-} to Cr^{3+}, as represented by the classical equation

$$K_2Cr_2O_7 \rightarrow Cr_2O_3 + K_2O + 3[O]$$

The rate of reduction increases markedly between 60 and 70 °C and is often accompanied by a decrease in the rate of adsorption because of the increase in pH resulting from K_2O. Since the reduction is accelerated by light, uneven illumination of treated cloth can lead to uneven results. The reduction to Cr^{3+} is not normally carried out as a separate operation but takes place simultaneously with dye adsorption. For some dyes which are prone to oxidation by acidic dichromate, reducing agents such as formic or oxalic acid are added to the bath. It has been suggested that addition of these reagents prevents oxidation of some of the cystine bonds. Certainly the use of these compounds leads to level chroming, but Gaunt [77] considers that their main function is not to reduce chromium(VI) to chromium(III) but, because they are themselves ligands, to compete with the carboxyl groups of wool for Cr^{3+} ions, thus promoting levelling. Thiocarbamide has recently been suggested for use in dyeing wool with chrome dyes [78]. It is found that it accelerates both the reduction of chromium(VI) to chromium(III) and the rate of penetration of the fibre, so that processing times and use of dichromate are both decreased.

Chelate formation between Cr^{3+} and the mordant dye (usually an *o,o'*-dihydroxyazo dye) takes place according to the principles already discussed, so that 1:1 and 1:2 chelates are possible and have been observed [76]. In chrome dyeing an excess of chromium over dye is always used, so that the formation of 1:1 chelates would be expected. Surprisingly, however, Race, Rowe and Speakman [79] found evidence for 1:2 chelates in chrome dyeings. They produced dyeings using stoichiometric quantities of dye and dichromate so as to obtain 1:2 chelates and found these to be brighter than dyeings produced normally. Furthermore, normal dyeings were brightened on extraction with 0.5 M oxalic acid, which removes unchelated Cr^{3+}, the amount removed corresponding with the residual Cr^{3+}–dye complex being 1:2. Oxalic acid is a chelating agent itself, however, and therefore removes unbound Cr^{3+}. It may therefore remove Cr^{3+} from some 1:1 Cr^{3+}–dye complex, thus allowing the dye which is then available to produce the 1:2 complex.

A question of considerable importance is whether or not, in the final chrome dyeing, any bonding exists between the metal and wool. The crux of the matter would seem to be the composition of the Cr^{3+}–dye chelate. If this is genuinely 1:2, as proposed by Race *et al.*, then the coordination number of Cr^{3+} is fully occupied by the dye and chelation of wool, e.g. by virtue of —COOH groups as discussed earlier, is not possible. Any 1:1 complex would, however, be able to chelate with suitable groups in the wool.

An interesting modern development of complex formation is in the transfer printing of wool. In this process the wool is padded with a solution of trivalent chromium followed, after drying, by transfer printing at *ca* 200 °C from a paper printed with sublimable low-r.m.m. monoazo dyes with hydroxyl groups in the *o,o'*- positions. Metal complexes with good fastness properties are produced in the printed areas [80]. Uneven application of chromium(III) is probably not important in this method, because the amount of complex finally produced is determined by the amount of azo dye transferred, which is a function of the transfer paper rather than the wool. Comparison of the complexes extracted from such prints with purified 1:1 and 1:2 chelates has shown that, here also, the 1:2 chelates predominate [81].

7.6.3 Preformed chelates

Two groups of metal-complex dyes are available for application to wool: the acid-dyeing 1:1 complexes and the neutral-dyeing 1:2 complexes. They are prepared by the dye manufacturer, so dye–metal chelate formation plays no part in their application and therefore need not be considered here, although it conforms entirely to the general principles already discussed.

The dyeing behaviour of these preformed chelates has not been studied as extensively as that of simple acid dyes, but the results that are available enable their practical dyeing behaviour to be explained.

The 1:1 complexes (e.g. Neolan and Palatine Fast dyes) are produced from selected simple acid dyes (mainly *o,o'*-dihydroxyazo compounds) which, because they contain one or two sulphonic acid groups per molecule, exhibit normal acid-dyeing behaviour. This leads to increasing exhaustion with decreasing pH, up to a maximum. Superimposed on this behaviour, however, is the effect of the chelated Cr^{3+}. Since the chelate is 1:1, three coordination positions in the chromium ion are still available for occupation, either by water or by groups in the wool such as ionised carboxyl (as discussed) and un-ionised amino groups. Lowering the pH will lead to gradual protonation of these groups to give —COOH and —$\overset{+}{N}H_3$, which no longer have a lone pair of electrons available for donation to chromium. Thus, below pH 3–4, chelation with wool will be less and substantivity due to this cause will decline. The combined effect, therefore, is that the affinity (as measured by degree of exhaustion) of 1:1 metal-complex dyes passes through a maximum in the pH range 3–4 [82]. This accounts for the practical observation that tippy dyeings are obtained with these dyes unless at least 8 per cent sulphuric acid is used to give a dyebath pH of *ca* 2.0, at which the exhaustion is well below its maximum value. Migration can therefore occur and tippiness be levelled. It must be pointed out, however, that this explanation does not account for the inferior wet-fastness obtained in dyeings containing smaller quantities of acid, where chelation to the wool ought to be at a maximum.

The 1:2 complexes (Irgalan, Cibalan, etc. dyes), unlike the 1:1 complexes, are applied to wool under essentially neutral conditions, as are other fast-to-milling dyes. Under these dyebath conditions the number of positive sites in wool (or any other basic fibre) is small and hence the ionic attraction for any anionic dye will also be small. Dyes having a high substantivity under these conditions therefore interact with wool mainly by forces other than ionic. The 1:2 complexes also are produced from *o,o′*-dihydroxyazo compounds, but they contain no sulphonic acid groups, water solubility being derived from groups such as methylsulphone, which can hydrogen-bond to water. A typical structure is shown below.

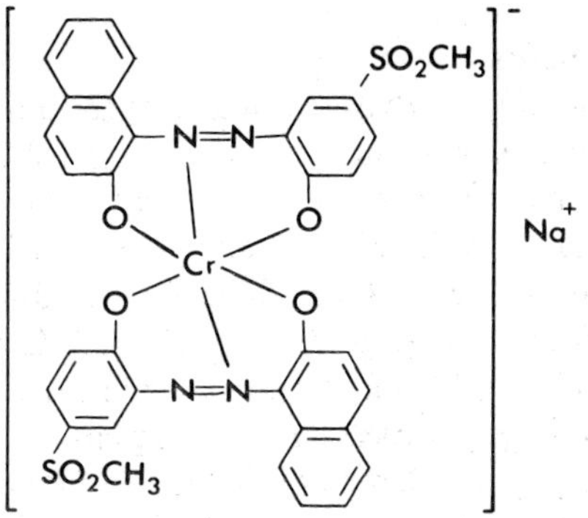

(X)

The complex bears one negative charge (the difference between four ionised hydroxyl groups and the Cr^{3+} ion), distributed over an r.m.m. approximately double that of a typical equalising acid dye. Usually also, groups such as $—NH_2$, which may be present in acid dyes, are absent. The result is a compound in which the hydrophobic–hydrophilic balance is the reverse of that of a typical equalising acid dye [83]. Extensive van der Waals attraction can thus be set up in the fibre, whilst at the same time the affinity for water is low (the dyes having only just adequate solubility), so the difference between the two (which is the observed affinity) is high. If the dyes were to be applied from a markedly acid bath, i.e. one which would lead to a high concentration of positive sites, dyeing would be extremely rapid but most likely uneven and it would not be possible to bring about levelling and penetration later. A useful review of the chemistry of metal-complex dyes, which elaborates many of the points made in this section, has been published recently [84].

REFERENCES

1. R and T Holliday, BP 2757/80.
2. H Zollinger, *Azo and diazo chemistry* (New York and London: Interscience, 1961).
3. E Bamberger, *Ber.*, **27** (1894) 1948.
4. A Hantzsch and M Schumann, *ibid.*, **32** (1899) 1691.
5. H Schmid, *Z. Elektrochem.*, **42** (1936) 579.
6. T W J Taylor, *J.C.S.* (1928) 1099, 1897.
7. L P Hammett, *Physical organic chemistry* (New York and London: McGraw Hill, 1940) 294.

8. J H Ridd, *Quart. Rev.*, **15** no. 4 (1961).
9. K H Saunders, *The aromatic diazo compounds*, 2nd edn (London: Arnold & Co., 1949).
10. A Hantzsch, *Ber.*, **27** (1894) 1702.
11. J B Conant and W D Peterson, *J. Amer. Chem. Soc.*, **52** (1930) 1220.
12. C R Hauser and D S Breslow, *ibid.*, **63** (1941) 418.
13. R Wistar and P D Bartlett, *ibid.*, **63** (1941) 413.
14. H Zollinger, *Helv. Chim. Acta.*, **36** (1953) 1723.
15. A F Hegarty, *The chemistry of diazonium and diazo groups* (ed. by S Patai) (Chichester: John Wiley & Sons, 1978) 551–592.
16. P Frey, *Textil. Praxis*, **36** (1981) 1119.
17. Griesheim Elektron., BP 6,379/12; 13,237/13; 23,732/13; 238,704.
18. A S Dunn, *Faraday Soc. Disc.*, **16** (1954) 230.
19. E H Daruwalla and R H Mehra, *Text. Research. J.*, **29** (1959) 595.
20. W J Marshall and R H Peters, *J.S.D.C.*, **63** (1947) 446.
21. A S Dunn, *ibid.*, **73** (1957) 92.
22. P Griess, *Annal.*, **137** (1866) 52.
23. Becker, German P 81,039, 86,367, 89,998.
24. C K Ingold, *Structure and mechanism in organic chemistry*, 2nd edn (London: G. Bell & Sons, 1969).
25. (a) A Johnson, MSc Tech thesis (University of Manchester, 1950); (b) G Siegrist, *Teintex*, **26** (1961) 887.
26. R J W Le Fevre, M F O'Dwyer and R L Werner, *Australian J. Chem.*, **6** (1953) 341.
27. R Dijkstra and J de Jonge, *Rec. Trav. Chim.*, **77** (1958) 538.
28. A K Gupta, *J.C.S.* (1952) 3473, 3579.
29. W J Marshall and R H Peters, *J.S.D.C.*, **68** (1952) 289.
30. A Geake and J T Lemon, *Trans. Faraday Soc.*, **34** (1938) 1395, 1409; **37** (1941) 68.
31. T Vickerstaff, *The physical chemistry of dyeing* (London: Oliver and Boyd, 1954), 291.
32. D Appleton and A Geake, *Trans. Faraday Soc.*, **37** (1941) 45.
33. P B Curzons, PhD thesis (University of Manchester, 1978).
34. G M Nabar and V A Shenai, *Text. Research J.*, **33** (1963) 471.
35. W J Marshall and R H Peters, *Bull. Inst. Text. France*, **30** (1952) 415.
36. *Idem*, *J.S.D.C.*, **69** (1953) 583.
37. E Atherton, A C Cooper and M R Fox, *ibid.*, **80** (1964) 521.
38. A H Liddell, D McKay and P J Weedall, *ibid.*, **90** (1974) 164.
39. M Bader, *Chim. et Ind.*, **449** (1924) 455.
40. E Baumann and F Thiemann, *Ber.*, **13** (1880) 408.
41. A von Baeyer, *ibid.*, **12** (1879) 1600.
42. M. Mader, *Amer. Dyestuff Rep.*, **27** (1938) 455.
43. *Idem*, *Rev. Gen. Mat. Col.*, **63** (1939) 241.
44. S Ainsworth and A Johnson, *J.S.D.C.*, **71** (1955) 592; **73** (1957) 41.
45. F Haber and J Weiss, *Proc. Roy. Soc.*, **A417** (1934) 332.
46. N Uri, *Chem. Rev.*, **50** (1952) 375.
47. W G Barb, J H Baxendale, P George and K B Hargrave, *Trans. Faraday Soc.*, **47** (1951) 462.
48. A M Dawson and A Johnson, *J.S.D.C.*, **82** (1966) 49.
49. A Johnson and A P Lockett, *J.S.D.C.*, **76** (1960) 412.
50. E Halfpenny and P L Robinson, *J.C.S.* (1952) 928, 940.
51. O H Claussen, A Johnson and P T Kamdar, *J.S.D.C.*, **81** (1965) 350.
52. P B Curzons, PhD thesis (University of Manchester, 1978).
53. R L M Allen, *Colour chemistry* (London: Nelson, 1971) Chapter 14.
54. G Kaufmann, *Melliand Textilber.*, **44** (1963) 1245.
55. H Luttringhaus, *Amer. Dyestuff Rep.*, **53** (1964) 728.
56. Y Minoura, *J. Soc. Rubber Ind. Japan*, **23** (1950) 213.
57. B Milligan and J M Swan, *Text. Research J.*, **31** (1961) 18.
58. F Osterloh, *Melliand Textilber*, **44** (1963) 57.
59. R L M Allen, ref. 53, Chapter 15.

60. M Wyler and ICI, BP 464,126.
61. H H Haddock, *Research*, **1** (1948) 685.
62. H Zollinger, see ref. 2.
63. C L Bird, *The theory and practice of wool dyeing*, 4th edn (Bradford: Society of Dyers and Colourists, 1972).
64. P Rys and H Zollinger, *Fundamentals of the chemistry and application of dyes* (London: Wiley, 1972).
65. J C Bailar, *The chemistry of the co-ordination compounds* (New York: Reinhold; London: Chapman and Hall, 1956).
66. G Schetty, *Amer. Dyestuff Rep.*, **54** (1965) 589.
67. R Price, in *Chemistry of synthetic dyes*, vol. 3 (ed. by K Venkataraman) (New York and London: Academic Press, 1970) 312–317.
68. A Johnson, M S Mort, R H Peters and M J Wadia, *Tex.*, **21** (1962) 543.
69. F A Snaveley, A C Fernelius and B E Douglas, *J.S.D.C.*, **79** (1957) 491.
70. A E Martell and M Calvin, *Chemistry of the metal chelate compounds* (New Jersey: Prentice Hall, 1956).
71. J Bjerrum and L G Sillen, *Stability constants of metal ion complexes*, Part I (London: The Chemical Society, sp. Pub. 17, 1964).
72. F A Snaveley, W C Fernelius and B R Block, *J. Amer. Chem. Soc.*, **79** (1957) 1028.
73. M Calvin and K W Wilson, *ibid.*, **67** (1945) 2003.
74. A Burawoy, A G Salem and A R Thompson, *J.C.S.* (1952) 4793.
75. J Keleman, S Moss, H Suater and T Winkler, *Dyes and Pigments*, **3** (1982) 27–47.
76. F R Hartley, *J.S.D.C.*, **85** (1969) 66.
77. J F Gaunt, *ibid.*, **79** (1954) 46.
78. I A Lednera and B N Mel'nikor, *ibid*, **98** (1982) 365.
79. E Race, F M Rowe and J B Speakman, *ibid.*, **62** (1946) 372.
80. P R Brady, P G Cookson, K W Fincher and D M Lewis, *ibid.*, **100** (1982) 398.
81. D M Lewis, M S Otterburn and D R Savage, *ibid.*, **100** (1984) 264.
82. I D Rattee, *ibid.*, **69** (1953) 288.
83. G Schetty, *ibid.*, **71** (1955) 705.
84. F Beffa and G Black, *Rev. Prog. Coloration*, **14** (1984) 33–41.

Index